T0211209

Galaxies in the Universe: An Introduction

Galaxies are the places where gas turns into luminous stars, powered by nuclear reactions that also produce most of the chemical elements. But the gas and stars are only the tip of an iceberg: a galaxy consists mostly of dark matter, which we know only by the pull of its gravity. The ages, chemical composition and motions of the stars we see today, and the shapes that they make up, tell us about each galaxy's past life. This book presents the astrophysics of galaxies since their beginnings in the early Universe. This Second Edition is extensively illustrated with the most recent observational data. It includes new sections on galaxy clusters, gamma ray bursts and supermassive black holes. Chapters on the large-scale structure and early galaxies have been thoroughly revised to take into account recent discoveries such as dark energy.

The authors begin with the basic properties of stars and explore the Milky Way before working out towards nearby galaxies and the distant Universe, where galaxies can be seen in their early stages. They then discuss the structures of galaxies and how galaxies have developed, and relate this to the evolution of the Universe. The book also examines ways of observing galaxies across the electromagnetic spectrum, and explores dark matter through its gravitational pull on matter and light.

This book is self-contained, including the necessary astronomical background, and homework problems with hints. It is ideal for advanced undergraduate students in astronomy and astrophysics.

LINDA SPARKE is a Professor of Astronomy at the University of Wisconsin, and a Fellow of the American Physical Society.

JOHN GALLAGHER is the W. W. Morgan Professor of Astronomy at the University of Wisconsin and is editor of the *Astronomical Journal.*

Galaxies in the Universe: An Introduction

Second Edition

Linda S. Sparke
John S. Gallagher III
University of Wisconsin, Madison

CAMBRIDGE
UNIVERSITY PRESS

CAMBRIDGE
UNIVERSITY PRESS

University Printing House, Cambridge CB2 8BS, United Kingdom

One Liberty Plaza, 20th Floor, New York, NY 10006, USA

477 Williamstown Road, Port Melbourne, VIC 3207, Australia

314-321, 3rd Floor, Plot 3, Splendor Forum, Jasola District Centre, New Delhi - 110025, India

79 Anson Road, #06-04/06, Singapore 079906

Cambridge University Press is part of the University of Cambridge.

It furthers the University's mission by disseminating knowledge in the pursuit of education, learning and research at the highest international levels of excellence.

www.cambridge.org
Information on this title: www.cambridge.org/9780521671866

First published 2000
Second edition 2007
14th printing 2018

A catalogue record for this publication is available from the British Library

ISBN 978-0-521-85593-8 Hardback
ISBN 978-0-521-67186-6 Paperback

Contents

Preface to the second edition

This text is aimed primarily at third- and fourth-year undergraduate students of astronomy or physics, who have undertaken the first year or two of university-level studies in physics. We hope that graduate students and research workers in related areas will also find it useful as an introduction to the field. Some background knowledge of astronomy would be helpful, but we have tried to summarize the necessary facts and ideas in our introductory chapter, and we give references to books offering a fuller treatment. This book is intended to provide more than enough material for a one-semester course, since instructors will differ in their preferences for areas to emphasize and those to omit. After working through it, readers should find themselves prepared to tackle standard graduate texts such as Binney and Tremaine's *Galactic Dynamics*, and review articles such as those in the *Annual Reviews of Astronomy and Astrophysics*.

Astronomy is not an experimental science like physics; it is a natural science like geology or meteorology. We must take the Universe as we find it, and deduce how the basic properties of matter have constrained the galaxies that happened to form. Sometimes our understanding is general but not detailed. We can estimate how much water the Sun's heat can evaporate from Earth's oceans, and indeed this is roughly the amount that falls as rain each day; wind speeds are approximately what is required to dissipate the solar power absorbed by the ground and the air. But we cannot predict from physical principles when the wind will blow or the rain fall. Similarly, we know why stellar masses cannot be far larger or smaller than they are, but we cannot predict the relative numbers of stars that are born with each mass. Other obvious regularities, such as the rather tight relations between a galaxy's luminosity and the stellar orbital speeds within it, are not yet properly understood. But we trust that they will yield their secrets, just as the color–magnitude relation among hydrogen-burning stars was revealed as a mass sequence. On first acquaintance galaxy astronomy can seem confusingly full of disconnected facts; but we hope to convince you that the correct analogy is meteorology or botany, rather than stamp-collecting.

We have tried to place material which is relatively more difficult or more intricate at the end of each subsection. Students who find some portions heavy going at a first reading are advised to move to the following subsection and return later to the troublesome passage. Some problems have been included. These aim mainly at increasing a reader's understanding of the calculations and appreciation of the magnitudes of quantities involved, rather than being mathematically demanding. Often, material presented in the text is amplified in the problems; more casual readers may find it useful to look through them along with the rest of the text.

Boldface is used for vectors; italics indicate concepts from physics, or specialist terms from astronomy which the reader will see again in this text, or will meet in the astronomical literature. Because they deal with large distances and long timescales, astronomers use an odd mixture of units, depending on the problem at hand; Appendix A gives a list, with conversion factors. Increasing the confusion, many of us are still firmly attached to the centimeter–gram–second system of units. For electromagnetic formulae, we give a parallel-text translation between these and units of the Système Internationale d'Unités (SI), which are based on meters and kilograms. In other cases, we have assumed that readers will be able to convert fairly easily between the two systems with the aid of Appendix A. Astronomers still disagree significantly on the distance scale of the Universe, parametrized by the Hubble constant H_0. We often indicate explicitly the resulting uncertainties in luminosity, distance, etc., but we otherwise adopt $H_0 = 75 \, \mathrm{km\,s^{-1}\,Mpc^{-1}}$. Where ages are required or we venture across a substantial fraction of the cosmos, we use the benchmark cosmology with $\Omega_\Lambda = 0.7$, $\Omega_m = 0.3$, and $H_0 = 70 \, \mathrm{km\,s^{-1}\,Mpc^{-1}}$.

We will use an equals sign ($=$) for mathematical equality, or for measured quantities known to greater accuracy than a few percent; approximate equality ($\approx$) usually implies a precision of 10%–20%, while $\sim$ (pronounced 'twiddles') means that the relation holds to no better than about a factor of two. Logarithms are to base 10, unless explicitly stated otherwise. Here, and generally in the professional literature, ranges of error are indicated by $\pm$ symbols, or shown by horizontal or vertical bars in graphs. Following astronomical convention, these usually refer to 1σ error estimates calculated by assuming a Gaussian distribution (which is often rather a bad approximation to the true random errors). For those more accustomed to 2σ or 3σ error bars, this practice makes discrepancies between the results of different workers appear more significant than is in fact the case.

This book is much the better for the assistance, advice, and warnings of our colleagues and students. Eric Wilcots test flew a prototype in his undergraduate class; our colleagues Bob Bless, Johan Knapen, John Mathis, Lynn Matthews, and Alan Watson read through the text and helped us with their detailed comments; Bob Benjamin tried to set us right on the interstellar medium. We are particularly grateful to our many colleagues who took the time to provide us with figures or the material for figures; we identify them in the captions. Bruno Binggeli, Dap Hartmann, John Hibbard, Jonathan McDowell, Neill Reid, and Jerry Sellwood re-analyzed, re-ran, and re-plotted for us, Andrew Cole integrated stellar energy outputs, Evan Gnam did orbit calculations, and Peter Erwin helped us out with some huge and complex images. Wanda Ashman turned our scruffy sketches into line drawings. For the second edition, Bruno Binggeli made us an improved portrait of the Local Group, David Yu helped with some complex plots, and Tammy Smecker-Hane and Eric Jensen suggested helpful changes to the problems. Much thanks to all!

Linda Sparke is grateful to the University of Wisconsin for sabbatical leave in the 1996–7 and 2004–5 academic years, and to Terry Millar and the University of Wisconsin Graduate School, the Vilas Foundation, and the Wisconsin Alumni Research Foundation for financial support. She would also like to thank the directors, staff, and students of the Kapteyn Astronomical Institute (Groningen University, Netherlands), the Mount Stromlo and Siding Spring Observatories (Australian National University, Canberra), and the Isaac Newton Institute for Mathematical Sciences (Cambridge University, UK) and Yerkes Observatory (University of Chicago), for their hospitality while much of the first edition was written. She is equally grateful to the Dominion Astrophysical Observatory of Canada, the Max Planck Institute for Astrophysics in Garching, Germany, and the Observatories of the Carnegie Institute of Washington (Pasadena, California) for refuge as we prepared the second edition. We are both most grateful to our colleagues in Madison for putting up with us during the writing. Jay Gallagher also thanks his family for their patience and support for his work on 'The Book'.

Both of us appear to lack whatever (strongly recessive?) genes enable accurate proofreading. We thank our many helpful readers for catching bugs in the first edition, which we listed on a website. We will do the same for this edition, and hope also to provide the diagrams in machine-readable form: please see links from our homepages, which are currently at www.astro.wisc.edu/~sparke and ~jsg.

Introduction

Galaxies appear on the sky as huge clouds of light, thousands of light-years across: see the illustrations in Section 1.3 below. Each contains anywhere from a million stars up to a million million (10^{12}); gravity binds the stars together, so they do not wander freely through space. This introductory chapter gives the astronomical information that we will need to understand how galaxies are put together.

Almost all the light of galaxies comes from their stars. Our opening section attempts to summarize what we know about stars, how we think we know it, and where we might be wrong. We discuss basic observational data, and we describe the life histories of the stars according to the theory of stellar evolution. Even the nearest stars appear faint by terrestrial standards. Measuring their light accurately requires care, and often elaborate equipment and procedures. We devote the final pages of this section to the arcana of stellar photometry: the magnitude system, filter bandpasses, and colors.

In Section 1.2 we introduce our own Galaxy, the Milky Way, with its characteristic 'flying saucer' shape: a flat disk with a central bulge. In addition to their stars, our Galaxy and others contain gas and dust; we review the ways in which these make their presence known. We close this section by presenting some of the coordinate systems that astronomers use to specify the positions of stars within the Milky Way. In Section 1.3 we describe the variety found among other galaxies and discuss how to measure the distribution of light within them. Only the brightest cores of galaxies can outshine the glow of the night sky, but most of their light comes from the faint outer parts; photometry of galaxies is even more difficult than for stars.

One of the great discoveries of the twentieth century is that the Universe is not static, but expanding; the galaxies all recede from each other, and from us. Our Universe appears to have had a beginning, the Big Bang, that was not so far in the past: the cosmos is only about three times older than the Earth. Section 1.4 deals with the cosmic expansion, and how it affects the light we receive from galaxies. Finally, Section 1.5 summarizes what happened in the first million years after the Big Bang, and the ways in which its early history has determined what we see today.

1.1 The stars

1.1.1 Star light, star bright . . .

All the information we have about stars more distant than the Sun has been deduced by observing their electromagnetic radiation, mainly in the ultraviolet, visible, and infrared parts of the spectrum. The light that a star emits is determined largely by its surface area, and by the temperature and chemical composition – the relative numbers of each type of atom – of its outer layers. Less directly, we learn about the star's mass, its age, and the composition of its interior, because these factors control the conditions at its surface. As we decode and interpret the messages brought to us by starlight, knowledge gained in laboratories on Earth about the properties of matter and radiation forms the basis for our theory of stellar structure.

The *luminosity* of a star is the amount of energy it emits per second, measured in watts, or ergs per second. Its *apparent brightness* or *flux* is the total energy received per second on each square meter (or square centimeter) of the observer's telescope; the units are $W\,m^{-2}$, or $erg\,s^{-1}\,cm^{-2}$. If a star shines with equal brightness in all directions, we can use the *inverse-square law* to estimate its luminosity L from the distance d and measured flux F:

$$F = \frac{L}{4\pi d^2}.\tag{1.1}$$

Often, we do not know the distance d very well, and must remember in subsequent calculations that our estimated luminosity L is proportional to d^2. The Sun's total or *bolometric* luminosity is $L_\odot = 3.86 \times 10^{26}\,W$, or $3.86 \times 10^{33}\,erg\,s^{-1}$. Stars differ enormously in their luminosity: the brightest are over a million times more luminous than the Sun, while we observe stars as faint as $10^{-4}L_\odot$.

Lengths in astronomy are usually measured using the *small-angle* formula. If, for example, two stars in a binary pair at distance d from us appear separated on the sky by an angle α, the distance D between the stars is given by

$$\alpha\ (\text{in radians}) = D/d.\tag{1.2}$$

Usually we measure the angle α in *arcseconds*: one arcsecond ($1''$) is $1/60$ of an arcminute ($1'$) which is $1/60$ of a degree. Length is often given in terms of the *astronomical unit*, Earth's mean orbital radius (1 AU is about 150 million kilometers) or in *parsecs*, defined so that, when $D = 1\,AU$ and $\alpha = 1''$, $d = 1\,pc = 3.09 \times 10^{13}\,km$ or 3.26 light-years.

The orbit of two stars around each other can allow us to determine their masses. If the two stars are clearly separated on the sky, we use Equation 1.2 to measure the distance between them. We find the speed of the stars as they orbit each other from the *Doppler shift* of lines in their spectra; see Section 1.2. Newton's equation for the gravitational force, in Section 3.1, then gives us the masses. The Sun's mass, as determined from the orbit of the Earth and other planets, is $\mathcal{M}_\odot = 2 \times 10^{30}$ kg, or 2×10^{33} g.

Stellar masses cover a much smaller range than their luminosities. The most massive stars are around $100\mathcal{M}_\odot$. A star is a nuclear-fusion reactor, and a ball of gas more massive than this would burn so violently as to blow itself apart in short order. The least massive stars are about $0.075\mathcal{M}_\odot$. A smaller object would never become hot enough at its center to start the main fusion reaction of a star's life, turning hydrogen into helium.

> **Problem 1.1** Show that the Sun produces 10 000 times less energy per unit mass than an average human giving out about 1 W kg^{-1}.

The radii of stars are hard to measure directly. The Sun's radius $R_\odot = 6.96 \times 10^5$ km, but no other star appears as a disk when seen from Earth with a normal telescope. Even the largest stars subtend an angle of only about 0.05″, 1/20 of an arcsecond. With difficulty we can measure the radii of nearby stars with an interferometer; in eclipsing binaries we can estimate the radii of the two stars by measuring the size of the orbit and the duration of the eclipses. The largest stars, the red supergiants, have radii about 1000 times larger than the Sun, while the smallest stars that are still actively burning nuclear fuel have radii around $0.1R_\odot$.

A star is a dense ball of hot gas, and its spectrum is approximately that of a *blackbody* with a temperature ranging from just below 3000 K up to 100 000 K, modified by the absorption and emission of atoms and molecules in the star's outer layers or *atmosphere*. A blackbody is an ideal radiator or perfect absorber. At temperature T, the luminosity L of a blackbody of radius R is given by the *Stefan–Boltzmann* equation:

$$L = 4\pi R^2 \sigma_{SB} T^4, \tag{1.3}$$

where the constant $\sigma_{SB} = 5.67 \times 10^{-8}$ W m^{-2} K^{-4}. For a star of luminosity L and radius R, we define an *effective temperature* T_{eff} as the temperature of a blackbody with the same radius, which emits the same total energy. This temperature is generally close to the average for gas at the star's 'surface', the *photosphere*. This is the layer from which light can escape into space. The Sun's effective temperature is $T_{eff} \approx 5780$ K.

> **Problem 1.2** Use Equation 1.3 to estimate the solar radius $R_\odot$ from its luminosity and effective temperature. Show that the gravitational acceleration g at the surface is about 30 times larger than that on Earth.

> **Problem 1.3** The red supergiant star Betelgeuse in the constellation Orion has $T_{\text{eff}} \approx 3500$ K and a diameter of $0.045''$. Assuming that it is 140 pc from us, show that its radius $R \approx 700 R_{\odot}$, and that its luminosity $L \approx 10^5 L_{\odot}$.

Generally we do not measure all the light emitted from a star, but only what arrives in a given interval of wavelength or frequency. We define the *flux per unit wavelength* F_{λ} by setting $F_{\lambda}(\lambda)\Delta\lambda$ to be the energy of the light received between wavelengths λ and $\lambda + \Delta\lambda$. Because its size is well matched to the typical accuracy of their measurements, optical astronomers generally measure wavelength in units named after the nineteenth-century spectroscopist Anders Ångström: 1 Å $= 10^{-8}$ cm or 10^{-10} m. The flux F_{λ} has units of W m^{-2} Å^{-1} or erg s^{-1} cm^{-2} Å^{-1}. The *flux per unit frequency* F_{ν} is defined similarly: the energy received between frequencies ν and $\nu + \Delta\nu$ is $F_{\nu}(\nu)\Delta\nu$, so that $F_{\lambda} = (\nu^2/c)F_{\nu}$. Radio astronomers normally measure F_{ν} in *janskys*: 1 Jy $= 10^{-26}$ W m^{-2} Hz^{-1}. The apparent brightness F is the integral over all frequencies or wavelengths:

$$F \equiv \int_0^{\infty} F_{\nu}(\nu)\,\mathrm{d}\nu = \int_0^{\infty} F_{\lambda}(\lambda)\,\mathrm{d}\lambda. \tag{1.4}$$

The hotter a blackbody is, the bluer its light: at temperature T, the peak of F_{λ} occurs at wavelength

$$\lambda_{\max} = [2.9/T \text{ (K)}]\,\text{mm}. \tag{1.5}$$

For the Sun, this corresponds to yellow light, at about 5000 Å; human bodies, the Earth's atmosphere, and the uncooled parts of a telescope radiate mainly in the infrared, at about 10 μm.

1.1.2 Stellar spectra

Figure 1.1 shows F_{λ} for a number of commonly observed types of star, arranged in order from coolest to hottest. The hottest stars are the bluest, and their spectra show absorption lines of highly ionized atoms; cool stars emit most of their light at red or infrared wavelengths, and have absorption lines of neutral atoms or molecules. Astronomers in the nineteenth century classified the stars according to the strength of the *Balmer* lines of neutral hydrogen H I, with A stars having the strongest lines, B stars the next strongest, and so on; many of the classes subsequently fell into disuse. In the 1880s, Antonia Maury at Harvard realized that, when the classes were arranged in the order O B A F G K M, the strengths of all the spectral lines, not just those of hydrogen, changed continuously along the sequence. The first large-scale classification was made at Harvard College Observatory between 1911 and 1949: almost 400 000 stars were included in the *Henry Draper Catalogue* and its supplements. We now know that Maury's spectral sequence lists the stars in order of decreasing surface temperature. Each of the classes has been subdivided

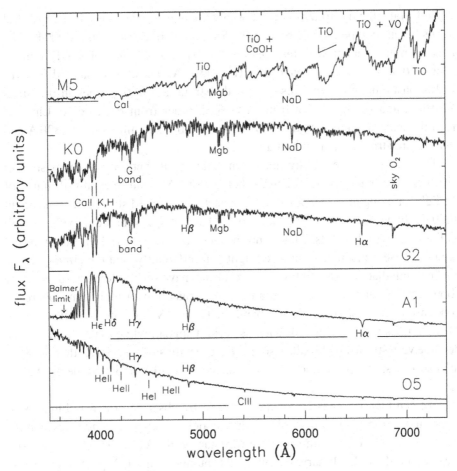

Fig. 1.1. Optical spectra of main-sequence stars with roughly the solar chemical composition. From the top in order of increasing surface temperature, the stars have spectral classes M5, K0, G2, A1, and O5 – G. Jacoby *et al.*, spectral library.

into subclasses, from 0, the hottest, to 9, the coolest: our Sun is a G2 star. Recently classes L and T have been added to the system, for the very cool stars discovered by infrared observers. Astronomers often call stars at the beginning of this sequence 'early types', while those toward the end are 'late types'.

The temperatures of O stars exceed 30 000 K. Figure 1.1 shows that the strongest lines are those of HeII (once-ionized helium) and CIII (twice-ionized carbon); the Balmer lines of hydrogen are relatively weak because hydrogen is almost totally ionized. The spectra of B stars, which are cooler, have stronger hydrogen lines, together with lines of neutral helium, HeI. The A stars, with temperatures below 11 000 K, are cool enough that the hydrogen in their atmospheres is largely neutral; they have the strongest Balmer lines, and lines of singly ionized metals such as calcium. Note that the flux decreases sharply at wavelengths less than 3800 Å, this is called the *Balmer jump*. A similar *Paschen jump* appears at wavelengths that are $3^2/2^2$ times longer, at around 8550 Å.

In F stars, the hydrogen lines are weaker than in A stars, and lines of neutral metals begin to appear. G stars, like the Sun, are cooler than about 6000 K. The most prominent absorption features are the 'H and K' lines of singly ionized calcium (CaII), and the G band of CH at 4300 Å. These were named in 1815 by Joseph Fraunhofer, who discovered the strong absorption lines in the Sun's spectrum, and labelled them from A to K in order from red to blue. Lines of neutral metals, such as the pair of D lines of neutral sodium (NaI) at 5890 Å and 5896 Å, are stronger than in hotter stars.

In K stars, we see mainly lines of neutral metals and of molecules such as TiO, titanium oxide. At wavelengths below 4000 Å metal lines absorb much of the light, creating the *4000 Å break*. The spectrum of the M star, cooler than about 4000 K, shows deep absorption bands of TiO and of VO, vanadium oxide, as well as lines of neutral metals. This is not because M stars are rich in titanium, but because these molecules absorb red light very efficiently, and the atmosphere is cool enough that they do not break apart. L stars have surface temperatures below about 2500 K, and most of the titanium and vanadium in their atmospheres is condensed onto dust grains. Hence bands of TiO and VO are much weaker than in M stars; lines of neutral metals such as cesium appear, while the sodium D lines become very strong and broad. T stars are those with surfaces cooler than 1400 K; their spectra show strong lines of water and methane, like the atmospheres of giant planets.

We can measure masses for these dwarfs by observing them in binary systems, and comparing with evolutionary models. Such work indicates a mass $\mathcal{M} \approx 0.15\mathcal{M}_\odot$ for a main-sequence M5 star, while $\mathcal{M} \approx 0.08\mathcal{M}_\odot$ for a single measured L0–L1 binary. Counting the numbers of M, L, and T dwarfs in the solar neighborhood shows that objects below $0.3\mathcal{M}_\odot$ contribute little to the total mass in the Milky Way's thin disk. 'Stars' cooler than about L5 have too little mass to sustain hydrogen burning in their cores. They are not true stars, but *brown dwarfs*, cooling as they contract slowly under their own weight. Over its first 100 Myr or so, a given brown dwarf can cool from spectral class M to L, or even T; the temperature drops only slowly during its later life.

The spectrum of a galaxy is *composite*, including the light from a mixture of stars with different temperatures. The hotter stars give out most of the blue light, and the lines observed in the blue part of the spectrum of a galaxy such as the Milky Way are typically those of A, F, or G stars. O and B stars are rare and so do not contribute much of the visible light, unless a galaxy has had a recent burst of star formation. In the red part of the spectrum, we see lines from the cooler K stars, which produce most of the galaxy's red light. Thus the blue part of the spectrum of a galaxy such as the Milky Way shows the Balmer lines of hydrogen in absorption, while TiO bands are present in the red region.

It is much easier to measure the strength of spectral lines relative to the flux at nearby wavelengths than to determine $F_\lambda(\lambda)$ over a large range in wavelength. Absorption and scattering by dust in interstellar space, and by the Earth's

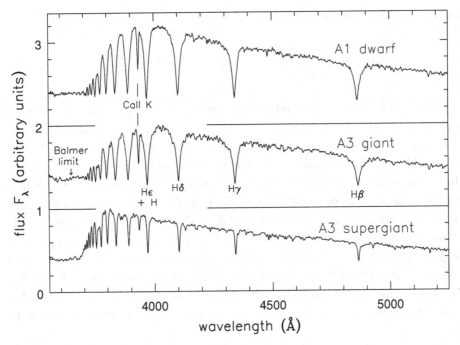

Fig. 1.2. Spectra of an A1 dwarf, an A3 giant, and an A3 supergiant: the most luminous star has the narrowest spectral lines – G. Jacoby *et al.*, spectral library.

atmosphere, affects the blue light of stars more than the red; blue and red light also propagate differently through the telescope and the spectrograph. In practice, stellar temperatures are often estimated by comparing the observed depths of absorption lines in their spectra with the predictions of a *model stellar atmosphere*. This is a computer calculation of the way light propagates through a stellar atmosphere with a given temperature and composition; it is calibrated against stars for which F_λ has been measured carefully.

The lines in stellar spectra also give us information about the surface gravity. Figure 1.2 shows the spectra of three stars, all classified as A stars because the overall strength of their absorption lines is similar. But the Balmer lines of the A dwarf are broader than those in the giant and supergiant stars, because atoms in its photosphere are more closely crowded together: this is known as the *Stark effect*. If we use a model atmosphere to calculate the surface gravity of the star, and we also know its mass, we can then find its radius. For most stars, the surface gravity is within a factor of three of that in the Sun; these stars form the *main sequence* and are known as *dwarfs*, even though the hottest of them are very large and luminous.

All main-sequence stars are burning hydrogen into helium in their cores. For any particular spectral type, these stars have nearly the same mass and luminosity, because they have nearly identical structures: the hottest stars are the most massive, the most luminous, and the largest. Main-sequence stars have radii between $0.1 R_\odot$

and about $25 R_\odot$: very roughly,

$$R \sim R_\odot \left(\frac{\mathcal{M}}{\mathcal{M}_\odot} \right)^{0.7} \quad \text{and} \quad L \sim L_\odot \left(\frac{\mathcal{M}}{\mathcal{M}_\odot} \right)^\alpha, \qquad (1.6)$$

where $\alpha \approx 5$ for $\mathcal{M} \lesssim \mathcal{M}_\odot$, and $\alpha \approx 3.9$ for $\mathcal{M}_\odot \lesssim \mathcal{M} \lesssim 10\mathcal{M}_\odot$. For the most massive stars with $\mathcal{M} \gtrsim 10\mathcal{M}_\odot$, $L \sim 50L_\odot(\mathcal{M}/\mathcal{M}_\odot)^{2.2}$. *Giant* and *supergiant* stars have a lower surface gravity and are much more distended; the largest stars have radii exceeding $1000 R_\odot$. Equation 1.3 tells us that they are much brighter than main-sequence stars of the same spectral type. Below, we will see that they represent later stages of a star's life. *White dwarfs* are not main-sequence stars, but have much higher surface gravity and smaller radii; a white dwarf is only about the size of the Earth, with $R \approx 0.01 R_\odot$. If we define a star by its property of generating energy by nuclear fusion, then a white dwarf is no longer a star at all, but only the ashes or embers of a star's core; it has exhausted its nuclear fuel and is now slowly cooling into blackness. A *neutron star* is an even smaller stellar remnant, only about 20 km across, despite having a mass larger than the Sun's.

Further reading: for an undergraduate-level introduction to stars, see D. A. Ostlie and B. W. Carroll, 1996, *An Introduction to Modern Stellar Astrophysics* (Addison-Wesley, Reading, Massachusetts); and D. Prialnik, 2000, *An Introduction to the Theory of Stellar Structure and Evolution* (Cambridge University Press, Cambridge, UK).

The strength of a given spectral line depends on the temperature of the star in the layers where the line is formed, and also on the abundance of the various elements. By comparing the strengths of various lines with those calculated for a hot gas, Cecelia Payne-Gaposhkin showed in 1925 that the Sun and other stars are composed mainly of hydrogen. The surface layers of the Sun are about 72% hydrogen, 26% helium, and about 2% of all other elements, by mass. Astronomers refer collectively to the elements heavier than helium as *heavy elements* or *metals*, even though substances such as carbon, nitrogen, and oxygen would not normally be called metals.

There is a good reason to distinguish hydrogen and helium from the rest of the elements. These atoms were created in the aftermath of the Big Bang, less than half an hour after the Universe as we now know it came into existence; the neutrons and protons combined into a mix of about three-quarters hydrogen, one-quarter helium, and a trace of lithium. Since then, the stars have burned hydrogen to form helium, and then fused helium into heavier elements; see the next subsection. Figure 1.3 shows the abundances of the commonest elements in the Sun's photosphere. Even oxygen, the most plentiful of the heavy elements, is over 1000 times rarer than hydrogen. The 'metals' are found in almost, but not

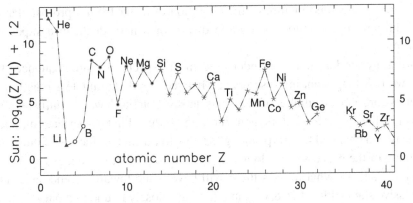

Fig. 1.3. Logarithm of the number of atoms of each element found in the Sun, for every 10^{12} hydrogen atoms. Hydrogen, helium, and lithium originated mainly in the Big Bang, the next two elements result from the breaking apart of larger atoms, and the remainder are 'cooked' in stars. Filled dots show elements produced mainly in quiescent burning; star symbols indicate those made largely during explosive burning in a supernova – M. Asplund *et al.*, astro-ph/0410214.

exactly, the same proportions in all stars. The small differences can tell us a lot about the history of the material that went into making a star; see Section 4.3.

The fraction by mass of the heavy elements is denoted Z: the Sun has $Z_\odot \approx$ 0.02, while the most metal-poor stars in our Galaxy have less than $1/10\,000$ of this amount. If we want to specify the fraction of a particular element, such as oxygen, in a star, we often give its abundance relative to that in the Sun. We use a logarithmic scale:

$$[A/B] \equiv \log_{10}\left\{\frac{(\text{number of A atoms/number of B atoms})_\star}{(\text{number of A atoms/number of B atoms})_\odot}\right\}, \qquad (1.7)$$

where $\star$ refers to the star and we again use $\odot$ for the Sun. Thus, in a star with $[\text{Fe/H}] = -2$, iron is 1% as abundant as in the Sun. A warning: $[\text{Fe/H}]$ is often used for a star's average heavy-element abundance relative to the Sun; it does not always refer to measured iron content.

1.1.3 The lives of the stars

Understanding how stars proceed through the different stages of their lives is one of the triumphs of astrophysics in the second half of the twentieth century. The discovery of nuclear-fusion processes during the 1940s and 1950s, coupled with the fast digital computers that became available during the 1960s and 1970s, has given us a detailed picture of the evolution of a star from a protostellar gas cloud through to extinction as a white dwarf, or a fiery death in a supernova explosion.

We are confident that we understand most aspects of main-sequence stars fairly well. A long-standing discrepancy between predicted nuclear reactions in the Sun's core and the number of neutrinos detected on Earth was recently resolved in favor of the stellar modellers: neutrinos are produced in the expected numbers, but many had changed their type along the way to Earth. Our theories falter at the beginning of the process – we do not know how to predict when a gas cloud will form into stars, or what masses these will have – and toward its end, especially for massive stars with $\mathcal{M} \gtrsim 8\mathcal{M}_\odot$, and for stars closely bound in binary systems. This remaining ignorance means that we do not yet know what determines the rate at which galaxies form their stars; the quantity of elements heavier than helium that is produced by each type of star; and how those elements are returned to the interstellar gas, to be incorporated into future generations of stars.

The mass of a star almost entirely determines its structure and ultimate fate; chemical composition plays a smaller role. Stars begin their existence as clouds of gas that become dense enough to start contracting under the inward pull of their own gravity. Compression heats the gas, making its pressure rise to support the weight of the exterior layers. But the warm gas then radiates away energy, reducing the pressure, and allowing the cloud to shrink further. In this *protostellar* stage, the release of gravitational energy counterbalances that lost by radiation. As a protostar, the Sun would have been cooler than it now is, but several times more luminous. This phase is short: it lasted only 50 Myr for the Sun, which will burn for 10 Gyr on the main sequence. So protostars do not make a large contribution to a galaxy's light.

The temperature at the center rises throughout the protostellar stage; when it reaches about 10^7 K, the star is hot enough to 'burn' hydrogen into helium by thermonuclear fusion. When four atoms of hydrogen fuse into a single atom of helium, 0.7% of their mass is set free as energy, according to Einstein's formula $\mathcal{E} = \mathcal{M}c^2$. Nuclear reactions in the star's *core* now supply enough energy to maintain the pressure at the center, and contraction stops. The star is now quite stable: it has begun its *main-sequence* life. Table 1.1 gives the luminosity and effective temperature for stars of differing mass on the *zero-age main sequence*; these are calculated from models for the internal structure, assuming the same chemical composition as the Sun. Each solid track on Figure 1.4 shows how those quantities change over the star's lifetime. A plot like this is often called a *Hertzsprung–Russell* diagram, after Ejnar Hertzsprung and Henry Norris Russell, who realized around 1910 that, if the luminosity of stars is plotted against their spectral class (or color or temperature), most of the stars fall close to a diagonal line which is the main sequence. The temperature increases to the left on the horizontal axis to correspond to the ordering O B A F G K M of the spectral classes. As the star burns hydrogen to helium, the mean mass of its constituent

Table 1.1 Stellar models with solar abundance, from Figure 1.4

Mass ($\mathcal{M}_\odot$)	L_{ZAMS} ($L_\odot$)	T_{eff} (K)	Spectral type	τ_{MS} (Myr)	τ_{red} (Myr)	$\int(L\,d\tau)_{MS}$ (Gyr $\times L_\odot$)	$\int(L\,d\tau)_{pMS}$ (Gyr $\times L_\odot$)
0.8	0.24	4860	K2	25 000		10	
1.0	0.69	5640	G5	9800	3200	10.8	24
1.25	2.1	6430		3900	1650	11.7	38
1.5	4.7	7110	F3	2700	900	16.2	13
2	16	9080	A2	1100	320	22.0	18
3	81	12 250	B7	350	86	38.5	19
5	550	17 180	B4	94	14	75.2	23
9	4100	25 150		26	1.7	169	40
15	20 000	31 050		12	1.1	360	67
25	79 000	37 930		6.4	0.64	768	145
40	240 000	43 650	O5	4.3	0.47	1500	112
60	530 000	48 190		3.4	0.43	2550	9
85	1 000 000	50 700		2.8		3900	
120	1 800 000	53 330		2.6		5200	

Note: L and T_{eff} are for the zero-age main sequence; spectral types are from Table 1.3; τ_{MS} is main-sequence life; τ_{red} is time spent later as a red star ($T_{eff} \lesssim 6000$ K); integrals give energy output on the main sequence (MS), and in later stages (pMS).

particles slowly increases, and the core must become hotter to support the denser star against collapse. Nuclear reactions go faster at the higher temperature, and the star becomes brighter. The Sun is now about 4.5 Gyr old, and its luminosity is almost 50% higher than when it first reached the main sequence.

Problem 1.4 What mass of hydrogen must the Sun convert to helium each second in order to supply the luminosity that we observe? If it converted all of its initial hydrogen into helium, how long could it continue to burn at this rate? Since it can burn only the hydrogen in its core, and because it is gradually brightening, it will remain on the main sequence for only about 1/10 as long.

Problem 1.5 Use Equation 1.3 and data from Table 1.1 to show that, when the Sun arrived on the main sequence, its radius was about $0.87 R_\odot$.

A star can continue on the main sequence until thermonuclear burning has consumed the hydrogen in its core, about 10% of the total. Table 1.1 lists the time τ_{MS} that stars of each mass spend there; it is most of the star's life. So, at any given time, most of a galaxy's stars will be on the main sequence. For an average value $\alpha \approx 3.5$ in Equation 1.6, we have

$$\tau_{MS} = \tau_{MS,\odot}\frac{\mathcal{M}/L}{\mathcal{M}_\odot/L_\odot} \sim 10\,\text{Gyr}\left(\frac{\mathcal{M}}{\mathcal{M}_\odot}\right)^{-2.5} = 10\,\text{Gyr}\left(\frac{L}{L_\odot}\right)^{-5/7}. \quad (1.8)$$

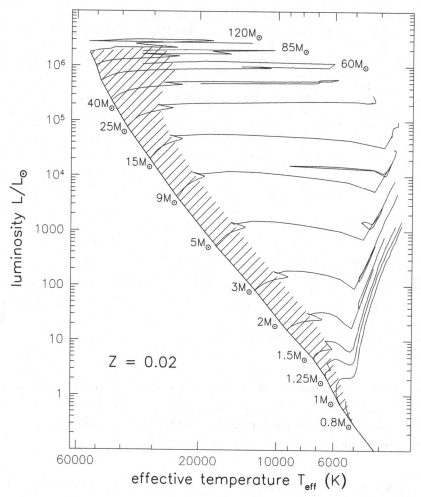

Fig. 1.4. Luminosity and effective temperature during the main-sequence and later lives of stars with solar composition: the hatched region shows where the star burns hydrogen in its core. Only the main-sequence track is shown for the $0.8\mathcal{M}_\odot$ star – Geneva Observatory tracks.

A better approximation is

$$\log(\tau_{MS}/10\,\text{Gyr}) = 1.015 - 3.49\log(\mathcal{M}/\mathcal{M}_\odot) + 0.83[\log(\mathcal{M}/\mathcal{M}_\odot)]^2. \quad (1.9)$$

The most massive stars will burn out long before the Sun. None of the O stars shining today were born when dinosaurs walked the Earth 100 million years ago, and all those we now observe will burn out before the Sun has made another circuit of the Milky Way. But we have not included any stars with $\mathcal{M} < 0.8\mathcal{M}_\odot$ in Figure 1.4, because none has left the main sequence since the Big Bang, $\sim$14 Gyr

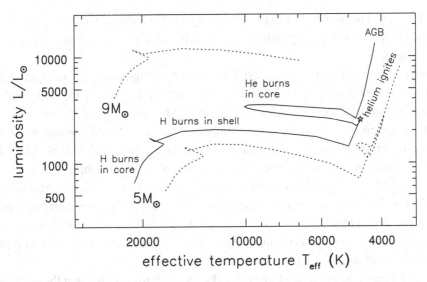

Fig. 1.5. Evolutionary tracks of a $5\mathcal{M}_\odot$ and a $9\mathcal{M}_\odot$ star with solar composition (dotted curves), and a metal-poor $5\mathcal{M}_\odot$ star with $Z = 0.001 \approx Z_\odot/20$ (solid curve). The metal-poor star makes a 'blue loop' while burning helium in its core; it is always brighter and bluer than a star of the same mass with solar metallicity – Geneva Observatory tracks.

in the past. Most of the stellar mass of galaxies is locked into these dim long-lived stars.

Decreasing the fraction of heavy elements in a star makes it brighter and bluer; see Figure 1.5. The 'metals' are a source of *opacity*, blocking the escape of photons which carry energy outward from the core through the interior and the atmosphere. If the metal abundance is low, light moves to the surface more easily; as a result, a metal-poor star is more compact, meaning that it is denser. So its core must be hotter, and produce more energy. Consequently, the star uses up its nuclear fuel faster.

In regions of a star where photons carry its energy out toward the surface, collisions between atoms cannot mix the 'ash' of nuclear burning with fresh material further out. The star, which began as a homogeneous ball of gas, develops strata of differing chemical composition. Convection currents can stir up the star's interior, mixing the layers. Our figures and table are computed for stars that do not spin rapidly on their axes. Fast rotation encourages mixing, and the fresh hydrogen brought into the star's core extends its life on the main sequence.

At the end of its main-sequence life, the star leaves the hatched area in Figure 1.4. Its life beyond that point is complex and depends very much on the star's mass. All stars below about $0.6\mathcal{M}_\odot$ stay on the main sequence for so long that none has yet left it in the history of the Universe. In *low-mass stars* with $0.6\mathcal{M}_\odot \lesssim M \lesssim 2\mathcal{M}_\odot$, the hydrogen-exhausted core gives out energy by shrinking; it becomes denser, while the star's outer layers puff up to a hundred times their former size. The star now radiates its energy over a larger area, so Equation 1.3 tells us that its surface temperature must fall; it becomes cool and red. This is the *subgiant* phase.

When the temperature just outside the core rises high enough, hydrogen starts to burn in a surrounding shell: the star becomes a *red giant*. Helium 'ash' is deposited onto the core, making it contract further and raising its temperature. The shell then burns hotter, so more energy is produced, and the star becomes gradually brighter. During this phase, the tracks of stars with $M \lesssim 2M_\odot$ lie close together at the right of Figure 1.4, forming the red giant branch. Stars with $M \lesssim 1.5M_\odot$ give out most of their energy as red giants and in later stages; see Table 1.1. By contrast with main-sequence stars, the luminosity and color of a red giant depend very little on its mass; so the giant branches in stellar systems of different ages can be very similar. Just as on the main sequence, stars with low metallicity are somewhat bluer and brighter.

As it contracts, the core of a red giant becomes dense enough that the electrons of different atoms interact strongly with each other. The core becomes *degenerate*; it starts to behave like a solid or a liquid, rather than a gas. When the temperature at its core has increased to about 10^8 K, helium ignites, burning to carbon; this releases energy that heats the core. In a gas, expansion would dampen the rate of nuclear reactions to produce a steady flow of energy. But the degenerate core cannot expand; instead, like a liquid or solid, its density hardly changes, so burning is explosive, as in an uncontrolled nuclear reactor on Earth. This is the *helium flash*, which occurs at the very tip of the red giant branch in Figure 1.4. In about 100 s, the core of the star heats up enough to turn back into a normal gas, which then expands.

On the red giant branch, the star's luminosity is set by the mass of its helium core. When the helium flash occurs, the core mass is almost the same for all stars below $\sim 2M_\odot$; so these stars should reach the same luminosity at the tip of the red giant branch. In any stellar population more than 2–3 Gyr old, stars above $2M_\odot$ have already completed their lives; if the metal abundance is below $\sim 0.5Z_\odot$, the red giants have almost the same color. So the apparent brightness at the tip of the red giant branch can be used to find the distance of a nearby galaxy.

Helium is now steadily burning in the core, and hydrogen in a surrounding shell. In Figure 1.4, we see that stars of $M_\odot$ to $2M_\odot$ stay cool and red during this phase; they are *red clump* stars. In Figure 2.2, showing the luminosity and color of stars close to the Sun, we see a concentration of stars in the red clump. Blue *horizontal branch* stars are in the same stage of burning. In these, little material remains in the star's outer envelope, so the outer gas is relatively transparent to radiation escaping from the hot core. Stars that are less massive or poorer in heavy elements than the red clump will become horizontal branch stars.

Helium burning provides less energy than hydrogen burning. We see from Table 1.1 that this phase lasts no more than 30% as long as the star's main-sequence life. Once the core has used up its helium, it must again contract, and the outer envelope again swells. The star moves onto the *asymptotic giant branch* (AGB); it now burns both helium and hydrogen in shells, and it is more luminous and cooler than it was as a red giant. This is as far as we can follow its evolution in Figure 1.4.

On the AGB, both of the shells undergo pulses of very rapid burning, during which the loosely held gas of the outer layers is lost as a *stellar superwind*. Eventually the hot naked core is exposed, as a *white dwarf*: its ultraviolet radiation ionizes the ejected gas, which is briefly seen as a *planetary nebula*. White dwarfs near the Sun have masses around $0.6\mathcal{M}_\odot$, meaning that at least half of the star's original material has been lost. The white dwarf core can do no further burning, and it gradually cools.

Stars of *intermediate mass*, from $2\mathcal{M}_\odot$ up to $6\mathcal{M}_\odot$ or $8\mathcal{M}_\odot$, follow much the same history, up to the point when helium ignites in the core. Because their central density is lower at a given temperature, the helium core does not become degenerate before it begins to burn. These stars also become red, but Figure 1.4 shows that they are brighter than red giants; their tracks lie above the place where those of the lower-mass stars come together. Once helium burning is under way, the stars become bluer; some of them become *Cepheid variables*, F- and G-type supergiant stars which pulsate with periods between one and fifty days. Cepheids are very useful to astronomers, because the pulsation cycle betrays the star's luminosity: the most massive stars, which are also the most luminous, have the longest periods. So once we have measured the period and apparent brightness, we can use Equation 1.1 to find the star's distance. Cepheids are bright enough to be seen far beyond the Milky Way. In the 1920s, astronomers used them to show that other galaxies existed outside our own.

Once the core has used up its helium, these stars become red again; they are asymptotic giant branch stars, with both hydrogen and helium burning in shells. Rapid pulses of burning dredge gas up from the deep interior, bringing to the surface newly formed atoms of elements such as carbon, and heavier atoms which have been further 'cooked' in the star by the *s-process*: the slow capture of neutrons. For example, the atmospheres of some AGB stars show traces of the short-lived radioactive element technetium. The stellar superwind pushes polluted surface gas out into the interstellar environment; these AGB stars are a major source of the elements carbon and nitrogen in the Galaxy.

An intermediate-mass star makes a spectacular planetary nebula, as its outer layers are shed and subsequently ionized by the hot central core. The core then cools to become a white dwarf. Stars at the lower end of this mass range leave a core which is mainly carbon and oxygen; remnants of slightly more massive stars are a mix of oxygen, neon, and magnesium. We know that white dwarfs cannot have masses above $1.4\mathcal{M}_\odot$; so these stars put most of their material back into the interstellar gas.

In *massive stars*, with $M \gtrsim 8\mathcal{M}_\odot$, the carbon, oxygen, and other elements left as the ashes of helium burning will ignite in their turn. The star Betelgeuse is now a red supergiant burning helium in its core. It probably began its main-sequence life 10–20 Myr ago, with a mass between $12 \mathcal{M}_\odot$ and $17 \mathcal{M}_\odot$. It will start to burn heavier elements, and finally explode as a supernova, within another 2 Myr. After their time on the main sequence, massive stars like Betelgeuse spend most of their time as blue or yellow supergiants; Deneb, the brightest star in the constellation Cygnus, is a yellow supergiant. Helium starts to burn in the core of a $25 \mathcal{M}_\odot$ star while it is a blue supergiant, only slightly cooler than it was on the main sequence. Once the core's helium is exhausted, this star becomes a red supergiant; but mass loss can then turn it once again into a blue supergiant before the final conflagration.

The later lives of stars with $\mathcal{M} \gtrsim 40 \mathcal{M}_\odot$ are still uncertain, because they depend on how much mass has been lost through strong *stellar winds*, and on ill-understood details of the earlier convective mixing. A star of about $50 \mathcal{M}_\odot$ may lose mass so rapidly that it never becomes a red supergiant, but is stripped to its nuclear-burning core and is seen as a blue *Wolf–Rayet* star. These are very hot stars, with characteristic strong emission lines of helium, carbon, and nitrogen coming from a fast stellar wind; the wind is very poor in hydrogen, since the star's outer layers were blown off long before. Wolf–Rayet stars live less than 10 Myr, so they are seen only in regions where stars have recently formed.

Once helium burning has finished in the core, a massive star's life is very nearly over. The carbon core quietly burns to neon, magnesium, and heavier elements. But this process is rapid, giving out little energy; most of that energy is carried off by *neutrinos*, weakly interacting particles which easily escape through the star's outer layers. A star that started on its main-sequence life with $10 \mathcal{M}_\odot \lesssim \mathcal{M} \lesssim 40 \mathcal{M}_\odot$ will burn its core all the way to iron. Such a core has no further source of energy. Iron is the most tightly bound of all nuclei, and it would *require* energy to combine its nuclei into yet heavier elements. The core collapses, and its neutrons are squeezed so tightly that they become degenerate. The outer layers of the star, falling in at a tenth of the speed of light, bounce off this suddenly rigid core, and are ejected in a blazing *Type II supernova*. Supernova 1987A which exploded in the Large Magellanic Cloud was of this type, which is distinguished by strong lines of hydrogen in its spectrum. The core of the star, incorporating the heavier elements such as iron, is either left as a *neutron star* or implodes as a *black hole*. The gas that escapes is rich in oxygen, magnesium, and other elements of intermediate atomic mass.

A star with an initial mass between $8 \mathcal{M}_\odot$ and $10 \mathcal{M}_\odot$ also ends its life as a Type II supernova, but by a slightly different process; the core probably collapses before it has burned to iron. After the explosion, a neutron star may remain, or the star may blow itself apart completely, like the Type Ia supernovae described below. A Wolf–Rayet star also becomes a supernova. Because its hydrogen has been lost, hydrogen lines are missing from the spectrum, and it is classified as Type Ic. These supernovae may be responsible for the energetic γ-ray bursts that we discuss in Section 9.2. We shall see in Section 2.1 that massive stars are only a tiny fraction of the total; but they are a galaxy's main producers of oxygen and heavier elements. Detailed study of their later lives can tell us how much of each element is returned to the interstellar gas by stellar winds or supernova explosions, and how much will be locked within a remnant neutron star or black hole. In Section 4.3 we will discuss what the abundances of the various elements may tell us about the history of our Galaxy and others.

Further reading: see the books by Ostlie and Carroll, and by Prialnik. For stellar life beyond the main sequence, see the graduate-level treatment of D. Arnett, 1996, *Supernovae and Nucleosynthesis* (Princeton University Press, Princeton, New Jersey).

1.1.4 Binary stars

Most stars are not found in isolation; they are in *binary or multiple star systems*. Binary stars can easily appear to be single objects unless careful measurements are made, and astronomers often say that 'three out of every two stars are in a binary'. Most binaries are widely separated, and the two stars evolve much like single stars. These systems cause us difficulty only because usually we cannot see the two stars as separate objects, even in nearby galaxies. When we observe them, we get a blend of two stars while thinking that we have only one.

In a close binary system, one star may remove matter from the other. It is especially easy to 'steal' gas from a red giant or an AGB star, since the star's gravity does not hold on strongly to the puffed-up outer layers. Then we can have some dramatic effects. For example, if one of the two stars becomes a white dwarf, hydrogen-rich gas from the companion can pour onto its surface, building up until it becomes dense enough to burn explosively to helium, in a sudden flash which we see as a *classical nova*. If the more compact star has become a neutron star or a black hole, gas falls onto its surface with such force that it is heated to X-ray-emitting temperatures.

A white dwarf in a binary can also explode as a *Type Ia supernova*. Such supernovae lack hydrogen lines in their spectra; they result from the explosive burning of carbon and oxygen. If the white dwarf takes enough matter from its binary companion, it can be pushed above the *Chandrasekhar limit* at about $1.4\mathcal{M}_\odot$. No white dwarf can be heavier than this; if it gains more mass, it is forced to collapse, like the iron core in the most massive stars. But unlike that core, the white dwarf still has nuclear fuel: its carbon and oxygen burn to heavier elements, releasing energy which blows it apart. There is no remnant: the iron and other elements are scattered to interstellar space. Much of the iron we now find in the Earth and in the Sun has been produced in these supernovae. Even though close binary stars are relatively rare, they make a significant difference to the life of their host galaxy.

A Type Ia supernova can be as bright as a whole galaxy, with a luminosity of $2 \times 10^9 L_\odot \lesssim L \lesssim 2 \times 10^{10} L_\odot$. The more luminous the supernova, the longer its light takes to fade. So, if we monitor its apparent brightness over the weeks following the explosion, we can estimate its intrinsic luminosity, and use Equation 1.1 to find the distance. Recently, Type Ia supernovae have been observed in galaxies more than 10^{10} light-years away; they are used to probe the structure of the distant Universe.

1.1.5 Stellar photometry: the magnitude system

Optical astronomers, and those working in the nearby ultraviolet and infrared regions, often express the apparent brightness of a star as an *apparent magnitude*. Originally, this was a measure of how much dimmer a star appeared to the eye in comparison with the bright A0 star α Lyrae (Vega). The brightest stars in the sky were of first magnitude, the next brightest were second magnitude, and so on: brighter stars have numerically smaller magnitudes. The apparent magnitudes m_1 and m_2 of two stars with measured fluxes F_1 and F_2 are related by

$$m_1 - m_2 = -2.5 \log_{10}(F_1/F_2). \tag{1.10}$$

So if $m_2 = m_1 + 1$, star 1 appears about 2.5 times brighter than star 2. The magnitude scale is close to that of natural logarithms: a change of 0.1 magnitudes corresponds to about a 10% difference in brightness.

Problem 1.6 Show that, if two stars of the same luminosity form a close binary pair, the apparent magnitude of the pair measured together is about 0.75 magnitudes brighter than either star individually.

We have referred glibly to 'measuring a star's spectrum'. But in fact, this is almost impossible. At far-ultraviolet wavelengths below $912\,\text{Å}$, even small amounts of hydrogen gas between us and the star absorb much of its light. The Earth's atmosphere blocks out light at wavelengths below $3000\,\text{Å}$, or longer than a few microns. In addition to the light pollution caused by humans, the night sky itself emits light. Figure 1.6 shows that the sky is relatively dim between $4000\,\text{Å}$ and $5500\,\text{Å}$; at longer wavelengths, emission from atoms and molecules in the Earth's atmosphere is increasingly intrusive. Taking high-resolution spectra of faint stars is also costly in telescope time. For all these reasons, we often settle instead for measuring the amount of light that we receive over various broad ranges of wavelength. Thus, our magnitudes and apparent brightness most often refer to a specific region of the spectrum.

We define standard *filter bandpasses*, each specified by the fraction of light $0 \leq \mathcal{T}(\lambda) \leq 1$ that it transmits at wavelength λ. When all the star's light is passed

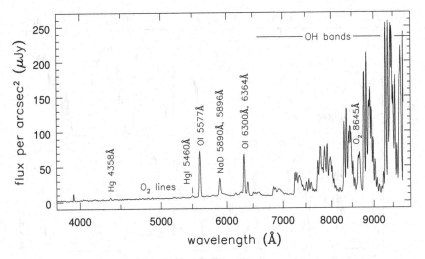

Fig. 1.6. Sky emission in the visible region, at La Palma in the Canary Islands – C. Benn.

by the filter then $\mathcal{T} = 1$, while $\mathcal{T} = 0$ means that no light gets through at this wavelength. The star's apparent brightness in the bandpass described by the filter $\mathcal{T}_{\mathrm{BP}}$ is then

$$F_{\mathrm{BP}} \equiv \int_0^\infty \mathcal{T}_{\mathrm{BP}}(\lambda) F_\lambda(\lambda) \mathrm{d}\lambda \approx F_\lambda(\lambda_{\mathrm{eff}}) \Delta\lambda, \qquad (1.11)$$

where the effective wavelength λ_{eff} and width $\Delta\lambda$ are defined in Table 1.2. The lower panel in Figure 1.7 shows one set of standard bandpasses for the optical and near-infrared part of the spectrum. The R and I bands are on the 'Cousins' system: the 'Johnson' system includes bands with the same names but at different wavelengths, so beware of confusion! In the visible region, these bands were originally defined by the transmission of specified glass filters and the sensitivity of photographic plates or photomultiplier tubes.

The upper curve in Figure 1.7 gives the transmission of the Earth's atmosphere. Astronomers refer to the wavelengths where it is fairly transparent, roughly from $3400\,\text{Å}$ to $8000\,\text{Å}$, as *visible* light. At the red end of this range, we encounter absorption bands of water and of atmospheric molecules such as oxygen, O_2. Between about $9000\,\text{Å}$ and $20\,\mu\text{m}$, windows of transparency alternate with regions where light is almost completely blocked. For $\lambda \gtrsim 20\,\mu\text{m}$ up to a few millimeters, the atmosphere is not only opaque; Figure 1.15 shows that it emits quite brightly. The standard infrared bandpasses have been placed in relatively transparent regions. The K' bandpass is very similar to K, but it has become popular because it blocks out light at the longer-wavelength end of the K band, where atmospheric molecules and warm parts of the telescope emit strongly. Magnitudes measured in these standard bands are generally corrected to remove the

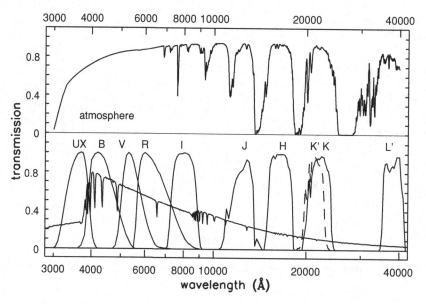

Fig. 1.7. Above, atmospheric transmission in the optical and near-infrared. Below, flux F_λ of a model A0 star, with transmission curves $\mathcal{T}(\lambda)$ for standard filters (from Bessell 1990 *PASP* **102**, 1181). UX is a version of the U filter that takes account of atmospheric absorption. For $JHK'KL'$, $\mathcal{T}(\lambda)$ describes transmission through the atmosphere and subsequently through the filter.

dimming effect of the Earth's atmosphere; they refer to the stars as we should observe them from space.

The lower panel of Figure 1.7 also shows the spectrum from a model A0 star. The Balmer jump occurs just at the blue edge of the B band, so the difference between the U and the B magnitudes indicates its strength; we can use it to measure the star's temperature. Because atmospheric transmission changes greatly between the short- and the long-wavelength ends of the U bandpass, the correction for it depends on how the star's flux $F_\lambda(\lambda)$ varies across the bandpass. So U-band fluxes are tricky to measure, and alternative narrower filters are often used instead. The R band includes the Balmer Hα line. Where many hot stars are present they ionize the gas around them, and Hα emission can contribute much to the luminosity in the R band.

The apparent magnitudes of two stars measured in the same bandpass defined by the transmission $\mathcal{T}_{\mathrm{BP}}(\lambda)$ are related by

$$m_{1,\mathrm{BP}} - m_{2,\mathrm{BP}} = -2.5 \log_{10} \left\{ \int_0^\infty \mathcal{T}_{\mathrm{BP}}(\lambda) F_{1,\lambda}(\lambda) \mathrm{d}\lambda \Big/ \int_0^\infty \mathcal{T}_{\mathrm{BP}}(\lambda) F_{2,\lambda}(\lambda) \mathrm{d}\lambda \right\}.$$

$$(1.12)$$

Table 1.2 Fluxes of a standard A0 star with $m = 0$ in bandpasses of Figure 1.7

	UX	B	V	R	I	J	H	K	L'
λ_{eff}	3660 Å	4360 Å	5450 Å	6410 Å	7980 Å	1.22 μm	1.63 μm	2.19 μm	3.80 μm
F_λ	4150	6360	3630	2190	1130	314	114	39.6	4.85
F_ν	1780	4050	3635	3080	2420	1585	1020	640	236
Zero point ZP_λ	−0.15	−0.61	0.0	0.55	1.27	2.66	3.76	4.91	7.18
Zero point ZP_ν	0.78	−0.12	0.0	0.18	0.44	0.90	1.38	1.89	2.97

Note: the bandpass UX is defined in Figure 1.7; data from Bessell *et al.* 1988 *AAp* **333**, 231 and M. McCall. For each filter, the effective wavelength $\lambda_{\text{eff}} \equiv \int \lambda \mathcal{T}_{\text{BP}} F_\lambda(\lambda)\,d\lambda / \int \mathcal{T}_{\text{BP}} F_\lambda(\lambda)\,d\lambda$, while the effective width $\Delta\lambda = \int \mathcal{T}_{\text{BP}}\,d\lambda$.
F_ν is in janskys, F_λ is in units of 10^{-12} erg s^{-1} cm^{-2} Å^{-1} or 10^{-11} W m^{-2} μm^{-1}.
Zero point ZP: $m = -2.5 \log_{10} F_\lambda + 8.90 - ZP_\lambda$ or $m = -2.5 \log_{10} F_\nu + 8.90 - ZP_\nu$ in these units.

Table 1.3 Photometric bandpasses used for the Sloan Digital Sky Survey

Bandpass	u	g	r	i	z
Average $\langle\lambda\rangle$	3551 Å	4686 Å	6165 Å	7481 Å	8931 Å
Width $\Delta\lambda$	580 Å	1260 Å	1150 Å	1240 Å	995 Å
Sun's magnitude: $M_\odot$	6.55	5.12	4.68	4.57	4.60

$\langle\lambda\rangle$ is the average wavelength; $\Delta\lambda$ is the full width at half maximum transmission, for point objects observed at an angle ZA to the zenith, where $1/\cos(ZA) = 1.3$ (1.3 airmasses); $M_\odot$ is the Sun's 'flux-based' absolute magnitude in each band: Data Release 4.

These 'in-band' magnitudes are generally labelled by subscripts: m_B is an apparent magnitude in the B bandpass of Figure 1.7, and m_R is the apparent magnitude in R. Originally, the star Vega was defined to have apparent magnitude zero in all optical bandpasses. Now, a set of A0 stars is used to define the zero point, and Vega has apparent magnitude 0.03 in the V band. Sirius, which appears as the brightest star in the sky, has $m_V \approx -1.45$; the faintest stars measured are near $m_V = 28$, so they are roughly 10^{12} times dimmer. Table 1.2 gives the *effective wavelength* – the mean wavelength of the transmitted light – for a standard A0 star viewed through those filters, and the fluxes F_λ and F_ν which correspond to apparent magnitude $m = 0$ in each filter.

At ultraviolet wavelengths, there is no well-measured set of standard stars to define the magnitude system, so 'flux-based' magnitudes were developed instead. The apparent magnitude m_{BP} in the bandpass specified by $\mathcal{T}_{\text{BP}}$ of a star with flux $F_\lambda(\lambda)$ is

$$m_{\text{BP}} = -2.5 \log_{10} \left(\frac{\langle F_{\text{BP}} \rangle}{\langle F_{V,0} \rangle} \right), \quad \text{where} \quad \langle F_{\text{BP}} \rangle \equiv \frac{\int \mathcal{T}_{\text{BP}}(\lambda) F_\lambda(\lambda)\,d\lambda}{\int \mathcal{T}_{\text{BP}}(\lambda)\,d\lambda}. \quad (1.13)$$

Here $\langle F_{V,0} \rangle \approx 3.63 \times 10^{-9}$ erg s^{-1} cm^{-2} Å^{-1}, the average value of F_λ over the V band of a star which has $m_V = 0$. Equivalently, when $\langle F_{BP} \rangle$ is measured in erg s^{-1} cm^{-2} Å^{-1}, we have

$$m_{BP} = -2.5 \log_{10} \langle F_{BP} \rangle - 21.1; \qquad\qquad (1.14)$$

the zero point ZP$_\lambda$ of Table 1.2 is equal to zero for all 'flux-based' magnitudes. Magnitudes on this scale do not coincide with those of the traditional system, except in the V band, and we no longer have $m_{BP} = 0$ for a standard A0 star. The Sloan Digital Sky Survey used a specially-built 2.5 m telescope to measure the brightness of 100 million stars and galaxies over a quarter of the sky, taking spectra for a million of them. The survey used 'flux-based' magnitudes in the filters of Table 1.3.

Non-astronomers often ask why the rather awkward magnitude system survives in use: why not simply give the apparent brightness in W m^{-2}? The answer is that, in astronomy, our relative measurements are often much more accurate than absolute ones. The relative brightness of two stars that are observed through the same telescope, with the same detector equipment, can be established to within 1%. The total (bolometric) luminosity of the Sun is well determined, but the apparent brightness of other stars can be compared with a laboratory standard no more accurately than within about 3%. One major problem is absorption in the Earth's atmosphere, through which starlight must travel to reach our telescopes. The fluxes in Table 1.2 were derived by using a model stellar atmosphere, which proves to be more precise than trying to correct for terrestrial absorption. At wavelengths longer than a few microns we do use physical units, because the response of the telescope is less stable. The power of a radio source is often known only to within 10%, so a comparison with terrestrial sources is as accurate as intercomparing two objects in the sky.

The color of a star is defined as the difference between the amounts of light received in each of two bandpasses. If one star is bluer than another, it will give out relatively more of its light at shorter wavelengths: this means that the difference $m_B - m_R$ will be *smaller* for a blue star than for a red one. Astronomers refer to this quantity as the '$B - R$ color' of the star, and often denote it just by $B - R$. Other colors, such as $V - K$, are defined in the same way. We always subtract the apparent magnitude in the longer-wavelength bandpass from that in the shorter-wavelength bandpass, so that a low or negative number corresponds to a blue star and a high one to a red star. Table 1.4 gives colors for main-sequence stars of each spectral type in most of the bandpasses of Figure 1.7.

Astronomers often try to estimate a star's spectral type or temperature by comparing its color in suitably chosen bandpasses with that of stars of known type. We can see that the blue color $B - V$ is a good indicator of spectral type for A, F, and G stars. But cool M stars, which emit most of their light at red and

Table 1.4 Average magnitudes and colors for main-sequence stars: class V (dwarfs)

	M_V	BC	$U-B$	$B-V$	$V-R$	$V-I$	$J-K$	$V-K$	T_{eff}
O3	−5.8	4.0	−1.22	−0.32					44 500
O5	−5.2	3.8	−1.19	−0.32	−0.14	−0.32	−0.25	−0.99	41 000
O8	−4.3	3.3	−1.14	−0.32	−0.14	−0.32	−0.24	−0.96	35 000
B0	−3.7	3.0	−1.07	−0.30	−0.13	−0.30	−0.23	−0.91	30 500
B3	−1.4	1.6	−0.75	−0.18	−0.08	−0.2	−0.15	−0.54	18 750
B6	−1.0	1.2	−0.50	−0.14	−0.06	−0.13	−0.09	−0.39	14 000
B8	−0.25	0.8	−0.30	−0.11	−0.04	−0.09	−0.06	−0.26	11 600
A0	0.8	0.3	0.0	0.0	0.0	0.0	0.0	0.0	9400
A5	1.8	0.1	0.08	0.19	0.13	0.27	0.08	0.38	7800
F0	2.4	0.1	0.06	0.32	0.16	0.33	0.16	0.70	7300
F5	3.3	0.1	−0.03	0.41	0.27	0.53	0.27	1.10	6500
G0	4.2	0.2	0.05	0.59	0.33	0.66	0.36	1.41	6000
Sun	4.83	0.07	0.14	0.65	0.36	0.72	0.37	1.52	5780
G5	4.93	0.2	0.13	0.69	0.37	0.73	0.41	1.59	5700
K0	5.9	0.4	0.46	0.84	0.48	0.88	0.53	1.89	5250
K5	7.5	0.6	0.91	1.08	0.66	1.33	0.72	2.85	4350
K7	8.3	1.0		1.32	0.83	1.6	0.81	3.16	4000
M0	8.9	1.2		1.41	0.89	1.80	0.84	3.65	3800
M2	11.2	1.7		1.5	1.0	2.2	0.9	4.3	3400
M4	12.7	2.7		1.6	1.2	2.9	0.9	5.3	3200
M6	16.5	4.3			1.9	4.1	1.0	7.3	2600

BC is the bolometric correction defined in Equation 1.16.

infrared wavelengths, all have similar values of $B - V$; the infrared $V - K$ color is a much better guide to their spectral type and temperature. The colors of giant and supergiant stars are slightly different from those of dwarfs; see Tables 1.5 and 1.6.

Optical and near-infrared colors are often more closely related to each other and to the star's effective temperature than to its spectral type. For example, stars very similar to the Sun, with the same colors and effective temperatures, can have spectral classification G1 or G3. The colors listed in Tables 1.4, 1.5, and 1.6 have been compiled from a variety of sources, and they are no more accurate than a few hundredths of a magnitude. But because the colors of different stars are measured in the same way, the *difference* in color between two stars can be found more accurately than either color individually.

We define the *absolute magnitude M* of a source as the apparent magnitude it would have at a standard distance of 10 pc. A star's absolute magnitude gives the same information as its luminosity. If there is no dust or other obscuring matter between us and the star, it is related by Equation 1.1 to the measured apparent magnitude m and distance d:

$$M = m - 5\log_{10}(d/10\,\text{pc}). \tag{1.15}$$

Table 1.5 Average magnitudes and colors for red giant stars: class III

	M_V	BC	$U-B$	$B-V$	$V-R$	$V-I$	$J-K$	$V-K$	T_{eff}
B0	−5.1	2.8							29 500
G5	0.9	0.3	0.50	0.88	0.48	0.93	0.57	2.10	5000
K0	0.7	0.4	0.90	1.02	0.52	1.00	0.63	2.31	4800
K5	0.3	1.1	1.87	1.56	0.84	1.63	0.95	3.60	3900
M0	−0.4	1.3	1.96	1.55	0.88	1.78	1.01	3.85	3850
M3	−0.6	1.8	1.83	1.59	1.10	2.47	1.13	4.40	3700
M5	−0.4	3	1.56	1.57	1.31	3.05	1.23	5.96	3400
M7	v	5	0.94	1.69	3.25	5.56	1.21	8.13	3100

Note: M7 stars of class III are often variable.

Table 1.6 Average magnitudes and colors for supergiant stars: class I

	M_V	BC	$U-B$	$B-V$	$V-R$	$V-I$	$V-K$	T_{eff}
O8	−6.3	3.2	−1.07	−0.24				33 000
O9.5	−6.3	2.9						30 500
B0	−6.3	2.8	−1.03	−0.22	−0.08	−0.2		29 000
B6	−6.2	1.0	−0.72	−0.09	−0.01	−0.07		13 500
A0	−6.3	0.2	−0.44	0.02	0.05	0.11	0.9	9600
F0	−6.6	−0.1	0.16	0.17	0.12	0.25		7700
G5	−6.2	0.4	0.84	1.02	0.44	0.82	3	4850
K5	−5.8	1.0	1.7	1.60	0.81	1.50		3850
M0	−5.6	1.4	1.9	1.71	0.95	1.91	4	3650

Note: supergiants have a large range in luminosity at any spectral type; Type Ia (luminous) and Ib (less luminous) supergiants can differ by 2 or 3 magnitudes.

As for apparent magnitudes, the bandpass in which the absolute magnitude of a star has been measured is indicated by a subscript. The Sun has absolute magnitudes $M_B = 5.48$, $M_V = 4.83$, $M_K = 3.31$; because it is redder than an A0 star, the absolute magnitude is numerically smaller in the longer-wavelength bandpasses. Supergiant stars have $M_V \sim -6$; they are over 10 000 times more luminous than the Sun in this band.

The absolute V-magnitudes listed in the tables are averages for each spectral subclass. For main-sequence stars near the Sun, the dispersion in M_V measured in magnitudes for each subclass ranges from about 0.4 for A and early F stars to 0.5 for late F and early G stars, decreasing to about 0.3 for late K and early M stars. This small variation arises because stars change their colors and luminosities as they age, and also differ in their metal content. But supergiants with the same spectral classification can differ in luminosity by as much as 2 or 3 magnitudes.

To compare observations with theoretical models, we need to find the total amount of energy coming from a star, integrated over all wavelengths; this is its *bolometric* luminosity L_{bol}. Because we cannot measure all the light of a star, we use models of stellar atmospheres to find how much energy is emitted in the regions that we do not observe directly. Then, we can define a *bolometric magnitude M_{bol}*. The zero point of the scale is set by fixing the Sun's absolute bolometric magnitude as $M_{\text{bol},\odot} = 4.75$. The second column in each of Tables 1.4, 1.5, and 1.6 gives the *bolometric correction*, the amount that must be *subtracted* from M_V to obtain the bolometric magnitude:

$$M_{\text{bol}} = M_V - \text{BC}. \qquad (1.16)$$

For the Sun, BC ≈ 0.07. Bolometric corrections are small for stars that emit most of their light in the blue–green part of the spectrum. They are large for hot stars, which give out most of their light at bluer wavelengths, and for the cool red stars. A warning: some astronomers prefer to define the bolometric correction with a + sign in Equation 1.16.

Finally, stellar and galaxy luminosities are often expressed as multiples of the Sun's luminosity. From near-ultraviolet wavelengths to the near-infrared range at a few microns, we say that a star has $L = 10L_\odot$ in a given bandpass if its luminosity in that bandpass is ten times that of the Sun *in the same bandpass*. But at frequencies at which the Sun does not emit much radiation, such as the X-ray and the radio, a source is generally said to have $L = 10L_\odot$ in a given spectral region if its luminosity there is ten times the Sun's *bolometric* luminosity. Occasionally, this latter definition is used for all wavebands.

Problem 1.7 A star cluster contains 200 F5 stars at the main-sequence turnoff, and 20 K0III giant stars. Use Tables 1.2 and 1.4 to show that its absolute V-magnitude $M_V \approx -3.25$ and its color $B - V \approx 0.68$. (These values are similar to those of the 4 Gyr-old cluster M67: see Table 2.2.)

Problem 1.8 After correcting for dust dimming (see Section 1.2), the star Betelgeuse has average apparent magnitude $m_V = 0$ and $V - K \approx 5$. (Like many supergiants, it is variable: m_V changes by roughly a magnitude over 100–400 days.) Taking the distance $d = 140\,\text{pc}$, find its absolute magnitude in V and in K.

Show that Betelgeuse has $L_V \approx 1.7 \times 10^4 L_{V,\odot}$ while at K its luminosity is much larger compared with the Sun: $L_K \approx 4.1 \times 10^5 L_{K,\odot}$. Using Table 1.4 to find a rough bolometric correction for a star with $V - K \approx 5$, show that $M_{\text{bol}} \approx -8$, and the bolometric luminosity $L_{\text{bol}} \approx 1.2 \times 10^5 L_{\text{bol},\odot}$. Looking back at Problem 1.3, show that the star radiates roughly 4.6×10^{31} W. (The magnitude system can sometimes be confusing.)

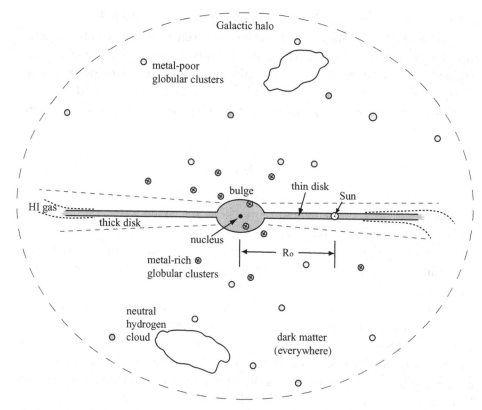

Fig. 1.8. A schematic side view of the Milky Way.

1.2 Our Milky Way

We are resident in the Milky Way, which is also called the Galaxy (with a capital G). Here, we have a close-up view of the stellar and gaseous content of a typical large spiral galaxy. This section gives a brief sketch of our Galaxy, and how we observe the gas and dust that lie between the stars. We also define some of the coordinate systems by which astronomers locate objects on the sky and within the Milky Way.

An external observer might see the Milky Way looking something like what is drawn in Figure 1.8. The Sun lies some way from the center, in the stellar disk that is the Milky Way's most prominent feature. As its name implies, the disk is thin and roughly circular; when we look out on a dark night, the disk stars appear as a luminous band stretching across the heavens. Dark patches in this band mark concentrations of dust and dense gas. In the southern sky, the bright central regions are seen as a bulge extending above and below the disk. At the center of the bulge is a dense *nucleus* of stars; this harbors a radio source, and a black hole with mass $\mathcal{M}_{BH} \approx 4 \times 10^6 \mathcal{M}_\odot$.

We generally measure distances within the Galaxy in *kiloparsecs*: 1 kpc is 1000 pc, or about 3×10^{16} km. The Milky Way's central bulge is a few kiloparsecs in radius, while the stellar disk stretches out to at least 15 kpc, with the Sun about 8 kpc from the center. The density n of stars in the disk drops by about a factor of e as we move out in radius R by one *scale length* h_R, so that $n(R) \propto e^{-R/h_R}$. Estimates for h_R lie in the range $2.5-4.5$ kpc.

The *thin disk* contains 95% of the disk stars, and all of the young massive stars. Its *scale height*, the distance we must move in the direction perpendicular to the disk to see the density fall by a factor of e, is 300–400 pc. The rest of the stars form the *thick disk*, which has a larger scale height of about 1 kpc. We will see in Chapter 2 that stars of the thick disk were made earlier in the Galaxy's history than those of the thin disk, and they are poorer in heavy elements. The gas and dust of the disk lie in a very thin layer; near the Sun's position, most of the neutral hydrogen gas is within 100 pc of the midplane. The thickness of the gas layer increases roughly in proportion to the distance from the Galactic center.

Both the Milky Way's disk and its bulge are rotating. Stars in the disk orbit the Galactic center at about $200 \, \mathrm{km \, s^{-1}}$, so the Sun takes roughly 250 Myr to complete its orbit. Disk stars follow nearly circular orbits, with small additional random motions amounting to a few tens of kilometers per second. Bulge stars have larger random velocities. We will see in Chapter 3 that this means they must orbit the center with a lower average speed, closer to $100 \, \mathrm{km \, s^{-1}}$. Stars and globular clusters of the *metal-poor halo* do not have any organized rotation around the center of the galaxy. Like comets in the solar system, their orbits follow random directions, and are often *eccentric*: the stars spend most of their time in the outer reaches of the Galaxy but plunge deeply inward at pericenter.

In all, the luminosity of the disk is about $(15-20) \times 10^9 L_\odot$, and the mass in stars is around $60 \times 10^9 \mathcal{M}_\odot$. For the bulge $L \approx 5 \times 10^9 L_\odot$, while the mass of stars is about $20 \times 10^9 \mathcal{M}_\odot$. The halo stars form only a small fraction of the Galaxy's mass, accounting for no more than about $10^9 \mathcal{M}_\odot$. When we measure the orbital speeds of gas, stars, and star clusters at large distances from the center of the Milky Way, and use Equation 3.20 to find the mass required to keep them in those orbits, we find that the total mass of the Galaxy must be more than just that present in the stars and gas. In particular, most of the Galaxy's mass appears to lie more than 10 kpc from the center, where there are relatively few stars. We call this the *dark matter* and usually assume, without a compelling reason, that it lies in a roughly spherical *dark halo*. The nature of the unseen material making up the dark halo of our Galaxy and others is one of the main fields of research in astronomy today.

1.2.1 Gas in the Milky Way

In the neighborhood of the Sun, we find about one star in every $10 \, \mathrm{pc^3}$. The diameter of a solar-type star is only about 10^{-7} pc, so most of interstellar space is empty of stars; but it is filled with gas and dust. This dilute material makes itself apparent both by absorbing radiation from starlight that travels through it and by its own emission. We receive radiation from gas within the Milky Way that is *ionized* (the atoms have lost one or more of their electrons), from neutral atoms, and from molecules. The radiation can be in the form of emission lines, or as *continuum* emission, a continuous spectrum without lines.

Atoms and ions radiate when one of their electrons makes a quantum jump to a lower energy level; the line photon carries off energy equal to the difference between the states. If m-times-ionized element X (written X^{+m}) captures an electron it becomes $X^{+(m-1)}$, which typically forms in an excited state. As this newly recombined ion relaxes to its ground state, a whole cascade of *recombination radiation* is emitted. Transitions between barely bound high levels produce radio-frequency photons, whereas electrons falling to lower levels give visible light: the Balmer lines of hydrogen correspond to transitions down to level $n = 2$. We observe these transitions in *H*II *regions* around hot stars, where hydrogen is almost completely ionized. Transitions to the lowest energy levels give rise to more energetic photons. When H^+ captures an electron to become neutral, and its electron drops from $n = 2$ into the ground state $n = 1$, it gives out an ultraviolet photon of the Lyman-α line at 1216 Å (10.2 eV). In heavier atoms, the lowest-level electrons are more tightly bound, and transitions to these states correspond to X-ray photons. From very hot gas we often see the 6.7 keV K line of Fe^{+24}, 24-times-ionized iron.

Gas can be *photoionized*: energetic photons liberate electrons from their atoms. O and B stars produce ultraviolet photons with wavelength below 912 Å, or energy above 13.6 eV, which are required to ionize hydrogen from its ground state. These stars develop HII regions, but cooler stars do not. Atoms can also be excited up to higher levels by collisions with electrons. *Collisional ionization* is important when the gas temperature T is high enough that the average particle energy $k_B T$ is comparable to νh_P, the energy of the emitted photon that corresponds to the difference between the levels. When atom A collides with atom B to form the excited state A*, we can have the reaction

$$A + B \longrightarrow A^* + B, \quad A^* \longrightarrow A + \nu h_P.$$

However, we see an emission line only if state A* decays before colliding yet again. Either the decay must be rapid, or the gas density quite low. *Forbidden lines* violate the quantum-mechanical rules that specify the most probable transitions (electric dipole) by which an atom could return to its ground state. These 'rule-breaking' transitions occur via less-probable slower pathways. They are not observed in dense laboratory plasmas because A* typically collides before it can decay. The electron of a hydrogen atom takes only 10^{-8} s to move from level $n = 3$ to $n = 2$ by radiating an Hα photon, but for forbidden lines this is typically 1 s or longer. At the *critical density* n_{crit} the line is close to its maximum strength;

Table 1.7 Some common optical and infrared forbidden lines of atoms

Atom	Transition	Wavelength	n_{crit} (cm^{-3})
CI	$^3P_1 \rightarrow {}^3P_0$	610 μm	(500)
	$^3P_2 \rightarrow {}^3P_1$	371 μm	(1 000)
CII (C$^+$)	$^2P_{3/2} \rightarrow {}^2P_{1/2}$	158 μm	(3 000)
NII (N$^+$)	$^1D_2 \rightarrow {}^3P_2$	6583 Å	66 000
	$^1D_2 \rightarrow {}^3P_1$	6548 Å	66 000
OI	$^1D_2 \rightarrow {}^3P_2$	6300 Å	2×10^6
	$^3P_1 \rightarrow {}^3P_2$	63.2 μm	(10^6)
	$^3P_0 \rightarrow {}^3P_1$	145.5 μm	(10^5)
OII (O$^+$)	$^2D_{5/2} \rightarrow {}^4S_{3/2}$	3729 Å	3 400
	$^2D_{3/2} \rightarrow {}^4S_{3/2}$	3726 Å	15 000
OIII (O^{++})	$^1D_2 \rightarrow {}^3P_2$	5007 Å	7×10^5
	$^1D_2 \rightarrow {}^3P_1$	4959 Å	7×10^5
	$^1S_0 \rightarrow {}^1D_2$	4363 Å	2×10^7
	$^3P_2 \rightarrow {}^3P_1$	51.8 μm	(4 000)
	$^3P_1 \rightarrow {}^3P_0$	88.4 μm	(2 000)
NeII (Ne$^+$)	$^2P_{1/2} \rightarrow {}^2P_{3/2}$	12.8 μm	7×10^5
NeIII (Ne^{++})	$^3P_1 \rightarrow {}^3P_2$	15.6 μm	2×10^5
	$^3P_0 \rightarrow {}^3P_1$	36.0 μm	3×10^4
NeV (Ne^{+4})	$^1D_2 \rightarrow {}^3P_2$	3426 Å	2×10^7
SII (S$^+$)	$^2D_{5/2} \rightarrow {}^4S_{3/2}$	6716 Å	2 000
	$^2D_{3/2} \rightarrow {}^4S_{3/2}$	6731 Å	2 000
SIII (S^{++})	$^3P_2 \rightarrow {}^3P_1$	18.7 μm	10 000
	$^3P_1 \rightarrow {}^3P_0$	33.5 μm	2 000
SiII (Si$^+$)	$^2P_{3/2} \rightarrow {}^2P_{1/2}$	34.8 μm	(3×10^5)
FeII (Fe^{++})	$a\,^4D_{7/2} \rightarrow a\,^4D_{9/2}$	1.64 μm	3×10^5

The line is close to maximum strength at the critical density n_{crit}. This density is calculated for collision with electrons at $T = 10\,000$ K, except quantities in () which refer to collisions with H atoms or molecules near 100 K.

in denser gas, collisions are so frequent that A* is most often knocked out of its excited state before it emits the photon.

Because their intensity depends so strongly on these quantities, forbidden lines often give us detailed information on the density and temperature of the emitting gas. Astronomers indicate them with square brackets; they refer, for example, to the [OIII] line at 5007 Å. Forbidden lines of ionized 'metals' such as OII, OIII (once- and twice-ionized oxygen, respectively), NII, and SII account for most of the energy radiated from HII regions. Table 1.7 lists some common forbidden lines.

Fine structure transitions reflect the coupling between the electron's orbital angular momentum and its spin. They correspond roughly to energy differences only $1/137^2$ times as large as between the main levels, so for neutral atoms or low ions, wavelengths lie in the far-infrared. The fine-structure lines of carbon, oxygen, and nitrogen are in the 10–300 μm range, and must be observed from aircraft in the stratosphere or from space. Because the energies of the excited states are less, these lines are important at lower temperatures. The line of singly ionized carbon CII at 158 μm, and lines of neutral oxygen at 63 μm and 145 μm, carry most of the energy radiated by the Milky Way's atomic gas at $T \sim 100\,\text{K}$. The line brightness is set by the rate at which C^+ ions collide with energetic-enough electrons or H_2 molecules. The energy of a photon at 158 μm corresponds to $T = 91.2\,\text{K}$; in cooler gas, collisions are too slow to excite CII into the upper state. In a dilute gas with n_e electrons and $n(C^+)$ carbon ions per cm^3, energy is radiated at the rate

$$\Lambda(C^+) = n_e n(C^+)\, T^{-1/2} \exp(-91.2\,\text{K}/T) \times 8 \times 10^{-20}\,\text{erg}\,\text{cm}^{-3}\,\text{s}^{-1}. \quad (1.17)$$

The exponential term arises because the electrons follow the *Maxwellian* energy distribution of Equation 3.58. The line brightness falls rapidly as the temperature drops below 100 K.

Hyperfine transitions are a consequence of coupling between the nuclear spin and the magnetic field generated by the orbiting electron. They have energy splittings 2000 times smaller yet than the fine-structure lines. The most important of these is in the hydrogen atom, as the spin of the electron flips from being parallel to the proton spin, to antiparallel. A photon in the 21 cm line carries off the small amount of energy released. The average hydrogen atom takes about 11 Myr to make this transition; but, since hydrogen is by far the most abundant element, 21 cm radio emission is ubiquitous in the Galaxy.

Problem 1.9 To find the strength of 21 cm emission, we first ask how often hydrogen atoms collide with each other. If the atoms have average speed v_{th}, each has energy of motion $m_p v_{th}^2/2 = 3k_B T/2$: show that $v_{th} \approx 2\,\text{km}\,\text{s}^{-1}$ at $T = 100\,\text{K}$. We regard each as a small sphere of mass m_p with the Bohr radius $a_0 = h_p^2/(4\pi^2 m_e e^2) = 0.529 \times 10^{-8}\,\text{cm}$. With n_H atoms per cm^3, the average time t_{coll} for one to run into another is

$$\frac{1}{t_{coll}} = n_H \pi a_0^2\, v_{th}, \quad \text{so}\ t_{coll} \approx 3000 \left(\frac{1\,\text{cm}^{-3}}{n_H}\right)\left(\frac{100\,\text{K}}{T}\right)^{1/2}\ \text{yr}. \quad (1.18)$$

In nearby neutral clouds $n_H \geq 0.1$ (see Table 2.4), so t_{coll} is much shorter than the time taken to emit a 21 cm photon. The energy difference between the excited and ground states is tiny, so after repeated collisions there will be equal numbers of atoms in each possible state. The excited atom with spins parallel (unit angular momentum) has three distinct states, the ground state (with opposed spins and zero angular momentum) only one. Thus each cubic centimeter emits $3n_H/4$ photons per 11 Myr.

Molecules can radiate like atoms, as electrons move between energy levels. But the atoms within the molecule also have quantized levels of vibration, and of rotation about the center of mass. We see radiation as the molecule jumps between these levels. The vibrational levels of common small molecules such as CO, CS, and HCN are separated by energies corresponding to emission at a few microns, whereas the rotational transitions lead to millimeter waves. The symmetric H_2 molecule has no dipole moment, so its rotational transitions are $\approx 137^2$ times slower. The least energetic transition of H_2 corresponds to emission at 20 μm, so cold H_2 hardly radiates at all; only shocked gas, with $T \sim 1000\,K$, gives strong emission. Ultraviolet photons below 2000 Å can excite a hydrogen molecule into a higher electronic level, where its energy is larger than that of two isolated hydrogen atoms. About 10% of these molecules *dissociate* into two H atoms; the rest emit an ultraviolet photon to return to the ground state. The ultraviolet spectra of hot stars show many absorption lines from cool H_2 between us and the star.

Table 1.8 lists some commonly observed molecular lines. H_2 is often excited by ultraviolet light; but other molecules are mainly excited by collisions, and we see their lines only when the typical thermal energy $k_B T$ exceeds the energy E_{upper} of the upper level. The decay rate for a line at frequency v is proportional to v^3, so the upper states are long-lived. The line strengths depend on the density, temperature and composition of the gas, and on the radiation shining on it. Unlike the atomic lines, the strong rotational transitions of molecules like CO, CS, and HCN are optically thick: a line photon trying to escape from a molecular cloud is likely to be absorbed by another molecule, putting it into an excited state. So we cannot simply define a critical density as we could for the atomic lines.

For CO, the most abundant molecule after H_2, the rotational transitions at 1.3 mm and 2.6 mm are normally strongest at densities $n(H_2) \sim 100\text{--}1000\,\text{cm}^{-3}$ and require only $T \sim 10\text{--}20\,K$ for excitation. Higher transitions at shorter wavelengths require larger densities and higher temperatures, so we can investigate the state of the emitting gas by observing multiple lines. Emission from NH_3, CS, and HCN is strongest at densities 10 to 100 times higher than for CO. We see SiO emission where silicate dust grains have been broken apart in shocks. Beyond the Milky Way, lines such as CO are usually weak in dwarf galaxies because their gas is poor in heavy elements. HCN is often strong in galaxies with active nuclei, where intense radiation shines on dense gas; but X-rays from the active nucleus easily destroy HCO^+.

Near luminous stars and protostars, and around active galactic nuclei (see Section 9.1), *masing* can occur in molecular transitions, such as those of OH at 1.7 GHz and of water, H_2O, at 22 GHz. Collisions with fast-moving molecules of hot H_2, and intense infrared radiation, can excite these molecules in a way that puts more of them into the higher of two energy states. Radiation corresponding to a transition down to the lower state is then amplified by stimulated emission. We see a masing spot when the emission happens to be beamed in our direction: the spot is very small, and the radiation intense.

We often use emission or absorption lines to measure the velocities of gas clouds or stars. If a light source is moving away from us, the wavelength λ_{obs} at

Table 1.8 Commonly observed lines from molecular clouds

Molecule	Transition	Frequency ν	Wavelength λ	$E_{\text{upper}}/k_{\text{B}}$ (K)	[a]Typical $n(H_2)$ (cm^{-3})
H_2	$v = 1-0$, S(1)	140 THz	2.128 μm		
H_2 ortho	$v = 0-0$, S(0)	11 THz	28.2 μm	510	
	$v = 0-0$, S(2)	24 THz	12.3 μm	1682	
H_2 para	$v = 0-0$, S(1)	18 THz	17.1 μm	1015	
CO	$J = 1 \rightarrow 0$	115.3 GHz	2.6 mm	5.5	~100
	$J = 2 \rightarrow 1$	230.5 GHz	1.3 mm	17	~1000
	$J = 3 \rightarrow 2$	345.8 GHz	0.87 mm	34	10^3–10^4
CS	$J = 1 \rightarrow 0$	49 GHz	6.1 mm	2.4	>5000
	$J = 5 \rightarrow 4$	244 GHz	1.2 mm	35	10^6
SiO	$J = 2 \rightarrow 1$	86.8 GHz	3.5 mm	6.3	[c]shocks
	$J = 5 \rightarrow 4$	217.1 GHz	1.4 mm	31.3	[c]shocks
	$J = 8 \rightarrow 7$	347.3 GHz	0.86 mm	75	[c]shocks
HCO^+	$J = 1 \rightarrow 0$	89 GHz	3.4 mm	4.3	>3000
	$J = 3 \rightarrow 2$	268 GHz	1.1 mm	26	>30000
HCN	$J = 1 \rightarrow 0$	89 GHz	3.4 mm	4.3	>10000
	$J = 3 \rightarrow 2$	266 GHz	1.1 mm	26	>10^5
HNC	$J = 1 \rightarrow 0$	91 GHz	3.3 mm	4.3	>10000
	$J = 3 \rightarrow 2$	272 GHz	1.1 mm	26	>10^5
NH_3 para	$(J, K) = (1, 1)-(1, 1)$	23.69 GHz	12.7 mm	23	2×10^3
	$(J, K) = (2, 2)-(2, 2)$	23.72 GHz	12.6 mm	64	2×10^3
NH_3 ortho	$(J, K) = (3, 3)-(3, 3)$	23.87 GHz	12.6 mm	122	
H_2CO ortho	$2_{12} \rightarrow 1_{11}$	140.8 GHz	2.1 mm	21.9	10^5
	$3_{12} \rightarrow 2_{11}$	225.7 GHz	1.3 mm	33.5	5×10^5
	$5_{33} \rightarrow 4_{32}$	364.3 GHz	0.82 mm	158.4	10^6
H_2CO para	$2_{02} \rightarrow 1_{01}$	145.6 GHz	2.1 mm	10.5	2×10^5
	$3_{22} \rightarrow 2_{21}$	218.5 GHz	1.4 mm	68.1	2×10^5
	$5_{23} \rightarrow 4_{22}$	365.4 GHz	0.82 mm	99.7	2×10^6
[b]OH	$^2\Pi_{3/2}$, $J = 3/2$	1.7 GHz	18 cm	0.1	10^4–10^6
[b]H_2O ortho	$6_{16} \rightarrow 5_{23}$	22.2 GHz	13.5 mm	640	10^7–10^9

[a] This density depends on cloud size, radiation field, etc.
[b] This line is often seen as a maser.
[c] This line indicates that shocks at speeds of 10–40 km s^{-1} have disrupted dust grains.

which we observe the line will be longer than the wavelength λ_e where it was emitted; if it moves toward us, we have $\lambda_{\text{obs}} < \lambda_e$. The *redshift* is the fractional change in wavelength $z = \lambda_{\text{obs}}/\lambda_e - 1$. For speeds well below that of light, we have the *Doppler* formula:

$$1 + z \equiv \frac{\lambda_{\text{obs}}}{\lambda_e} = 1 + \frac{V_r}{c}; \tag{1.19}$$

V_r is the *radial velocity*, the speed at which the source moves away from us, and c is the speed of light. Radio telescopes routinely measure wavelengths and velocities to about one part in 10^6, while optical telescopes normally do no better than one part in 10^5. Astronomers correct for the motion of the Earth, as it varies during the year; quoted velocities are generally *heliocentric*, measured relative to the Sun.

A diffuse ionized gas also produces continuum radiation. *Free–free* radiation (also called *bremsstrahlung*, or braking radiation) is produced when the electrical forces from ions deflect free electrons onto curved paths, so that they radiate. Hot gas in the center of the Milky Way, and in clusters and groups of galaxies, has $T \sim 10^7$–10^8 K; its free–free radiation is mainly X-rays. Ionized gas in HII regions around hot stars, with $T \sim 10^4$ K, can be detected via free–free emission at radio wavelengths which penetrates the surrounding dusty gas. Strong magnetic fields also force electrons onto curved paths; if they are moving at nearly the speed of light, they give out strongly polarized *synchrotron* radiation. This process powers the radio emission of supernova remnants, and the radio source at the Milky Way's nucleus. If the electrons have very high energy, synchrotron radiation can be produced at optical and even X-ray energies.

About 1% of the mass of the interstellar material consists of dust particles, mainly silicates and forms of carbon, smaller than $\sim 1\,\mu\text{m}$. These grains scatter and absorb radiation efficiently at wavelengths less than their own dimensions. Dust heated by diffuse stellar radiation has $T \sim 10$–20 K and glows at $\sim 200\,\mu\text{m}$; dust near bright stars is hotter.

When dust is spread uniformly, light loses an equal fraction of its power for every parsec that it travels through the dusty gas. Then, if two observers at x and $x + \Delta x$ look at a distant star in the negative-x direction, it will appear brighter to the closer observer. We can write the apparent brightness at wavelength λ as

$$F_\lambda(x + \Delta x) = F_\lambda(x)[1 - \kappa_\lambda \, \Delta x], \qquad (1.20)$$

where the *opacity* κ represents the rate at which light is absorbed. If the distant star is at position $x_0 < x$, we have

$$\frac{\mathrm{d}F_\lambda}{\mathrm{d}x} = -\kappa_\lambda F_\lambda, \quad \text{so} \quad F_\lambda(x) = F_{\lambda,0}e^{-\kappa_\lambda(x-x_0)} = F_{\lambda,0}e^{-\tau_\lambda}, \qquad (1.21)$$

where $F_{\lambda,0}$ is the apparent brightness that we would have measured without the dust, and τ_λ is the *optical depth* of the dust layer. Blue and ultraviolet radiation is more strongly scattered and absorbed by dust than red light, so dust between us and a star makes it appear both dimmer and redder. For interstellar dust, we often make the approximation $\kappa_\lambda \propto 1/\lambda$ at visible wavelengths in the range $3000\,\text{Å} < \lambda < 1\,\mu\text{m}$.

Problem 1.10 When a source is dimmed by an amount $e^{-\tau_\lambda}$, show that, according to Equation 1.10, its apparent magnitude *increases* by an amount $A_\lambda = 1.086\tau_\lambda$. A_λ is called the *extinction* at wavelength λ.

Problem 1.11 Near the Sun, the diffuse interstellar gas has a density of about one atom cm^{-3}. Show that you would need to compress a cube of gas 30 km on a side into $1\,cm^3$ to bring it to Earth's normal atmospheric density and pressure (6×10^{23} atoms in 22.4 liters: a cube 10 cm on a side has a volume of one liter). Interstellar gas is about 10^{10} times more rarefied than a good laboratory vacuum, which is itself $\sim 10^{10}$ times less dense than Earth's atmosphere.

Assume that each dust grain is a sphere of radius $0.1\,\mu m$, and the gas contains one grain for every 10^{12} hydrogen atoms. Show that, as light travels through a 1 cm layer of the compressed gas in the previous problem, about 1% of it will be intercepted. Show that $\kappa = 0.015\,cm^{-1}$, so that a layer about 70 cm thick would block a fraction $1/e$ of the rays ($\tau_\lambda = 1$). If the air around you were as dusty as interstellar space you could see for less than a meter, as in the London fogs described by Charles Dickens.

Problem 1.12 Assuming that the Milky Way's luminosity $L \approx 2 \times 10^{10} L_\odot$, and making the very rough approximation that it is a sphere 5 kpc in radius, use Equation 1.3 to show that, if it radiated as a blackbody, $T_{eff} \approx 5\,K$. Near the Sun, starlight heats interstellar dust to 15–20 K.

On scales larger than ~ 1 pc, dust is fairly evenly mixed with the Galactic gas. Looking 'up' toward the north Galactic pole, we see distant objects dimmed by an average of 0.15 magnitudes or 13% in the V band. If the dusty layer had constant thickness z_0, objects beyond the Milky Way seen at an angle b from the Galactic pole are viewed through a length $z_0/\cos b$ of dusty gas. So τ_λ, and the increase in the star's apparent magnitude, proportional to $1/\cos b$. Since the dusty gas is quite clumpy, at high latitudes we can make a better estimate of this *Galactic extinction* by using 21 cm emission to measure the neutral hydrogen, and assuming the amount of dust to be proportional. Roughly, the extinction A_V in the V band is related to the number N_H of hydrogen atoms per square centimeter by

$$N_H \approx 1.8 \times 10^{21}\,cm^{-2} \times A_V (\text{magnitudes}). \qquad (1.22)$$

We will discuss our own Galaxy's interstellar gas further in Section 2.4, and that of other galaxies in Chapters 4, 5, and 6.

1.2.2 What's where in the Milky Way: coordinate systems

Just as we use latitude and longitude to specify the position of a point on Earth, we need a way to give the positions of stars on the sky. Often we use *equatorial* coordinates, illustrated in Figure 1.9. We imagine that the stars lie on the *celestial sphere*, a very large sphere centred on the Earth, and define the *celestial poles* as the points that are directly overhead at the Earth's north and south poles. The

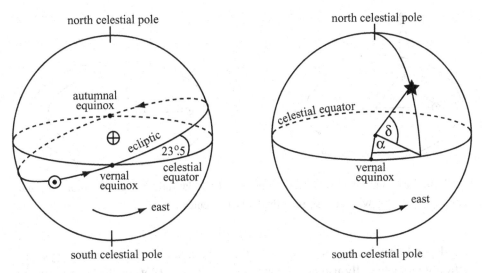

Fig. 1.9. The celestial sphere, showing the ecliptic: right ascension α is measured eastward from the vernal equinox, and declination δ from the celestial equator.

celestial equator is the great circle on the celestial sphere that runs directly above the Earth's equator.

A star's *declination*, akin to latitude on Earth, is the angle between its position on the celestial sphere, and the nearest point on the celestial equator. An object at the north celestial pole has declination $\delta = 90°$, whereas at the south celestial pole it has $\delta = -90°$. During the night, the Earth turns anticlockwise on its axis as seen from above the north pole, so the stars appear to rise in the east and move westward across the sky, circling the celestial poles. Each star rises where the circle of 'latitude' corresponding to its declination cuts the eastern horizon, and sets where that circle intersects the horizon in the west.

Throughout the year, the Sun appears to move slowly from west to east against the background of the stars; it lies north of the celestial equator in June and south of it in January, following the great circle of the *ecliptic*. The ecliptic is inclined by 23°27′ to the celestial equator, intersecting it at the *vernal equinox* and the *autumnal equinox*. So the Sun crosses the equator twice a year: at the vernal equinox in the spring, usually on March 21st, and at the autumnal equinox around September 23rd. To define a longitude on the sky, we use the vernal equinox as a zero point, like the Greenwich meridian on Earth. The 'longitude' of a star is its *right ascension*, denoted by α. Right ascension is measured eastward from the vernal equinox in hours, with 24 hours making up the complete circle.

The direction of the Earth's rotation axis changes slowly because of *precession*: the celestial poles and equator do not stay in fixed positions on the sky. The vernal equinox moves westward along the ecliptic at about 50″ per year, so the *tropical year* from one vernal equinox to the next is about 20 minutes shorter than the Earth's orbital period, the *sidereal year*. Hence the coordinates α, δ of a star

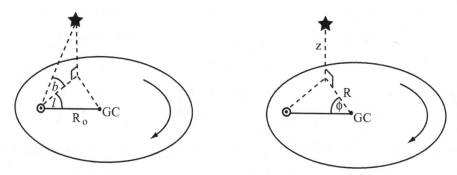

Fig. 1.10. Left, Sun-centred Galactic longitude l and latitude b; right, cylindrical polar coordinates R, ϕ, z with the origin at the Galactic center.

will depend on which year we take as the reference for our coordinate system. Astronomers generally use coordinates relative to the 1950 or 2000 equinox, or the equinox of the current year. Computer programs easily convert between these systems. But more than one astronomer has pointed a telescope in the wrong direction by forgetting to specify the equinox.

The stars that have reached their highest point in the sky at any moment all lie on a great circle passing through the celestial poles and through the *zenith*, the point directly overhead. These stars all have the same right ascension. On the date of the autumnal equinox, in September, the position of the vernal equinox, and all stars with right ascension zero, are highest in the sky at midnight. Three hours later, when the Earth has made an eighth of a turn on its axis, the stars further east, with right ascension 3^h, will be at their highest; and so on through the night. So the positions of the stars can be used to tell the time. At any moment, the right ascension of the stars that are at their highest gives the local star time, or *sidereal* time: all observatories have clocks telling sidereal time, as well as the usual *civil* time.

Problem 1.13 Draw a diagram to show that, as the Earth circles the Sun during the course of a year, relative to the stars it makes $366\frac{1}{4}$ rotations on its axis. The number of sidereal days in a year is one more than the number of solar days, which are measured from midnight to midnight. So a sidereal day lasts only 23^h56^m.

To give the positions of stars as we see them in relation to the Milky Way, we use the Sun-centred system of Galactic latitude and longitude shown in the left panel of Figure 1.10. The center of the Galaxy lies in the direction $\alpha = 17^h42^m24^s$, $\delta = -28°55'$ (equinox 1950). Galactic longitude l is measured in the plane of the disk from the Sun–center line, defined as $l = 0$, toward the direction of the Sun's rotation, $l = 90°$. The region $0 < l < 180°$ is sometimes called the 'northern' half of the Galaxy, because it is visible from the Earth's northern hemisphere, while $180° < l < 360°$ is the 'southern' Galaxy. The latitude b

gives the angle of a star away from the plane of the disk; b is measured positive toward the *north Galactic pole* at $\alpha = 12^h49^m$, $\delta = 27°24'$ (1950). The north Galactic pole is just the pole of the disk that is visible from the Earth's northern hemisphere. Inconveniently, the Earth's rotation axis and that of the Milky Way are at present about 120° apart, so the Milky Way's rotation is clockwise when seen from 'above' the north Galactic pole; its spin axis points closer to the direction of the south Galactic pole.

To specify the positions of stars in three-dimensional space, we can use a system of Galactocentric cylindrical polar coordinates R, ϕ, z (Figure 1.10). The radius R measures the distance from the Galactic center in the disk plane; the height 'above' the midplane is given by z, with $z > 0$ in the direction of the north Galactic pole. The azimuthal angle ϕ is measured from the direction toward the Sun, so that it is positive in the direction toward $l = 90°$. For motions near the Sun, we sometimes use Cartesian coordinates x, y, z with x pointing radially outwards and y in the direction of the Sun's rotation.

We take a more detailed look at the Milky Way in Chapters 2 and 3.

1.3 Other galaxies

This section introduces the study of galaxies other than our own Milky Way. We discuss how to classify galaxies according to their appearance in optical light, and how to measure the amount of light that they give out. Although big galaxies emit most of the light, the most common type of galaxy is a tiny dim dwarf.

The existence of other galaxies was established only in the 1920s. Before that, they were listed in catalogues of *nebulae*: objects that appeared fuzzy in a telescope and were therefore not stars. Better images revealed stars within some of these 'celestial clouds'. Using the newly opened 100″ telescope on Mount Wilson, Edwin Hubble was able to find variable stars in the Andromeda 'nebula' M31. He showed that their light followed the same pattern of changing brightness as Cepheid variable stars within our Galaxy. Assuming that all these stars were of the same type, with the same luminosities, he could find the relative distances from Equation 1.1. He concluded that the stars of Andromeda were at least 300 kpc from the Milky Way, so the nebula must be a galaxy in its own right. We now know that the Andromeda galaxy is about 800 kpc away.

Hubble set out his scheme for classifying the galaxies in a 1936 book, *The Realm of the Nebulae*. With later additions and modifications, this system is still used today; see Figure 1.11. Hubble recognized three main types of galaxy: ellipticals, lenticulars, and spirals, with a fourth class, the irregulars, for galaxies that would not fit into any of the other categories.

Elliptical galaxies are usually smooth, round, and almost featureless, devoid of such photogenic structures as spiral arms and conspicuous dust lanes. Ellipticals are generally lacking in cool gas and consequently have few young blue stars.

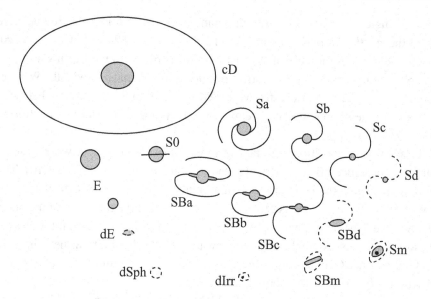

Fig. 1.11. Galaxy classification: a modified form of Hubble's scheme.

Though they all appear approximately elliptical on the sky, detailed study shows that large bright ellipticals have rather different structures from their smaller and fainter counterparts.

Ellipticals predominate in rich clusters of galaxies, and the largest of them, the *cD galaxies*, are found in the densest parts of those clusters. Around an elliptical core, the enormous diffuse envelope of a cD galaxy may stretch for hundreds of kiloparsecs; these systems can be up to 100 times more luminous than the Milky Way. Normal or *giant* ellipticals have luminosities a few times that of the Milky Way, with characteristic sizes of tens of kiloparsecs. The stars of these bright ellipticals show little organized motion, such as rotation; their orbits about the galaxy center are oriented in random directions. The left panel of Figure 1.12 shows a giant elliptical, which has an active nucleus (see Section 9.1) that is a bright compact radio source.

In less luminous elliptical galaxies, the stars have more rotation and less random motion. Often there are signs of a disk embedded within the elliptical body. The very faintest ellipticals, with less than $\sim 1/10$ of the Milky Way's luminosity, split into two groups. The first comprises the rare compact ellipticals, like the nearby system M32. The other group consists of the faint diffuse *dwarf elliptical* (dE) galaxies, and their even less luminous cousins the *dwarf spheroidal* (dSph) galaxies, which are so diffuse as to be scarcely visible on sky photographs. The right panel of Figure 1.12 shows a dwarf elliptical satellite of M31. The dE and dSph galaxies show almost no ordered rotation.

Lenticular galaxies show a rotating disk in addition to the central elliptical bulge, but the disk lacks any spiral arms or extensive dust lanes. These galaxies are labelled *S0* (pronounced 'ess-zero'), and they form a transition class between

Fig. 1.12. Left, the giant elliptical galaxy NGC 3998; brightness rapidly increases toward the center, which is over-exposed. Almost all the faint compact objects are globular clusters. Right, nearby dwarf elliptical NGC 147 in the V band; we see individual stars in its outer parts. The brightest images are foreground stars of the Milky Way – WIYN telescope.

ellipticals and spirals. They resemble ellipticals in lacking extensive gas and dust, and in preferring regions of space that are fairly densely populated with galaxies; but they share with spirals the thin and fast-rotating stellar disk. The left panel of Figure 1.13 shows an SB0 galaxy, with a central linear bar.

Spiral galaxies (Figure 1.14) are named for their bright spiral arms, which are especially conspicuous in the blue light that was most easily recorded by early photographic plates. The arms are outlined by clumps of bright hot O and B stars, and the compressed dusty gas out of which these stars form. About half of all spiral and lenticular galaxies show a central linear bar: the barred systems SB0, SBa, ..., SBd form a sequence parallel to that of the unbarred galaxies. Along the sequence from Sa spirals to Sc and Sd, the central bulge becomes less important relative to the rapidly rotating disk, while the spiral arms become more open and the fraction of gas and young stars in the disk increases. Our Milky Way is probably an Sc galaxy, or perhaps an intermediate Sbc type; M31 is an Sb. On average, Sc and Sd galaxies are less luminous than the Sa and Sb systems, but some Sc galaxies are still brighter than a typical Sa spiral.

At the end of the spiral sequence, in the Sd galaxies, the spiral arms become more ragged and less well ordered. The Sm and SBm classes are *Magellanic spirals*, named after their prototype, which is our Large Magellanic Cloud; see Section 4.1. In these, the spiral is often reduced to a single stubby arm. As the galaxy luminosity decreases, so does the speed at which the disk rotates; dimmer galaxies are less massive. The Large Magellanic Cloud rotates at only $80 \, \mathrm{km \, s^{-1}}$, a third as fast as the Milky Way. Random stellar motions are also diminished in

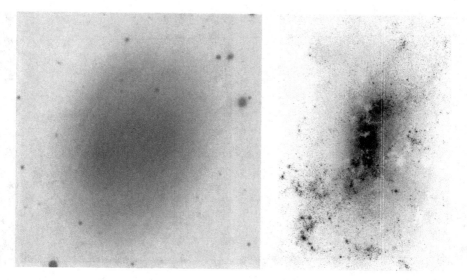

Fig. 1.13. Negative images of two disk galaxies. Left, NGC 936, a luminous barred S0 with $L \approx 2 \times 10^{10} L_\odot$; the smooth disk has neither dust lanes nor spiral arms. Luminous regions appear darkest in this negative image – CTIO Blanco telescope. Right, NGC 4449, classified as irregular or SBm; this is a small gas-rich galaxy with $L \approx 4 \times 10^9 L_\odot$. Bright star-forming knots are strewn about the disk – A. Aloisi, F. Annibali, and J. Mack; Hubble Space Telescope/NASA/ ESA.

the smaller galaxies, but even so, ordered rotational motion forms a less important part of their total energy. We indicate this in Figure 1.11 by placing these galaxies to the left of the Sd systems.

The terms 'early type' and 'late type' are often used to describe the position of galaxies along the sequence from elliptical galaxies through S0s to Sa, Sb, and Sc spirals. Some astronomers once believed that this progression might describe the life cycle of galaxies, with ellipticals turning into S0s and then spirals. Although this hypothesis has now been discarded, the terms live on. Confusingly, 'early-type' galaxies are full of 'late-type' stars, and vice versa.

Hubble placed all galaxies that did not fit into his other categories in the *irregular* class. Today, we use that name only for small blue galaxies which lack any organized spiral or other structure (Figure 1.13). The smallest of the irregular galaxies are called *dwarf irregulars*; they differ from the dwarf spheroidals by having gas and young blue stars. It is possible that dwarf spheroidal galaxies are just small dwarf irregulars which have lost or used up all of their gas. Locally, about 70% of moderately bright galaxies are spirals, 30% are elliptical or S0 galaxies, and 3% are irregulars.

Other galaxies that Hubble would have called irregulars include the *starburst galaxies*. These systems have formed many stars in the recent past, and their disturbed appearance results in part from gas thrown out by supernova explosions.

Fig. 1.14. Our nearest large neighbor, the Andromeda galaxy M31; north is to the right, east is upward. Note the large central bulge of this Sb galaxy, and dusty spiral arms in the disk. Two satellites are visible: M32 is round and closer to the center, NGC 205 is the elongated object to the west – O. Nielsen.

Interacting galaxies, in which two or more systems have come close to each other, and galaxies that appear to result from the merger of two or more smaller systems, would also have fallen into this class. We have come to realize that galaxies are not 'island universes', but affect each other's development throughout their lives. Chapters 4, 5, and 6 of this book discuss the structures of nearby galaxies, while Chapter 7 considers how galaxies interact in groups and clusters.

We usually refer to galaxies by their numbering in a catalogue. Charles Messier's 1784 catalogue lists 109 objects that look 'fuzzy' in a small telescope; it includes the Andromeda galaxy as M31. The *New General Catalogue* of more than 7000 nonstellar objects includes clusters of stars and gaseous nebulae as well as galaxies. This catalogue, published by J. L. E. Dreyer in 1888, with additions in 1895 and 1908, was based largely on the work of William Herschel (who discovered the planet Uranus), his sister Caroline, and his son John Herschel. The Andromeda galaxy is NGC 224.

Modern catalogues of bright galaxies include the *Third Reference Catalogue of Bright Galaxies*, by G. and A. de Vaucouleurs and their collaborators (1991; Springer, New York), which includes all the NGC galaxies, and the

Uppsala General Catalogue of Galaxies, by P. Nilson (1973; Uppsala Observatory), with its southern extension, the *ESO/Uppsala Survey of the ESO(B) Atlas*, by A. Lauberts (1982; European Southern Observatory). Galaxies that emit brightly in the radio, X-rays, etc., also appear in catalogues of those sources. Many recent catalogues are published electronically; for example, NASA's Extragalactic Database (http://ned.ipac.caltech.edu).

Further reading: E. Hubble, 1936, *The Realm of the Nebulae* (Yale University Press; reprinted by Dover, New York); for pictures to illustrate Hubble's classification, see A. Sandage, 1961, *The Hubble Atlas of Galaxies* (Carnegie Institute of Washington; Washington, DC). A recent graduate text on galaxy classification is S. van den Bergh, 1998, *Galaxy Morphology and Classification* (Cambridge University Press, Cambridge, UK).

1.3.1 Galaxy photometry

Unlike stars, galaxies do not appear as points of light; they extend across the sky. Turbulence in the Earth's atmosphere has the effect of blurring galaxy images; this is known as *seeing*. Because of it, a ground-based optical telescope rarely shows details smaller than about $1/3''$. For sharper images, we must use a telescope in space or resort to techniques such as interferometry or adaptive optics.

Although the classification of galaxies is still based on their appearance in optical images, most work on galaxies is quantitative, measuring how much light, at what wavelengths, is emitted by the different regions. The *surface brightness* of a galaxy $I(\mathbf{x})$ is the amount of light per square arcsecond on the sky at a particular point $\mathbf{x}$ in the image. Consider a small square patch of side D in a galaxy that we view from a distance d, so that it subtends an angle $\alpha = D/d$ on the sky. If the combined luminosity of all the stars in this region is L, its apparent brightness F is given by Equation 1.1; then the surface brightness is

$$I(\mathbf{x}) \equiv \frac{F}{\alpha^2} = \frac{L/(4\pi d^2)}{D^2/d^2} = \frac{L}{4\pi D^2}. \tag{1.23}$$

I is usually given in $\mathrm{mag\,arcsec^{-2}}$ (the apparent magnitude of a star that appears as bright as one square arcsecond of the galaxy's image), or in $L_\odot\,\mathrm{pc^{-2}}$. The surface brightness at any point does not depend on distance unless d is so large that the expansion of the Universe has the effect of reducing $I(\mathbf{x})$; we discuss this further in Section 8.3. Contours of constant surface brightness on a galaxy image are called *isophotes*. Equation 1.23 shows that the position of an isophote within the galaxy is independent of the observer's distance.

We generally measure surface brightness in a fixed wavelength band, just as for stellar photometry. The centers of galaxies reach only $I_B \approx 18\,\mathrm{mag\,arcsec^{-2}}$ or $I_R \approx 16\,\mathrm{mag\,arcsec^{-2}}$, and the stellar disks are much fainter. Galaxies do not have sharp edges, so we often measure their sizes within a fixed isophote. One popular choice is the 25th-magnitude isophote in the blue B band, denoted R_{25}. This is about 1% of the sky level on an average night; before CCD photometry (see Section 5.1), it was close to the limit of what could be measured reliably. Another option is the *Holmberg radius* at $I_B(\mathbf{x}) = 26.5$ mag arcsec^{-2}. To find the luminosity of the whole galaxy, we measure how the amount of light coming from within a given radius grows as that radius is moved outward, and we extrapolate to reach the total.

> **Problem 1.14** In a galaxy at a distance of d Mpc, what would be the apparent B-magnitude of a star like our Sun? In this galaxy, show that $1''$ on the sky corresponds to $5d$ pc. If the surface brightness $I_B = 27$ mag arcsec^{-2}, how much B-band light does one square arcsecond of the galaxy emit, compared with a star like the Sun? Show that this is equivalent to $L_\odot$ pc^{-2} in the B band, but that $I_I = 27$ mag arcsec^{-2} corresponds to only $0.3L_\odot$ pc^{-2} in the I band.

Table 1.9 gives the surface brightness of the *night sky* measured in the bandpasses of Figure 1.7. These are approximate average values, since the sky brightness depends on solar activity (the sunspot cycle), the observatory's location on Earth, and the direction in the sky. Typically, the sky is brighter than all but the inner core of a galaxy, and on a moonlit night even the center can disappear into the bright sky. From Earth's surface, optical observations of galaxies must generally be made during the dark of the moon.

If our eyes could perceive colors at such low light levels, we would see the sky glowing red with emission in the bands of atmospheric molecules. In the near-infrared at $2\,\mu$m, from most observatory sites the sky is over a thousand times brighter than it would be in space. Figure 1.15 shows how steeply its emission rises at longer wavelengths, in the *thermal infrared*. The high cold South Pole is the best Earth-based site at these wavelengths, but the sky is still enormously brighter than from space. Standard infrared filters are chosen to lie where the atmosphere is most transparent. Between these regions, we see a blackbody spectrum corresponding to the temperature of the opaque layers.

We can cut down the sky light by designing our filters to exclude some of the strongest lines; using the K' filter instead of K blocks out about two-thirds of the emission. But Table 1.9 makes clear that, when we observe from the ground, the infrared sky is *always* brighter than the galaxy. To find the surface brightness accurately, we must measure the brightness of a patch of blank sky as it changes throughout the night just as accurately as we measure the galaxy-plus-sky; the small difference between the two gives $I(\mathbf{x})$. A telescope in space gives us a much darker sky at red and infrared wavelengths. We can also observe in the near-ultraviolet, where the sky brightness is yet lower.

Table 1.9 Average sky brightness in ultraviolet, optical, and infrared wavebands

Band	Wavelength	Full moon (mag arcsec^{-2})	Dark sky (mag arcsec^{-2})	From space (mag arcsec^{-2})	From space (μJy arcsec^{-2})	South Pole (μJy arcsec^{-2})
	1500 Å			25.0		
	2000 Å			26.0		
	2500 Å			25.6		
U	3700 Å		22.0	23.2		
B	4400 Å	19.4	22.7	23.4	1.8	
V	5500 Å	19.7	21.8	22.7		
R	6400 Å	19.9	20.9	22.7		
I	8000 Å	19.2	19.9	22.2	3.2	
J	1.2 μm	15.0	15.0	20.7	2.4	300–600
H	1.6 μm	13.7	13.7	20.9	4.4	800–2 000
K	2.2 μm	12.5	12.5	21.3	1.9	300–700
K′	2.2 μm	13.7	13.7	21.3	1.9	500
L	3.3 μm				1.1	10^5
M	4.9 μm				8.0	10^6
N	10.6 μm				220	4×10^7
Q	19 μm				400	3×10^8

The two columns headed 'from space' differ only in their units.

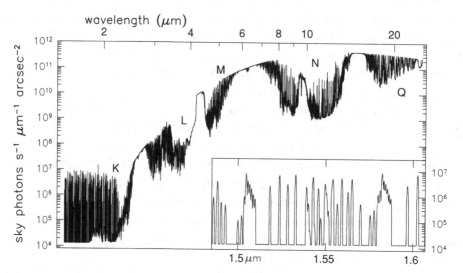

Fig. 1.15. Sky emission on Mauna Kea, Hawaii, at 4000 m elevation; standard infrared bandpasses are indicated. The inset shows that the sky background consists mainly of closely spaced emission lines – Gemini telescope project.

There are many more small dim galaxies than large bright ones. Figure 1.16 shows the number of galaxies measured at absolute magnitude $M(B_J)$, in the 2dF survey from the Anglo-Australian Observatory. Notice that most of the very luminous galaxies are red; these are elliptical and S0 galaxies. Most of the dim galaxies are spirals or irregulars, which are blue because they contain recently born

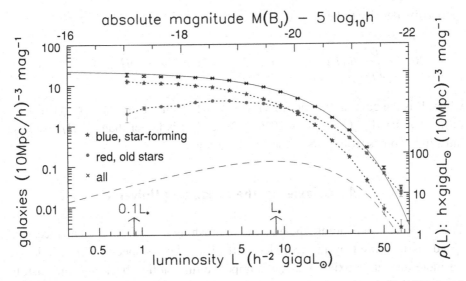

Fig. 1.16. Number of galaxies per 10 Mpc cube between absolute magnitude $M(B_J)$ and $M(B_J) + 1$ (crosses). Dotted lines show numbers of blue (stars) and red (filled dots) galaxies making up this total; vertical bars indicate errors. The solid line shows the luminosity function of Equation 1.24; the dashed line gives $\Phi(M) \times L/L_\star$, the light from galaxies in each interval of absolute magnitude. The blue bandpass B_J is matched to the photographic plates used to select the galaxies – 2dF survey, D. Croton.

massive stars. Although spirals and irregulars are far more numerous, elliptical galaxies contain about half the total mass in stars.

The solid curve in Figure 1.16 shows what is expected if the number of galaxies $\Phi(L)\Delta L$ per Mpc^3 between luminosity L and $L + \Delta L$ is given by

$$\Phi(L)\Delta L = n_\star \left(\frac{L}{L_\star}\right)^\alpha \exp\left(-\frac{L}{L_\star}\right)\frac{\Delta L}{L_\star}; \qquad (1.24)$$

this is the *Schechter function*. According to this formula, the number of galaxies brighter than the luminosity $L_\star$ drops very rapidly. We often use the criterion $L \gtrsim 0.1L_\star$ to define a 'bright' or 'giant' galaxy, as opposed to a dwarf. The solid curve is for $L_\star \approx 9 \times 10^9 h^{-2} L_\odot$, corresponding to $M_\star(B_J) = -19.7 + 5\log_{10}h$; as explained in the next section, the parameter h measures the rate at which the Universe expands. Taking $h = 0.7$, we find $L_\star \approx 2 \times 10^{10} L_\odot$, roughly the Milky Way's luminosity.

The number of galaxies in each unit interval in absolute magnitude is almost constant when $L < L_\star$; the curve is drawn for $n_\star = 0.02h^3\,\mathrm{Mpc}^{-3}$ and $\alpha = -0.46$. The Schechter formula overestimates the density of very faint galaxies; for $\alpha \leq -1$, it even predicts that the total number of galaxies $\int_L^\infty \Phi(L)\mathrm{d}L$ should increase without limit as $L \to 0$. But the dashed line shows that most of the light comes from galaxies close to $L_\star$. Integrating Equation 1.24, we estimate the total luminosity density to be

$$\rho_{\rm L}(B_J) = \int_0^\infty \Phi(L)L\,{\rm d}L = n_\star L_\star \Gamma(\alpha+2) \approx 2\times 10^8 h L_\odot\,{\rm Mpc}^{-3}, \quad (1.25)$$

where Γ is the gamma function; $\Gamma(j+1) = j!$ when j is an integer. In the near-infrared K band $\rho_{\rm L}(K) \approx 6\times 10^8 h L_\odot\,{\rm Mpc}^{-3}$; it is larger than $\rho_{\rm L}(B_J)$ because most light comes from stars redder than the Sun.

1.4 Galaxies in the expanding Universe

The Universe is expanding; the galaxies are rushing away from us. The recession speed, as measured by the Doppler shift of a galaxy's spectral lines, is larger for more distant galaxies. We can extrapolate this motion back into the past, to estimate when the Universe had its beginning in the *Big Bang*. Doing this, we link the recession speed or *redshift* that we measure for a galaxy with the time after the Big Bang at which its light was given out; the redshift becomes a measure of the galaxy's age when it emitted that light.

In 1929, on the basis of only 22 measurements of radial velocities for nearby galaxies, and some distance estimates which turned out to be wrong by about a factor of ten, Hubble claimed that the galaxies are moving away from us with speeds V_r proportional to their distance d:

$$V_r \approx H_0 d. \quad (1.26)$$

Subsequent work proved him right, and this relation is now known as *Hubble's law*. Current estimates for the parameter H_0, the *Hubble constant*, lie between 60 and 75 km s^{-1} Mpc^{-1}. Figure 1.17 shows that galaxies that recede faster are indeed fainter, as expected if they all have roughly the same luminosity, but are progressively more distant.

We often use Hubble's law to estimate the distances of galaxies from their measured velocities. It is common to indicate the uncertainty in the Hubble constant explicitly, by writing h for the value of H_0 in units of 100 km s^{-1} Mpc^{-1}. Then Equation 1.26 implies

$$d = h^{-1}\left[V_r\,({\rm km\,s}^{-1})/100\right]\,{\rm Mpc}. \quad (1.27)$$

When the distance of a galaxy is found from its radial velocity V_r, the derived luminosity $L \propto h^{-2}$. This is why the parameter $L_\star$ of Equation 1.24 has a value proportional to h^{-2}; similarly, the density $n_\star \propto h^3$. If we estimate the mass $\mathcal{M}$ of a galaxy by using Equation 1.2 with a distance from Equation 1.27, together with

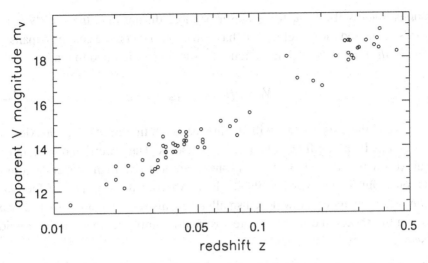

Fig. 1.17. Apparent magnitude in the V band for the brightest galaxies in rich galaxy clusters. The magnitude increases proportionally to the logarithm of the redshift z, as we expect if the galaxy's distance is proportional to its recession speed cz – data from J. E. Gunn and J. B. Oke 1975 *ApJ* **195**, 255.

Newton's equation for the gravitational force (see Section 3.1), then we have that $\mathcal{M} \propto h^{-1}$.

If the average speeds of the galaxies had always remained constant, they would have been on top of each other at a time t_H before the present, where

$$t_H = \frac{1}{H_0} = 9.78h^{-1}\,\text{Gyr} = 15\,\text{Gyr} \times \frac{67\,\text{km s}^{-1}\,\text{Mpc}^{-1}}{H_0}. \qquad (1.28)$$

This is called the *Hubble time*; we can use it as a rough estimate of the age of the Universe, the time since the Big Bang.

Problem 1.15 If a galaxy has absolute magnitude M, use Equations 1.1 and 1.27 to show that its apparent magnitude m is related to the redshift $z = V_r/c$ of Equation 1.19 by $m = M + 5\log_{10} z + C$, where C is a constant, the same for all objects. Draw an approximate straight line through the points in Figure 1.17; check that its slope is roughly what you would expect if the brightest galaxy in a rich cluster always had the same luminosity.

Using Hubble's law to find approximate distances for galaxies, we can examine their distribution in space: Figure 8.3 shows the region of the 2dF survey of Figure 1.16. We do not see galaxies spread uniformly through space, but concentrated into groups and clusters. Within rich clusters, the galaxies' orbits give them peculiar velocities up to $1500\,\text{km s}^{-1}$. So, if we use Equation 1.27 to find their positions, they will appear to be closer or more distant than they really are. A galaxy's measured radial velocity V_r has two components: the cosmic expansion, and a *peculiar velocity* V_{pec}. Equation 1.26 should be modified to read

$$V_r = H_0 d + V_{pec}. \tag{1.29}$$

Between the clusters, we will see in Chapter 8 that individual galaxies and small groups lie along filaments or in large sheets. The groups and associations of galaxies within these filaments and sheets are less rich than clusters, but more numerous. Our Milky Way and its neighbor Andromeda form part of the *Local Group*, which includes a few dozen smaller systems within a radius of 1–2 Mpc. Between the sheets and filaments are vast nearly empty regions; in these voids, we see only a few isolated galaxies.

1.4.1 Densities and ages

In Section 8.2 we will examine the dynamics of the cosmic expansion, and how it is related to the density of matter and energy in the Universe. If the average density is now greater than the *critical density*, the expansion can in future reverse to a contraction; if it is less, the galaxies continue to recede forever. The critical density is

$$\rho_{crit}(\text{now}) = \frac{3H_0^2}{8\pi G} = 1.9 \times 10^{-29} h^2 \, \text{g cm}^{-3}$$
$$= 2.8 \times 10^{11} h^2 \mathcal{M}_\odot \, \text{Mpc}^{-3}. \tag{1.30}$$

For $H_0 = 67 \, \text{km s}^{-1} \, \text{Mpc}^{-1}$, this is equivalent to a good-sized galaxy in each megaparsec cube, or about five hydrogen atoms per m^3. If matter in the Universe has exactly this density, the time t_0 from the Big Bang to the present day is

$$t_0 = \frac{2}{3H_0} \approx 10 \, \text{Gyr} \times \left(\frac{67 \, \text{km s}^{-1} \, \text{Mpc}^{-1}}{H_0} \right). \tag{1.31}$$

If the average density exceeds ρ_{crit}, the Universe is younger than this, whereas if the density is less, it is older. We shall see that the density is unlikely to be greater than the critical value, so the time since the Big Bang is at least that given by Equation 1.31. The present age t_0 can be larger than t_H only if the equations of General Relativity are modified by including *dark energy*, which pushes the galaxies away from each other.

We will see in the following section that normal matter makes up only about 4% of the critical density. In the current *benchmark* model for the cosmic expansion, the total density has exactly the critical value, and $H_0 = 70 \, \text{km s}^{-1}$ Mpc^{-1}. Matter makes up 30% of ρ_{crit}, but most of it is *dark*. The dark matter probably consists of particles that, like neutrinos, are weakly interacting – or we should have seen them – and have some small but nonzero mass. Collectively, these are known as weakly interacting massive particles, or WIMPs. Dark energy accounts for the remainder; we have little idea about its nature. The present age t_0 of the benchmark model is $0.964t_H$ or 13.5 Gyr.

Problem 1.16 Use Equation 1.25 to show that, for the Universe to be at the critical density, the average ratio of mass to luminosity $\mathcal{M}/L$ would have to be approximately $1700h\mathcal{M}_\odot/L_\odot$ in blue light.

1.4.2 Galaxies in the Universe

Why is the history of the Universe relevant to our study of galaxies? First, as we will see in Section 2.2, the Hubble time t_H is very close to the ages that we estimate for the oldest stars in our own Galaxy and others. The galaxies, and the stars in them, can be no older than the Universe. To understand how galaxies came into existence, we must know how much time it took to form the earliest stars, and to build up the elements heavier than helium. The atmospheres of old low-mass stars in galaxies are fossils from the early Universe, preserving a record of the abundances of the various elements in the gas out of which they formed. Our knowledge of stellar evolution provides us with a clock measuring in gigayears how long ago these stars began their lives on the main sequence. The redshifts of distant galaxies tell the time by a different clock, giving information on how long after the Big Bang their light set off on its journey to us. To relate times measured by these two clocks, we must know how the scale of the Universe has changed with time. In Section 8.2 we will see how to calculate the *scale length* $\mathcal{R}(t)$, which grows proportionally to the distance between the galaxies; the Hubble constant H_0 is given by $\dot{\mathcal{R}}(t_0)/\mathcal{R}(t_0)$. For the simplest models, $\mathcal{R}(t)$ depends only on the value of H_0 and the present density $\rho(t_0)$.

The expansion of the Universe also affects the light that we receive from galaxies. Consider two galaxies separated by a distance d, separating at speed $V_r = H_0 d$ according to Equation 1.26. If one of these emits light of wavelength λ_e, an observer in the other galaxy will receive it at a time $\Delta t = d/c$ later, with a longer wavelength $\lambda_{obs} = \lambda_e + \Delta\lambda$. If the galaxies are fairly close, so that $V_r \ll c$, we can use the Doppler formula of Equation 1.19 to show that the ratio λ_{obs}/λ_e is

$$1 + \frac{\Delta\lambda}{\lambda_e} \approx 1 + \frac{H_0 d}{c} = 1 + H_0 \Delta t = 1 + \left[\frac{1}{\mathcal{R}(t)} \frac{d\mathcal{R}(t)}{dt} \right]_{t_0} \Delta t. \qquad (1.32)$$

We can rewrite this as an equation for the wavelength λ as a function of time:

$$\frac{1}{\lambda} \frac{d\lambda}{dt} = \frac{1}{\mathcal{R}(t)} \frac{d\mathcal{R}(t)}{dt}. \qquad (1.33)$$

Integrating gives the formula for the *cosmological redshift z:*

$$1 + z \equiv \frac{\lambda_{\text{obs}}}{\lambda_{\text{e}}} = \frac{\mathcal{R}(t_0)}{\mathcal{R}(t_{\text{e}})}, \qquad (1.34)$$

which holds for large redshifts as well as small. Since the wavelength of light expands proportionally to $\mathcal{R}(t)$, its frequency decreases by a factor of $1 + z$. All processes in a distant galaxy appear stretched in time by this factor; when we observe the distant Universe, we see events taking place in slow motion.

We will discuss galaxy groups and clusters in Chapter 7. The other topics of this section are treated in greater depth in Chapters 8 and 9.

1.5 The pregalactic era: a brief history of matter

Here, we sketch what we know of the history of matter in the Universe before the galaxies formed. When a gas is compressed, as in filling a bicycle tire, it heats up; when it is allowed to expand, as in using a pressurized spray can, its temperature drops. The gas of the early Universe was extremely hot and dense, and it has been cooling off during its expansion. This is the *Big Bang* model for the origin of the Universe: the cosmos came into existence with matter at a very high temperature, and expanding rapidly. The physics that we have developed in laboratories on Earth then predicts how this fireball developed into the cosmos that we know today. Two aspects of the early hot phase are especially important for our study of galaxies.

First, the abundance of the lightest elements, hydrogen, deuterium (heavy hydrogen), helium, and lithium, was largely determined by conditions in the first half-hour after the Big Bang. The observed abundance of helium is, amazingly, quite closely what is predicted by the Big Bang model. The measured fraction of deuterium, ^{3}He, and lithium can tell us how much matter the Universe contains. In later chapters, we will compare this figure with the masses that we measure in and around galaxies.

Second, the cosmic microwave background radiation, a relic of the pregalactic Universe, allows us to find our motion relative to the rest of the cosmos. The Milky Way's speed through the cosmic microwave background is our *peculiar velocity*, as defined in Equation 1.29. It turns out to be surprisingly large, indicating that huge concentrations of distant matter have exerted a strong pull on our Local Group.

Further reading: for an undergraduate-level introduction, see B. Ryden, 2003, *Introduction to Cosmology* (Addison Wesley, San Francisco, USA); and A. Liddle, 2003, *An Introduction to Modern Cosmology*, 2nd edition (John Wiley & Sons, Chichester, UK).

1.5.1 The hot early Universe

For at least the first hundred thousand years after the Big Bang, most of the energy in the Universe was that of the blackbody radiation emitted from the hot matter, and of *relativistic* particles: those moving at nearly the speed of light, so rapidly that they behave much like photons. During the expansion, Equation 1.34 tells us that wavelengths grow proportionally to the scale length $\mathcal{R}(t)$. By Equation 1.5, the radiation temperature T varies inversely as the wavelength $\lambda_{\max}$ at which most light is emitted, and the temperature drops as $T \propto 1/\mathcal{R}(t)$; see the problem below.

Problem 1.17 If photons now fill the cosmos uniformly with number density $n(t_0)$, show that, at time t, the density $n(t) = n(t_0)\mathcal{R}^3(t_0)/\mathcal{R}^3(t)$. Use Equation 1.34 to show that the energy density of radiation decreases as $1/\mathcal{R}^4(t)$. For blackbody radiation at temperature T, the number density of photons with energy between ν and $\nu + \Delta\nu$ is

$$n(\nu)\Delta\nu = \frac{2\nu^2}{c^3}\frac{\Delta\nu}{\exp[h\nu/(k_{\mathrm{B}}T)] - 1}. \qquad (1.35)$$

Show that, if the present spectrum is that of blackbody radiation at temperature T_0, then at time t the expansion transforms this exactly into blackbody radiation at temperature $T(t) = T_0\mathcal{R}(t_0)/\mathcal{R}(t)$.

During the first three minutes of its life, the Universe was full of energetic γ-rays, which would smash any atomic nuclei apart into their constituent particles. When the temperature of the radiation field is high enough, pairs of particles and their antiparticles can be created out of the vacuum. Because photons are never at rest, two of them are required to produce a particle pair. A typical photon of radiation at temperature T carries energy $\mathcal{E} = 4k_{\mathrm{B}}T$, where k_{B} is Boltzmann's constant; so proton–antiproton pairs could be produced when $k_{\mathrm{B}}T \gtrsim m_{\mathrm{p}}c^2$, where m_{p} is the proton mass. We usually measure these energies in units of an *electron volt*, the energy that an electron gains by moving through a potential difference of 1 volt: $1\,\mathrm{eV} = 1.6 \times 10^{-19}\,\mathrm{J}$ or $1.6 \times 10^{-12}\,\mathrm{erg}$. In these units, $m_{\mathrm{p}}c^2 \approx 10^9\,\mathrm{eV}$ or 1 GeV, so pairs of protons and antiprotons were created freely in the first 10^{-4} s, when

$$k_{\mathrm{B}}T \gg m_{\mathrm{p}}c^2, \quad \text{or} \quad T \gg 10^{13}\,\mathrm{K}. \qquad (1.36)$$

As the expansion continued, the temperature fell, and the photons had too little energy to make a proton–antiproton pair; almost all the antiprotons met with a proton and annihilated to leave a pair of γ-rays. We do not understand why this was so, but in the early Universe there were slightly more protons: about $10^9 + 1$ protons for every 10^9 antiprotons. The small excess of matter over antimatter was left over to form the galaxies. The photons produced in the annihilation are seen today as the cosmic microwave background.

Electrons are about 2000 times less massive than protons: their rest energy $m_e c^2$ is only 0.5 MeV. So the radiation still produced pairs of electrons and anti-electrons (*positrons*, e^+), until the temperature dropped a thousandfold, to $T \sim 10^{10}$ K. Before this time, the reaction

$$e^- + e^+ \longleftrightarrow \nu_e + \bar{\nu}_e$$

could produce *electron neutrinos* ν_e and their antiparticles $\bar{\nu}_e$. The great abundance of electrons, positrons, and neutrinos allowed neutrons to turn into protons, and vice versa, through reactions such as

$$e^- + p \longleftrightarrow n + \nu_e, \quad \bar{\nu}_e + p \longleftrightarrow n + e^+, \quad n \longleftrightarrow p + e^- + \bar{\nu}_e.$$

In equilibrium at temperature T, there would have been slightly fewer neutrons than protons, since the neutron mass m_n is larger. The ratio of neutrons to protons was given by

$$n/p = e^{-Q/(k_B T)}, \quad \text{where} \quad Q = (m_n - m_p)c^2 = 1.293 \, \text{MeV}. \quad (1.37)$$

Neutrinos are very weakly interacting particles; from the Sun, 10^{15} of them fly harmlessly through each square meter of the Earth's surface every second. Only in the extremely hot material of supernova cores, or in the early Universe, do they have an appreciable chance of reacting with other particles. While electron–positron pairs were still numerous, the density of neutrinos was high enough to keep the balance between neutrons and protons at this equilibrium level. But later, once $k_B T \lesssim 0.8$ MeV or $t \gtrsim 1$ s, expansion had cooled the matter and neutrinos so much that a neutron or proton was very unlikely to interact with a neutrino. The neutrons *froze out*, with $n/p \approx 1/5$.

1.5.2 Making the elements

Neutrons can survive if they are bound up in the nuclei of atoms, but free neutrons are not stable; they decay exponentially into a proton, an electron, and an anti-neutrino $\bar{\nu}_e$. After a time $\tau_n = 886 \pm 1$ s, which is also the mean lifetime, the number is reduced by a factor $1/e$. Very few neutrons would now be left, if they had not combined with protons to form *deuterium*, a nucleus of 'heavy hydrogen' containing a neutron and a proton, by the reaction

$$n + p \rightarrow D + \gamma;$$

here γ represents a photon, a γ-ray carrying away the 2.2 MeV of energy set free in the reaction. This reaction also took place at earlier times, but any deuterium that managed to form was immediately torn apart by photons in the blackbody radiation.

After the electron–positron pairs were gone, at $T \lesssim 3 \times 10^9$ K, the energy density in the Universe was almost entirely due to blackbody radiation. General Relativity tells us that the temperature fell according to

$$t = \left(\frac{3c^2}{32\pi G a_B T^4} \right)^{1/2} \approx 230\,\text{s} \left(\frac{10^9\,\text{K}}{T} \right)^2 ; \qquad (1.38)$$

here $a_B = 7.56 \times 10^{-16}\,\text{J m}^{-3}\,\text{K}^{-4}$ is the blackbody constant. About a quarter of the neutrons had decayed before the temperature fell to about 10^9 K, when they could be locked into deuterium; this left one neutron for every seven protons. The excess protons, which became the nuclei of hydrogen atoms, accounted for roughly 75% of the total mass.

Deuterium easily combines with other particles to form ^{4}He, a helium nucleus with two protons and two neutrons. Essentially all the neutrons, and so about 25% of the total mass of neutrons and protons, ended up in ^{4}He. Only a little deuterium and some ^{3}He (with two protons and one neutron) remained. Traces of boron and lithium were also formed, but the Universe expanded too rapidly to build up heavier nuclei. The amount of helium produced depends on the half-life of the neutron, but hardly at all on the density of matter at that time; almost every neutron could find a proton and form deuterium, and almost every deuterium nucleus reacted to make helium. The observed abundance of helium is between 22% and 24%, in rough accord with this calculation. If, for example, we had found the Sun to contain 10% of helium by weight, that observation would have been very hard to explain in the Big Bang cosmology.

Problem 1.18 Deuterium can become abundant only when $k_B T \lesssim 70$ keV. Use Equation 1.38 to show that this temperature is reached at $t \approx 365$ s, by which time about 35% of the free neutrons have decayed. The mean lifetime τ_n is hard to measure; until recently, laboratory values varied from 700 s to 1400 s. If the mean lifetime had been 750 s, show that the predicted fraction of helium would be about 2% lower, whereas if it had been 1100 s, we would expect to find close to 2% more helium.

By contrast, the small fraction of deuterium left over is very much dependent on the density of neutrons and protons, collectively known as *baryons*. If there had been very little matter, many of the deuterium nuclei would have missed the chance to collide with other particles, before reactions ceased as the Universe became too dilute. If the present number of baryons had been as low as $n_B = 10^{-8}$ cm^{-3}, then as many as 1% of the deuterium nuclei would remain. If the density were now as high as $n_B = 2 \times 10^{-6}\,\text{cm}^{-3}$, we would expect to find less than one deuterium nucleus for each 10^9 atoms of hydrogen.

Deuterium also burns readily to helium inside stars. So to measure how much was made in the Big Bang, we must look for old metal-poor stars that have not burned the deuterium in their outer layers, or at intergalactic clouds of gas that have not yet formed many stars. Our best measurements show one deuterium nucleus for every 20 000 or 30 000 atoms of hydrogen. Along with measurements of ^{3}He and lithium, these show that the combined density of neutrons and protons today is

$$n_B = (2.5 \pm 0.5) \times 10^{-7}\,\text{cm}^{-3}, \quad \text{or} \quad \rho_B = (5-7) \times 10^9 \mathcal{M}_\odot\,\text{Mpc}^{-3}. \quad (1.39)$$

This is much less than the critical density of Equation 1.30: the ratio is

$$0.02h^{-2} \lesssim \rho_B/\rho_{\text{crit}} \lesssim 0.025h^{-2}, \quad (1.40)$$

where h is Hubble's constant H_0 in units of $100\,\text{km s}^{-1}\,\text{Mpc}^{-1}$. Observations seem to require $h \gtrsim 0.6$, so neutrons and protons cannot make up more than about 7% of ρ_{crit}. Since $h \lesssim 0.75$, baryons account for no less than 3% of the critical density. In the benchmark model, $\rho_B = 0.045\rho_{\text{crit}}$. We will find in Section 5.3 that this is more mass than we can see as the gas and luminous stars of galaxies; at least part of their 'dark stuff' must consist of normal matter.

Problem 1.19 The Universe must contain at least as much matter as that of the neutrons and protons: use Equations 1.39 and 1.25 to show that the average mass-to-light ratio must exceed $50h^{-1}\mathcal{M}_\odot/L_\odot$.

1.5.3 Recombination: light and matter uncoupled

The next few hundred thousand years of the Universe's history were rather boring. Its density had dropped too low for nuclear reactions, and the background radiation was energetic enough to ionize hydrogen and disrupt other atoms. The cosmos was filled with glowing gas, like the inside of a fluorescent light. Photons could not pass freely through this hot plasma, but they were scattered by free electrons. Matter would not collapse under its own gravity to form stars or other dense objects, because the pressure of the radiation trapped inside was too high.

The density of radiation decreases with the scale length $\mathcal{R}(t)$ as $T^4 \propto \mathcal{R}^{-4}(t)$. So after some time it must drop below that of matter, which falls only as $\mathcal{R}^{-3}(t)$. In the benchmark model, at this time of *matter–radiation equality* radiation had

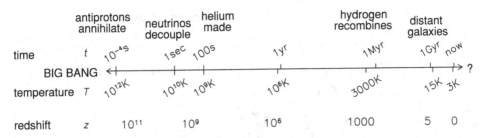

Fig. 1.18. Important moments in the history of the Universe.

cooled to $T \approx 10\,000\,\text{K}$. As measured by $\mathcal{R}(t)$, the Universe was then about 1/3600 of its present size. Later on, photons of the blackbody radiation lacked the energy to remove the electron from a hydrogen atom. During its subsequent expansion, hydrogen atoms *recombined*, the gas becoming neutral and transparent as it is today. By the time that $\mathcal{R}(t)/\mathcal{R}(t_0) \approx 1/1100$, photons of the background radiation were able to escape from the matter. Their outward pressure no longer prevented the collapse of matter, into the galaxies and clusters that we now observe. The most distant galaxies so far observed are at redshifts $z \gtrsim 6$; when their light left them, the Universe was less than 1 Gyr old. Figure 1.18 presents a brief summary of cosmic history up to that time.

The radiation coming to us from the period of recombination has been red-shifted according to Equation 1.34; it now has a much longer wavelength. Its temperature $T = 2.728 \pm 0.002\,\text{K}$, so it is known as the *cosmic microwave background*. There are about 420 of these photons in each cm^3 of space, so, according to Equation 1.40, we have $(2\text{--}4) \times 10^9$ photons for every neutron or proton. The energy density of the background radiation is about equal to that of starlight in the outer reaches of the Milky Way. It is given by $a_B T^4 = 4.2 \times 10^{-14}\,\text{J}\,\text{m}^{-3}$; so from each steradian of the sky we receive $ca_B T^4/(4\pi) \approx 10^{-6}\,\text{W}\,\text{m}^{-2}$.

Problem 1.20 Using Equation 1.25, show that, even if we ignore the energy loss that goes along with the redshift, it would take more than 100 Gyr for all the galaxies, at their present luminosity, to emit as much energy as is in the microwave background today.

Figure 1.19 shows the extragalactic background radiation, estimated by observing from our position in the Milky Way and attempting to subtract local contributions. The energy of the cosmic background is far larger than that in the infrared, visible, and ultraviolet spectral regions. It would be very difficult to explain such enormous energy as coming from any other source than the Big Bang. Radiation from the submillimeter region through to the ultraviolet at $\lesssim 0.1\,\text{keV}$ comes from stars and active galactic nuclei, either directly or after re-radiation

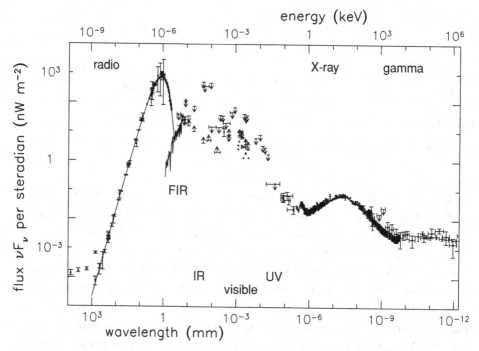

Fig. 1.19. Extragalactic background radiation: the vertical *logarithmic* scale shows energy density per decade in frequency or wavelength. Arrows show upper and lower limits. The curve peaking at $\lambda \sim 1$ mm is the cosmic microwave background; the far-infrared background is the light of stars and active galactic nuclei, re-radiated by dust – T. Ressell and D. Scott.

by heated dust. The high-energy 'tail' in X-rays and γ-rays is mainly from active nuclei. Since photons lose energy in an expanding Universe, almost all this radiation, aside from that in the microwave background, must have been emitted over the past ~10 Gyr, at times corresponding to redshifts $z \lesssim 3$.

The microwave background is now very close to a blackbody spectrum; it is also extremely uniform. Between different parts of the sky, we see small irregularities in its temperature that are just a few parts in 100 000 – with only one exception. In the direction $l = 265°$, $b = 48°$ the peak wavelength is shorter than average, and the temperature higher, by a little more than 0.1%. In the opposite direction, the temperature is lower by the same amount. This difference reflects the Sun's motion through the background radiation. If T_0 is the temperature measured by an observer at rest relative to the backgound radiation, then an observer moving with relative speed $V \ll c$ would measure a temperature $T(\theta)$ at an angle θ to the direction of motion, given by

$$T(\theta) \approx T_0(1 + V \cos\theta/c). \qquad (1.41)$$

For the Sun, $V = 370\,\mathrm{km\,s^{-1}}$. Taking into account the Sun's orbit about the Milky Way, and the Milky Way's motion relative to nearby galaxies, we find that our Local Group has a peculiar motion of $V_{\mathrm{pec}} \approx 600\,\mathrm{km\,s^{-1}}$ relative to the background radiation and to the Universe as a whole. The Local Group's motion is unexpectedly and troublingly large: we discuss it further in Chapter 8.

2

Mapping our Milky Way

Our position in the Milky Way's disk gives us a detailed and close-up view of a fairly typical large spiral galaxy. We begin this chapter by looking at the Sun's immediate vicinity. Examining the closest stars gives us a sample of the disk stuff, and we can ask how many stars of each luminosity, mass, composition, and age are present. Combining this information with theories of stellar evolution, we investigate the star-forming past of the solar neighborhood.

In Section 2.2 we venture further afield. Measuring stellar distances allows us to map out the Milky Way's structure: the thin and thick disks, the metal-poor halo, and the central bulge. Star clusters, where all the members were born together, with the same initial composition, are especially useful; comparing their color–magnitude diagrams with the predictions of stellar models yields the age and composition jointly with the cluster's distance. We find that the youngest stars belong to the disk, and are relatively rich in elements heavier than helium, whereas the stars of the metal-poor halo are extremely old.

Most of the Milky Way's stars and almost all of its gas lie in the disk, orbiting the center like the planets around the Sun. Section 2.3 deals with the Galaxy's rotation: how we measure it, and how we use it to find the distribution of gas within the disk. By contrast with motions in the solar system, the rotational speed of material in the furthest part of the disk is nearly the same as that for gas at the Sun's position. To prevent this distant gas from flying off into intergalactic space, a large amount of mass must be concealed in the outer reaches of the Milky Way, in a form that emits very little light, or none at all: the *dark matter*.

Measured by mass, the Milky Way has only a tenth as much gas as stars; but the gas has profound effects. The densest gas clouds collapse to make new stars; and at the end of its life a star pollutes the gas with dust and heavy elements produced by its nuclear burning. Section 2.4 discusses the complex processes that heat, ionize, and push around the Milky Way's gas, and how these affect the pace at which stars are born.

2.1 The solar neighborhood

In this section, we consider the closest stars, the Sun's immediate neighbors. We ask what kinds of stars are present, and in what numbers? How many are on the main sequence, and how many are in the later stages of their lives? How many were formed recently, and how many are very old? To answer these questions, we must first find out how far away the stars are.

The distances of astronomical objects are in general extremely difficult to measure, but they are essential for our understanding of their nature. The luminosities of stars and galaxies are always derived by using the inverse-square law (Equation 1.1). We usually find linear dimensions from a measured angular size on the sky and an estimated distance, and our calculations of masses usually depend on those size estimates. Many astronomical disputes come down to an argument over how far away something is; so astronomers have had good reason to develop a wide range of inventive and sometimes bizarre techniques for measuring distances.

Triangulation, or *trigonometric parallax*, allows us to measure distances only for the nearest stars. We then compare more distant stars with similar stars close at hand, assuming that stars with similar spectra have the same luminosity and finding the relative distance from their relative apparent brightness. Those distant stars in turn are used to estimate distances to the nearest galaxies, those close enough that we can pick out individual stars. This *cosmic distance ladder* is then extended to more distant galaxies by comparing them with nearby systems. The relative distances at each stage of comparison are often quite well determined – in astronomy, this means to within a few percent – but the accumulation of errors at each stage can leave extragalactic distances uncertain by as much as a factor of two. Occasionally we are lucky enough to find a way of circumventing some of the lower rungs of the ladder, to measure directly the distance to an object beyond the reach of trigonometric parallax. These opportunities are much prized; we discuss some of them in Section 2.2 below.

In general, the *relative* distances of two stars or galaxies can be found far more accurately than the *absolute* distance (in meters, or light-years) to either one. The *parsec* was adopted in 1922 by the International Astronomical Union to specify stellar distances in units of the Earth's mean distance from the Sun, the *astronomical unit*. With interplanetary spacecraft, and measurements of the reflection times for light or radio waves bounced from the surfaces of planets, we now know the scale of the solar system to within one part in a million; we retain the parsec for historical reasons. Within the Galaxy, distances are sometimes given in units of R_0, the Galactocentric radius of the Sun.

2.1.1 Trigonometric parallax

Within a few hundred parsecs, we can use *trigonometric parallax* to find stellar distances. As the Earth orbits the Sun, our viewing position changes, and closer

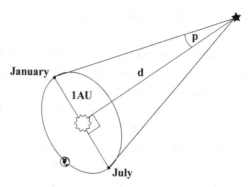

Fig. 2.1. Trigonometric parallax: in the course of a year, the star appears to move in an ellipse with a major axis of $2p$.

stars appear to move relative to more distant objects. In the course of a year, a nearby star traces out an elliptical path against the background of distant stars (Figure 2.1). The angle p is the *parallax*; it is always small, so, for a star at distance d, we have

$$\frac{1\,\text{AU}}{d} = \tan p \approx p \text{ (in radians)}. \tag{2.1}$$

One astronomical unit (1 AU) is the Earth's mean distance from the Sun, about 150 million kilometers or 8.3 light-minutes. The *parsec* (pc) is defined as the distance d at which a star has a parallax of $1''$, one *second of arc*: $1'' = 1/60 \times 1/60$ of $1°$. One radian is roughly $206\,265''$, so a parsec is $206\,265$ AU or about 3.26 light-years. The stars are so distant that none has a parallax even as large as $1''$. Proxima Centauri, the nearest star, has $p = 0.8''$, so its distance is 1.3 pc or 4.3 light-years.

The European Space Agency's Hipparcos satellite (1989–93) repeatedly measured the apparent motions across the sky of $120\,000$ bright stars, to an accuracy of a milli-arcsecond. The Hipparcos database gives us distances, and hence accurate luminosities, for the stars within a few hundred parsecs, as well as their motions through space, and orbits for the binary or multiple stars. For this fabulous sample of stars, the accurate distances allow us to determine many of the basic stellar parameters that we discussed in Section 1.1.

Problem 2.1 To determine a star's trigonometric parallax, we need at least three measurements of its position relative to much more distant objects: why? (What else could change its position on the sky?)

Often, we express our distance as a *distance modulus*, which is defined as the difference between the apparent magnitude m and the absolute magnitude M

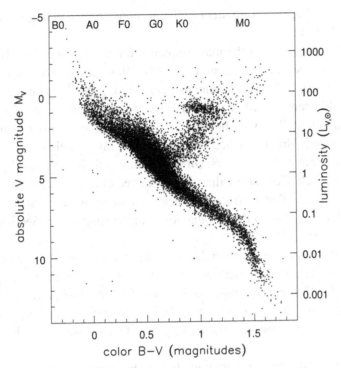

Fig. 2.2. A color–magnitude diagram and approximate spectral types for 15 630 stars within 100 pc of the Sun, for which Hipparcos measured the trigonometric parallax to <10%, and the color $B - V$ to within 0.025 magnitudes – M. Perryman.

given by Equation 1.15. We write it as

$$(m - M)_0 \equiv 5 \log_{10}\left(\frac{d}{10 \, \text{pc}}\right) = 5 \log_{10}\left(\frac{0.1''}{p}\right), \tag{2.2}$$

where the subscript 0 indicates that the apparent magnitude has been adjusted to compensate for dimming by interstellar dust, which absorbs and scatters the starlight.

Problem 2.2 Show that an error or uncertainty of 0.1 magnitudes in the distance modulus is roughly equivalent to a 5% error in d.

If space were transparent, with no dust, it would be easy to find a star's absolute magnitude by combining a measurement of its parallax with the observed apparent magnitude. For nearby stars, the effect of dust is small, and we can generally neglect it. Figure 2.2 shows the $B - V$ color for all the stars measured by Hipparcos within 100 pc of the Sun, and the absolute V-magnitude M_V derived by using Equation 2.2. This is a *color–magnitude* or *Hertzsprung–Russell* diagram; we can compare it with Figure 1.4, to identify the portions corresponding to the various stages of stellar life.

We can easily pick out the main sequence for stars fainter than $M_V \approx -4$; brighter stars are so rare that none falls within this sphere. The main sequence is quite broad. For stars more luminous than the Sun, this is mainly because the stars are of different ages. For dimmer stars, variations in the metal abundance are more important. The very bright stars at the right are the red giants, which have used up the hydrogen in their central cores. Many stars lie near $M_V \approx 1$ and $B-V \approx 1$; these are *red clump* stars, burning helium in their cores, and hydrogen in a surrounding shell. Comparing the color of the red clump with the predictions of stellar models shows that the fraction of heavy elements in these stars lies between the solar value and about a third of that level: in the language of Section 1.1, this means $-0.5 \lesssim [\text{Fe}/\text{H}] \lesssim 0$.

Few of the stars brighter than $M_V \approx 2$ lie in the triangle between the main sequence and the red stars with $B - V \gtrsim 0.9$, corresponding to temperatures $T_{\text{eff}} \lesssim 5500\,\text{K}$. This nearly empty region is called the *Hertzsprung gap*. Once their main-sequence life is over, stars more massive than about $2\mathcal{M}_\odot$ swell very rapidly to become luminous red stars; they spend very little time at intermediate temperatures.

A few stars fall about a magnitude below the main sequence; most of these are metal-poor *subdwarfs*, which are bluer than stars of the same mass with solar composition. Four dim *white dwarfs* are seen at the lower left. Very dim stars are hard to find; even within 100 pc, it is likely that more than half of the white dwarfs and the dimmest main-sequence stars have so far escaped our searches.

2.1.2 Luminosity functions and mass functions

Most of the stars in Figure 2.2 are more luminous than absolute magnitude $M_V \approx 8$; but the less luminous stars are in fact more common. The *luminosity function* $\Phi(M_V)$ describes how many stars of each luminosity are present in each pc^3: $\Phi(M_V)\Delta M_V$ is the density of stars with absolute V-magnitude between M_V and $M_V + \Delta M_V$. To compute $\Phi(M_V)$, we must know the volume of space within which we have observed stars of each luminosity. It is common to select for observation those stars or galaxies that appear brighter than some fixed apparent magnitude. The Hipparcos observing list included almost all the stars with $m_V \lesssim 8$, as well as some that were fainter. A star like the Sun with $M_V = 4.83$ might not have been included if its distance modulus had been larger than $8 - 4.83 = 3.17$, corresponding to $d \approx 43\,\text{pc}$. The solid dots in Figure 2.3 show an approximate luminosity function, in one-magnitude bins, calculated by using the formula

$$\Phi(x) = \frac{\text{number of stars with } M_V - 1/2 < x < M_V + 1/2}{\text{volume } \mathcal{V}_{\text{max}}(M_V) \text{ over which these could be seen}}. \qquad (2.3)$$

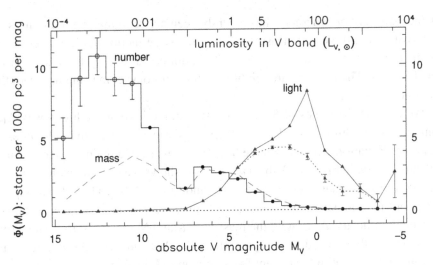

Fig. 2.3. The histogram shows the luminosity function $\Phi(M_V)$ for nearby stars: solid dots from stars of Figure 2.2, open circles from Reid *et al.* 2002 *AJ* **124**, 2721. Lines with triangles show $L_V\Phi(M_V)$, light from stars in each magnitude bin; the dotted curve is for main-sequence stars alone, the solid curve for the total. The dashed curve gives $\mathcal{M}\Phi_{MS}(M_V)$, the mass in main-sequence stars. Units are $L_\odot$ or $\mathcal{M}_\odot$ per 10 pc cube; vertical bars show uncertainty, based on numbers of stars in each bin.

Problem 2.3 Show that the volume in Equation 2.3 is $V_{max}(M) \approx 4\pi d_{max}^3/3$, where d_{max} is the smaller of 100 pc and 10 pc $\times 10^{0.2(8-M)}$. Using Table 1.4 for M_V, find d_{max} for an M4 dwarf. Why are you surprised to see such faint stars in Figure 2.2?

It is quite difficult to determine the faint end of the luminosity function, since dim stars are hard to find. The bright end of $\Phi(M_V)$ also presents problems; because luminous stars are rare, we will not find enough of them unless we survey a volume larger than our 100 pc sphere. But stars are not spread out uniformly in space. For example, their density falls as we go further out of the Milky Way's disk in the direction of the Galactic poles. So, if we look far afield for luminous stars, the average density in our search region is lower than it is near the Sun's position. Finally, many stars are in binary systems so close that they are mistaken for a brighter single star. Despite these uncertainties, it is clear that dim stars are overwhelmingly more numerous than bright ones.

Figure 2.3 also shows how much of the V-band light is emitted by stars of each luminosity: stars in the range from M_V to $M_V + \Delta M_V$ contribute an amount $L_V\Phi(M_V)\Delta M_V$ of the total. Almost all the light comes from the brighter stars, mainly A and F main-sequence stars and K giants. Rare luminous stars such as main-sequence O and B stars, and bright supergiants, contribute more light than all the stars dimmer than the Sun; so the total luminosity of a galaxy depends strongly on whether it has recently been active in making these massive short-lived stars. If we had measured our luminosity function at ultraviolet wavelengths rather than in the V band at 5500 Å, almost all the light would be from O and B stars. The near-infrared light, at a few microns, comes mainly from the luminous red stars.

We can use the tables of Section 1.1 to find the bolometric luminosity L_{bol} from the V-band luminosity L_V, and then use Table 1.1, or the mass–luminosity relation of Equation 1.6, to calculate the average mass of a main-sequence star at each luminosity. In Figure 2.3, the dashed curve shows the *mass* in main-sequence stars with absolute magnitude between M_V and $M_V + \Delta M_V$. The red giants make a tiny contribution, since they are even less massive than main-sequence stars of the same luminosity. Almost all the mass is in K and M dwarfs, stars so faint that we cannot see them in galaxies beyond the Milky Way and its satellites. The stars that emit most of the light account for hardly any of the mass.

The luminosity function of Figure 2.3 corresponds to about 65 stars in each $1000 \, \mathrm{pc}^3$, with a luminosity equivalent to $40 L_\odot$ in the V band. About 75% of the light comes from main-sequence stars, which have mass totalling about $30 \mathcal{M}_\odot$. The averaged *mass-to-light* ratio $\mathcal{M}/L$ gives a measure of the proportion of massive luminous stars to dim stars. We find $\mathcal{M}/L_V \approx 1$ for the main-sequence stars alone, and $\mathcal{M}/L_V \approx 0.74$ for all the stars, when we measure $\mathcal{M}$ in solar masses and L_V as a fraction of the Sun's V-band luminosity. Even including white dwarfs and the interstellar gas, the mass-to-light ratio $\mathcal{M}/L_V \lesssim 2$ locally. In Section 2.3, we will see that the ratio of mass to light for the Milky Way as a whole is much larger than that for this sample of stars near the Sun. The outer Galaxy contains unseen mass, in a form other than the stars and interstellar gas found near the Sun.

Using models of stellar evolution, we can work backward from the present-day *stellar population* to find how many stars were born with each mass. We define the *initial luminosity function*, $\Psi(M_V)$, such that $\Psi(M_V)\Delta M_V$ is the number of stars formed that had absolute magnitude between M_V and $M_V + \Delta M_V$ when they were on the main sequence. Stars less massive than the Sun have main-sequence lives of 10 Gyr or longer. Counting the numbers of faint white dwarfs indicates that stars have been forming in the local disk only for a time $\tau_{gal} \approx$ 8–10 Gyr, so the first-born K dwarfs have evolved very little. For these low-mass stars, $\Psi(M_V)$ is almost the same as the present-day luminosity function $\Phi(M_V)$. But O, B, and A stars burn out rapidly, and only those most recently born are still on the main sequence.

We can calculate the initial luminosity function if we assume that the disk has been forming stars at a uniform rate throughout its history. If $\Phi_{MS}(M_V)$ is the present-day luminosity function for main-sequence stars alone, and a star remains on the main sequence with absolute magnitude M_V for a time $\tau_{MS}(M_V)$, then

$$\Psi(M_V) = \Phi_{MS}(M_V) \qquad \text{for } \tau_{MS}(M_V) \geq \tau_{gal},$$
$$= \Phi_{MS}(M_V) \times \frac{\tau_{gal}}{\tau_{MS}(M_V)} \quad \text{when } \tau_{MS}(M_V) < \tau_{gal}. \qquad (2.4)$$

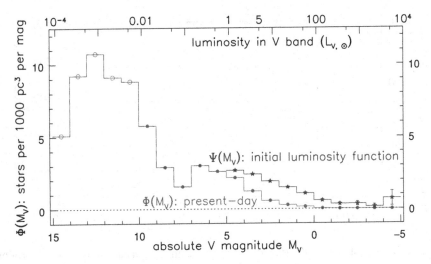

Fig. 2.4. Circles show the luminosity function $\Phi_{MS}(M_V)$ for main-sequence stars as in Figure 2.3. The histogram gives the initial luminosity function $\Psi(M_V)$, assuming that stars were born at a constant rate over the past 10 Gyr. Both functions have a minimum, the *Wielen dip*, at $M_V \approx 8$. This V-band luminosity corresponds to only a tiny range of stellar mass $\mathcal{M}$. The mass function $\xi(\mathcal{M})$ probably has no dip or inflection at this mass.

Figure 2.4 shows a crude estimate for $\Psi(M_V)$ calculated according to this formula, assuming $\tau_{gal} = 10$ Gyr. Massive stars are formed more rarely than dim low-mass stars, but the disproportion is not so great as in the present-day luminosity function $\Phi(M_V)$.

Problem 2.4 Suppose that stars are born at a constant rate. Assuming $\tau_{gal} = 10$ Gyr and using Table 1.1 for stellar lifetimes, show that only 11% of all the $2\mathcal{M}_\odot$ stars ever made are still on the main sequence today. What fraction of all the $3\mathcal{M}_\odot$ stars are still there? What fraction of all the $0.5\mathcal{M}_\odot$ stars? Now suppose that star formation slows with time t as $e^{-t/t_\star}$, with $t_\star = 3$ Gyr. Show that now only 1.6% of all $2\mathcal{M}_\odot$ stars survive, and merely 0.46% of stars of $3\mathcal{M}_\odot$.

For these stars, explain why $\Psi(M_V)$ is larger for a given observed $\Phi(M_V)$ when starbirth declines with time (the $\star$ and $\bullet$ points in Figure 2.4 must be further apart) than if it stays constant. How much larger must $\Psi(M_V)/\Phi(M_V)$ become for stars of $2\mathcal{M}_\odot$? How would a gradual slowdown change the inferred $\Psi(M_V)$ for stars longer-lived than the Sun?

For an accelerating rate of starbirth $t_\star < 0$, in what sense would this affect our estimates of the initial luminosity function $\Psi(M_V)$?

($\Phi(M_V)$ is a fairly smooth function, and we have no reason to expect that $\Psi(M_V)$ will have a kink or change in slope near the Sun's luminosity. Together, these imply that star formation locally has not slowed or speeded up by more than a factor of two over the past few gigayears.)

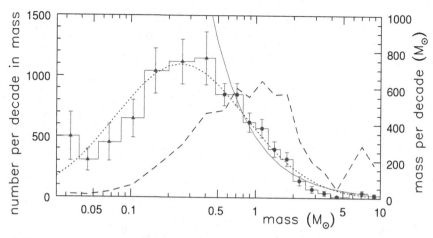

Fig. 2.5. Masses of stars in the Pleiades cluster: the number in each mass range is proportional to the area under the histogram. The smooth curve shows the Salpeter initial mass function, the dotted curve is a lognormal function. The dashed line shows mass: stars near $0.25\mathcal{M}_\odot$ are most numerous, but those of $(1–2)\mathcal{M}_\odot$ account for most of the cluster's mass – E. Moreau.

We can convert the initial luminosity function $\Psi(\mathcal{M})$ into an *initial mass function*: $\xi(\mathcal{M})\Delta\mathcal{M}$ is the number of stars that have been born with masses between $\mathcal{M}$ and $\mathcal{M} + \Delta\mathcal{M}$. Near the Sun, a good approximation for stars more massive than $\sim 0.5\mathcal{M}_\odot$ is

$$\xi(\mathcal{M})\Delta\mathcal{M} = \xi_0(\mathcal{M}/\mathcal{M}_\odot)^{-2.35}(\Delta\mathcal{M}/\mathcal{M}_\odot), \qquad (2.5)$$

where the constant ξ_0 sets the local stellar density; this is called the *Salpeter* initial mass function. Figure 2.5 shows the observed numbers of stars at each mass in the Pleiades cluster, shown in Figure 2.11 below. This cluster is only 100 Myr old, so for masses below $5\mathcal{M}_\odot$ the initial mass function is identical to what we observe at the present day. The Salpeter function overestimates the number of stars with masses below $0.5\mathcal{M}_\odot$, but otherwise it gives a good description. Observations in very different parts of our Galaxy and the nearby Magellanic Clouds show that $\xi(\mathcal{M})$ is surprisingly uniform, from dense stellar clusters to diffuse associations of stars. If we understood better how stars form, we might be able to predict the initial mass function.

Problem 2.5 Suppose that Equation 2.5 describes stars formed within a 100 pc cube with masses between $\mathcal{M}_\ell$ and an upper limit $\mathcal{M}_u \gg \mathcal{M}_\ell$. Write down and solve the integrals that give (a) the number of stars, (b) their total mass, and (c) the total luminosity, assuming that Equation 1.6 holds with $\alpha \approx 3.5$. Show that the number and mass of stars depend mainly on the mass $\mathcal{M}_\ell$ of the smallest stars, while the luminosity depends on $\mathcal{M}_u$, the mass of the largest stars.

Taking $\mathcal{M}_\ell = 0.3\mathcal{M}_\odot$ and $\mathcal{M}_u \gg 5\mathcal{M}_\odot$, show that only 2.2% of all stars have $\mathcal{M} > 5\mathcal{M}_\odot$, while these account for 37% of the mass. The Pleiades cluster has $\mathcal{M} \approx 800\mathcal{M}_\odot$: show that it has about 700 stars. Taking $\mathcal{M}_u = 10\mathcal{M}_\odot$ (see Figure 2.5), show that the few stars with $\mathcal{M} > 5\mathcal{M}_\odot$ should contribute over 80% of the light. Why do we see so few stars in Figure 2.11 compared with the number in Figure 2.13?

2.2 The stars in the Galaxy

Most stars do not have measurable parallaxes, which tells us that the Galaxy is much larger than 500 pc across. To estimate distances of stars further away, we rely on the cosmic distance ladder; we measure their distances relative to stars that are close enough to show parallaxes. Occasionally we can use information from velocities to obtain distances without this intermediate step. In this section, we first explore some of these opportunities, then discuss the distribution of stars and star clusters in the Milky Way, which reveals its basic structure. Our position in the Galactic disk gives us a unique and detailed three-dimensional view of one spiral galaxy.

2.2.1 Distances from motions

Radial velocities V_r, toward or away from an observer, are measured using the *Doppler shift* of emission or absorption lines in the spectra of stars or gas (Equation 1.19). *Tangential velocities* V_t are found from the angular rate at which a star appears to move across the sky: this *proper motion* μ is so small that it is often measured in milli-arcseconds per year. The tangential velocity is the product of distance and proper motion:

$$V_t = \mu \text{ (radians/time)} \times d, \text{ or } \mu\,(0.001''/\text{year}) = \frac{V_t\,(\text{km s}^{-1})}{4.74 \times d\,(\text{kpc})}. \quad (2.6)$$

If we know how V_r and V_t are related for a particular object, then, by measuring V_r and the proper motion, we can find its distance.

Our current best estimate of the Milky Way's size comes from the proper motions of stars around a very massive black hole that is believed to mark the exact center. The stellar orbits are shown in Figure 2.17. The problem below shows how observations of their position and velocity yield not only the mass $\mathcal{M}_{BH}$ of the black hole, but also its distance: the Sun is 7.6 ± 0.3 kpc from the Galactic center.

Problem 2.6 Using an 8-meter telescope to observe the Galactic center regularly over two decades, you notice that one star moves back and forth across the sky in a straight line: its orbit is edge-on. You take spectra to measure its radial velocity

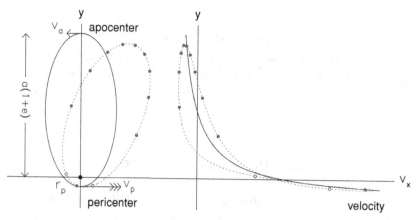

Fig. 2.6. Left, orbits of two stars around a point mass at the origin. Filled dots are at equal time intervals $P/10$, open dots at $\pm 0.01P$ from pericenter. Right, velocity in the horizontal direction. A distant observer looking along the x-axis sees the radial velocity v_x repeat exactly when the orbit is symmetric about the plane $x = 0$ of the sky (solid line); this does not happen if the orbit is misaligned (dotted line).

V_r, and find that this repeats exactly each time the star is at the same point in the sky. You are in luck: the furthest points of the star's motion on the sky are also when it is closest to the black hole (pericenter) and furthest from it (apocenter), as in Figure 2.6. You measure the separation s of these two points on the sky, and the orbital period P. Assuming that the black hole provides almost all the gravitational force, follow these steps to find both the mass $\mathcal{M}_{BH}$ of the black hole and its distance d from us.

From the definition below Equation 2.1, show that the orbit's semi-major axis $a = 0.5\,\mathrm{AU} \times (s/1'')(d/1\,\mathrm{pc})$. You observe $s = 0.248''$: what would a be at a distance of 8 kpc? At the two extremes of its motion across the sky, the star's radial velocity is $V_a = 473\,\mathrm{km\,s^{-1}}$ and $V_p = 7\,326\,\mathrm{km\,s^{-1}}$: at which point is it closest to the black hole? The orbit's eccentricity is e; explain why the conservation of angular momentum requires that $V_p(1 - e) = V_a(1 + e)$, and show that, here, $e = 0.876$.

At distance r from the black hole moving at speed V, the star has kinetic energy $\mathcal{KE} = m_\star V^2/2$ and potential energy $\mathcal{PE} = -Gm_\star\mathcal{M}_{BH}/r$. Since the total energy $\mathcal{KE} + \mathcal{PE}$ does not change during the orbit, show that

$$V_p^2 - V_a^2 = \frac{G\mathcal{M}_{BH}}{a} \times \frac{4e}{1 - e^2}. \tag{2.7}$$

Measuring v in $\mathrm{km\,s^{-1}}$, $\mathcal{M}_{BH}$ in $\mathcal{M}_\odot$, and a in parsecs, $G = 4.3 \times 10^{-3}$. Convert a to AU to show that $\mathcal{M}_{BH}/\mathcal{M}_\odot = 3822(a/1\,\mathrm{AU})$. Because $m_\star \ll \mathcal{M}_{BH}$, we can use Kepler's third law: P^2 (in years) $= a^3$ (in AU)$/\mathcal{M}_{BH}$ (in $\mathcal{M}_\odot$). You measure $P = 15.24\,\mathrm{yr}$; use Equation 2.7 to eliminate $\mathcal{M}_{BH}/a$ and show that $a = 942\,\mathrm{AU}$ and $\mathcal{M}_{BH} = 3.6 \times 10^6\mathcal{M}_\odot$. What is the distance to the Galactic center?

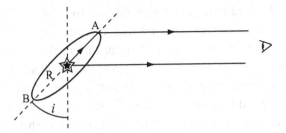

Fig. 2.7. The 'light echo' from a gas ring around supernova 1987a is seen by an observer to the right of the figure.

In February 1987, a supernova was seen to explode in the Large Magellanic Cloud (LMC). Shortly afterward, narrow emission lines of highly ionized carbon, nitrogen and oxygen appeared in the spectrum; the lines were so narrow that they must have come from cool gas, surrounding the star. When the supernova faded, an ellipse of glowing gas was seen around it. This is probably a circular ring of material, thrown off in the plane of the star's equator during the red-giant phase, which is tipped so as to appear elliptical (Figure 2.7).

The narrow emission lines started to become bright only about 85 days after the supernova had exploded. So it must have taken 85 days longer for light to reach the nearer edge of the ring and to ionize the gas, and then for the gas emission to reach us, than for light to come to us directly from the explosion. From this, we can find the radius of the ring in light-days; see Problem 2.7 below. Together with its measured size in arcseconds, this gives us a distance to the LMC between 50 kpc and 53 kpc. Information from two spacecraft observatories was used to make this measurement: the gas emission lines were at ultraviolet wavelengths that cannot penetrate the Earth's atmosphere, and the ring was so small that its image in a ground-based telescope would have been too badly blurred by atmospheric turbulence.

Problem 2.7 The ring around supernova 1987a measures about $1.62'' \times 1.10''$ across on the sky; if its true shape is circular, show that the ring is inclined at $i \approx 43°$ to face-on. If the ring radius is R, use Figure 2.7 to explain why light travelling first to the point A and then to us is delayed by a time $t_- = R(1-\sin i)/c$ relative to light coming straight to us from the supernova. Thus we see a *light echo*. If t_+ is the time delay for light reaching us by way of point B, show that $R = c(t_- + t_+)/2$. The measured values are $t_- = 83$ days, $t_+ = 395$ days: find the radius R in light-days, and hence the distance d to the supernova. At its brightest, the supernova had apparent magnitude $m_V \approx 3$; show that its luminosity was $L_V \approx 1.4 \times 10^8 L_\odot$. (Most Type II supernovae are even more luminous.)

2.2.2 Spectroscopic parallax: the vertical structure of the disk

As we discussed in Section 1.1, the width and depth of lines in a star's spectrum depend on its luminosity. We can use this fact as the basis of a technique for finding stellar distances. For example, a star with the spectrum of an F2 main-sequence star is roughly as luminous as other F2 dwarfs with about the same chemical composition. If we can measure a parallax for one of these, then Equation 2.1 tells us its distance, and hence its luminosity and that of all similar F2 stars. So if we then measure our more distant star's apparent brightness and can compensate for the dimming caused by interstellar dust, Equation 1.1 will tell us its distance. This method is called *spectroscopic parallax* because it should give the same information as that we gain by measuring the trigonometric parallax: namely, the star's distance.

Spectroscopic parallax works well for some types of star and poorly for others. The luminosities of main-sequence stars can often be found to within 10%, leading to 5% uncertainties in their distance. But in K giants, the temperature in the atmosphere is almost the same, no matter what the luminosity; the giant branch is almost vertical in Figure 1.4. The best we can hope for is to determine the luminosity to within 0.5 in the absolute magnitude, and hence the distance to 25%.

Taking high-quality spectra for many faint stars demands long hours of observation on large telescopes. A 'poor man's variant' is to estimate the spectral type of a star from its color, and to rely on other indications to establish whether it is a dwarf or a giant. This method of *photometric parallax* can be reasonably successful if the measured color changes substantially with the luminosity of the observed stars, and if we can correct for the effects of interstellar dust. Results are best when we observe a cluster of stars; with measurements of many stars, we can estimate both the cluster's distance and the reddening caused by dust.

When we look out perpendicular to the Galactic disk and select the sharp stellar images from among the fuzzy shapes of distant galaxies, almost all the red stars fainter than apparent magnitude $m_V \approx 14$ are K and M dwarfs. (At a distance of 1 kpc, what would be the apparent magnitude of a K or M giant with $M_V \approx 0$?) There is little dust in this direction to dim and redden the stars, so the $V - I$ color should be a good indication of the spectral type. The stellar densities in Figure 2.8 were compiled using photometric distances determined by measuring the $V - I$ colors of 12 500 stars with apparent magnitude $m_V < 19$ in the direction of the south Galactic pole. The late G and early K dwarfs are bluer than red giants, and there are very few giant stars of the same color to mislead us.

Using the Galaxy-centred spherical polar coordinates R, ϕ, z of Figure 1.10, we often approximate the density $n(R, z, S)$ of stars of a particular type S by a double-exponential form

$$n(R, z, S) = n(0, 0, S)\exp[-R/h_R(S)]\exp[-|z|/h_z(S)], \qquad (2.8)$$

where h_R is called the *scale length* of the disk and h_z is the *scale height*. Figure 2.8 shows that, near the midplane, $h_z \approx 300$–350 pc for K dwarfs, while for more massive and shorter-lived stars, such as the A dwarfs, it is smaller, $h_z \lesssim 200$ pc.

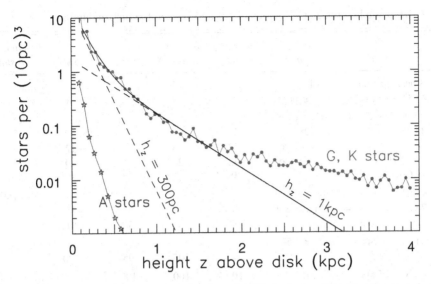

Fig. 2.8. Looking toward the south Galactic pole, filled circles show the density of stars with $5 < M_V < 6$; these are late G and early K dwarfs. Sloping dashed lines show $n(z) \propto \exp(-z/300\,\mathrm{pc})$ (thin disk) and $n(z) \propto \exp(-z/1\,\mathrm{kpc})$ (thick disk); the solid curve is their sum. At $z \gtrsim 2\,\mathrm{kpc}$, most stars belong to the metal-poor halo. A dwarfs (star symbols) lie in a very thin layer – N. Reid and J. Knude.

Gas in the disk, and the dust that is mixed with it, is confined to an even thinner layer. Near the Sun, $h_z < 150\,\mathrm{pc}$ for most of the neutral hydrogen gas, and no more than 60–70 pc for the cold clouds of molecular gas from which stars are born. Table 2.1 lists values of the scale height h_z for various types of stars and for gas; these are only approximate, since the density does not exactly follow Equation 2.8. The scale length is probably in the range $2.5\,\mathrm{kpc} \lesssim h_R \lesssim 4.5\,\mathrm{kpc}$.

Problem 2.8 By integrating Equation 2.8, show that at radius R the number of stars per unit area (the surface density) of type S is $\Sigma(R, S) = 2n(0, 0, S)h_z(S)$ $\exp[-R/h_R(S)]$. If each has luminosity $L(S)$, the surface brightness $I(R, S) = L(S)\Sigma(R, S)$. Assuming that h_R and h_z are the same for all types of star, show that the disk's total luminosity $L_D = 2\pi I(R = 0)h_R^2$.

For the Milky Way, taking $L_D = 1.5 \times 10^{10}L_\odot$ in the V band and $h_R = 4\,\mathrm{kpc}$, show that the disk's surface brightness at the Sun's position 8 kpc from the center is $\sim 20L_\odot\,\mathrm{pc}^{-2}$. We will see in Section 3.4 that the mass density in the disk is about $(40$–$60)\mathcal{M}_\odot\,\mathrm{pc}^{-2}$, so we have $\mathcal{M}/L_V \sim 2$–3. Why is this larger than $\mathcal{M}/L_V$ for stars within 100 pc of the Sun? (Which stars are found only close to the midplane?)

Assuming that the solar neighborhood is a typical place, we can estimate how fast the Milky Way's disk is currently making stars. Taking $\mathcal{M}/L_V \approx 2$, the

Table 2.1 Scale heights and velocities of gas and stars in the disk and halo

Galactic component	h_z or shape	$\sigma_x = \sigma_R$ (km s^{-1})	$\sigma_y = \sigma_\phi$ (km s^{-1})	σ_z (km s^{-1})	$\langle v_y \rangle$ (km s^{-1})	Fraction of local stars
HI gas near the Sun	130 pc		≈ 5	≈ 7	Tiny	
Local CO, H_2 gas	65 pc		4		Tiny	
Thin disk: $Z > Z_\odot/4$	(Figure 2.9)					90%
$\tau < 3$ Gyr	≈ 280 pc	27	17	13	-10	
$3 < \tau < 6$ Gyr	≈ 300 pc	32	23	19	-12	
$6 < \tau < 10$ Gyr	≈ 350 pc	42	24	21	-19	
$\tau > 10$ Gyr		45	28	23	-30	
Thick disk	0.75–1 kpc					5%–15%
$\tau > 7$ Gyr, $Z < Z_\odot/4$	(Figure 2.9)	68	40	32	-32	
$0.2 \lesssim Z/Z_\odot \lesssim 0.6$		63	39	39	-51	
Halo stars near Sun	$b/a \approx 0.5$–0.8					$\sim 0.1\%$
$Z \lesssim Z_\odot/50$		140	105	95	-190	
Halo at $R \sim 25$ kpc	Round	100	100	100	-215	

Note: gas velocities are measured looking up out of the disk (σ_z of HI), or at the tangent point (σ_ϕ for HI and CO); velocities for thin-disk stars refer to Figure 2.9. For thick disk and halo, abundance Z, shape, and velocities refer to particular samples of stars. Velocity $\langle v_y \rangle$ is in the direction of Galactic rotation, relative to the *local standard of rest*, a circular orbit at the Sun's radius R_0, assuming $v_{y,\odot} = 5.2$ km s^{-1}.

disk's luminosity $L_V \sim 1.5 \times 10^{10} L_\odot$ corresponds to $3 \times 10^{10} \mathcal{M}_\odot$ in stars. If stars are produced with the same initial mass function that we measure locally, roughly half of their material is returned to the interstellar gas as they age. So, to build the disk over 10 Gyr, the Milky Way must produce $(3-5)\mathcal{M}_\odot$ of new stars each year. We will see in Section 2.4 that there is $(5-10) \times 10^9 \mathcal{M}_\odot$ of cool gas in the disk – so this rate of starbirth can be sustained for at least a few gigayears.

Even if we cannot measure enough stars to find their distribution in space, we can still use the volume $\mathcal{V}_{max}$ of Equation 2.3 to test whether they are uniformly spread: this is the $\mathcal{V}/\mathcal{V}_{max}$ *test*. Suppose that we choose a sample of stars by some well-determined rule (e.g., all brighter than a given apparent magnitude), and find their distance d and absolute magnitude M_V (equivalently, the luminosity L_V). For each one, we find the largest distance d_{max} and volume $\mathcal{V}_{max}$ to which we could have included it in our sample, and compare that with $\mathcal{V}(<d)$, the volume closer to us than the star. If the stars are equally common everywhere, then on average $\mathcal{V}(<d) = \mathcal{V}_{max}/2$: the star is equally likely to be in the nearer or the further half of the volume $\mathcal{V}_{max}$. A smaller value for this average indicates that the stars become less common further away from us.

Problem 2.9 Suppose that we look at G dwarfs brighter than $m_V = 15$ within 5° of the north Galactic pole. Assuming that they all have $M_V = 5$, to what height z_{max} can we see them? Then $\mathcal{V}_{max} = \Omega z_{max}^3/3$, where $\Omega/(4\pi)$ is the fraction of

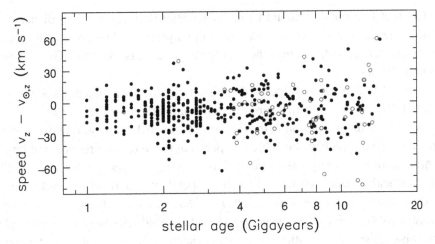

Fig. 2.9. For nearby main-sequence F and G stars, velocity $v_z - v_{z,\odot}$ is perpendicular to the Galactic plane, measured relative to the Sun. Open circles show stars with less than a quarter of the Sun's iron abundance. Older stars tend to move faster; the average velocity is negative, showing that the Sun moves 'upward' at $7\,\mathrm{km\,s^{-1}}$ – B. Nordström *et al.* 2004 *AAp* **418**, 98.

the sky covered by our $5°$ circle. If there are $n(z)$ stars per cubic parsec, show that the number N that we see is

$$N = \Omega \int_0^{z_{max}} n(z)z^2\,\mathrm{d}z, \quad \text{while} \quad \left\langle \frac{\mathcal{V}}{\mathcal{V}_{max}} \right\rangle = \frac{1}{N z_{max}^3} \int_0^{z_{max}} n(z)z^2 \cdot \Omega z^3\,\mathrm{d}z.$$

When $n(z)$ is constant, show that $\mathcal{V}/\mathcal{V}_{max} = 0.5$. Suppose that $n(z) = 1$ for $z < 800\,\mathrm{pc}$ and is zero further away: show that $\mathcal{V}/\mathcal{V}_{max} = 0.26$. Historically, this test was used to show that quasars were either more luminous or more common in the past.

The older stars have larger scale heights because the Galactic disk is lumpy. As they orbit, the stars feel the gravitational force of giant clouds of molecular gas, which can have masses up to $10^7 \mathcal{M}_\odot$, and clumps of stars and gas in the spiral arms. Over time, their orbits are disturbed by random pulls from these concentrations of matter, which increase both their in-and-out motion in radius and their vertical speed. Figure 2.9 shows velocities perpendicular to the Galactic disk for nearby F stars; clearly, older stars are more likely to be fast moving.

In Table 2.1, we give the *velocity dispersion* σ_z for different groups of stars. This quantity measures the spread of vertical velocities v_z:

$$\sigma_z^2 \equiv \left\langle v_z^2 - \langle v_z \rangle^2 \right\rangle, \tag{2.9}$$

where the angle-brackets denote an average over all the stars. For the F stars of Figure 2.9, we see that σ_z increases steadily with the age of the stars. Groups of stars that live for only a short time never attain a large velocity dispersion. Main-sequence A stars live no more than a gigayear; for them, σ_z is only a few kilometers per second, whereas the average for G dwarfs like our 5 Gyr-old Sun is about 30 km s^{-1}.

Table 2.1 also gives velocity dispersions in the disk plane: as shown in Figure 1.10, $\sigma_x = \sigma_R$ is measured radially outward from the Galactic center, and $\sigma_y = \sigma_\phi$ refers to motion in the direction of the disk's rotation. Older stars of a given spectral type have both higher speeds in the vertical direction and larger random motions in the plane of the disk than younger stars. The more rapid the stars' vertical motion, the more time they spend further from the midplane of the disk; accordingly, the scale height h_z is larger. Generally, $\sigma_R \gtrsim \sigma_\phi \gtrsim \sigma_z$. The velocity $\langle v_y \rangle$ gives the *asymmetric drift*, the speed relative to a circular orbit in the disk at the Sun's position. It is systematically negative. Groups of stars with larger velocity dispersion lag most strongly: we will see why in Section 3.3.

The stars within about 400 pc of the midplane belong mainly to the *thin disk* of the Galaxy. At greater heights, Figure 2.8 shows that the density of K dwarfs does not decrease as fast as predicted by Equation 2.8; these 'extra' stars belong to the *thick disk*. The density of the thick disk is often described by Equation 2.8, with $h_z \approx 1$ kpc, but the true vertical distribution is not very well known. (In fact, the problem of finding scale heights for both the thin and the thick disk simultaneously from a single measured run of $n(z)$ is *ill posed*: small errors or random variations in $n(z)$ cause huge changes in the values of h_z that we infer.)

Stars of the thick disk make up 10% of the total near the Sun at $z \approx 0$, and the number of thick-disk stars per square parsec is only 30% of that of stars in the thin disk. Unlike the thin disk, which is still forming stars, the thick disk includes no O, B, or A stars, so it must be older than about 3 Gyr. Because the luminous young stars are absent, the thick disk will have a higher mass-to-light ratio $\mathcal{M}/L$ than the thin disk. Thus its contribution to the luminosity will be somewhat below 30%.

Problem 2.10 Thin-disk stars make up 90% of the total in the midplane while 10% belong to the thick disk, but h_z for the thin disk is roughly three times smaller than for the thick disk. Starting from Equation 2.8, show that the surface density of stars per square parsec follows $\Sigma(R, \text{thin disk}) \approx 3\Sigma(R, \text{thick disk})$.

Most stars in the thin disk have heavy-element abundances between the solar level and about half-solar, though some are more metal-poor. The spectra of thick-disk stars generally show a smaller fraction of heavy elements, with most having $Z_\odot/10 \lesssim Z \lesssim Z_\odot/2$. When we see stars with ages and compositions characteristic of the thick disk near the Sun, they have rapid vertical motions, with $\sigma_z \gtrsim 40$ km s^{-1}, so they have enough energy to travel a kiloparsec or more above the midplane.

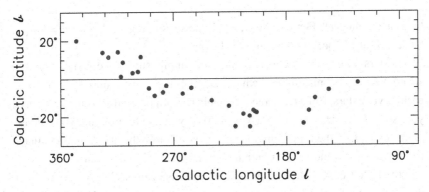

Fig. 2.10. Short-lived bright stars with $M_V < 3$, at distances $100\,\text{pc} < d < 500\,\text{pc}$ from the Sun, taken from the Hipparcos catalogue. Most of these B and A stars lie in a plane tilted by about 20° to the midplane of the disk.

Our present thick disk may be the remains of an early thin disk in the young Milky Way. If a small satellite galaxy had collided with our Milky Way and merged into it, the disk would have been shaken up, and the energy of the impact largely transferred into increased random motions of the stars. The gas would have fallen to the midplane of the disk; stars which later formed from it would make up the thin disk that we observe today.

Disk stars are born in clusters and associations, where a gas cloud has come together that is large enough to collapse under its own gravity. The Sun lies inside a partial ring or disk of young stars known as Gould's Belt. Within 500 pc, stars younger than about 30 Myr are not found in the plane of the disk, but rather in a layer tilted by 20° about a line roughly along the Sun's orbit at $l = 90°$, with stars nearer to the Galactic center lying above the midplane (Figure 2.10). Clouds of hydrogen gas form a similarly tilted ring, which is expanding outward at 1–$2\,\text{km s}^{-1}$ from a point about 150 pc away from us. By the time the Sun has made a few orbits of the Galaxy, the stars of Gould's Belt will have dispersed into the disk. But measuring their velocities will show that they follow very similar orbits; such a collection of stars is called a 'moving group'.

Problem 2.11 Here you make a numerical model describing both the distribution of stars and the way we observe them, to explore the *Malmquist bias*. If we observe stars down to a fixed apparent brightness, we do not get a fair mixture of all the stars in the sky, but we include more of the most luminous stars. This method of 'Monte Carlo simulation' is frequently used when a mathematical analysis would be too complex.

(a) Your model sky consists of G-type stars in regions A ($70\,\text{pc} < d < 90\,\text{pc}$), B ($90\,\text{pc} < d < 110\,\text{pc}$), and C ($110\,\text{pc} < d < 130\,\text{pc}$). If the density is uniform, and you have ten stars in region B, how many are in regions A and C (round to the nearest integer)? For simplicity, let all the stars in region A be at $d = 80\,\text{pc}$, those in B at $100\,\text{pc}$, and those in C at $120\,\text{pc}$.

G stars do not all have exactly the same luminosity; if the variation corresponds to about 0.3 magnitudes, what fractional change in luminosity is this? For each of your stars, roll a die, note the number N_1 on the upturned face, and give your star $M_V = M_{V,\odot} + 0.2 \times (N_1 - 3.5)$. If you like to program, you can use more stars, place them randomly in space, and choose the absolute magnitudes from a Gaussian random distribution, with mean $M_{V,\odot}$ and variance 0.3.

(b) To 'observe' your sky, use a 'telescope' that can 'see' only stars brighter than apparent magnitude $m_V = 10$; these stars are your *sample*. How different is their mean absolute magnitude from that for all the stars that you placed in your sky?

What is the average distance of all the stars in your sample? Suppose you assumed that your sample stars each had the average luminosity for all the stars in your sky, and then calculated their distances from their apparent magnitudes: what would you find for their average distance? In which sense would you make an error?

(c) Metal-poor main-sequence stars are bluer for a given luminosity, so they must be fainter at a particular spectral type; if the star's fraction by weight of heavy elements is Z, then $\Delta M_V \approx -0.87 \log_{10}(Z/Z_\odot)$. For each of your stars, roll the die again, note the number N_2, set $Z/Z_\odot = (N_2 + 0.5)/6$ and change its absolute magnitude from part (a) by ΔM_V.

Observe them again with the same telescope. For your sample of stars, calculate the average Z of those that fall in regions B and C. Are these more or less metal-rich than all the stars that you placed in your sky?

(Errors in measurement have the same kind of effect as a spread in the true luminosity of a class of stars or galaxies. You can make corrections if you know what your measurement errors are; but most people are too optimistic and underestimate their errors!)

2.2.3 Distances to star clusters

If we can observe not just a single star, but an entire cluster of stars that are all at the same distance and that were formed together out of the same gaseous raw material, we can make a much more accurate estimate of the distance. The Pleiades cluster, shown in Figure 2.11, contains about 700 stars that appear brighter than $m_V = 17$.

The color–magnitude diagram of Figure 2.12 shows that most of the stars are still on the main sequence, but those brighter than $m_V \approx 5$, that can be seen without a telescope or binoculars, have left it to become blue giants. The main sequence is much narrower than it is in Figure 2.2, because all the stars are the same age and have the same abundance of heavy elements. Stars with the lowest masses have not yet reached the main sequence, and some of them are still

Fig. 2.11. The central region of the Pleiades open cluster; the brightest few stars easily outshine the rest of the cluster – NOAO.

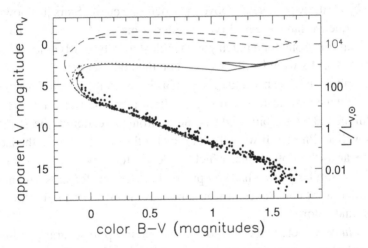

Fig. 2.12. Measured apparent magnitude m_V and color $B - V$ for stars in the Pleiades cluster; points show observed stars, and the solid line is an isochrone for stars 100 Myr old. The dotted line shows the same isochrone without correction for dust reddening; the dashed line is an isochrone for age 16 Myr – J.-C. Mermilliod.

partially hidden by dust. The gas and dust still present around this cluster shows that no stars in it have yet exploded as supernovae; these would have swept it clean of gas.

The solid line in Figure 2.12 is an *isochrone*, showing where stars of different masses, but all of the same age, would appear if they were moved to the distance modulus corresponding to the Pleiades: $(m - M)_0 = 5.6$. It has been calculated with the same stellar models as Figure 1.4, for stars with the same chemical

Table 2.2 Some open clusters in the Milky Way

Cluster	d (pc)	[Fe/H]	M_V (mag)	L_V ($10^3 L_\odot$)	Color $(B-V)$	r_c (pc)	σ_r (km s^{-1})	Age	Mass ($\mathcal{M}_\odot$)	$\mathcal{M}/L$ ($\mathcal{M}_\odot/L_\odot$)
NGC 3603	6500	–	–	20 000	−0.3	0.5	–	<5 Myr	7000	0.001
Pleiades (M45)	132	0.0	−4.3	4.5	−0.05	3	0.5	0.125 Gyr	800	0.2
NGC 6705 (M11)	1900	0.14	−6.0	22	0.18	1	–	0.25 Gyr	–	–
Hyades	46	0.14	−2.7	1.0	0.40	3	0.3	0.625 Gyr	460	0.4
NGC 7789	2000	−0.26	−5.7	17	0.98	5	0.8	1.5 Gyr	–	–
NGC 2682 (M67)	860	−0.05	−3.3	1.8	0.78	1	0.8	4 Gyr	2000	1
NGC 6791	4000	0.3	−3.5	2.1	1.02	3	–	8–10 Gyr	–	–

Note: d is distance from the Sun; [Fe/H] $= \log_{10}(Z/Z_\odot)$; M_V and $B - V$ are corrected for the obscuring effect of dust; r_c is the core radius measured for stars of roughly the Sun's luminosity; σ_r is the dispersion in the radial velocity V_r of stars in the cluster's central region. Masses are found from the stellar luminosity function, using infrared observations to find faint or obscured stars.

composition as the Sun, at an age of 100 Myr. Most of the stars lie very close to the isochrone, but almost a third of the stars in the cluster's central region are in binary or multiple star systems that are so close that the stars appear as one brighter object; these points fall above the isochrone curve. Stars just now leaving the main sequence have spectral type B8 and a mass $\mathcal{M} \approx 5\mathcal{M}_\odot$. The dashed line shows an isochrone for a much younger cluster, only 16 Myr old; at that time, even stars of $10\mathcal{M}_\odot$ would still have been on the main sequence.

To determine the distance and the age of a cluster by this method, we must take account of interstellar dust, which scatters light from the stars and makes them appear dimmer. Because blue light is more strongly scattered, the stars appear redder as well as dimmer. If we did not allow for the effects of dust, the isochrone would appear as the dotted line, which is clearly too blue at $m_V \approx 5$ to fit the observations. Similarly, if we had computed our isochrone for stars with much less than the solar metal abundance, it would not have been a good fit to the observed brightness and colors.

Open clusters such as the Pleiades can contain up to several hundred stars, bound together by their mutual gravitational attraction. Table 2.2 shows that cluster luminosities generally fall in the range $100 L_\odot$ to $30\,000 L_\odot$. The central density can be as high as $100 L_\odot$ pc^{-3} but is frequently less; the *core radius*, where the surface brightness falls to half its central value, is typically a few parsecs. The stars have small random speeds, below 1 km s^{-1}. The mass-to-light ratio is generally below unity; it increases with the cluster's age as the luminous massive stars die. This ratio is hard to determine, because the diffuse outskirts of the cluster contain disproportionately many of the dim stars that contribute most of the cluster's mass, but little of its light. We will see in Section 3.2 how this *mass segregation* can develop.

Open clusters are quite often surrounded by gas and dust, and they always lie close to the plane of the disk. The youngest cluster in Table 2.2, NGC 3603, is near the Galactic center. Only ∼1% of its visible light can reach us, so we must study it in the infrared. In other galaxies, we see open clusters in the spiral arms. Unfortunately we cannot see most of the Milky Way's open clusters, because visible light does not travel well through the dust in the disk. We can make a fairly complete map of the clusters only within about 5 kpc of the Sun.

Only about 5% of the 1200 or so known open clusters are more than 1 Gyr old, and most have ages less than 300 Myr; the nearby Hyades cluster is about 600 Myr old. Because most of a cluster's light comes from its brightest stars, the color of its total or *integrated* light gives us a rough estimate of its age. For example, the Pleiades cluster has $B - V = -0.05$, close to the color of its most luminous stars that are still on the main sequence; see Table 1.4. As the massive stars die, and more of the cluster's light comes from red giants, its color becomes redder. We can use observations of nearby clusters, and the theory of stellar evolution, to relate a cluster's color to its age and metal abundance. When we observe star clusters in other galaxies, generally we measure only the integrated light. Comparing a cluster's color with clusters in the Milky Way gives us an estimate of its age and the composition of its stars.

Outside Gould's Belt, most of the known open clusters younger than 300 Myr lie within 50 pc of the Galactic midplane. The older clusters, like the old stars of the thin disk, have a larger scale height, $h_z \approx 375$ pc. Clusters older than a gigayear are found mainly in the outer Galaxy, beyond the Sun's orbit, where the gravitational forces are weaker. They are also relatively well populated and compact. Younger clusters are scattered more uniformly in Galactocentric radius; many have larger sizes or fewer members, and so are more fragile and prone to fall apart. Since the gravitational attraction of open clusters does not bind them together very strongly, they tend to be pulled apart as they pass through the spiral arms of the disk, and also to dissolve because of the gravitational pull of the cluster stars on each other. The old clusters that we see today are probably the robust survivors of a much larger population.

NGC 3603 is a very young and massive cluster; in the central parsec are several O3 stars, each with $L \sim 30\,000 L_\odot$. The stars are no more than 5 Myr old, and may be younger. Because its most massive stars are still shining, it is hugely more luminous than the other clusters in Table 2.2. It is a smaller version of the *super-star-clusters* formed in *starburst* galaxies. Here, starbirth is so vigorous that it will consume the galaxy's supply of cool gas within ∼300 Myr. (Recall from Section 2.2 that the Milky Way will not run out of gas for at least 2–3 Gyr.) We will discuss starbursts further in Section 7.1.

Since elements heavier than helium are produced in the interiors of stars, they become more abundant as the Universe ages. Figure 2.9 showed that F stars older than 10 Gyr are much more likely to be iron-poor than are the younger stars. So we might expect that old open clusters, which formed earlier in the life of the Milky Way, should have lower metal abundances than the younger clusters. Surprisingly, this is not so; the old cluster NGC 6791 is as metal-rich as the Sun.

Table 2.3 Some globular clusters in the Milky Way, and one belonging to the Fornax dwarf spheroidal galaxy

Cluster		d (kpc)	[Fe/H]	M_V (mag)	L_V ($10^3 L_\odot$)	r_c (pc)	r_t (pc)	σ_r (km s^{-1})
NGC 5139	ω Cen	5.2	-1.6^a	-10.2	1100	4	70	20
NGC 104	47 Tuc	4.5	-0.71	-9.5	500	0.5	50	11
NGC 7078	M15	10.8	-2.15	-9.3	440	$<0.01^b$	85	12
NGC 6341	M92	8.5	-2.15	-8.3	180	0.5	35	5
NGC 7099	M30	9.1	-2.13	-7.6	95	$<0.1^b$	45	5
NGC 6121	M4	1.73	-1.2	-7.2	60	0.5	25	4
	Pal 13	24.3	-1.9	-3.8	3	0.5	>50	0.6–0.9
NGC 1049	Fornax 3	140	-2.0	-7.8	100	1.6	>50	9

Note: d is distance from the Sun; [Fe/H] $= \log_{10}(Z/Z_\odot)$; r_c is the core radius, r_t is the tidal or truncation radius; and σ_r is the dispersion in the radial velocity V_r of stars in the central region.
[a] 20%–30% of the stars of ω Centauri's core are more metal-rich.
[b] A *collapsed core*: see Section 3.2.

There is considerable scatter in the chemical composition of open clusters at all ages. The buildup of metals in the Galaxy must have proceeded quite unevenly: some regions even recently were relatively poor in these elements. Open clusters further from the Galactic center are more likely to be metal-poor; the outer Milky Way seems to be enriching itself more slowly than the inner parts.

The *globular clusters* are very different from open clusters; they contain far more stars, much more tightly packed. The brightest of the Milky Way's 150 or so known globulars is ω Centauri; with $L \approx 10^6 L_\odot$ it contains about a million stars. In Figure 2.13 we see 47 Tucanae, another luminous cluster; Table 2.3 shows that dimmer clusters range down to $10^4 L_\odot$. The stellar density is roughly constant inside the core radius: $r_c \approx 5$ pc. The stars of globular clusters have higher random speeds σ_r than those of stars in the open clusters of Table 2.2. Globular clusters around Fornax, a small nearby dwarf galaxy, are very similar to those in the Milky Way. At some outer radius r_t, usually beyond 30 pc, the density of stars drops sharply toward zero. This is the *tidal radius* or *truncation radius*. Stars beyond this point are so loosely bound to the cluster that they can be swept away by the 'tidal' gravitational force of the Milky Way as the cluster orbits around it. We will study this process further in Section 4.1.

None of the Milky Way's globular clusters is younger than several gigayears, and most are much older. The two color–magnitude diagrams of Figure 2.14 show no young stars at all. The globular cluster 47 Tucanae is more than 10 Gyr old. Its stars have only 15% of the solar abundance of metals, so the main sequence ends with stars a little bluer and more luminous than the Sun. As in the Pleiades, the main sequence is very narrow because all the stars are the same age and have the same composition. Brighter stars fall on a narrow red giant branch, or on the *horizontal branch* with $M_V \approx 0$. Like the local disk stars that populate the red clump

Fig. 2.13. The luminous globular cluster 47 Tucanae – Southern African Large Telescope.

in Figure 2.2, horizontal branch stars burn helium in their cores and hydrogen in a surrounding shell. The stars of the most metal-poor clusters appear to be 12–15 Gyr old. Even taking account of likely observational errors, and uncertainties in the theory of stellar evolution, their ages are unlikely to be less than 11–12 Gyr. This is close to the estimate t_H for the age of the Universe given by Equation 1.31.

Why has the Milky Way ceased production of globular clusters, when it continues to make open clusters? We will see in Problem 3.14 that star formation must be more efficient in globular clusters than it normally is today, to allow them to remain dense. As the galaxies were assembled, extremely high gas density and pressure may have produced denser and more massive star clusters than can form today, scaled-up versions of the open cluster NGC 3603, that could survive as today's globular clusters.

Globular clusters are lacking in heavy elements. The stars of metal-rich globulars have 1/3 to 1/10 of the solar metallicity, while metal-poor clusters contain as little as 1/300 of the Sun's proportion of these elements. The color–magnitude diagram gives a clue about the cluster's abundance. In a metal-poor globular cluster, the horizontal branch can lie well to the blue of the red giant branch. It shifts redward in younger clusters and those richer in metals. Horizontal branch stars of the metal-poor cluster M92 are as blue as spectral type A0 ($B - V = 0$), while in 47 Tucanae they are about as red as the Sun, with $B - V \approx 0.6$ (Figure 2.14). The main sequence and giant branch of M30 are also much bluer than those in 47 Tucanae.

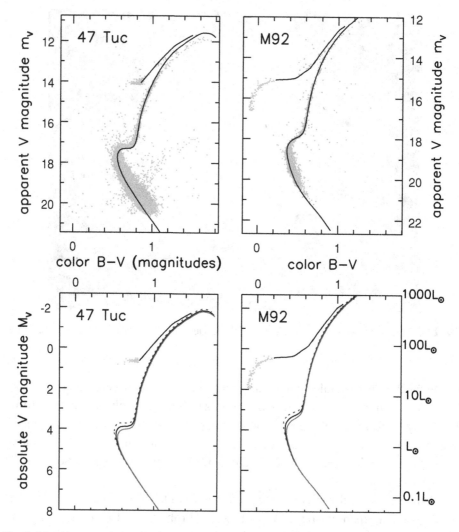

Fig. 2.14. Above, color–magnitude diagrams for globular clusters 47 Tucanae and M92; all vertical scales coincide in luminosity. Above left, the star sequence crossing the main sequence near $B - V, m_V = (0.8, 19.5)$ is the red giant branch of the Small Magellanic Cloud, seen in the background. The model curve shows stars that are 12 Gyr old. Above right, the model curve is for metal-poor stars of age 13 Gyr. Below, the central isochrones match those above; the dotted lines show stars 2 Gyr younger, and the lighter lines those 2 Gyr older – P. Stetson; models from BaSTI at Teramo Observatory.

No globular cluster lies close enough to us that its distance can be measured by trigonometric parallax. Instead, we compare the observed color–magnitude diagram with the predictions of models for stellar evolution like those in Figure 2.14. We adjust the assumed distance, age, and fraction of heavy elements to obtain the best correspondence. Another way to find the distance is from *RR Lyrae stars*, low-mass stars on the horizontal branch. These stars pulsate; as the radius grows

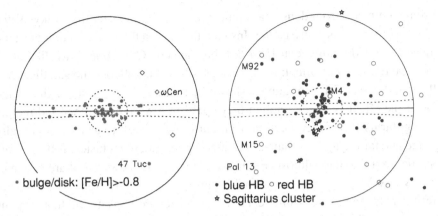

Fig. 2.15. Left, positions on the sky of the Milky Way's metal-rich 'disk' globular clusters (filled dots), and unusual objects, perhaps remnants of disrupted dwarf galaxies (open diamonds). Right, metal-poor clusters with [Fe/H] < −0.8. Those of the Sagittarius dwarf (stars) fall in a great circle on the sky. Clusters with a blue horizontal branch (filled dots) are more concentrated to the center than are those with a red horizontal branch (open circles). Circles mark 20° and 90° from the direction to the Galactic center; the solid line is the Galactic equator. Between the dashed lines at $b = \pm 5°$, clusters may easily hide in the dusty disk – D. Mackey.

alternately larger and smaller, their brightness varies regularly with periods of 0.2–1 day, which makes them easy to find. RR Lyrae stars all have about the same luminosity, $L \approx 50 L_\odot$; so, if we can measure their apparent brightness, Equation 1.1 gives the distance.

Unlike open clusters, most globular clusters are high above the midplane, so they are not hidden from us in the dusty disk. We see many more globular clusters when we look toward the Galactic center than we do in the anticenter direction, showing that the Sun lies several kiloparsecs out in the disk. Figure 2.15 shows that metal-rich globulars are more numerous in the inner Galaxy and closer to the midplane; they may have been formed together with the bulge and thick disk. The metal-poor clusters form a rough sphere around the center of the Galaxy. Among them, red horizontal branches may be a sign of relative youth. Those are less concentrated toward the Galaxy center than are the clusters with blue horizontal branches.

We can measure the radial velocities of stars in globular clusters from the Doppler shift of absorption lines in their spectra. These tell us that the most metal-poor globular clusters do not follow circular orbits. They plunge deep into the Galaxy, but spend most of their time at large distances. The orbits of halo clusters are oriented almost randomly, so the cluster system as a whole does not rotate about the Galactic center like the disk. The system of metal-rich globular clusters does rotate; the clusters follow orbits much like those of stars in the thick disk.

Stars are made from dense gas clouds: so we might expect the Milky Way's oldest stars to be in its dense center. Instead, they are in the halo globular clusters, its most extended component. How did this happen? Our Milky Way, like other sizable galaxies, is a cannibal: it has eaten its closest neighbors and satellites. We will see in Section 7.1 how gravity slows the motion of two galaxies that come close; if the effect is strong enough they spiral in toward each other and merge. The Sagittarius dwarf (see Section 4.1), our closest satellite galaxy, is partially digested and falling apart. A half-dozen metal-poor globular clusters follow orbits so similar to it that they almost certainly belonged to this dwarf, but are now part of the Milky Way's halo. The globular cluster ω Centauri contains stars with differing metal abundances, unlike a true star cluster. Perhaps it is all that remains of a dense dwarf galaxy after its outer stars were torn away. The Magellanic Clouds will share the Sagittarius dwarf's fate within 3–5 Gyr. Some astronomers believe that the 'blue' metal-poor clusters in Figure 2.15 joined the Milky Way when it swallowed their parent galaxies.

The *metal-poor halo* of the Milky Way consists of the halo globular clusters, and roughly 100 times more individual metal-poor stars, some with less than 10^{-5} of the Sun's abundance. Locally, only one star in a thousand belongs to the metal-poor halo. They follow orbits similar to those of metal-poor globulars, and so are moving very fast relative to the Sun. We can pick out a few nearby halo stars by their unusually high proper motions. The blue horizontal branch stars, distinctive variable stars such as the RR Lyrae, and red giants stand out clearly from among the more numerous foreground disk stars. Metal-poor globular clusters and stars of the metal-poor halo have been found as far as 100 kpc from the Galactic center.

Most halo stars seem to be as old as the metal-poor globular clusters. As with open clusters, the globular clusters that we see today are the few survivors of a larger population. Some halo stars must be the remains of globular clusters that dissolved or were torn apart by the Milky Way's gravity (see Sections 3.2 and 4.1). The globular cluster Palomar 5 has probably lost 90% of its stars; two spectacular stellar tails extend away from it more than $10°$ across the sky. Many globular clusters in the far reaches of the Milky Way, like Palomar 13 in Table 2.3, have relatively few members. If clusters like this had been formed closer to the Galactic center, they would have been torn apart long ago. The metal-poor halo also contains various 'moving groups' of stars that follow a common orbit, and are probably the remains of captured satellite systems. A stream of carbon stars and M giants apparently stripped from the Sagittarius dwarf galaxy stretches around the sky.

The most useful approach to mapping the metal-poor halo has been to guess at the luminosity function and the density of stars in the disk, bulge, and halo, and then use Monte Carlo techniques to calculate how many stars of each spectral type one expects to see in a particular area of the sky with a given apparent brightness. We then adjust the guesses to achieve the best correspondence with

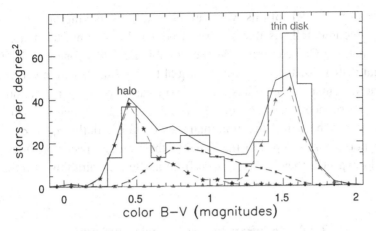

Fig. 2.16. Numbers of stars at each $B - V$ color with apparent V magnitude $19 < m_V < 20$, per square degree near the north Galactic pole. The solid line shows the prediction of a model: thin-disk stars (triangles) are red, halo stars (stars) are blue, and thick-disk stars (squares) have intermediate colors – N. Reid.

what is observed. Figure 2.16 shows the number of stars per square degree at the north Galactic pole at apparent brightness $19 < m_V < 20$, compared with the predictions of a model in which 0.15% of the stars near the Sun belong to the metal-poor halo, and the density of halo stars drops with radius approximately as r^{-3}. Almost all the blue stars in this brightness range are halo stars, while nearly all the red ones are in the thin disk (see the following problem), so we can separate the various components. The total mass of the metal-poor halo stars is only about $10^9 \mathcal{M}_\odot$, much less than that of stars in the disk or bulge.

Problem 2.12 The range in apparent magnitude for Figure 2.16 was chosen to separate stars of the thin disk cleanly from those in the halo. To see why this works, use Figure 2.2 to represent the stars of the local disk, and assume that the color–magnitude diagram for halo stars is similar to that of the metal-poor globular cluster M92, in Figure 2.14.

(a) What is the absolute magnitude M_V of a disk star at $B - V = 0.4$? How far away must it be to have $m_V = 20$? In M92, the bluest stars still on the main sequence have $B - V \approx 0.4$. Show that, if such a star has apparent magnitude $m_V = 20$, it must be at $d \approx 20$ kpc.

(b) What absolute magnitudes M_V could a disk star have, if it has $B - V = 1.5$? How far away would that star be at $m_V = 20$? In M92, what is M_V for the reddest stars, with $B - V \approx 1.2$? How distant must these stars be if $m_V = 20$?

(c) Explain why the reddest stars in Figure 2.16 are likely to belong to the disk, while the bluest stars belong to the halo.

Comparing models of this type with observations looking out of the disk in various directions tells us that the metal-poor halo is somewhat flattened, but rounder than the Galactic bulge. We see in Table 2.1 that for the halo stars the tangential drift velocity $\langle v_y \rangle$ is almost equal to the Sun's rotation speed around the Galactic center; the halo has little or no rotation of its own. The outer halo seems to be rounder than the inner part. Table 2.1 shows that outer-halo stars move slower in the radial direction than they do tangentially: $\sigma_R < \sigma_{\theta,\phi}$. These stars do not plunge deep into the Galaxy; their radial speeds are low, just as we would expect if they had been torn from a satellite spiralling into the Milky Way.

2.2.4 An infrared view: the bulge and nucleus

Studies of the Galactic disk have always been hampered by interstellar dust. By scattering and absorbing at visible and ultraviolet wavelengths, the dust denies us a clear view of distant stars in the disk. By comparing the numbers of stars in different parts of the sky, William Herschel was able to show in 1800 that the Milky Way is a disk. But since he saw roughly the same number of stars in every direction within the disk, he erroneously concluded that the Sun must be near the center of the Galaxy. In fact, dust in the disk hid most of the stars.

The best way to map the Milky Way's central bulge is to use infrared light, which travels more freely than visible light through the dusty disk. These observations indicate that, both for the thin and for the thick disk, the scale length h_R of Equation 2.8 probably lies in the range $2.5\,\mathrm{kpc} < h_R < 4.5\,\mathrm{kpc}$. Beyond a radius $R_{\mathrm{max}} \approx 15\,\mathrm{kpc}$, the density of disk stars appears to drop rapidly toward zero. We will find in Section 5.1 that an 'edge' of this kind is also seen in the stellar disks of some other galaxies.

The near-infrared image on the front cover of this book shows the Milky Way as seen at wavelengths of 1.25, 2.5, and 3.5 μm. We clearly see a flattened central bulge, which accounts for roughly 20% of the Galaxy's total light; most of that comes from within ~1 kpc of the center. The bulge appears pear-shaped, larger on one side than the other. It is probably a central *bar*, extending 3–4 kpc from the center; the end at $l > 0$ appears larger because it is closer to us. In Hubble's classification, the Milky Way is probably an Sbc or Sc galaxy. It is not so strongly barred as to be an SBbc or SBc, although some astronomers might place it in a category intermediate between barred and unbarred spirals, labelled SAB.

The density of the stellar halo rises toward the Galactic center, and it is natural to wonder whether the galactic bulge is just the dense inner portion of the halo. It is not. Although the bulge stars are several gigayears old, they are not metal-poor like the halo. The average metal fraction is at least half of the solar value, and some stars have up to three times the solar abundance of heavy elements. The bulge is more flattened than the inner halo, and the bulge stars circle the Galactic center in the same direction as the disk does. The averaged rotation speed in the bulge is about $100 \, \text{km s}^{-1}$, somewhat slower than in the disk; the bulge stars have larger random motions. In Section 4.3 below, we will discuss how the bulge, halo, and disk of the Milky Way might have been formed as they are. Very close to the Galactic center, we find dense gas and young stars. About 150 pc away, near-infrared observations reveal a huge dense star cluster, Sagittarius B2, which is making stars at a furious rate. Then, 30–50 pc from the center, the Quintuplet and Arches clusters are each more luminous than $10^6 L_\odot$, containing several very massive stars. At the heart of the Galaxy is a torus of hot dense molecular clouds, about 2 pc in radius with $10^6 \mathcal{M}_\odot$ of gas. It surrounds the Milky Way's *stellar nucleus*, an extraordinary concentration of stars.

At optical wavelengths the nucleus is invisible, because surrounding dust absorbs and scatters the light; in the V band, it is dimmed by 31 magnitudes! It is best seen in the near-infrared, at $\lambda \sim 5$–7 µm; at longer wavelengths the warmed dust radiates strongly, overwhelming the starlight. In mass and size, the stellar nucleus is not so different from a massive globular cluster, with $3 \times 10^7 \mathcal{M}_\odot$ of stars within a central cusp of radius $10''$ or 0.2 pc. The density of stars reaches $3 \times 10^7 \mathcal{M}_\odot \, \text{pc}^{-3}$ within $1''$ from the center. But, unlike in globular clusters, at least 30 massive stars have formed here over the last 2–7 million years. Star-cluster nuclei are common in giant spiral galaxies; they are by far the densest regions of these systems. Unlike true star clusters, they contain stars with a range of ages and composition.

The innermost young stars are less than 0.05 pc from the Milky Way's central radio source. Figure 2.17 shows some of their orbits, calculated from the observed radial velocities and proper motions. These stars follow almost the same Keplerian motion as the planets in the solar system, and we saw in Problem 2.6 how to use them to measure the mass of the central compact object. This is almost certainly a black hole: we cannot otherwise understand how $4 \times 10^6 \mathcal{M}_\odot$ can fit into such a small volume.

Radio maps of the inner region show narrow filaments, tens of parsecs long but only a fraction of a parsec wide, reaching up out of the Galactic plane. They are highly polarized, which tells us that this is synchrotron emission; the radiating electrons are probably held inside the filaments by magnetic fields. At the position of the black hole is a central pointlike source, Sagittarius A⋆, which varies its brightness so rapidly that it must be less than 10 light-minutes, or 1.3 AU, across. Sagittarius A⋆ may be a small version of the spectacular nuclear radio and X-ray sources found in active galactic nuclei, which we will discuss in Section 9.1. If so, it is a very weak example. Its power is no more than a few thousand times the Sun's total energy output, while, in Seyfert galaxies and quasars, the nucleus alone can outshine the rest of the galaxy.

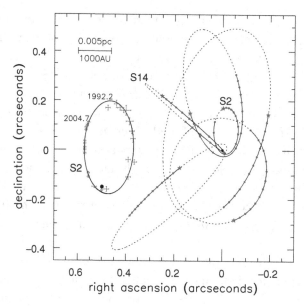

Fig. 2.17. Star symbols show the positions of stars in the Galactic nucleus in 1992.25, with small dots at one-year intervals along the orbits; the stars move faster near the black hole (filled dot at the origin). The inset shows coordinates measured for the O8/O9 star S2, which has made almost a complete orbit during 12.5 years of observation – F. Eisenhauer, MPE Galactic Center Team.

Problem 2.13 Here we make a crude model to estimate how many stars you could see with your unaided eye, if you observed from the center of the Galaxy. Naked-eye stars are those brighter than apparent magnitude $m_V \approx 5$; from Earth, we see about 7000 of them. Assume that the Milky Way's nucleus is a uniform sphere of stars with radius 3 pc, and ignore the dimming effects of dust. What is the luminosity L_{eye} of a star that is seen 3 pc away with $m_V = 5$? For a main-sequence star, use Equation 1.6 to show that L_{eye} corresponds to $\mathcal{M} \approx 0.6 \mathcal{M}_\odot$.

In our simple model, almost all stars that spend less than 3 Gyr on the main sequence have now died; according to Table 1.1, what stellar mass $\mathcal{M}_u$ does this correspond to? Approximate the number $\xi(\mathcal{M})\Delta\mathcal{M}$ of main-sequence stars with masses between $\mathcal{M}$ and $\mathcal{M} + \Delta\mathcal{M}$ by Equation 2.5: $\xi(\mathcal{M}) \propto \mathcal{M}^{-2.35}$ for $\mathcal{M} \gtrsim 0.2\mathcal{M}_\odot$, with few stars of lower mass. Find the total number and total mass of main-sequence stars with $\mathcal{M} < \mathcal{M}_u$, in terms of the parameter ξ_0. How do we know that red giants will contribute little mass? Taking the total mass as $10^7 \mathcal{M}_\odot$, find ξ_0; show that the nucleus contains $N_{eye} \sim 4 \times 10^6$ main-sequence stars with $L \geq L_{eye}$. How do we know that many fewer red giants will be visible? (For advanced students: stars with $L < L_{eye}$ will be seen as naked-eye stars if they are close enough to the observer. Show that these make little difference to the total.)

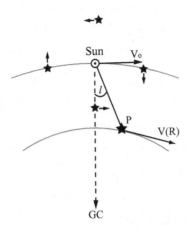

Fig. 2.18. Galactic rotation: stars closer to the Galactic center (GC) pull ahead of us in their orbits, while those further out are left behind. A star at the same Galactocentric radius moves sideways relative to us.

Further reading: F. Melia, 2003, *The Black Hole at the Center of our Galaxy* (Princeton University Press, Princeton, New Jersey) is written for the general reader.

2.3 Galactic rotation

To a good approximation, the stars and gas in the disk of our Milky Way move in near-circular paths about the Galactic center. We can take advantage of this orderly motion to map out the distribution of galactic gas, from its measured velocities in each direction. From the observed speeds, we can calculate how much inward force is needed to keep the gas of the outer Galaxy in its orbit; it turns out to be far more than expected. Additional mass, the *dark matter*, is required in addition to that of the luminous stars and gas.

Stars closer to the Galactic center complete their orbits in less time than do those further out. This *differential rotation* was first discovered by considering the proper motions of nearby stars. Looking inward, we see stars passing us in their orbits; their motion relative to us is in the same direction as the Sun's orbital velocity V_0. Looking outward, we see stars falling behind us, so they have proper motions in the opposite direction (Figure 2.18). Stars at the same Galactocentric radius orbit at the same rate as the Sun, so they maintain a fixed distance and have a 'sideways' motion. So, for stars close to the Sun, the proper motion μ has a component that varies with Galactic longitude l as $\mu \propto \cos(2l)$. This pattern had been noticed already by 1900; Dutch astronomer Jan Oort explained it in 1927 as an effect of Galactic rotation. By the 1920s, photographic plates had become more sensitive, and could record stellar spectra well enough to determine radial velocities accurately; these are now more useful for measuring differential rotation in the Galaxy.

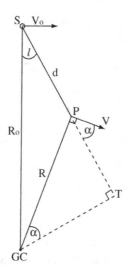

Fig. 2.19. Galactic rotation: a star or gas cloud at P with longitude l and Galactocentric radius R, at distance d from the Sun, orbits with speed $V(R)$. The line of sight to P is closest to the Galactic center at the *tangent point* T.

The Sun does not lie exactly in the Galactic midplane, but about 15 pc above it, and its path around the Galactic center is not precisely circular. The *local standard of rest* is defined as the average motion of stars near the Sun, after correcting for the asymmetric drift $\langle v_y \rangle$ of Table 2.1. Relative to this average, the Sun is moving 'in' toward the Galactic center at $10 \, \mathrm{km \, s^{-1}}$, and travels faster in the direction of rotation by about $5 \, \mathrm{km \, s^{-1}}$; its 'upward' speed toward the north Galactic pole is 7–$8 \, \mathrm{km \, s^{-1}}$. Published velocities of stars and gas are frequently given with respect to this standard.

Usually (but not always; see Problem 2.16 below), we assume that the local standard of rest follows a circular orbit around the Galactic center. In 1985 the International Astronomical Union (IAU) recommended the values $R_0 = 8.5 \, \mathrm{kpc}$, for the Sun's distance from the Galactic center, and $V_0 = 220 \, \mathrm{km \, s^{-1}}$, for its speed in that circular orbit. To allow workers to compare their measurements, astronomers often compute the distances and speeds of stars by using the IAU values, although current estimates are closer to $R_0 \approx 8 \, \mathrm{kpc}$ and $V_0 \approx 200 \, \mathrm{km \, s^{-1}}$.

Problem 2.14 Using the IAU values for R_0 and V_0, show that it takes the Sun about 240 Myr to complete one orbit about the Galactic center. This period is sometimes called a 'Galactic year'.

2.3.1 Measuring the Galactic rotation curve

We can calculate the radial velocity V_r of a star or gas cloud, assuming that it follows an exactly circular orbit; see Figure 2.19. At radius R_0 the Sun (or more

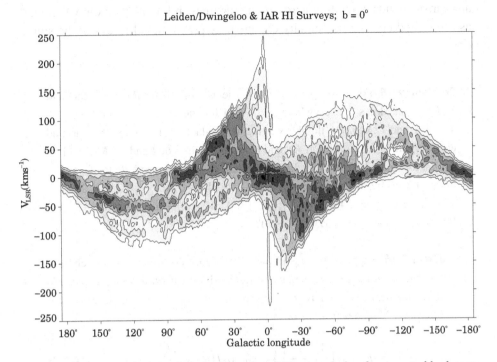

Leiden/Dwingeloo & IAR HI Surveys; b = 0°

Fig. 2.20. In the plane of the disk, the intensity of 21 cm emission from neutral hydrogen gas moving toward or away from us with velocity V_{LSR}, measured relative to the local standard of rest – D. Hartmann and W. Burton.

precisely, the local standard of rest) orbits with speed V_0, while a star P at radius R has orbital speed $V(R)$. The star moves away from us at speed

$$V_r = V \cos\alpha - V_0 \sin l. \tag{2.10}$$

Using the sine rule, we have $\sin l / R = \sin(90° + \alpha)/R_0$, and so

$$V_r = R_0 \sin l \left(\frac{V}{R} - \frac{V_0}{R_0} \right). \tag{2.11}$$

If the Milky Way rotated rigidly like a turntable, the distances between the stars would not change, and V_r would always be zero. In fact, stars further from the center take longer to complete their orbits; the angular speed V/R drops with radius R. Then Equation 2.11 tells us that V_r is positive for nearby objects in directions $0 < l < 90°$, becoming negative for stars on the other side of the Galaxy that are so distant that $R > R_0$. For $90° < l < 180°$, V_r is always negative, while for $180° < l < 270°$ it is always positive; in the sector $270° < l < 360°$, the pattern of the first quadrant is repeated with the sign of V_r reversed. Figure 2.20 shows the intensity of 21 cm line emission from neutral hydrogen gas in the disk of the Galaxy: as expected, there is no gas with positive velocities in the second quadrant ($90° < l < 180°$), or with negative velocities in the third quadrant. The dark narrow bands extending across many degrees in longitude show where gas has been piled up, and its velocity changed, by gravitational forces in the spiral arms.

Problem 2.15 For a simple model of the Galaxy with $R_0 = 8\,\mathrm{kpc}$ and $V(R) = 220\,\mathrm{km\,s^{-1}}$ everywhere, find $V_r(l)$ for gas in circular orbit at $R = 4, 6, 10$, and $12\,\mathrm{kpc}$. Do this by varying the Galactocentric azimuth ϕ around each ring; find d for each (ϕ, R), and hence the longitude l and V_r. Make a plot similar to Figure 2.20 showing the gas on these rings. In Figure 2.20 itself, explain where the gas lies that corresponds to $(l \sim 50°, V > 0)$; $(l \sim 50°, V < 0)$; $(l \sim 120°, V < 0)$; $(l \sim 240°, V > 0)$; $(l \sim 300°, V > 0)$; and $(l \sim 300°, V < 0)$. Where is the gas at $(l \sim 120°, V > 0)$?

Problem 2.16 Suppose that gas in the Galaxy does not follow exactly circular orbits, but in addition has a velocity $U(R, l)$ radially outward from the Galactic center; stars near the Sun have an outward motion U_0. Show that gas at point P in Figure 2.19 recedes from us at speed

$$V_r = R_0 \sin l \left(\frac{V}{R} - \frac{V_0}{R_0} \right) - R_0 \cos l \left(\frac{U}{R} - \frac{U_0}{R_0} \right) + d \frac{U}{R}. \qquad (2.12)$$

Suppose now that the Sun is moving outward with speed $U_0 > 0$, but that gas in the rest of the Galaxy follows circular orbits; how should velocities measured in the direction $l = 180°$ differ from zero? For gas at a given radius R, in which direction are the extrema (maxima or minima) of V_R shifted away from $l = 90°$ and $l = 270°$? Use Figure 2.20 to show that the Sun and the local standard of rest are probably moving outward from the Galactic center.

When our star or gas cloud is close to the Sun, so that $d \ll R$, we can neglect terms in d^2; using the cosine rule for triangle S–P–GC then gives $R \approx R_0 - d \cos l$. The radial velocity of Equation 2.11 becomes

$$V_r \approx R_0 \sin l \left(\frac{V}{R} \right)' (R - R_0) \approx d \sin(2l) \left[-\frac{R}{2} \left(\frac{V}{R} \right)' \right]_{R_0} \equiv d A \sin(2l), \quad (2.13)$$

where we use the prime for differentiation with respect to R. The constant A, named after Oort, is measured as $14.8 \pm 0.8\,\mathrm{km\,s^{-1}\,kpc^{-1}}$.

The proper motion of a star at P relative to the Sun can be calculated in a similar way. From Figure 2.19, the tangential velocity is

$$V_t = V \sin \alpha - V_0 \cos l. \qquad (2.14)$$

Noting that $R_0 \cos l = R \sin \alpha + d$, we have

$$V_t = R_0 \cos l \left(\frac{V}{R} - \frac{V_0}{R_0} \right) - V \frac{d}{R}. \qquad (2.15)$$

Close to the Sun, we can substitute $R_0 - R \approx d \cos l$, to show that V_t varies almost linearly with the distance d:

$$V_t \approx d \cos(2l) \left[-\frac{R}{2} \left(\frac{V}{R} \right)' \right]_{R_0} - \frac{d}{2} \left[\frac{1}{R}(RV)' \right]_{R_0} \equiv d[A \cos(2l) + B], \quad (2.16)$$

where the constant $B = -12.4 \pm 0.6 \, \text{km s}^{-1} \, \text{kpc}^{-1}$. In Section 3.3, we will see another method to estimate B. The *Oort constants* A and B measure respectively the local shear, or deviation from rigid rotation, and the local vorticity, or angular-momentum gradient in the disk.

Problem 2.17 Show that $A + B = -V'$, while $A - B = V_0/R_0$. Show that the IAU values for V_0 and R_0 imply $A - B = 26 \, \text{km s}^{-1} \, \text{kpc}^{-1}$. Do the measured values of A and B near the Sun correspond to a rising or a falling rotation curve? What effects might cause us to measure $A + B \neq 0$ near the Sun, even though the Milky Way's rotation speed is roughly constant at that radius?

If we could measure the speed V_r for stars of known distance scattered throughout the disk, we could work backward to find $V(R)$, the *rotation curve* of the Milky Way. Unfortunately, visible light from disk stars and clusters is blocked by dust. Radio waves can travel through dust, and we receive emission in the 21 cm hyperfine transition of atomic hydrogen from gas almost everywhere in the Galaxy. But, in general, we have no way of knowing the distance to the emitting gas.

For the inner Galaxy ($R < R_0$), the *tangent-point method* circumvents this difficulty and allows us to find the rotation curve. The angular speed V/R drops with radius. So Equation 2.11 tells us that, when we look out in the disk along a fixed direction with $0 < l < 90°$, the radial speed $V_r(l, R)$ is greatest at the tangent point T in Figure 2.19, where the line of sight passes closest to the Galactic center. Here, we have

$$R = R_0 \sin l \qquad \text{and} \qquad V(R) = V_r + V_0 \sin l. \qquad (2.17)$$

Thus, if there is emitting gas at virtually every point in the disk, we can find $V(R)$ by measuring in Figure 2.20 the largest velocity at which emission is seen for each longitude l; Figure 2.21 shows the results. The gravitational pull of the extra mass in spiral arms can easily change the velocity of gas passing through them by

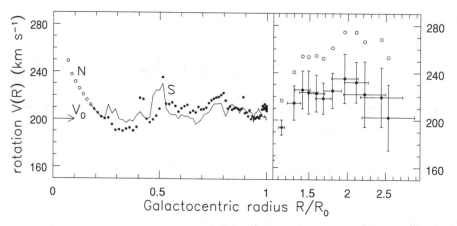

Fig. 2.21. Left, the Milky Way's rotation from the tangent-point method, taking $V_0 = 200\,\mathrm{km\,s^{-1}}$; dots show velocities of northern HI gas with $l > 270°$; the curve gives results from southern gas at $l < 90°$. The tangent-point method fails at $R \lesssim 0.2R_0$ (open circles) because this gas follows oval orbits in the Galactic bar. Right, the rotation speed of the outer Galaxy, calculated for $V_0 = 200\,\mathrm{km\,s^{-1}}$ (filled circles) and for $V_0 = 220\,\mathrm{km\,s^{-1}}$ (open circles); crosses show estimated errors – W. B. Burton and M. Honma.

$10–20\,\mathrm{km\,s^{-1}}$. If the tangent point falls close to an arm, then the rotation speed found by using Equation 2.17 will differ from the average speed of an orbit at that Galactocentric radius.

Measuring rotation speeds in the outer Galaxy is harder. We must first find the distances to associations of young stars by the methods of spectroscopic or photometric parallax. Their velocity V_r is then measured from the emission lines of hot or cool gas around the stars. The stellar distances are often not very well determined, but they are good enough to tell us that the rotation speed $V(R)$ does not decline much in the outer Galaxy, and may even rise further.

2.3.2 Dark matter in the Milky Way

In Section 3.1 we will see that, for a spherical system, the speed V in a circular orbit at radius R is related to the mass $\mathcal{M}(<R)$ interior to that radius by the exact equation

$$\mathcal{M}(<R) = RV^2/G. \tag{2.18}$$

When R is measured in parsecs, time in megayears, and $\mathcal{M}$ in $\mathcal{M}_\odot$, Newton's gravitational constant $G = 4.5 \times 10^{-3}$. For orbits in a flattened disk, this formula gives $\mathcal{M}(< R)$ to within $10\% - 15\%$. Since $V(R)$ does not decline, the mass of the Milky Way must increase almost linearly with radius, even far beyond the Sun where there are very few stars. Later, we will see that this discrepancy between the distributions of light and mass is generally present in spiral galaxies. Astronomers often refer to it as the 'missing mass' or 'dark matter' problem.

Galaxies presumably contain a large amount of matter that gives out virtually no light; this nonluminous mass is assumed to lie in a *dark halo*.

Problem 2.18 Use Equation 2.18 to find the mass $\mathcal{M}(< R_0)$ within a sphere of radius R_0 about the Galactic center. What is the average density within that sphere, in $\mathcal{M}_\odot\,\mathrm{pc}^{-3}$? Show that this is about 10^5 times larger than the critical density of Equation 1.30.

Taking $h_R = 4\,\mathrm{kpc}$ in Equation 2.8, show that 60% of the Milky Way's disk lies within the Sun's orbit. Taking $L_V = 5 \times 10^9 L_\odot$ for the bulge and $15 \times 10^9 L_\odot$ for the disk, show that the mass-to-light ratio $\mathcal{M}/L(<R_0) \approx 5$. Using the result of Problem 2.8, explain why we believe that no more than half of the mass within R_0 is dark matter.

Problem 2.19 The Galaxy's HI disk extends outward to about $2.5R_0$. From Figure 2.21, show that the mass $\mathcal{M}(<2.5R_0) \approx 2 \times 10^{11} \mathcal{M}_\odot$, so that the mass-to-light ratio $\mathcal{M}/L_V \gtrsim 10$. How does this compare with what we found in Problem 2.8? Where is most of the Milky Way's dark matter?

Problem 2.20 Consider the spherical density distribution $\rho_H(r)$ with

$$4\pi G\rho_H(r) = \frac{V_H^2}{r^2 + a_H^2}, \tag{2.19}$$

where V_H and a_H are constants; what is the mass $\mathcal{M}(<r)$ contained within radius r? Use Equation 2.18 to show that the speed $V(r)$ of a circular orbit at radius r is given by

$$V^2(r) = V_H^2[1 - (a_H/r)\arctan(r/a_H)], \tag{2.20}$$

and sketch $V(r)$ as a function of radius. This density law is often used to represent the mass of a galaxy's dark halo – why?

2.4   Milky Way meteorology: the interstellar gas

Between the Milky Way's stars lies the gas from which they were made, and to which they return the heavy elements produced by their nuclear burning. Almost all of the gas lies in the disk. Although its mass is less than 10% of that in stars, it gives the Galaxy many of its distinctive properties. Without gas, the Milky Way would be an S0 galaxy, not a spiral: our disk would have no hot young stars and no spiral pattern. Only about half of the Galaxy's starlight escapes freely into intergalactic space; dusty interstellar gas absorbs the rest. The front cover shows the Milky Way's gas, dust and stars observed at wavelengths from radio to γ-rays.

Like the stellar disk, the Galaxy's interstellar material is subject to gravity, which ultimately causes the densest gas to collapse into new stars. But we must also consider other forces, which can safely be ignored when discussing stellar motions: gas pressure, magnetic forces, and the pressure of cosmic rays. The gas is heated and ionized by stellar radiation; it is shocked and set into motion by fast stellar winds, violent supernova explosions, and passage through the spiral arms. Like the Earth's atmosphere, the interstellar medium is in complex motion.

2.4.1 Mapping the gas layer

Unlike stars, gas does not come in units of standard size. The mass of a clump of gas is not directly linked to its temperature, or any other quantity that we could measure independently of its distance. So the distances of gas clouds are very uncertain, except in special cases: for example, when we know that the gas surrounds a star. When we see absorption lines from interstellar gas in the spectrum of a star, we know that the gas is closer to us than the star. For gas in circular orbits in the Milky Way's disk, we can calculate a *kinematic distance*. From its spectral line emission, we find how much gas is moving at each radial velocity V_r in the direction at longitude l. Then, from the rotation curve $V(R)$, we can use Equation 2.11 to estimate its distance. Thus we can build up a picture of how the Milky Way's gas is distributed.

When all of its radiation reaches us without being absorbed, the mass of gas moving with a particular velocity is proportional to the intensity of the radiation at the corresponding wavelength. Visible light is strongly absorbed by the interstellar dust, whereas radio waves can travel through the dusty gas. But, in the disk, we often look through enough material that radio waves from distant gas are partially absorbed by gas closer to us: the emission is *optically thick*. The 21 cm line of neutral hydrogen, HI, is optically thick in the inner parts of the Galaxy. Dense cool clouds, where the gas is largely molecular, are often traced by the millimeter-wavelength lines of ^{12}CO, which are nearly always optically thick. Thus much of the molecular gas may be hidden from our direct view.

Recall from Section 1.2 that we do not see emission lines from cool molecular hydrogen, because H_2 is a symmetric molecule. The next most abundant molecule in the dense gas is carbon monoxide, with roughly one CO molecule for every 10^4 of H_2. On large scales, we can measure the total amount of atomic and molecular gas by comparing the numbers of energetic *cosmic rays* (see below), detected as they zip through the Earth's atmosphere, with the brightness in γ-rays, which are produced as cosmic rays hit the gas atoms in the Galactic plane. By comparing in turn with observations in CO, we can find the average ratio X_{CO} of the column density of molecular hydrogen to CO emission in the lowest rotational transition. In the Milky Way X_{CO} is fairly uniform in the disk. It is higher in the central few hundred parsecs and in *starbursts*, where very vigorous star formation gives rise to a strong radiation field.

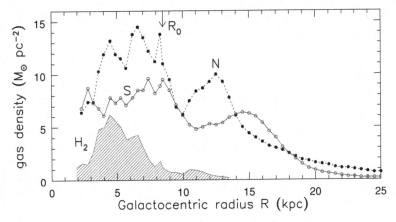

Fig. 2.22. The surface density of neutral hydrogen, as estimated separately for the northern ($0 < l < 180°$; filled dots) and southern ($180° < l < 360°$; open circles) halves of the Galaxy. Within the solar circle, the density is sensitive to corrections for optical thickness; outside, it depends on what is assumed for $V(R)$. The shaded region shows the surface density of molecular hydrogen, as estimated from the intensity of CO emission – W. Burton and T. Dame.

Table 1.8 of Section 1.2 lists spectral lines from some common interstellar molecules. Most of the gas in molecular clouds is cold, with $T \approx 10$–$20\,\mathrm{K}$, and emits most strongly in CO. Even colder gas is detected when by chance it lies in front of a distant radio galaxy or quasar, and absorbs radiation in the appropriate spectral lines. We will see later that the densest gas is in the small warm cores of molecular clouds, where clusters and associations of young stars are born. Here we observe molecules such as NH_3, CN, and H_2CO.

Figure 2.22 shows how atomic and molecular gases are distributed in the Galaxy. It is based on kinematic distances, using CO to trace the H_2. The Milky Way probably contains $(4$–$8) \times 10^9 \mathcal{M}_\odot$ of HI, and about half that amount of molecular gas. Almost all of the H_2, but less than half of the HI, lies within the *solar circle*, at radius R_0. Molecular gas is piled up in a ring of radius 4 kpc. Inside this ring we find little molecular or atomic gas, except in the central few hundred parsecs. On the cover, we see that atomic hydrogen spreads to much larger radii than the molecular gas or the stars. It also forms a thicker layer than the molecular gas. Near the Sun, CO-emitting clouds lie mainly within 80 pc of the midplane, while the HI disk is about twice as thick. The HI layer puffs up even more strongly further out.

As in other spiral galaxies, the dense molecular gas, dust, and young stars are concentrated into the spiral arms: see Section 5.5. The Sun lies just outside the Sagittarius–Carina arm, which can be traced for almost a full turn around the disk. Its pitch angle is about 10°, which is typical for an Sbc or Sc galaxy. Looking away from the Galactic center, we see a short nearby arm-spur in the constellation of Orion. Roughly 2 kpc further out lies the Perseus arm, which shows up in

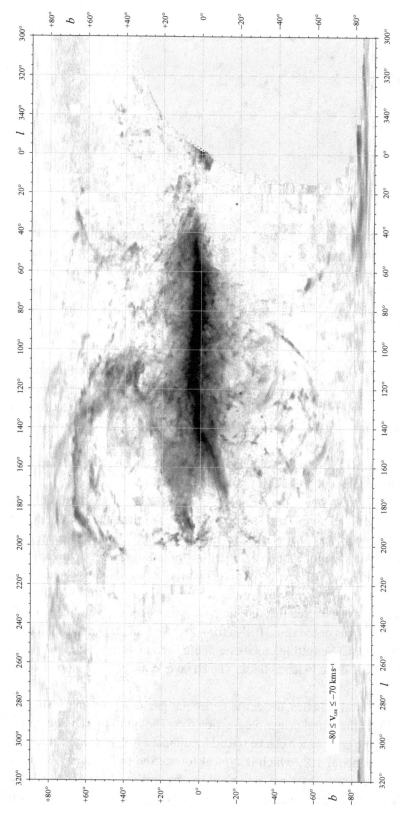

Fig. 2.23. Neutral hydrogen at velocities $-80\,\mathrm{km\,s^{-1}} < V_r < -70\,\mathrm{km\,s^{-1}}$; the Galactic plane $b = 0$ runs horizontally through the middle of the figure. Note the high-latitude streamers of gas. Empty areas of the plot could not be observed from the telescope in Dwingeloo (Netherlands) – D. Hartmann and W. Burton.

Figure 2.20 as a dark ridge at $l > 90°$, $V_{LSR} \approx 50 \, \text{km s}^{-1}$. However, we do not have a picture of our whole spiral pattern. Dust blocks our view through the disk in visible light, and kinematic distances for the gas are unreliable, because the gravitational tug of the spiral arms pulls it out of circular rotation.

Figure 2.23 shows atomic gas far above the Galactic plane. Looking away from the Galactic center toward $l = 180°$, from Equation 2.11 we expect that gas in circular orbit has $V_r \sim 0$, moving neither toward us nor away. Instead, the gas near $b = +70°$ is approaching at $>70 \, \text{km s}^{-1}$. The *high-velocity clouds* of HI rain down on the disk even faster, at over $100 \, \text{km s}^{-1}$. Some of these may be disk material that has been thrown up above the midplane by supernovae or winds from hot massive stars, so that it is now falling back. Others, like the gas in the Magellanic Stream (Section 4.1), come from beyond the Milky Way. We do not generally know the distances to these clouds and so cannot estimate their masses. If they resemble those found around our neighbor M31 (see Section 4.2), most lie within 50 kpc of the disk, with masses $10^4 \mathcal{M}_\odot$ to $10^7 \mathcal{M}_\odot$. Infalling high-velocity clouds may add metal-poor gas to the disk, potentially solving the *G-dwarf problem* which we discuss in Section 4.3.

Beyond 1–2 kpc from the Galactic plane, almost all the gas is hot or warm. We can probe clouds of warm gas in the halo of our Galaxy and others by looking at the absorption lines they produce in the ultraviolet spectra of distant stars or quasars: see Section 9.3. In dense warm gas, light is absorbed by low ions such as MgII. The more diffuse clouds are highly ionized; we see lines of CIV and NV. We also see absorption in lines of OVI, which arises in gas at $\sim 3 \times 10^5 \, \text{K}$, where Figure 2.25 shows that cooling is most rapid. This material must lie at the boundaries between the warm clouds and the hottest gas at $\sim 10^6 \, \text{K}$.

Within about 3 kpc of the Galactic center, inside the ring of dense gas, the average surface density of H_2 drops below $5 \mathcal{M}_\odot \, \text{pc}^{-2}$. At these radii, both atomic and molecular gas lie in a disk tilted $10°$–$20°$ from the plane $b = 0$, with gas at positive longitudes lying below that plane and that at $l < 0$ above it. We do not know what caused the tilt, nor why it persists. The tilted-disk gas is not in circular rotation. Instead, emission is observed at negative velocities for $l > 0$, and at positive velocities at negative longitudes, which by Equation 2.11 would not be allowed for circular motion. The emission at 'forbidden' velocities can be explained by the barlike Galactic bulge that we discussed in Section 2.2. Under the bar's gravitational force, the gas must follow oval orbits that take it alternately toward and away from the Galactic center.

The central 200 pc of the Galactic bulge is a gas-rich region which is actively forming stars; it harbors at least $10^8 \mathcal{M}_\odot$ of molecular gas, or about 10% of the total in the Milky Way. The material in the outer part of this region forms the inner edge of the tilted disk, its density now so high that the gas has become predominantly molecular. Further in, dense gas is again found close to the plane $b = 0$; here it is drawn out into long arcs and filaments as it orbits the central black hole.

There is so much dust in the Milky Way's disk that we cannot see optical and ultraviolet light from this central region. We observe X-rays from a hot plasma with $T \approx 10^7$ K, filling the spaces around the denser and cooler clouds. The Galaxy's gravity is too weak to hold onto such hot gas; it may be escaping as a wind into intergalactic space.

2.4.2 A physical picture

The interstellar gas is a multiphase medium. On scales between 1 pc and about 1 kpc, the hot, warm, and cool phases are all mixed together. Most of the atoms are in the dense cool phases, but diffuse warm and hot gas occupy most of the volume. The cold molecular and atomic material is over 10 000 times denser than the hot plasma – a contrast larger than that between air and water. So we can often think of clouds of cool gas moving with little hindrance through the more diffuse medium.

The densest gas is in a few thousand *giant molecular clouds* in the spiral arms. Any cloud with $N_H \gtrsim 10^{20}$ H atoms cm^{-2}, or roughly $\mathcal{M}_\odot$ pc^{-2}, becomes largely molecular, since the ultraviolet photons that can break up H_2 molecules do not penetrate to greater depth. (Unshielded molecules are highly vulnerable: near the Sun, photons of ambient starlight destroy them in only a few hundred years.) The clouds can be larger than 20 pc, with masses above $10^5 \mathcal{M}_\odot$, and densities $\gtrsim 200$ H_2 molecules cm^{-3}, rising above 10^4 cm^{-3} in their cores. They are surrounded by cool HI, forming large complexes up to ~ 100 pc across with $10^7 \mathcal{M}_\odot$ of gas. Between the arms, the clouds are smaller: typically $\mathcal{M} \sim 40 \mathcal{M}_\odot$, sizes are ~ 2 pc, and densities hardly rise above $n_H \sim 100$ cm^{-3}. In the central 200 pc, the clouds are denser ($n_H \approx 10^4$ cm^{-3}) and hotter (typically around 70 K) than those near the Sun.

The cool atomic hydrogen is less dense than molecular clouds, with $n_H \sim 25$ cm^{-3} and $T \leq 80$ K. Near the Sun, about half of the HI is much warmer, with $T \sim 8000$ K and $n_H \sim 0.3$ cm^{-3}. This neutral gas is mixed with warm ionized gas, with the same temperature and pressure. Clouds of warm gas are themselves enveloped in hot diffuse plasma with $n_H \sim 0.002$ cm^{-3} and $T \sim 10^6$ K. At these temperatures hydrogen and helium are almost completely ionized by violent collisions with fast-moving electrons. Our Sun itself is moving through a warm cloud, a parsec in size and roughly 50% ionized, which lies within an irregularly shaped and expanding *local bubble* of hot gas, 100 pc across. At 1–2 kpc above the midplane, we still find clouds of HI, but the proportions of warm ionized gas and hot plasma are larger. Table 2.4 gives a summary of the various phases of the interstellar gas. Notice that the product $n_H T = p/k_B$ is approximately the same for the cool HI clouds, for the warm gas, and for the hot plasma: they are in pressure balance. We will see why below, and discuss why the hot, warm, and cool phases have the temperatures that they do.

Only O and B stars emit many photons above 13.6 eV, the energy required to ionize hydrogen from its ground state. When one of these begins to shine, its ultraviolet light first breaks up the surrounding H_2 molecules into atomic hydrogen,

Table 2.4 A 'zeroth-order' summary of the Milky Way's interstellar medium (after J. Lequeux)

Component	Description	Density (cm^{-3})	Temperature (K)	Pressure (p/k_B)	Vertical extent	Mass ($\mathcal{M}_\odot$)	Filling factor
Dust grains						10^7–10^8	Tiny
large $\lesssim 1\,\mu$m	Silicates, soot		~20		150 pc		
small ~ 100 Å	Graphitic C		30–100				
PAH < 100 atoms	Big molecules				80 pc		
Cold clumpy gas	Molecular: H$_2$	> 200	< 100	Big	80 pc	(2) $\times 10^9$	<0.1%
	Atomic: HI	25	50–100	2 500	100 pc	3 $\times 10^9$	2%–3%
Warm diffuse gas	Atomic: HI	0.3	8 000	2 500	250 pc	2 $\times 10^9$	35%
	Ionized: HII	0.15	8 000	2 500	1 kpc	10^9	20%
HII regions	Ionized: HII	1–10^4	~10 000	Big	80 pc	5 $\times 10^7$	Tiny
Hot diffuse gas	Ionized: HII	~0.002	~10^6	2 500	~5 kpc	(10^8)	45%
Gas motions	$\frac{3}{2}\langle\rho_{HI}\rangle\sigma_r^2$	$\langle n_H\rangle$ ~ 0.5	10 km s^{-1}	8 000			
Cosmic rays	Relativistic	1 eV cm^{-3}		8 000	~3 kpc	Tiny	
Magnetic field	B ~ 5 μG	1 eV cm^{-3}		8 000	~3 kpc		
Starlight	$\langle \nu h_P \rangle$ ~ 1 eV	1 eV cm^{-3}			~500 pc		
UV starlight	11–13.6 eV	0.01 eV cm^{-3}					

Note: ( ) denotes a very uncertain value. Pressures and filling factors refer to the disk midplane near the Sun; notice that the pressures from cosmic rays, in magnetic fields, and the turbulent motions of gas clouds are roughly equal.

then *photoionizes* the gas to create an *H*II *region*. The zone where this happens is called the *photodissociation region*. As the newly-ionized gas is heated to ~10 000 K, its pressure suddenly shoots up to roughly 1000 times that of the surrounding cloud. The HII region expands, pushing the cold gas outward supersonically. It often pierces the molecular cloud, escaping as a *champagne flow*.

The Milky Way's entire gas layer is also threaded by a tangled magnetic field. Its strength is about 0.5 nT or 5 μG near the Sun; it is higher at smaller Galactocentric radii and falls to about half its local value at $2R_0$. All but the coldest dense gas is sufficiently ionized to be a good electrical conductor, so the field is *frozen* into it, moving along with the gas.

The expanding remnant of a supernova explosion sweeps up this magnetic field along with the gas, and the moving field accelerates protons and heavier atomic nuclei to near-light speeds, as *cosmic rays*. Cosmic rays with energy above 10^9 GeV or $10^9 m_p c^2$ can escape from the Galaxy's magnetic field, while those of lower energy are trapped within it. Cosmic rays penetrate into even dense molecular clouds, keeping them partially ionized. Both cosmic rays and magnetic field resist attempts to squeeze the gas, effectively adding to the gas pressure.

Expanding supernova remnants also accelerate electrons to relativistic speeds. We observe their *synchrotron radiation* at radio wavelengths, as they spiral in the Galactic magnetic field. This emission is brightest near the Galactic plane, but the topmost image on the front cover shows that a diffuse *radio halo* extends many kiloparsecs above and below it. This shows that both field and fast particles can escape from the dense disk gas.

The interstellar gas is in motion on large and small scales. Like stars, interstellar gas clouds do not follow exactly circular orbits about the Galactic center. They also have random motions: typically about $5\,\mathrm{km\,s^{-1}}$ for molecular clouds and $8-10\,\mathrm{km\,s^{-1}}$ for clouds of atomic gas. Even within molecular clouds, gas in the denser cores must be in motion: the observed width of CO lines corresponds to velocities of $1-10\,\mathrm{km\,s^{-1}}$, while the thermal speeds for CO molecules are below $0.1\,\mathrm{km\,s^{-1}}$ (see the problem below). Motions that are faster than the local speed of sound, which is roughly the thermal speed of atoms or molecules in the gas, give rise to *shocks*. Here the gas velocity changes sharply across a narrow region, and energy of motion is converted into heat.

We do not know exactly how these motions arise, just as we cannot predict the Earth's weather in detail even though we know that winds derive their energy from sunlight heating the air and the ground. The interstellar gas may be *turbulent*, with energy passed successively from motions on larger scales to smaller. Models of subsonic turbulence predict that random speeds σ measured inside a cloud of size L should increase as $\sigma \propto L^{1/3}$. This is approximately true for the atomic and molecular gas.

Problem 2.21 For molecules of H_2 in a cloud with $T = 20\,\mathrm{K}, n(H_2) = 200\,\mathrm{cm^{-3}}$, calculate the sound speed c_s: $k_B T = \mu m_H c_s^2$, where μm_H is the molecular mass. What is the sound speed for CO molecules in this same cloud? From Table 2.4, what is the sound speed for cool HI and for warm neutral gas? Show that cool HI clouds move through the warm interstellar medium at speeds close to the sound speed of that warm gas. An HII region cannot expand into the surrounding gas faster than this sound speed: how fast is that when $T \approx 10^4\,\mathrm{K}$? If its temperature does not change, show that an HII region 1 pc across with $n_H = 10^3\,\mathrm{cm^{-3}}$ would take 1 Myr to expand until its pressure balances that of the warm diffuse gas.

The Milky Way's dust absorbs nearly half of its ultraviolet and optical starlight. Most of this energy warms the grains, which radiate in the infrared, as we see in the fifth and sixth panels on the front cover. Cool interstellar gas contains about one grain of dust per 10^{12} hydrogen atoms: on average, one grain per 100-meter cube. Most obviously, dust absorbs the visible light of stars, which would otherwise travel through the gas. Absorption by dust diminishes sharply toward the infrared. Since dust most efficiently absorbs light of wavelengths smaller than the grain size, this tells us that most grains have sizes $\lesssim 0.3\,\mu\mathrm{m}$.

Problem 2.22 Use the blackbody equations of Section 1.1 to find the approximate temperature of a dust grain that radiates mainly at 150 μm. Show that a large dust grain 1 pc from an O star with $L = 10^6 L_\odot$ will be heated to roughly this temperature. (Remember that a grain of radius r_g absorbs starlight over an area πr_g^2, but emits from its whole surface.) The galaxy M82 is undergoing a

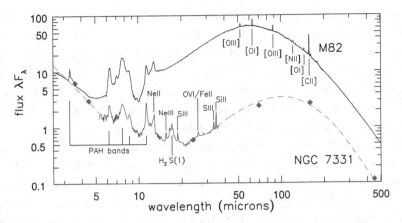

Fig. 2.24. Infrared spectra of Sb spiral galaxy NGC 7331 and starburst galaxy M82. Equal heights in λF_λ correspond to equal energy radiated per decade in frequency. For wavelengths of atomic lines, see Table 1.7; PAHs emit shortward of 30 μm; emission from large dust grains peaks near 100 μm – J. D. Smith, D. Dale, and A. Li: Spitzer Space Telescope; G. Lagache: ISO.

burst of star formation: at what wavelength does it emit most of its energy? Show that its large dust grains have $T \sim 50\,\text{K}$. How far from the star would the dust have to be to reach an average temperature of 150 K, so that it radiates at 30 μm?

Figure 2.24 shows an infrared spectrum of the star-forming ring of NGC 7331, a galaxy much like the Milky Way or our neighbor M31. We see that its dust grains are a complex mixture. Large grains, with temperatures $\sim 30\,\text{K}$, are responsible for the emission peaking at 100 μm. Radiation at around 30 μm requires hotter grains at $T \gtrsim 100\,\text{K}$. Few grains are so close to stars that they reach such a high average temperature. Instead, we believe that this emission comes from grains smaller than 10 nm, with fewer than 10^6 carbon atoms. These are so tiny that absorbing a single ultraviolet photon raises their temperature above 100 K.

Probably 10%–20% of the mass of interstellar dust is in the tiniest particles, the *polycyclic aromatic hydrocarbons* (PAHs), with 100 carbon atoms or fewer. Their carbon atoms are arranged in rings that make up a flat sheet or even a round 'buckyball'. These behave like large molecules rather than amorphous solids. Stretching of their C—C and C—H bonds gives rise to the strong emission lines in the 3–20 μm region. When a PAH molecule absorbs an ultraviolet photon, about 10% of the time it will throw out a fast-moving *photoelectron*, which loses its energy as it collides with electrons in the gas. This is probably the main way that the atomic gas is heated.

Dust grains consist largely of magnesium and iron silicates, from the oxygen-rich atmospheres of red giant stars, and carbon in various forms: amorphous soot, graphite, and PAHs. In dense cold clouds, mantles of water ice, methane, and ammonia condense out onto the larger grains. Dust makes up about 1% of the mass of interstellar material, and more in denser gas. In gas of approximately solar composition, elements heavier than helium hold only 2% of the mass. Thus, in dense clouds, almost all of the carbon, oxygen, magnesium, etc. must be in dust, leaving the gas *depleted* of those elements. Dust grains are continually knocked apart as they collide with fast-moving atoms and with each other, and built up by absorbing atoms of interstellar gas onto their surfaces. The material now present in a grain has probably been there for less than 500 Myr.

The surface of grains is the main site where hydrogen molecules are made. These form only slowly in a gas, because the atoms rarely encounter each other and then find it difficult to lose energy to become bound in a molecule, since the process is strongly forbidden because of symmetry. At typical densities $n_H \sim 100 \, \mathrm{cm}^{-3}$, atomic hydrogen would take 10^{13} yr to form H_2. When the atoms can be absorbed onto grains to 'find' each other there, and then transfer excess energy to the grain to release a bound molecule, H_2 forms $\sim 10^8$ times faster.

Problem 2.23 In a very simple model, H atoms that collide with a grain stick to it for long enough to find a partner; the pair departs as a molecule of H_2. In a cloud at $T = 50 \, \mathrm{K}$, show that the thermal speed of Problem 1.9 is $v_{th} \approx 1 \, \mathrm{km \, s}^{-1}$. From Problem 1.11, take the grain radius $a = 0.1 \mu \mathrm{m}$ and number density $n_g = 10^{-12} n_H$. Show that an H atom collides with a grain after an average time $(n_g \pi a^2 v_{th})^{-1}$ or $10 \, \mathrm{Myr} \times (1 \, \mathrm{cm}^{-3}/n_H)$. The Sun, and gas orbiting along with it, takes 5%–10% of a 'Galactic year' (see Problem 2.14) to pass through a spiral arm. Show that this is long enough for an H I cloud with $n_H = 100 \, \mathrm{cm}^{-3}$ to become largely molecular.

The interstellar gas is an 'open' system: it needs a continuous energy supply. A star like our Sun will not change its orbit significantly within a Hubble time, unless the Milky Way has a near-collision with another galaxy: the stellar disk and bulge are close to equilibrium. By contrast, the gas layer is like a pan of boiling water; unless energy is supplied to it, the gas will cool rapidly, and the random motions of the clouds will dissipate. Energy is added to the interstellar gas by stellar radiation, by collisions with cosmic rays, and mechanically by supernovae, stellar winds, and Galactic rotation which stretches the magnetic field.

As one example, we can ask how quickly ionized hydrogen reverts to a neutral state. In each cubic centimeter, neutral atoms are produced at a rate which increases as the number of electrons, n_e, times the number of protons, $n_p \approx n_e$, times the rate at which they encounter each other and recombine, which depends on the temperature T_e of the (lighter and faster-moving) electrons. Thus electrons recombine at the rate

$$-\frac{dn_e}{dt} = n_e^2 \alpha(T_e) \quad \text{with} \quad \alpha(T_e) \approx 2 \times 10^{-13} \left(\frac{T_e}{10^4 \, \mathrm{K}} \right)^{-3/4} \mathrm{cm}^3 \, \mathrm{s}^{-1}.$$

$$(2.21)$$

Here, the function $\alpha(T_e)$ hides the physics of encounters with a range of relative speed; we have taken it from Equation 5.6 of the book by Dyson and Williams. The approximation is good for $5000\,K \lesssim T \lesssim 20\,000\,K$. The recombination time t_{rec} is given by the number of electrons, divided by the rate at which they disappear:

$$t_{rec} = \frac{n_e}{|dn_e/dt|} = \frac{1}{n_e\alpha(T_e)} \approx 1500\,\text{yr} \times \left(\frac{T_e}{10^4\,K}\right)^{3/4}\left(\frac{100\,\text{cm}^{-3}}{n_e}\right). \quad (2.22)$$

When the gas is hotter, electrons and protons collide more frequently but are less likely to stick together, so t_{rec} is longer.

Within HII regions, t_{rec} is only a few thousand years. The ionized gas rapidly recombines once the star no longer provides ultraviolet photons. In the warm ionized interstellar gas the density is only $\sim 0.1\,\text{cm}^{-3}$, and recombination takes $\sim 2\,\text{Myr}$. But, because there is so much of this gas, it must absorb at least 25% of the ultraviolet radiation from all the O and B stars in the disk in order to maintain its ionized state. These energetic photons must find their way to $\sim 1\,\text{kpc}$ above the midplane, between the clouds of neutral gas.

Problem 2.24 We can estimate the size of an HII region around a massive star that radiates $S_\star$ photons with energy above 13.6 eV each second. Assume that the gas within radius $r_\star$ absorbs all these photons, becoming almost completely ionized, so that $n_e \approx n_H$, the density of H nuclei. In a steady state atoms recombine as fast as they are ionized, so the star ionizes a mass of gas $\mathcal{M}_g$, where

$$S_\star = (4r_\star^3/3)n_H^2\alpha(T_e) = (\mathcal{M}_g/m_p)n_H\alpha(T_e).$$

Use Equation 2.21 to show that a mid-O star radiating $S_\star = 10^{49}\,\text{s}^{-1}$ into gas of density $10^3\,\text{cm}^{-3}$ creates an HII region of radius 0.67 pc, containing $\sim 30\mathcal{M}_\odot$ of gas (assume that $T_e = 10^4$ K). What is $r_\star$ if the density is ten times larger? Show that only a tenth as much gas is ionized. How large is the HII region around a B1 star with $n_H = 10^3\,\text{cm}^{-3}$ but only $S_\star = 3 \times 10^{47}\,\text{s}^{-1}$?

The *cooling time* t_{cool} measures how fast the gas radiates away its thermal energy. When there are n atoms cm^{-3}, the energy in each cubic centimeter is proportional to nT; if it radiates with luminosity L, $t_{cool} \propto nT/L$. When the gas is optically thin, we have a formula like Equations 1.17 and 2.21: the number of photons from that volume is proportional to n^2. We can write

$$L = n^2\Lambda(T), \quad \text{so} \quad t_{cool} \propto T/[n\Lambda(T)]. \quad (2.23)$$

$\Lambda(T)$ depends only on the temperature, so denser gas cools more rapidly.

Table 2.5 Main processes that cool the interstellar gas

Temperature	Cooling process	Spectral region
$> 10^7$ K	Free–free	X-ray
10^7 K $< T < 10^8$ K	Iron resonance lines	X-ray
10^5 K $< T < 10^7$ K	Metal resonance lines	UV, soft X-ray
8000 K $< T < 10^5$ K	C, N, O, Ne forbidden lines	IR, optical
Warm neutral gas: ~ 8000 K	Lyman-α, [OI]	1216 Å, 6300 Å
100 K $< T < 1000$ K	[OI], [CII], H_2	Far IR: 63 µm, 158 µm
$T \sim 10-50$ K	CO rotational transitions	Millimeter-wave

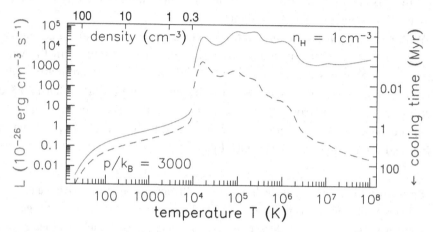

Fig. 2.25. For gas of solar composition, luminosity L, from each cubic centimeter (solid curve), and cooling time t_{cool} (broken curve). Above 10 000 K we set $n_H = 1$ cm^{-3}; the gas is optically thin, and $L = n^2 \Lambda(T)$. Below 10 000 K the thermal pressure $p/k_B = 3000$, and cosmic-ray and ultraviolet fluxes are as measured near the Sun; we set $N_H = 10^{19}$ cm^{-2}, so almost all H is atomic – M. Wolfire and G. Hensler.

Figure 2.25 shows the cooling curve for gas of roughly solar composition, and Table 2.5 lists the main processes that carry away energy. Above about 10^7 K, almost all the atoms are fully ionized, and the gas cools by free–free radiation (see Section 1.2). Roughly $\Lambda(T) \propto \sqrt{T}$, so $t_{cool} \propto \sqrt{T}/n$; we can see from Figure 2.25 that hotter gas needs *longer* to cool. At lower temperatures the resonance lines of iron and other metals become the main coolants. They are very efficient, so $\Lambda(T)$ rises steeply between 10^6 and 10^4 K; gas cannot remain long at temperatures between that of the hot and warm phases of the interstellar medium. In neutral gas below about 8000 K, the energy loss drops sharply. It also depends on the column density N_H, which determines how far ultraviolet photons penetrate to ionize atoms and dissociate molecules. Almost all of the cooling below 10^7 K depends on elements heavier than hydrogen and helium; so, in the metal-poor gas of the first galaxies, it would have been much weaker than it is today.

Various processes heat the Milky Way's gas, replacing the lost energy. The diffuse hot gas was heated to $\sim 10^6$ K, as it passed through the shock caused by the expanding remnant of a supernova. At the densities observed near the Sun, Figure 2.25 shows that it cools rapidly, within 10^4–10^5 yr, condensing into cooler clouds. Near the Sun's position, a given region is crossed by a supernova shock about once per 1–5 Myr, reheating the gas. Far from the midplane the hot gas is less dense, and cooling times can reach 1 Gyr.

Clouds of warm and cool HI gas are warmed by photoelectric heating, as ultraviolet light of stars falls onto the smallest dust grains. They are cooled by far-infrared lines of oxygen and carbon. When the gas temperature falls below ~ 100 K, collisions among gas atoms are not energetic enough to excite the far-infrared atomic lines. In molecular clouds, energy is lost mainly in the millimeter-wavelength lines of CO. Table 1.8 shows that the lowest rotational level lies only 5.5 K above the ground state, so it is excited by collisions in gas near that temperature or above. The main source of heat is cosmic rays, which penetrate right through the clouds; they strip electrons from gas atoms, which then share their energy as they bump into electrons in the gas. Molecular clouds are dark; only far-infrared and longer wavelengths of light can reach the interior. The infrared light warms the dust grains, which in turn prevent even the densest and darkest clouds from cooling below ~ 10 K (see Problem 1.12).

Unlike stars, the warm and cool gas clouds are large enough that they occasionally collide with one another. So, like molecules in a gas, they exert pressure. The random speeds of HI clouds are typically $\sigma_r \sim 10\,\mathrm{km\,s}^{-1}$, and the volume-averaged density $\langle n_{\mathrm{HI}} \rangle \sim 0.5\,\mathrm{cm}^{-3}$ near the Sun; so the density of kinetic energy $3\langle \rho_{\mathrm{HI}} \rangle \sigma_r^2 / 2$ is equivalent to $p/k_{\mathrm{B}} \sim 8000$. Table 2.4 shows that this is much larger than the thermal pressure of the gas, but about the same as the pressures contributed by magnetic fields and cosmic rays.

This rough equality is no accident. The Milky Way's magnetic field takes most of its energy from differential rotation, which tends to pull gas at small radii ahead of that further out. The magnetic field is frozen into the gas, so field lines connecting clouds at different radii are stretched out as the Galaxy rotates, strengthening the field. Random motions of the gas clouds, the pressure of cosmic rays, and disturbances from stellar winds and supernova explosions also stretch and tangle the field. The strength of the field depends on the vigor of these processes.

As they collide, much of the clouds' bulk motion is converted to heat, which is radiated away. The clouds' random motion would cease within 10–30 Myr if they were not shaken about by supernova explosions, winds from HII regions, the pull of magnetic fields, and passage through spiral arms. Thus we see that energy is continually transferred among cloud motions, magnetic field, and cosmic rays. In very complex processes, often the energies of the various motions are driven to be equal on average, just as kinetic energy is on average distributed equally among the colliding molecules of gas in a room. Here, each component exerts roughly equal pressure.

New stars are born in the Milky Way's dense molecular clouds. These clouds are at much higher pressure than the surrounding atomic gas, because they must resist the inward pull of their own gravity. We will see in Section 8.5 that gravity will cause a gas cloud of density ρ and temperature T to collapse on itself, if its diameter exceeds the *Jeans length*

$$\lambda_J = c_s \sqrt{\frac{\pi}{G\rho}};$$

here c_s is the sound speed $c_s^2 = k_B T/(\mu m_H)$, and μm_H is the mean molecular mass. The mass $\mathcal{M}_J$ within this sphere is the *Jeans mass*:

$$\mathcal{M}_J \equiv \frac{\pi}{6}\lambda_J^3 \rho = \left(\frac{1}{\mu m_H}\right)^2 \left(\frac{k_B T}{G}\right)^{3/2} \left(\frac{4\pi n}{3}\right)^{-1/2} \frac{\pi^3}{3\sqrt{3}}$$

$$\approx 20\left(\frac{T}{10\,\mathrm{K}}\right)^{3/2} \left(\frac{100\,\mathrm{cm}^{-3}}{n}\right)^{1/2} \mathcal{M}_\odot. \qquad (2.24)$$

If gas pressure is not enough to prevent it, the cloud collapses after approximately a free-fall time (see Equation 3.23 in Section 3.1):

$$t_{ff} = \sqrt{\frac{1}{G\rho}} \approx \frac{10^8}{\sqrt{n_H}}\,\mathrm{yr}. \qquad (2.25)$$

When do we expect collapse to be so rapid? A gas heats up as it is compressed, so the Jeans mass increases according to Equation 2.24. Unless it can radiate away this heat, gas pressure will slow the collapse. Thus the cooling time must be short: $t_{cool} \ll t_{ff}$. If T does not grow, the Jeans mass decreases as the density rises, and the original cloud can break into smaller fragments which themselves collapse independently. This continues until the densest fragments become optically thick; they heat up and begin to shine as *protostars*.

Table 2.4 shows that the Galaxy has $(1–2) \times 10^9 \mathcal{M}_\odot$ in molecular clouds, at densities above $100\,\mathrm{cm}^{-3}$, and $T \sim 10–20\,\mathrm{K}$. According to Equations 2.24 and 2.25, any of these clouds larger than $60\mathcal{M}_\odot$ should collapse within about 10 Myr. Converting all of the Galaxy's molecular material to stars in this way yields $\sim100\mathcal{M}_\odot\,\mathrm{yr}^{-1}$ of new stars – far more than the $(3–5)\mathcal{M}_\odot\,\mathrm{yr}^{-1}$ of new stars that we observe. Either a collapsing molecular cloud turns very little of its mass into stars, or something – perhaps 'frozen' magnetic fields or turbulent motions – must slow the collapse.

Just as water passes between solid, liquid, and vapor phases in Earth's atmosphere, so interstellar material passes continually between different phases. As HI gas cools or is compressed in a spiral arm, more of it converts to the dense cold phase. When atomic clouds become dense enough that ultraviolet light cannot penetrate their interiors, H_2 forms on dust grains. The molecular clouds lose heat and gradually contract, forming new stars if they are not first disrupted. As new massive stars shine on the remains of the dense cloud in which they were born, their ultraviolet photons split H_2 molecules apart, then ionize the atoms to form an HII region. This expands, breaking out of the molecular cloud to mix with the warm ionized medium.

Near the end of their lives, we saw in Section 1.1 that low-mass stars become red giants and supergiants, shedding dusty gas enriched in heavy elements produced by their nuclear burning. Supernova explosions also release dust and heavy elements. Even though the energy in these explosions is only $\sim 1\%$ of that in starlight, supernovae are the main source of the Galaxy's hot gas and cosmic rays. Their shock waves heat surrounding gas to over a million degrees, sweep up and so strengthen the magnetic field, and accelerate cosmic rays.

Once massive stars have destroyed their natal molecular cloud with ultraviolet radiation, stellar winds, and supernova explosions, no further stars can be born there until the gas has had time to cool and become dense again. If a galaxy undergoes a *starburst* (see Section 7.1), turning most of its cool gas into stars within $\lesssim 300$ Myr, repeated supernova explosions in a small volume can heat up so much gas that it forces its way out of the galaxy as a *superwind*. The average stellar birthrate in the Milky Way is set by this *feedback loop*: too-vigorous star formation in a particular region inhibits further starbirth. In a large galaxy like ours, an expanding supernova compresses cool gas in nearby regions of the disk, and can trigger collapse of the densest parts to make new stars. Thus star formation can 'spread like a disease' across the face of the galaxy. We will see in Section 4.4 that a dwarf galaxy is more likely to have episodes of rapid starbirth across the entire system, interspersed with quiet periods.

Further reading: two undergraduate texts are J. E. Dyson and D. A. Williams, 1997, *The Physics of the Interstellar Medium*, 2nd edition; and D. C. B. Whittet, 1992, *Dust in the Galactic Environment* (both from Institute of Physics Publishing, London and Bristol, UK). On the origin of cosmic rays, see M. S. Longair, 1994, *High Energy Astrophysics*, 2nd edition, Chapters 17–21 of Volume 2, *Stars, the Galaxy and the Interstellar Medium* (Cambridge University Press, Cambridge, UK). On the graduate level, see J. Lequeux, *The Interstellar Medium* (English translation, 2004; Springer, Berlin and Heidelberg, Germany).

3

The orbits of the stars

Stars travel around the Galaxy, and galaxies orbit within their groups and clusters, under the force of gravity. Stars are so much denser than the interstellar gas through which they move that neither gas pressure nor the forces from embedded magnetic fields can deflect them from their paths. If we know how mass is distributed, we can find the resulting gravitational force, and from this we can calculate how the positions and velocities of stars and galaxies will change over time.

But we can also use the stellar motions to tell us where the mass is. As we discovered in Chapter 2, much of the matter in the Milky Way cannot be seen directly. Its radiation may be absorbed, as happens for the visible light of stars in the dusty disk. Some material simply emits too weakly: dense clouds of cold gas do not show up easily in radio-telescope maps. The infamous dark matter still remains invisibly mysterious. But, since the orbits of stars take them through different regions of the galaxies they inhabit, their motions at the time we observe them have been affected by the gravitational fields through which they have travelled earlier. So we can use the equations for motion under gravity to infer from observed motions how mass is distributed in those parts of galaxies that we cannot see directly.

Newton's law of gravity, and methods for computing the gravitational forces, are introduced in Section 3.1. Usually we can consider the stars as point masses, because their sizes are small compared with the distances between them. Since galaxies contain anywhere between a million stars and 10^{12} of them, we usually want to look at the average motion of many stars, rather than following the individual orbit of each one. We prove the *virial theorem*, relating average stellar speeds to the depth of the gravitational potential well in which they move. Orbital dynamics and the virial theorem are our tools to find masses of star clusters and galaxies.

The gravitational potential of a galaxy or star cluster can be regarded as the sum of a smooth component, the average over a region containing many stars, and the very deep potential well around each individual star. In Section 3.2, we will see that the motion of stars within a galaxy is determined almost entirely by the smooth part of the force. *Two-body encounters*, transferring energy between individual stars, can be important within dense star clusters. We discuss how these encounters change the cluster's structure, eventually causing it to disperse or 'evaporate'.

Section 3.3 covers the *epicycle* theory, which is a way to simplify the calculation of motions for stars like the Sun, that follow very nearly circular orbits within a galaxy's disk. Using epicycles, we can explain the observed motions of disk stars near the Sun. Section 3.4 is the most technical of the book: it introduces the *collisionless Boltzmann equation*, linking the number of stars moving with given velocity at each point in space to the gravitational force acting on them. We survey a few of its many uses, such as finding the mass density in the Galactic disk near the Sun. We remind readers of Plato's warning (*Timaeus*, 91d): the innocent and lightminded, who believe that astronomy can be studied by looking at the heavens without knowledge of mathematics, will return in the next life as birds.

Symbols. We use boldface to indicate a vector quantity. The energy and angular momentum of a star are given by $\mathcal{E}$ and $\mathcal{L}$; E is energy per unit mass, while $\mathbf{L}$ and its components denote angular momentum per unit mass. F is used both for a force and for force per unit mass, depending on the context.

3.1 Motion under gravity: weighing the Galaxy

Newton's law of gravity tells us that a point mass $\mathcal{M}$ attracts a second mass m separated from it by distance $\mathbf{r}$, causing the velocity $\mathbf{v}$ of m to change according to

$$\frac{d}{dt}(m\mathbf{v}) = -\frac{Gm\mathcal{M}}{r^3}\mathbf{r}, \tag{3.1}$$

where G is Newton's gravitational constant. In a cluster of N stars with masses m_α ($\alpha = 1, 2, \ldots, N$), at positions $\mathbf{x}_\alpha$, we can add the forces on star α from all the other stars:

$$\frac{d}{dt}(m_\alpha \mathbf{v}_\alpha) = -\sum_{\substack{\beta \\ \alpha \neq \beta}} \frac{Gm_\alpha m_\beta}{|\mathbf{x}_\alpha - \mathbf{x}_\beta|^3}(\mathbf{x}_\alpha - \mathbf{x}_\beta). \tag{3.2}$$

The mass m_α cancels out of this equation, so the acceleration $d\mathbf{v}_\alpha/dt$ is independent of the star's mass: light and heavy objects fall equally fast. This is the *principle of equivalence* between gravitational and inertial mass, which is the basis for the general theory of relativity. We can write the force from the cluster on a star of mass m at position $\mathbf{x}$ as the gradient of the *gravitational potential* $\Phi(\mathbf{x})$:

$$\frac{d}{dt}(m\mathbf{v}) = -m\,\nabla\Phi(\mathbf{x}), \quad \text{with } \Phi(\mathbf{x}) = -\sum_\alpha \frac{Gm_\alpha}{|\mathbf{x} - \mathbf{x}_\alpha|} \text{ for } \mathbf{x} \neq \mathbf{x}_\alpha, \qquad (3.3)$$

where we have chosen an arbitrary integration constant so that $\Phi(\mathbf{x}) \to 0$ at large distances. If we think of a continuous distribution of matter in a galaxy or star cluster, the potential at point $\mathbf{x}$ is given by an integral over the density $\rho(\mathbf{x}')$ at all other points:

$$\Phi(\mathbf{x}) = -\int \frac{G\rho(\mathbf{x}')}{|\mathbf{x} - \mathbf{x}'|}\,d^3\mathbf{x}', \qquad (3.4)$$

and the force $\mathbf{F}$ per unit mass is

$$\mathbf{F}(\mathbf{x}) = -\nabla\Phi(\mathbf{x}) = -\int \frac{G\rho(\mathbf{x}')(\mathbf{x} - \mathbf{x}')}{|\mathbf{x} - \mathbf{x}'|^3}\,d^3\mathbf{x}'. \qquad (3.5)$$

The integral relation of Equation 3.4 can be turned into a differential equation. Applying ∇^2 to both sides, we have

$$\nabla^2\Phi(\mathbf{x}) = -\int G\rho(\mathbf{x}')\nabla^2\left(\frac{1}{|\mathbf{x} - \mathbf{x}'|}\right)d^3\mathbf{x}'. \qquad (3.6)$$

In three dimensions, differentiating with respect to the variable $\mathbf{x}$ gives, for $\mathbf{x} \neq \mathbf{x}'$ (check by trying it in Cartesian coordinates),

$$\nabla\left(\frac{1}{|\mathbf{x} - \mathbf{x}'|}\right) = -\frac{\mathbf{x} - \mathbf{x}'}{|\mathbf{x} - \mathbf{x}'|^3}, \quad \text{and} \quad \nabla^2\left(\frac{1}{|\mathbf{x} - \mathbf{x}'|}\right) = 0. \qquad (3.7)$$

So the integrand on the right-hand side of Equation 3.6 is zero outside a small sphere $S\epsilon(\mathbf{x})$ of radius ϵ centred on $\mathbf{x}$. If we take ϵ small enough that the density ρ is almost constant inside the sphere, we have

$$\nabla^2\Phi(\mathbf{x}) \approx -G\rho(\mathbf{x})\int_{S\epsilon(\mathbf{x})} \nabla^2\left(\frac{1}{|\mathbf{x} - \mathbf{x}'|}\right)d^3\mathbf{x}'$$

$$= -G\rho(\mathbf{x})\int_{S\epsilon(\mathbf{x})} \nabla^2_{\mathbf{x}'}\left(\frac{1}{|\mathbf{x} - \mathbf{x}'|}\right)dV'; \qquad (3.8)$$

in the last step, $\nabla^2_{\mathbf{x}'}$ means that the derivative is taken with respect to the variable $\mathbf{x}'$, instead of $\mathbf{x}$. (Check in Cartesian coordinates that the two ∇^2s are equal for any function of $|\mathbf{x} - \mathbf{x}'|$.)

Now we can use the *divergence theorem*: for any smooth-enough function f, the volume integral of $\nabla^2_{\mathbf{x}'}f$ over the interior of any volume is equal to the integral of $\nabla_{\mathbf{x}'} f \cdot d\mathbf{S}'$ over the surface. We also have $\nabla_{\mathbf{x}'} f = -\nabla f$ for any function $f(\mathbf{x} - \mathbf{x}')$. Setting $f = 1/|\mathbf{x} - \mathbf{x}'|$, Equation 3.7 tells us that, on the surface of the sphere $S\epsilon(\mathbf{x})$, the gradient $\nabla_{\mathbf{x}'} f$ is a vector of length ϵ^{-2} pointing in toward the point $\mathbf{x}$. The surface area is $4\pi\epsilon^2$, so the integral of $\nabla_{\mathbf{x}'} f \cdot d\mathbf{S}'$ in Equation 3.8 is -4π. We have *Poisson's equation*:

$$\nabla^2 \Phi(\mathbf{x}) = 4\pi G\rho(\mathbf{x}). \tag{3.9}$$

This can be a more convenient relationship between the potential $\Phi(\mathbf{x})$ and the corresponding density than the integral in Equation 3.4. To choose an approximation for the density $\rho(\mathbf{x})$ of a star cluster or galaxy, we can select a mathematically convenient form for the potential $\Phi(\mathbf{x})$, and then calculate the corresponding density. We must take care that $\rho(\mathbf{x}) \geq 0$ everywhere for our chosen potential; various apparently friendly potentials turn out to imply $\rho(\mathbf{x}) < 0$ in some places. The problems below deal with some commonly used potentials.

Problem 3.1 Use Equation 3.1 to show that, at distance r from a point mass $\mathcal{M}$, the gravitational potential is

$$\Phi(r) = -\frac{G\mathcal{M}}{r}. \tag{3.10}$$

Problem 3.2 The *Plummer sphere* is a simple if crude model for star clusters and round galaxies. Its gravitational potential

$$\Phi_P(r) = -\frac{G\mathcal{M}}{\sqrt{r^2 + a_P^2}} \tag{3.11}$$

approaches that of a point mass at $\mathbf{x} = 0$ when $r \gg a_P$. What is its total mass? (Hint: look ahead to Equation 3.22.) Show that its density is

$$\rho_P(r) = \frac{1}{4\pi G}\frac{1}{r^2}\frac{d}{dr}\left(r^2\frac{d\Phi_P}{dr}\right) = \frac{3a_P^2}{4\pi}\frac{\mathcal{M}}{\left(r^2 + a_P^2\right)^{5/2}}. \tag{3.12}$$

In Section 3.4 we will see that $\rho_P(r)$ describes a *polytropic* system, where the number of stars at each energy E is proportional to a power of $(-E)$.

When the Plummer sphere is viewed from a great distance along the axis z, show that the surface density at distance R from the center is

$$\Sigma_P(R) = \int_{-\infty}^{\infty} \rho_P(\sqrt{R^2 + z^2})dz = \frac{\mathcal{M}}{\pi}\frac{a_P^2}{\left(a_P^2 + R^2\right)^2}. \tag{3.13}$$

Check that the *core radius* r_c, where $\Sigma_P(R)$ drops to half its central value, is at $r_c \approx 0.644 a_P$.

Problem 3.3 The potential for the 'dark halo' mass distribution of Equation 2.19 cannot be written in a simple form, except in the limit that $a_H \to 0$. Show that the potential corresponding to the density

$$\rho_{SIS}(r) = \frac{\rho(r_0)}{(r/r_0)^2} \quad \text{is} \quad \Phi_{SIS}(r) = V_H^2 \ln(r/r_0), \quad (3.14)$$

where r_0 is a constant and $V_H^2 = 4\pi G r_0^2 \rho(r_0)$: this is the *singular isothermal sphere*. The density has a *cusp*: it grows without limit at the center. Show that both Φ_{SIS} and the mass within radius r have no finite limit as $r \to \infty$, and that the speed in a circular orbit is V_H at all radii. The singular isothermal sphere describes a system in which the number of stars at each energy E is proportional to $\exp[-E/(2V_H^2)]$.

Problem 3.4 A simple disk model potential is that of the *Kuzmin disk*: in cylindrical polar coordinates R, z,

$$\Phi_K(R, z) = -\frac{G\mathcal{M}}{\sqrt{R^2 + (a_K + |z|)^2}}. \quad (3.15)$$

Irrespective of whether z is positive or negative, this is the potential of a point mass $\mathcal{M}$ at $R = 0$, displaced by a distance a_K along the z axis, on the opposite side of the plane $z = 0$. Show that $\nabla^2 \Phi = 0$ everywhere except at $z = 0$; use the divergence theorem to show that there the surface density is

$$\Sigma_K(R) = \frac{a_K}{2\pi} \frac{\mathcal{M}}{(R^2 + a_K^2)^{3/2}}. \quad (3.16)$$

For a spherical galaxy or star cluster, Newton proved two useful theorems about the gravitational field. The first states that *the gravitational force inside a spherical shell of uniform density is zero*. In Figure 3.1, the star at S experiences a gravitational pull from the material at A within a narrow cone of solid angle $\Delta\Omega$, and a force in the opposite direction from mass within the same cone at B. By symmetry, the line AB makes the same angle with the normal OA to the surface at A as it does with OB at B. Thus the ratio of the mass enclosed is just $(SA/SB)^2$; by the inverse-square law, the forces are exactly equal, and cancel each other out. Thus there is no force on the star, and *the potential $\Phi(\mathbf{x})$ must be constant within the shell*.

The second theorem says that, *outside any spherically symmetric object, the gravitational force is the same as if all its mass had been concentrated at the center*. If we can show that this is so for a uniform spherical shell, it must be true for any spherically symmetric object built from those shells. To find the potential $\Phi(\mathbf{x})$ at a point P lying outside a uniform spherical shell of mass $\mathcal{M}$ and radius a,

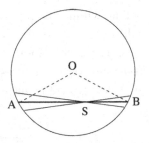

Fig. 3.1. The gravitational force inside a uniform hollow sphere with its center at O.

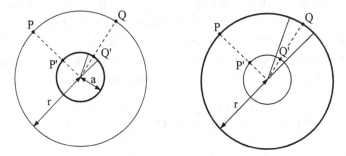

Fig. 3.2. The gravitational potential outside a uniform spherical shell.

at distance r from the center, we can add the contributions $\Delta\Phi$ from small patches of the shell. On the left of Figure 3.2, the mass in a narrow cone of opening solid angle $\Delta\Omega$ around Q' contributes

$$\Delta\Phi[\mathbf{x}(P)] = -\frac{G\mathcal{M}}{|\mathbf{x}(P) - \mathbf{x}(Q')|}\frac{\Delta\Omega}{4\pi}. \tag{3.17}$$

Now think of the potential Φ' at point P', lying at distance a from the center inside a sphere of the same mass $\mathcal{M}$, but now with radius r. On the right in Figure 3.2, we see that the contribution $\Delta\Phi'$ from material in the same cone, which cuts the larger sphere at Q, is

$$\Delta\Phi'[\mathbf{x}(P')] = -\frac{G\mathcal{M}}{|\mathbf{x}(P') - \mathbf{x}(Q)|}\frac{\Delta\Omega}{4\pi}. \tag{3.18}$$

But, because $PQ' = P'Q$, this is equal to $\Delta\Phi[\mathbf{x}(P)]$. So, when we integrate over the whole sphere,

$$\Phi[\mathbf{x}(P)] = \Phi'[\mathbf{x}(P')] = \Phi'[\mathbf{x} = 0] = -\frac{G\mathcal{M}}{r}; \tag{3.19}$$

the potential and force at P are exactly the same as if all the mass of the sphere with radius a had been concentrated at its center.

These two theorems tell us that, within any spherical object with density $\rho(r)$, the gravitational force toward the center is just the sum of the inward forces from all the matter inside that radius. The acceleration V^2/r of a star moving with speed $V(r)$ in an orbit of radius r about the center must be provided by the inward gravitational force $-F_r(r)$. So, if $\mathcal{M}(<r)$ is the mass within radius r, we have

$$\frac{V^2(r)}{r} = -F_r(r) = \frac{G\mathcal{M}(<r)}{r^2}. \tag{3.20}$$

We already used this equation in Section 2.3 to find the mass of the Milky Way. Whenever we can find gas or stars in near-circular orbit within a galaxy, this is by far the simplest and most reliable way to estimate the mass.

For a point mass, we have $V(r) \propto r^{-1/2}$; in a spherical galaxy the rotation speed can never fall more rapidly than this. The potential $\Phi(r)$ is

$$\Phi(r) = -\left[\frac{G\mathcal{M}(<r)}{r} + 4\pi G \int_r^\infty \rho(r')r'\,dr'\right] \tag{3.21}$$

(check that differentiating this with respect to r gives you back Equation 3.20). We see that $\Phi(r)$ is *not* equal to $-G\mathcal{M}(<r)/r$, unless all the mass lies within radius r. But Equation 3.4 implies that, at a great distance from any system with finite mass $\mathcal{M}_{tot}$,

$$\Phi(\mathbf{x}) \longrightarrow -\frac{G\mathcal{M}_{tot}}{|\mathbf{x}|}. \tag{3.22}$$

Problem 3.5 Use Equation 3.20 to show that, if the density ρ in a spherical galaxy is constant, then a star following a circular orbit moves so that its *angular speed* $\Omega(r) = V(r)/r$ is constant. Show that a star moving on a radial orbit, i.e., in a straight line through the center, would oscillate harmonically in radius with period

$$P = \sqrt{\frac{3\pi}{G\rho}} \sim 3t_{ff}, \qquad \text{where } t_{ff} \equiv \sqrt{1/G\rho}. \tag{3.23}$$

The *free-fall time* t_{ff} is roughly the time that a gas cloud of density ρ would take to collapse under its own gravity, if it is not held up by pressure. Show that, if you bored a hole through the center of the Earth to the other side, and dropped an egg down it, then (ignoring air resistance, outflows of molten lava, etc.) you could return about an hour and a half later to retrieve the egg as it returned to its starting point.

Problem 3.6 From the Sun's orbital speed of $200\,\text{km s}^{-1}$, find the mass within its orbit at $r = 8\,\text{kpc}$. Show that the average density inside a sphere of this radius around the Galactic center is $\sim 0.03 \mathcal{M}_\odot \, \text{pc}^{-3}$, so that $t_{\text{ff}} \sim 100 \, \text{Myr}$. This is a typical density for the inner parts of a galaxy. Processes such as bursts of star formation that involve large parts of the galaxy happen on roughly this timescale, because gravitational forces cannot move material any faster through the galaxy.

Problem 3.7 The *Navarro–Frenk–White* (NFW) model describes the halos of cold dark matter that form in simulations like that of Figure 7.16. Show that the potential corresponding to the density

$$\rho_{\text{NFW}}(r) = \frac{\rho_{\text{N}}}{(r/a_{\text{N}})(1 + r/a_{\text{N}})^2} \quad \text{is} \quad \Phi_{\text{NFW}}(r) = -\sigma_{\text{N}}^2 \frac{\ln(1 + r/a_{\text{N}})}{(r/a_{\text{N}})}, \tag{3.24}$$

where $\sigma_{\text{N}}^2 = 4\pi G \rho_{\text{N}} a_{\text{N}}^2$. The density rises steeply at the center, but less so than in the singular isothermal sphere; at large radii $\rho(r) \propto r^{-3}$. Show that the speed V of a circular orbit at radius r is given by

$$V^2(r) = \sigma_{\text{N}}^2 \left[\frac{\ln(1 + r/a_{\text{N}})}{(r/a_{\text{N}})} - \frac{1}{(1 + r/a_{\text{N}})} \right]. \tag{3.25}$$

When finding the orbit of a single star moving through a galaxy, we will see in Section 3.2 that we can usually ignore the effect which that star has in attracting all the other stars, and thus changing the gravitational potential. If the mass distribution is static (the galaxy is not, for example, collapsing, or colliding with something), the potential at position $\mathbf{x}$ does not depend on time. Then as the star moves with velocity $\mathbf{v}$, the potential $\Phi(\mathbf{x})$ at its location changes according to $d\Phi/dt = \mathbf{v} \cdot \nabla\Phi(\mathbf{x})$. Taking the scalar product of Equation 3.3 with $\mathbf{v}$, we have

$$\mathbf{v} \cdot \frac{d}{dt}(m\mathbf{v}) + m\mathbf{v} \cdot \nabla\Phi(\mathbf{x}) = 0 = \frac{d}{dt}\left[\frac{1}{2}m\mathbf{v}^2 + m\Phi(\mathbf{x})\right]. \tag{3.26}$$

Thus

$$\mathcal{E} \equiv \frac{1}{2}m\mathbf{v}^2 + m\Phi(\mathbf{x}) = \text{constant along the orbit.} \tag{3.27}$$

The star's *energy* $\mathcal{E}$ is the sum of its *kinetic energy* $\mathcal{KE} = m\mathbf{v}^2/2$ and the *potential energy* $\mathcal{PE} = m\Phi(\mathbf{x})$. The kinetic energy cannot be negative, and Equation 3.22 tells us that, far from an isolated galaxy or star cluster, $\Phi(\mathbf{x}) \to 0$. So a star at position $\mathbf{x}$ can escape only if it has $\mathcal{E} > 0$; it must be moving faster than the local *escape speed* v_{e}, given by

$$v_{\text{e}}^2 = -2\Phi(\mathbf{x}). \tag{3.28}$$

Problem 3.8 The Sun moves in a near-circular orbit about the Galactic center at radius $R_0 \approx 8\,\text{kpc}$, with speed $V_0 \approx 200\,\text{km s}^{-1}$. If all the mass of the Milky Way were concentrated at its center, show that its total mass would be about $7 \times 10^{10}\,\mathcal{M}_\odot$, and that a nearby star would escape from the Galaxy if it moved faster than $\sqrt{2}V_0$. In fact, we see local stars with speeds as large as $500\,\text{km s}^{-1}$; explain why this tells us that the Galaxy contains appreciable mass outside the Sun's orbit.

The star's *angular momentum* $\mathcal{L} = \mathbf{x} \times m\mathbf{v}$ changes according to

$$\frac{d\mathcal{L}}{dt} = \mathbf{x} \times \frac{d}{dt}(m\mathbf{v}) = -m\mathbf{x} \times \nabla\Phi \qquad (3.29)$$

(why can we leave out the term $d\mathbf{x}/dt \times m\mathbf{v}$?). If a galaxy is spherically symmetric about $\mathbf{x} = 0$, the force $\nabla\Phi$ points toward the center, and $\mathcal{L}$ does not change. For a star moving in an axisymmetric galaxy, we will see in Section 3.3 below that only the component of angular momentum parallel to the symmetry axis remains constant.

Problem 3.9 Calculate the energy $\mathcal{E}$ and angular momentum $\mathcal{L}$ of a star of mass m moving in a circular orbit of radius r in the Plummer potential of Equation 3.11. Show that the circular speed $V(r)$ increases with radius near the center and falls further out, and that $d\mathcal{L}/dr > 0$ everywhere, while the angular speed $\Omega(r) = V/r$ is always decreasing.

Problem 3.10 For a particle in circular orbit in a potential $\Phi(r) = -Kr^{-\alpha}$, where K and α are positive constants, show that $V^2(r) = -\alpha\Phi(r)$. Two gas clouds, of masses m_1 and m_2, follow circular orbits at radii r_1 and r_2, with $r_1 < r_2$. What is the total energy $\mathcal{E}$ and the angular momentum $\mathcal{L}$?

The gas clouds are now displaced to different circular orbits at radii $r_1 + \Delta r_1$ and $r_2 + \Delta r_2$. How must Δr_1 and Δr_2 be related so that $\mathcal{L}$ remains unchanged? Assuming that Δr_1 and Δr_2 are small, what is the energy change $\Delta\mathcal{E}$? Show that, if $\alpha < 2$, the angular momentum $rV(r)$ of a circular orbit increases with r. We will see in Section 3.3 that this condition is met whenever the circular orbit is stable. Show that the second state then has lower energy than the initial energy if $\Delta r_1 < 0$. Processes that couple different regions of a rotating disk, such as viscosity or spiral structure, can extract energy from the rotation by moving mass inward and angular momentum outward.

In a cluster of stars, the gravitational potential will change as the stars move: $\Phi = \Phi(\mathbf{x}, t)$. The energy of each star is no longer conserved, only the total for the cluster as a whole. To show this, we take the scalar product of Equation 3.2 with $\mathbf{v}_\alpha$, and sum over all the stars. The left-hand side gives the derivative of the total kinetic energy $\mathcal{KE}$:

$$\sum_\alpha \mathbf{v}_\alpha \cdot \frac{d}{dt}(m_\alpha \mathbf{v}_\alpha) = \frac{d}{dt}\mathcal{KE} = -\sum_{\substack{\alpha,\beta \\ \alpha \neq \beta}} \frac{Gm_\alpha m_\beta}{|\mathbf{x}_\alpha - \mathbf{x}_\beta|^3}(\mathbf{x}_\alpha - \mathbf{x}_\beta) \cdot \mathbf{v}_\alpha. \tag{3.30}$$

But we could have started with the equation for the force on star β, and taken the scalar product with $\mathbf{v}_\beta$ to find

$$\frac{1}{2}\sum_\beta \frac{d}{dt}(m_\beta \mathbf{v}_\beta \cdot \mathbf{v}_\beta) = -\sum_{\substack{\alpha,\beta \\ \alpha \neq \beta}} \frac{Gm_\alpha m_\beta}{|\mathbf{x}_\alpha - \mathbf{x}_\beta|^3}(\mathbf{x}_\beta - \mathbf{x}_\alpha) \cdot \mathbf{v}_\beta. \tag{3.31}$$

Adding the right-hand sides of the last two equations gives

$$-\sum_{\substack{\alpha,\beta \\ \alpha \neq \beta}} \frac{Gm_\alpha m_\beta}{|\mathbf{x}_\alpha - \mathbf{x}_\beta|^3}(\mathbf{x}_\alpha - \mathbf{x}_\beta) \cdot (\mathbf{v}_\alpha - \mathbf{v}_\beta) = \sum_{\substack{\alpha,\beta \\ \alpha \neq \beta}} \frac{d}{dt}\left(\frac{Gm_\alpha m_\beta}{|\mathbf{x}_\alpha - \mathbf{x}_\beta|} \right). \tag{3.32}$$

The cluster's potential energy $\mathcal{PE}$ is the sum of contributions from pairs of stars:

$$\mathcal{PE} = -\frac{1}{2}\sum_{\substack{\alpha,\beta \\ \alpha \neq \beta}} \frac{Gm_\alpha m_\beta}{|\mathbf{x}_\alpha - \mathbf{x}_\beta|} = \frac{1}{2}\sum_\alpha m_\alpha \Phi(\mathbf{x}_\alpha) \quad \text{or} \quad \frac{1}{2}\int \rho(\mathbf{x})\Phi(\mathbf{x})dV; \tag{3.33}$$

dividing by two means that each pair contributes only one term to the sum. On adding Equations 3.30 and 3.31, we see that

$$2\frac{d}{dt}\left[\mathcal{KE} - \frac{1}{2}\sum_{\substack{\alpha,\beta \\ \alpha \neq \beta}} \frac{Gm_\alpha m_\beta}{|\mathbf{x}_\alpha - \mathbf{x}_\beta|} \right] = 0. \tag{3.34}$$

Thus the total energy $\mathcal{E} = \mathcal{KE} + \mathcal{PE}$ of the cluster is constant.

Problem 3.11 Show that, at radius r inside a uniform sphere of density ρ, the radial force $F_r = -4\pi G\rho r/3$. If the density is zero for $r > a$, show that

$$\Phi(r) = -2\pi G\rho\left(a^2 - \frac{r^2}{3} \right) \quad \text{for } r \leq a, \tag{3.35}$$

so that the potential energy is related to the mass $\mathcal{M}$ by

$$\mathcal{PE} = -\frac{16\pi^2}{15}G\rho^2 a^5 = -\frac{3}{5}\frac{G\mathcal{M}^2}{a}. \tag{3.36}$$

Taking $a = R_\odot$, the solar radius, and the mass $\mathcal{M} = \mathcal{M}_\odot$, show that $\mathcal{PE} \sim L_\odot \times 10^7$ yr; approximately this much energy was set free as the Sun contracted from a diffuse cloud of gas to its present size. Since the Earth is about 4.5 Gyr old, and the Sun has been shining for at least this long, it clearly has another energy source – nuclear fusion.

Problem 3.12 Show that, for the Plummer sphere of Equation 3.12,

$$\mathcal{PE} = -\frac{3\pi}{32} \frac{G\mathcal{M}^2}{a_P}.$$

(3.37)

We will use this result to find the masses of star clusters.

According to Equation 3.34, the stars in an isolated cluster can change their kinetic and potential energies, as long as the sum of these remains constant. As they move further apart, their potential energy increases, and their speeds must drop so that the kinetic energy can decrease. If the stars moved so far apart that their speeds dropped to zero, and then just stayed there, the system could still satisfy this equation. But star clusters cannot remain in this state: Equation 3.2 makes clear that the stars are accelerated into motion. The *virial theorem* tells us how, on average, the kinetic and potential energies are in balance.

To prove this theorem, we return to Equation 3.2, but we now add an external force $\mathbf{F}_{\text{ext}}$; this might represent, for example, the gravitational pull of a galaxy on a star cluster within it. We take the scalar product with $\mathbf{x}_\alpha$ and sum over all the stars to find

$$\sum_\alpha \frac{\mathrm{d}}{\mathrm{d}t}(m_\alpha \mathbf{v}_\alpha) \cdot \mathbf{x}_\alpha = -\sum_{\substack{\alpha,\beta \\ \alpha \neq \beta}} \frac{Gm_\alpha m_\beta}{|\mathbf{x}_\alpha - \mathbf{x}_\beta|^3}(\mathbf{x}_\alpha - \mathbf{x}_\beta) \cdot \mathbf{x}_\alpha + \sum_\alpha \mathbf{F}_{\text{ext}}^\alpha \cdot \mathbf{x}_\alpha. \quad (3.38)$$

We would have had a similar equation if we had started with the β force:

$$\sum_\beta \frac{\mathrm{d}}{\mathrm{d}t}(m_\beta \mathbf{v}_\beta) \cdot \mathbf{x}_\beta = -\sum_{\substack{\alpha,\beta \\ \alpha \neq \beta}} \frac{Gm_\alpha m_\beta}{|\mathbf{x}_\alpha - \mathbf{x}_\beta|^3}(\mathbf{x}_\beta - \mathbf{x}_\alpha) \cdot \mathbf{x}_\beta + \sum_\beta \mathbf{F}_{\text{ext}}^\beta \cdot \mathbf{x}_\beta. \quad (3.39)$$

The left-hand sides of these two equations are the same; each is equal to

$$\frac{1}{2}\sum_\alpha \frac{\mathrm{d}^2}{\mathrm{d}t^2}(m_\alpha \mathbf{x}_\alpha \cdot \mathbf{x}_\alpha) - \sum_\alpha m_\alpha \mathbf{v}_\alpha \cdot \mathbf{v}_\alpha = \frac{1}{2}\frac{\mathrm{d}^2 I}{\mathrm{d}t^2} - 2\mathcal{KE}, \quad (3.40)$$

where I is the *moment of inertia* of the system:

$$I \equiv \sum_\alpha m_\alpha \mathbf{x}_\alpha \cdot \mathbf{x}_\alpha.$$

(3.41)

By averaging Equations 3.38 and 3.39, we find (compare with Equation 3.32) that the first term on the right-hand side is the potential energy $\mathcal{PE}$: so

$$\frac{1}{2}\frac{d^2 I}{dt^2} - 2\mathcal{KE} = \mathcal{PE} + \sum_\alpha \mathbf{F}^\alpha_{\text{ext}} \cdot \mathbf{x}_\alpha. \qquad (3.42)$$

Now we average this equation over the time interval $0 < t < \tau$:

$$\frac{1}{2\tau}\left[\frac{dI}{dt}(\tau) - \frac{dI}{dt}(0)\right] = 2\langle\mathcal{KE}\rangle + \langle\mathcal{PE}\rangle + \sum_\alpha \langle\mathbf{F}^\alpha_{\text{ext}} \cdot \mathbf{x}_\alpha\rangle, \qquad (3.43)$$

where the angle brackets are used to represent this long-term average. As long as all the stars are bound to the cluster, the products $|\mathbf{x}_\alpha \cdot \mathbf{v}_\alpha|$, and hence $|dI/dt|$, never exceed some finite limits. Thus the left-hand side of this equation must tend to zero as $\tau \to \infty$, giving

$$2\langle\mathcal{KE}\rangle + \langle\mathcal{PE}\rangle + \sum_\alpha \langle\mathbf{F}^\alpha_{\text{ext}} \cdot \mathbf{x}_\alpha\rangle = 0. \qquad (3.44)$$

This is the virial theorem, one of the fundamental results of dynamics.

The virial theorem is our tool for finding the masses of star clusters and galaxies where the orbits are far from circular. The process is straightforward if the star cluster or galaxy is nearly spherical and has no strong rotation; otherwise, we must use the *tensor virial theorem* of Section 6.2. Unless the system is actively colliding with another, or is still forming by collapse, we assume that it is close to a steady state so that the virial theorem applies. Generally we start by assuming that the ratio of mass to luminosity $\mathcal{M}/L$ is the same everywhere in the system, so that the measured surface brightness $I(\mathbf{x})$ indicates the density of mass. We measure the stellar radial velocities V_r relative to the cluster's mean motion, and find the *velocity dispersion* σ_r. This is defined by $\sigma_r^2 = \langle V_r^2\rangle$, where the angle brackets represent an average over the stars of the cluster. For example, in globular clusters V_r can be measured with a precision of $0.5\,\mathrm{km\,s^{-1}}$, and σ_r is typically 5–$15\,\mathrm{km\,s^{-1}}$; see Table 3.1.

Many star clusters are so distant that tangential motions are very hard to measure (what proper motion μ corresponds to $10\,\mathrm{km\,s^{-1}}$ at $d = 30\,\mathrm{kpc}$?). We often assume that the average motions are *isotropic*, equal in all directions. Then, $\langle\mathbf{v}_\alpha \cdot \mathbf{v}_\alpha\rangle \approx 3\sigma_r^2$, and the cluster's kinetic energy is $\mathcal{KE} \approx (3\sigma_r^2/2)(\mathcal{M}/L)L_{\text{tot}}$. (Proper-motion studies of a few globular clusters have shown that the orbits of stars in the outer parts are highly elongated; motions toward and away from the center are on average larger than those in the perpendicular directions. Taking this anisotropy into account modifies the derived masses slightly.) To estimate the potential energy $\mathcal{PE}$, we set $\mathcal{M} = L_{\text{tot}} \times \mathcal{M}/L$. Often, we take the cluster to be

Table 3.1 Dynamical quantities for globular and open clusters in the Milky Way

Cluster		σ_r (km s^{-1})	$\log_{10} \rho_c$ ($\mathcal{M}_\odot$ pc^{-3})	r_c (pc)	$t_{relax,c}$ (Myr)	Mass ($10^3 \mathcal{M}_\odot$)	$\mathcal{M}/L_V$ ($\mathcal{M}_\odot/L_\odot$)
NGC 5139	ω Cen	20	3.1	4	5000	2600	2.5
NGC 104	47 Tuc	11	4.9	0.5	50	800	1.5
NGC 7078	M15	12	>7	<0.1	<1	900	2
NGC 6341	M92	5	5.2	0.5	2	200	1
NGC 6121	M4	4	4–5	0.5	30	60	1
	Pal 13	~0.8	2	1.7	10	3	3–7
NGC 1049	Fornax 3	9	3.5	1.6	600	400	~3
Open cluster	Pleiades	0.5	0.5	3	100	0.8	0.2

Note: σ_r is the dispersion in radial velocity V_r in the cluster core; ρ_c is central density; $t_{relax,c}$ is the relaxation time at the cluster's center found using Equation 3.55 with $V = \sqrt{3}\sigma_r$, $\langle m_* \rangle = 0.3\mathcal{M}_\odot$, and $\Lambda = r_c/1$ AU. Clusters with upper limits to r_c probably have collapsed cores.

spherically symmetric. Then, from its surface brightness, we can find the volume density of stars and hence the potential energy: see the problem below.

Problem 3.13 For a random sample of stars in a globular cluster, $\sigma_r \approx 10$ km s^{-1}; the surface brightness can be fit approximately by the Plummer model of Equation 3.13 with $a_P = 10$ pc. Assuming that the cluster is spherical and contains no unseen dark matter, use the virial theorem to show that its mass $\mathcal{M} \approx 2 \times 10^6 \mathcal{M}_\odot$. What did you assume about the stars' random motions?

Another tactic is to measure the cluster's core radius r_c, where the measured surface brightness $I(\mathbf{x})$ has fallen to half its central value. The core radii of globular clusters are generally about a parsec, but can be larger (Table 2.3). We can write the potential energy as $\mathcal{PE} = -G\mathcal{M}^2/(2\eta r_c)$ for some constant $\eta \sim 1$. For example, the homogeneous sphere of Problem 3.1 has $r_c = \sqrt{3}a/2$, and using Equation 3.36 yields $\eta = 5/(3\sqrt{3}) \approx 0.96$. For the Plummer model of Problem 3.12, $\eta \approx 2.6$. Ignoring external forces from the rest of the Galaxy, Equation 3.44 gives the cluster's mass as

$$\mathcal{M} \approx 6\eta\sigma_r^2 r_c/G. \tag{3.45}$$

Sometimes it is easier to measure the cluster's central surface brightness $I(\mathbf{x} = 0)$ than its total luminosity. If we write $L_{tot} \approx 4\pi r_c^2 I(0)/3$, and for the velocity dispersion we take $\sigma_r(0)$, the observed central value, then

$$\frac{\mathcal{M}}{L} \approx \frac{9}{2\pi} \frac{\sigma_r^2(0)}{G I(0) r_c}. \tag{3.46}$$

This equation is correct to within a few percent in a spherical system where σ_r is exactly constant everywhere (the *isothermal* model of Section 3.4). It is a fairly good approximation for a wide range of cluster-like and galaxy-like potentials (see Richstone and Tremaine 1986 *AJ* **92**, 72).

Most globular clusters have masses in the range $10^4 \mathcal{M}_\odot$ to $10^6 \mathcal{M}_\odot$. Table 3.1 shows that typically the mass-to-light ratio $1 \lesssim \mathcal{M}/L \lesssim 4$. It is larger than that measured in Section 2.1 for the immediate solar neighborhood, which is not surprising since all the bright massive stars in these old clusters have now died. But it is much less than what we found in Problem 2.18 for the Milky Way as a whole: globular clusters do not seem to contain much, if any, 'dark matter'. Open clusters are less dense than globular clusters, and their stars have smaller random speeds. The mass-to-light ratios in Table 2.2 are even lower than those in globular clusters because luminous stars are still on the main sequence.

Problem 3.14 Suppose that an isolated cluster of stars is initially in equilibrium. If a fraction f of its mass is suddenly removed so that the mass m_α of each becomes $(1-f)m_\alpha$, by what factor do the kinetic and potential energies change? If the initial potential energy is $\mathcal{PE}_0$, show that the total energy $\mathcal{E}_0 = \mathcal{PE}_0/2$, and afterward it is $\mathcal{E}_1 = (1-f)(1-2f)\mathcal{E}_0$. So, if $f > 0.5$, the stars are no longer bound together. Use the virial theorem to show that, when the stars come to a new equilibrium, the average distance between them is larger by a factor $(1-f)/(1-2f)$.

The fast winds from massive stars, and their final explosion as supernovae, blow away any cluster gas that was not initially converted into stars. In the Milky Way's disk, we see many clumps of young stars that were born together but are now dissolving; they became unbound as leftover gas was expelled. Globular clusters must have formed in a different way, using up almost all their gas, or else the clusters would not have remained as dense as they now are.

To find the average motion of one star within an unchanging cluster potential, we can regard the gravity of all the other stars as giving rise to an external force. For each star, the virial theorem now tells us that

$$\langle \mathbf{v}^2 \rangle = \langle \nabla \Phi(\mathbf{x}) \cdot \mathbf{x} \rangle. \tag{3.47}$$

(For a star following a circular orbit in a spherically symmetric potential $\Phi(r)$, check that this gives the same orbital speed $V(r)$ as Equation 3.20.) In the outer parts of any cluster, the magnitude of $\nabla \Phi \cdot \mathbf{x}$ begins to decrease with radius. Just as for planets in the solar system, stars in the central parts of a cluster or galaxy generally move faster than those spending most of their time in the outskirts.

Problem 3.15 The Milky Way's satellite galaxies orbit at distances of 60–80 kpc: see Table 4.1 for a list. Their radial velocities $V_r(\odot)$ measured relative to the Sun are typically around $100\,\mathrm{km\,s^{-1}}$. Looking back at Figure 2.19, explain why you must add $V_0 \sin l \cos b$ to find the motion relative to the Galactic center.

Simplify the problem by taking the Galaxy to be spherically symmetric, with its mass entirely within the satellite orbits, so you can treat it as a central point mass $\mathcal{M}_G$. The mutual potential energy of the satellites is small compared with the external force term $-G\mathcal{M}_G/r^2$, and the satellite's Galactocentric radius r is almost equal to d, its distance from the Sun. Assume that on average the satellites have equal speeds in all directions; for each in turn, use Equation 3.47 to estimate $\mathcal{M}_G$. Average the results, to show that the Milky Way's mass exceeds $10^{12}\mathcal{M}_\odot$ so that $\mathcal{M}/L_V \gtrsim 50$.

Further reading: H. Goldstein, C. Poole, and J. Safko, 2002, *Classical Mechanics*, 3rd edition (Addison-Wesley, San Francisco), Chapters 1–3; and J. Binney and S. Tremaine, 1987, *Galactic Dynamics* (Princeton University Press, Princeton, New Jersey), Sections 2.1, 2.2, and 4.3; these are both graduate texts.

3.2 Why the Galaxy isn't bumpy: two-body relaxation

Given enough time, molecules of air or scent, or small particles of smoke, will spread themselves out evenly within a room. This happens because particles can exchange energy and momentum during 'collisions': two of them come so close that the forces between them are much stronger than the force that each feels from all the other molecules together. At an average room temperature and normal atmospheric pressure, each molecule of oxygen or nitrogen has about 10^{11} such encounters every second.

Similarly, Figure 3.3 shows how we can think of the gravitational potential of the Galaxy as the sum of two parts: a smooth component, averaged over a region containing many stars, and the remainder, which includes the very deep potential well around each star. The successive tugs of individual stars on each other, described by the sharply varying part of the potential, cause them to deviate from the courses they would have taken if just the smooth part of the force had been present: we can think of these sharp pulls as 'collisions' between stars.

We will see in this section that stars in a galaxy behave quite differently from air molecules. The cumulative effect of the small pulls of distant stars is more important in changing the course of a star's motion than the huge forces generated as stars pass very near to each other. But, except in dense star clusters, even these distant collisions have little effect over the lifetime of the Galaxy in randomizing or 'relaxing' the stellar motions. For example, the smooth averaged part of the Galactic potential almost entirely determines the motion of stars like the Sun.

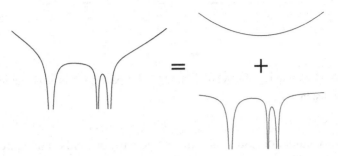

Fig. 3.3. The potential $\Phi(\mathbf{x})$ of a stellar system, represented here by vertical height, can be split into a smoothly varying averaged component and a steep potential well near each star.

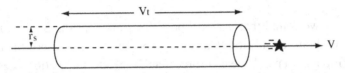

Fig. 3.4. During time t, this star will have a strong encounter with any other star lying within the cylinder of radius r_s.

3.2.1   Strong close encounters

We can calculate the average time between *strong encounters*, in which one star comes so near to another that the collision completely changes its speed and direction of motion, as follows. Suppose that the stars all have mass m and move in random directions with average speed V. For the moment, we neglect the gravitational force from the rest of the galaxy or cluster. Then, if two stars approach within a distance r, the sum of their kinetic energies must increase to balance the change in potential energy. When they are a long way apart, their mutual potential energy is zero. We say that they have a *strong encounter* if, at their closest approach, the change in potential energy is at least as great as their starting kinetic energy. This requires

$$\frac{Gm^2}{r} \gtrsim \frac{mV^2}{2}, \quad \text{which means } r \lesssim r_s \equiv \frac{2Gm}{V^2}; \tag{3.48}$$

we call r_s the *strong-encounter radius*. Near the Sun, stars have random speeds of $V \approx 30\,\mathrm{km\,s^{-1}}$, and taking $m = 0.5\mathcal{M}_\odot$ gives $r_s \approx 1\,\mathrm{AU}$.

How often does this happen? We know that the Sun has not had a strong encounter in the past 4.5 Gyr; if another star had come so near, it would have disrupted the orbits of the planets. As the Sun moves relative to nearby stars at speed V for a time t, it has a strong encounter with any other stars within a cylinder of radius r_s, and volume $\pi r_s^2 Vt$ centred on its path; see Figure 3.4. If there are n

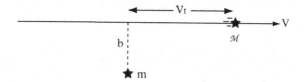

Fig. 3.5. A weak encounter: star $\mathcal{M}$ moves at speed V past the stationary star m, approaching to within distance b.

stars per unit volume, our Sun will on average have one close encounter in a time t_s such that $n\pi r_s^2 Vt = 1$, so the mean time between strong encounters is

$$t_s = \frac{V^3}{4\pi G^2 m^2 n} \approx 4 \times 10^{12}\,\mathrm{yr}\left(\frac{V}{10\,\mathrm{km\,s^{-1}}}\right)^3 \left(\frac{m}{\mathcal{M}_\odot}\right)^{-2} \left(\frac{n}{1\,\mathrm{pc^{-3}}}\right)^{-1}. \quad (3.49)$$

In Section 2.1 we found that $n \approx 0.1\,\mathrm{pc^{-3}}$ for stars near the Sun; so $t_s \sim 10^{15}$ years. This is about ten million 'Galactic years', and it far exceeds the age of the Universe. Gravity is a much weaker force than the electromagnetic forces between atoms, and, even though stars are by terrestrial standards very massive, they still do not often come close enough for the gravitational attraction of one to cause a large change in another's orbit. Strong encounters are important only in the dense cores of globular clusters, and in galactic nuclei.

3.2.2 Distant weak encounters

For molecules in the air, the electric and magnetic forces of distant particles will tend to cancel each other out, averaging to zero. Thus strong close encounters are overwhelmingly more important in changing their speeds and direction of motion. But gravity is always an attractive force; a star is pulled toward all other stars, however far away. In this section we will see that the cumulative pull of distant stars is more effective over time in changing a star's direction of motion than are single close encounters.

   In a distant encounter, the force of one star on another is so weak that the stars hardly deviate from their original paths while the encounter takes place. So we can use the *impulse approximation*, calculating the forces that the stars would feel as they move along the paths they would follow if they had not been disturbed. We start with a star of mass $\mathcal{M}$ in Figure 3.5, moving at speed V along a path that will take it within distance b of a stationary star of mass m. The motion of $\mathcal{M}$ is approximately along a straight line; the pull of m gives it a small motion $V_\perp$ perpendicular to that path. If we measure time from the point of closest approach, the perpendicular force is

$$\mathbf{F}_\perp = \frac{Gm\mathcal{M}b}{(b^2 + V^2 t^2)^{3/2}} = \mathcal{M}\frac{\mathrm{d}V_\perp}{\mathrm{d}t}. \quad (3.50)$$

Upon integrating over time, we find that, long after the encounter, the perpendicular speed of $\mathcal{M}$ is

$$\Delta V_\perp = \frac{1}{\mathcal{M}} \int_{-\infty}^{\infty} \mathbf{F}_\perp(t)\mathrm{d}t = \frac{2Gm}{bV}; \tag{3.51}$$

the faster $\mathcal{M}$ flies past m, the smaller the velocity change. In this approximation, the speed V of $\mathcal{M}$ along its orginal direction is unaffected; the force pulling it forward at times $t < 0$ exactly balances that pulling it back when $t > 0$. So the path of $\mathcal{M}$ is bent through an angle

$$\alpha = \frac{\Delta V_\perp}{V} = \frac{2Gm}{bV^2}. \tag{3.52}$$

Setting $V = c$ here shows that, according to Newtonian gravity, light should be bent by exactly *half* the angle that General Relativity predicts in Equation 7.13.

Momentum in the direction of $\mathbf{F}_\perp$ must be conserved, so after the encounter m is moving toward the path of $\mathcal{M}$ at a speed $2G\mathcal{M}/(bV)$. The impulse approximation is valid only if the perpendicular motion does not change the relative positions of $\mathcal{M}$ and m significantly over the time $\Delta t \sim b/V$ during which most of the velocity change takes place. The perpendicular velocity of approach must be small compared with V, so we need

$$b \gg \frac{2G(m + \mathcal{M})}{V^2}. \tag{3.53}$$

So a weak encounter requires b to be much larger than r_s, the strong-encounter radius of Equation 3.48.

As star $\mathcal{M}$ proceeds through the Galaxy many stars m will tug at it, each changing its motion by an amount $\Delta V_\perp$, but in different directions. If the forces are random, then we should add the *squares* of the perpendicular velocities to find the expected value of $\Delta V_\perp^2$. During time t, the number of stars m passing $\mathcal{M}$ with separations between b and $b + \Delta b$ is just the product of their number density n and the volume $Vt \cdot 2\pi b\,\Delta b$ in which these encounters can take place. Multiplying by $\Delta V_\perp^2$ from Equation 3.51 and integrating over b gives the expected squared speed: after time t,

$$\langle \Delta V_\perp^2 \rangle = \int_{b_{\min}}^{b_{\max}} nVt\left(\frac{2Gm}{bV}\right)^2 2\pi b\,\mathrm{d}b = \frac{8\pi G^2 m^2 nt}{V} \ln\left(\frac{b_{\max}}{b_{\min}}\right). \tag{3.54}$$

After a time t_{relax} such that $\langle \Delta V_\perp^2 \rangle = V^2$, the star's expected speed perpendicular to its original path becomes roughly equal to its original forward speed;

the 'memory' of its initial path has been lost. Defining $\Lambda \equiv (b_{max}/b_{min})$, we find that this *relaxation time* is much shorter than the strong-encounter time t_s of Equation 3.49:

$$t_{relax} = \frac{V^3}{8\pi G^2 m^2 n \ln \Lambda} = \frac{t_s}{2\ln \Lambda}$$

$$\approx \frac{2 \times 10^9 \,\text{yr}}{\ln \Lambda} \left(\frac{V}{10\,\text{km s}^{-1}}\right)^3 \left(\frac{m}{\mathcal{M}_\odot}\right)^{-2} \left(\frac{n}{10^3\,\text{pc}^{-3}}\right)^{-1}. \quad (3.55)$$

It is not clear what value we should take for Λ. Our derivation is certainly not valid if $b < r_s$, and we usually take $b_{min} = r_s$ and b_{max} to be equal to the size of the whole stellar system. For stars near the Sun, $r_s = 1$ AU, and $300\,\text{pc} \lesssim b_{max} \lesssim 30\,\text{kpc}$, giving $\ln \Lambda \approx 18$–22; the exact values of b_{min} and b_{max} are clearly not important. Although the many weak pulls of distant stars change the direction of motion of a star like the Sun more rapidly than do the very infrequent close encounters, *the time required is still* $\sim 10^{13}$ yr, much longer than the age of the Universe. So, when calculating the motion of stars like the Sun, we can ignore the pulls of individual stars, and consider all the stars to move in the smoothed-out potential of the entire Galaxy. We will take advantage of this fact in the next section, where we examine the orbits of stars in the Milky Way's disk.

Table 3.1 gives the average random speed σ_r and the relaxation time at the centers of a number of Galactic globular clusters. In ω Centauri, the largest, t_{relax} is about 5 Gyr. This is much longer than the time $t_{cross} \approx 0.5$ Myr that a star takes to move across the core. We can safely calculate the path of a star over a few orbits by using only the smoothed part of the gravitational force. But, to understand how globular clusters change throughout the lifetime of the Galaxy, we must take account of energy exchanges between individual stars. The central parts of most clusters have been affected by relaxation.

> **Problem 3.16** Assuming an average stellar mass of $0.5\mathcal{M}_\odot$ and $\Lambda = r_c/1$ AU, use the information in Table 3.1 to find the relaxation time t_{relax} at the center of the globular cluster 47 Tucanae. Show that the crossing time $t_{cross} \approx 2r_c/\sigma_r \sim 10^{-3} t_{relax}$.

The open clusters are comparable in size to globular clusters, but they have much lower densities, typically $n \sim 10\,\text{pc}^{-3}$ or less, and the stars move more slowly, $\sigma_r \sim 1\,\text{km s}^{-1}$. For an average stellar mass of $0.5\mathcal{M}_\odot$ Equation 3.55 predicts $t_{relax} \sim 50$ Myr, while for $r_c = 2$–3 pc the crossing time is about 5 Myr. So, within ten crossing times, the cumulative effect of weak encounters can change the stellar orbits radically. It is exceptionally difficult to calculate how the structure of an open cluster should develop over time. We cannot simply follow the orbits of stars in the smoothed part of the cluster potential; this would give inaccurate results after only a few orbits. But a gravitational N-body simulation, integrating Equation 3.2 accurately to follow the stars through close encounters where their gravitational forces are strong and rapidly varying, would take far too long on a standard computer. A further complication is that the relaxation time is close to the lifetime of a $5\mathcal{M}_\odot$ star, and mass lost from aging stars is likely to escape from the cluster. Some progress is being made with specially built computer hardware.

In an isolated cluster consisting of N stars with mass m moving at average speed V, the average separation between stars is roughly half the size R of the system. Equation 3.44 then tells us that

$$\frac{1}{2}NmV^2 \sim \frac{G(Nm)^2}{2R}, \quad \text{so } \Lambda = \frac{R}{r_s} \sim \frac{GmN}{V^2} \cdot \frac{V^2}{2Gm} \sim \frac{N}{2}. \quad (3.56)$$

The crossing time $t_{\text{cross}} \sim R/V$; since $N = 4n\pi R^3/3$, we have

$$\frac{t_{\text{relax}}}{t_{\text{cross}}} \sim \frac{V^4 R^2}{6NG^2m^2 \ln \Lambda} \sim \frac{N}{6\ln(N/2)}. \quad (3.57)$$

In a galaxy with $N \sim 10^{11}$ stars, relaxation will be important only after about 10^9 crossing times, much longer than the age of the Universe. Globular clusters contain about 10^6 stars, so for the cluster as a whole $t_{\text{relax}} \sim 10^4 t_{\text{cross}} \sim 10^{10}$ yr. In an open cluster with $N = 100$, as we saw above, the two timescales are almost equal.

Gravitational N-body simulations of galaxies generally use between 10^4 and 10^6 'stars' attracting each other by their gravity, according to Equation 3.2. Galaxies are centrally concentrated, and, in the dense inner regions, crossing times are only 10^6–10^7 years. Equation 3.57 shows that, if the 'stars' are treated as point masses, particles are pulled right off their original orbits on timescales $t_{\text{relax}} \lesssim 10^3 t_{\text{cross}} \sim 10^{10}$ yr. These computations cannot be trusted to behave like a real galaxy for longer than a gigayear or two; beyond that, relaxation is important. We can extend this time limit if we can somehow reduce Λ. A common tactic is to *soften* the potential, reducing the attractive force when 'stars' come very near each other. For example, we could substitute the potential of a Plummer sphere from Equation 3.11 for that of each point mass. The attractive pull of a 'star' of mass $\mathcal{M}$ is limited to $G\mathcal{M}/a_p^2$, and so $b_{\text{min}} \approx a_P$. But we pay the price that our model galaxy becomes 'fuzzy'; we cannot properly include any structures smaller than a few times a_P.

Problem 3.17 Gravitational N-body simulations of galactic disks often confine all the particles to a single plane: instead of n stars per unit volume we have $\mathcal{N}$ per unit area. The term $2\pi b\, db$ in Equation 3.54 is replaced by $2\, db$ – why? Show that now t_{relax} does not depend on Λ, but only on b_{min}, and that taking $b_{min} = r_s$ yields $t_{relax}/t_{cross} = V^2/(4GRm\mathcal{N})$. If the mass density $m\mathcal{N}$ is fixed, this ratio is independent of the number of simulation particles.

3.2.3 Effects of two-body relaxation

While a star moves in the smoothed potential of a star cluster, Equation 3.27 tells us that its orbit does not depend on whether it is heavy or light, but only on its position or velocity. If the smoothed potential $\Phi(\mathbf{x})$ does not change with time, the energy of the star remains constant. By contrast, two-body 'collisions' allow two stars to exchange energy and momentum in a way that depends on both their masses; this is known as *two-body relaxation.*

Just as for the air molecules in a room, the exchanges on average will shift the velocities of the stars toward the most probable way of sharing the available energy: this is a *Maxwellian* distribution. The fraction f of stars with velocities v between v and $v + \Delta v$ is given by $f_M(\mathcal{E})\, 4\pi v^2\, \Delta v$, where

$$f_M(\mathcal{E}) \propto \exp\left(\frac{-\mathcal{E}}{k_B T}\right) = \exp\left\{-\left[m\Phi(\mathbf{x}) + \frac{mv^2}{2}\right]\Big/(k_B T)\right\}, \qquad (3.58)$$

where k_B is Boltzmann's constant. The 'temperature' T depends on the energy of the system: it is higher when the stars are moving faster. The problem below shows that, for stars of mass m, T is related to the average kinetic energy by

$$\frac{1}{2}m\langle\mathbf{v}^2(\mathbf{x})\rangle = \frac{3}{2}k_B T. \qquad (3.59)$$

Just as oxygen molecules in the Earth's atmosphere move less rapidly than the lighter hydrogen molecules, heavier stars in a Maxwellian distribution move on average more slowly than the less massive ones.

Problem 3.18 Explain why the velocity dispersion is given by

$$\langle\mathbf{v}^2(\mathbf{x})\rangle = \int_0^\infty v^2 \exp\left(-\frac{mv^2}{2k_B T}\right)4\pi v^2\, dv \Big/ \int_0^\infty \exp\left(-\frac{mv^2}{2k_B T}\right)4\pi v^2\, dv.$$

Write both integrals as multiples of $\int_0^\infty x^2 e^{-x^2}\, dx$ to show that Equation 3.59 holds.

As it pushes their velocity distribution toward the Maxwellian form, two-body relaxation causes stars to *evaporate* from the cluster. The distribution $f_M(\mathcal{E})$ includes a small number of stars with arbitrarily high energy; but any stars moving faster than the escape speed v_e given by Equation 3.28 are not bound to the cluster and will escape. In a cluster of N stars with masses m_α at positions $\mathbf{x}_\alpha$, Equation 3.33 tells us that the average kinetic energy needed for escape is

$$\left\langle \frac{1}{2}mv_e^2(\mathbf{x}) \right\rangle = -\frac{1}{N}\sum_\alpha m_\alpha \Phi(\mathbf{x}_\alpha) = -\frac{2}{N}\mathcal{PE} = \frac{4}{N}\mathcal{KE}, \qquad (3.60)$$

where $\mathcal{PE}$ and $\mathcal{KE}$ are the potential and kinetic energy of the cluster as a whole; we have used Equation 3.44, the virial theorem, in the last step. The average kinetic energy needed for escape is just four times the average for each star, or $6k_B T$, so the fraction of escaping stars in the Maxwellian distribution f_M is

$$\int_{\sqrt{12k_B T/m}}^{\infty} f_M(\mathcal{E})v^2 \, dv \Bigg/ \int_0^\infty f_M(\mathcal{E})v^2 \, dv = 0.0074 \approx \frac{1}{136}. \qquad (3.61)$$

These stars leave the cluster; after a further time t_{relax}, new stars are promoted above the escape energy, and depart in their turn. The cluster loses a substantial fraction of its stars over an *evaporation time*

$$t_{evap} \sim 136 t_{relax}. \qquad (3.62)$$

In the observed globular clusters, t_{evap} is longer than the age of the Universe; any clusters with very short evaporation times presumably dissolved before we could observe them. For open clusters t_{evap} is only a few gigayears. In practice, these clusters fall apart even more rapidly, since evaporation is helped along by the repeated gravitational tugs from the spiral arms and from giant clouds of molecular gas in the disk.

Two-body relaxation also leads to *mass segregation*. Heavier stars congregate at the cluster center, while lighter stars are expelled toward the periphery; we see the result in Figure 3.6. If initially the cluster stars are thoroughly mixed, with similar orbital speeds, the more massive stars will have larger kinetic energy. But, in a Maxwellian distribution, their kinetic energies must be equal. Thus, on average, a massive star will be moving slower after a 'collision' than it did before. It then sinks to an orbit of lower energy; the cluster center fills up with stars that have too little energy to go anywhere else. But, as the cluster becomes centrally concentrated, these tightly bound stars must move faster than those further out, increasing their tendency to give up energy.

Meanwhile, the upwardly mobile lighter stars have gained energy from their encounters, but spend it in moving out to the suburbs. Their new orbits require slower motion than before, so they have become even poorer in kinetic energy.

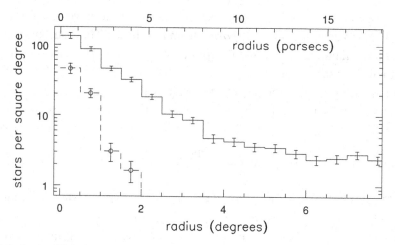

Fig. 3.6. In the Pleiades open cluster, stars with masses above $\mathcal{M}_\odot$ (dashed histogram) are more concentrated toward the center than stars with $\mathcal{M} < \mathcal{M}_\odot$ (solid histogram) – J. D. Adams.

Mass segregation is a runaway process: the lightest stars are pushed outward into an ever-expanding diffuse outer halo, while the heavier stars form an increasingly dense core at the center. Almost all star clusters have been affected by mass segregation. The smallest and least luminous stars, that carry most of the cluster's mass (recall Figure 2.3), are dispersed far from the center. So we must be careful to trace them when estimating the cluster's mass or the stellar mass function. Pairs of stars bound in a tight binary will effectively behave like a single more-massive star, sinking to the core. The X-ray sources in globular clusters are binaries in which a main-sequence star orbits a white dwarf or neutron star; they are all found near the cluster center.

Even if all the stars in a cluster have exactly the same mass, stars on low-energy orbits close to the center have higher orbital speeds than do those further out. So the inner stars tend to lose energy, while the outer stars gain it. Over time, some stars are expelled from the cluster core into the expanding halo, while the remaining core contracts. The core becomes denser, while the outer parts puff up and become more diffuse. Calculations for clusters of equal-mass stars predict that, after $(12–20)t_{\text{relax}}$, the core radius shrinks to zero, as the central density increases without limit: this is *core collapse*. A cluster that is near this state should have a small dense core and a diffuse halo, as we see for M15 in Figure 3.7.

What happens to a cluster after core collapse? In the dense core, binary stars become important sources of energy. Just as two-body 'collisions' tend to remove energy from fast-moving stars, so encounters between single stars and a tight binary pair will on average take energy from the binary. The energy is transferred to the single star, while the binary is forced closer. Depending on how many are present, binaries may supply so much energy to the stars around them that the core of the cluster starts to re-expand.

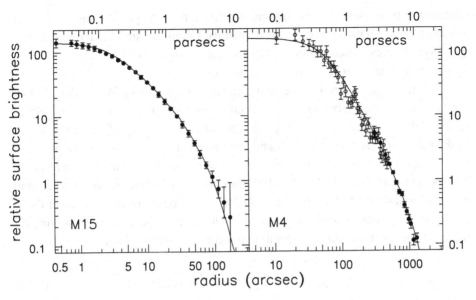

Fig. 3.7. Surface brightnesses of two globular clusters. Left, M15: the constant-density core is absent, or too small to measure. Right, M4: the surface brightness is nearly constant at small radii, dropping almost to zero at the truncation radius $r_t \approx 3000''$. The solid lines show a *King model* (Section 3.4) – A. Pasquali, G. Fahlman, and C. Pryor.

Problem 3.19 With the temperature T defined in Equation 3.59, find the kinetic energy of a system with N stars each of mass m, and use the virial theorem to show that its energy $\mathcal{E}$ satisfies

$$\frac{d\mathcal{E}}{dT} = -\frac{3}{2}Nk_{\mathrm{B}} < 0 \; (!) \tag{3.63}$$

The specific heat of a gravitating system is negative – removing energy makes it hotter. (As a mundane example, think of an orbiting satellite subject to the frictional drag of the Earth's atmosphere; as it loses energy, the orbit shrinks, and its speed increases.)

Further reading: graduate texts covering this material are J. Binney and S. Tremaine, 1987, *Galactic Dynamics* (Princeton University Press, Princeton, New Jersey), Sections 8.0, 8.2, and 8.4; and L. Spitzer, 1987, *Dynamical Evolution of Globular Clusters* (Princeton University Press, Princeton, New Jersey).

3.3 Orbits of disk stars: epicycles

We showed in the last section that the orbits of stars in a galaxy depend almost entirely on the *smooth* part of the gravitational field, averaged over a region containing many stars. From now on, when we refer to gravitational forces or potentials, we will mean this averaged quantity. Often, the smoothed potential has some symmetries which simplify the orbit calculations. In this section, we look at the orbits of stars in an *axisymmetric* galaxy.

Like the planets circling the Sun, the stars in the Milky Way's disk follow orbits that are nearly, but not quite, circular, and lie almost in the same plane. In the Galactocentric cylindrical polar coordinates (R, ϕ, z) of Section 1.2, the midplane of the disk is at $z = 0$ and the center at $R = 0$. If we are prepared to overlook non-axisymmetric structures such as an inner bar, the spiral arms, and local features such as Gould's Belt (Section 2.2), the smoothed gravitational potential is independent of ϕ. Thus $\partial \Phi / \partial \phi = 0$, and there is no force in the ϕ direction; a star conserves its angular momentum about the axis z. On writing L_z for the z angular momentum per unit mass, for each star we have

$$\frac{\mathrm{d}}{\mathrm{d}t}(R^2 \dot{\phi}) = 0, \quad \text{so } L_z \equiv R^2 \dot{\phi} = \text{constant.} \tag{3.64}$$

Since the potential does not change with time, $\Phi = \Phi(R, z)$. We can write the equation of motion in the radial direction as

$$\ddot{R} = R\dot{\phi}^2 - \frac{\partial \Phi}{\partial R} = -\frac{\partial \Phi_{\text{eff}}}{\partial R}, \quad \text{where } \Phi_{\text{eff}} \equiv \Phi(R, z) + \frac{L_z^2}{2R^2}. \tag{3.65}$$

The *effective potential* $\Phi_{\text{eff}}(R, z; L_z)$ behaves like a potential energy for the star's motion in R and z. By the same reasoning as that which led us to Equation 3.27, multiplying Equation 3.65 by $\dot{R}$ and integrating shows that, for a star moving in the midplane $z = 0$,

$$\frac{1}{2}\dot{R}^2 + \Phi_{\text{eff}}(R, z = 0; L_z) = \text{constant.} \tag{3.66}$$

Figure 3.8 shows $\Phi_{\text{eff}}(R, z = 0; L_z)$ for the Plummer potential of Equation 3.11. Since $\dot{R}^2 \geq 0$, the L_z^2 term in Φ_{eff} acts as an 'angular-momentum barrier', preventing a star with $L_z \neq 0$ from coming closer to the axis $R = 0$ than some perigalactic radius where $\dot{R} = 0$. Unless it has enough energy to escape from the Galaxy, each star must remain within some apogalactic outer limit.

The star's vertical motion is given by

$$\ddot{z} = -\frac{\partial \Phi}{\partial z}(R, z) = -\frac{\partial \Phi_{\text{eff}}}{\partial z}(R, z). \tag{3.67}$$

If the 'top' and 'bottom' halves of the disk are mirror images of each other, then $\Phi(R, z) = \Phi(R, -z)$, and the z force is zero in the plane $z = 0$. Let R_{g} be the average value of R for the star's orbit; we will define it more precisely below. Expanding $\Phi(R, z)$ in a Taylor series around $(R_{\text{g}}, 0)$, we make fractional errors

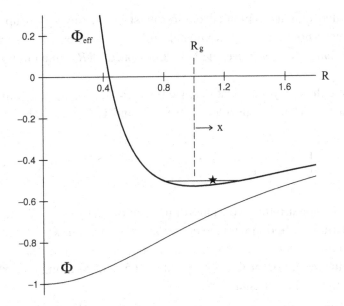

Fig. 3.8. The effective potential Φ_{eff} (upper curve) for a star with angular momentum $L_z = 0.595$, orbiting in a Plummer potential Φ_P (lower curve). The scale length $a_P = 1$; L_z is in units of $\sqrt{G\mathcal{M}/a_P}$; units for Φ and Φ_{eff} are $G\mathcal{M}/a_P$. The vertical dashed line marks the guiding center R_g; the star oscillates about R_g between inner and outer limiting radii.

only as large as z^2/R^2 or $(R - R_g)^2/R^2$ by keeping the leading term alone. So, for these nearly circular orbits,

$$\ddot{z} \approx -z\left[\frac{\partial^2\Phi}{\partial z^2}(R_g, z)\right]_{z=0} \equiv -\nu^2(R_g)z; \tag{3.68}$$

motion in z is almost independent of that in R, ϕ. So this is the equation of a harmonic oscillator with angular frequency ν; $z = Z\cos(\nu t + \theta)$, for some constants Z and θ. In a flattened galaxy, $\nu(R)$ is larger than the angular speed $\Omega(R)$ in a circular orbit.

A star with angular momentum L_z can follow an exactly circular orbit with $\dot{R} = 0$ only at the radius R_g where the effective potential Φ_{eff} is stationary with respect to R. There, Equation 3.65 tells us that

$$\frac{\partial\Phi}{\partial R}(R_g, z = 0) = \frac{L_z^2}{R_g^3} = R_g\Omega^2(R_g), \tag{3.69}$$

where $\Omega(R)$ is the angular speed of the circular orbit in the plane $z = 0$. If the effective potential has a minimum at the radius R_g, a circular path is the orbit with least energy for the given angular momentum L_z. The circular orbit is stable, and any star with the same L_z must oscillate around it. As that star moves radially in and out, its azimuthal motion must alternately speed up and slow down. We can show that it approximately follows an elliptical *epicycle* around its *guiding center*, which moves with angular speed $\Omega(R_g)$ in a circular orbit of radius R_g.

To derive the epicyclic equations, we set $R = R_g + x$ in Equation 3.65. We assume that $x \ll R$ and neglect terms in z^2/R^2 and x^2/R^2, to find

$$\ddot{x} \approx -x\left[\frac{\partial^2 \Phi_{\text{eff}}}{\partial R^2}\right]_{R_g} \equiv -\kappa^2(R_g)x, \qquad \text{so } x \approx X\cos(\kappa t + \psi), \quad (3.70)$$

where X and ψ are arbitrary constants of integration. When $\kappa^2 > 0$, this equation describes harmonic motion with the *epicyclic frequency* κ. If $\kappa^2 < 0$, the circular orbit is unstable, and the star moves away from it at an exponentially increasing rate. From the definition of Φ_{eff} in Equation 3.65, and recalling that $R\Omega^2(R) = \partial\Phi(R, z = 0)/\partial R$ in a circular orbit,

$$\kappa^2(R) = \frac{d}{dR}[R\Omega^2(R)] + \frac{3L_z^2}{R^4} = \frac{1}{R^3}\frac{d}{dR}[(R^2\Omega)^2] = -4B\Omega, \quad (3.71)$$

where B is Oort's constant, defined in Section 2.3. Locally, $B < 0$, so κ^2 is positive and near-circular orbits like that of our Sun are, fortunately, stable. The angular momentum on a circular orbit is $R^2\Omega(R)$; we see that, if it increases outward at radius R, the circular orbit there is stable. This condition always holds for circular orbits in galaxy-like potentials. Near a static black hole of mass $\mathcal{M}$, however, the last stable circular orbit is at $R = 6G\mathcal{M}/c^2$; those at smaller radii are unstable.

Problem 3.20 Effective potentials have many uses. The motion of a star around a non-rotating black hole of mass $\mathcal{M}_{\text{BH}}$ is given by

$$\left(\frac{dr}{d\tau}\right)^2 = E^2 - \left(c^2 - \frac{2G\mathcal{M}_{\text{BH}}}{r}\right)\left(1 + \frac{L^2}{c^2r^2}\right) \equiv E^2 - 2\Phi_{\text{eff}}(r); \quad (3.72)$$

we can interpret r as distance from the center, and τ as time. (More precisely, r is the usual Schwarzschild radial coordinate, τ is proper time for a static observer at radius r, and E and L are, respectively, the energy and angular momentum per unit mass as measured by that observer.) Show that there are no circular orbits at $r < 3G\mathcal{M}_{\text{BH}}/c^2$, and that the stable circular orbits lie at $r > 6G\mathcal{M}_{\text{BH}}/c^2$ with $L > 2\sqrt{3}G\mathcal{M}_{\text{BH}}/c$.

Further reading: S. L. Shapiro and S. A. Teukolsky, 1983, *Black Holes, White Dwarfs and Neutron Stars* (Wiley, New York).

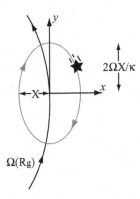

Fig. 3.9. The star moves in an elliptical epicycle around its guiding center at ($x = 0$, $y = 0$), which is carried around the Galactic center with angular speed $\Omega(R_g)$.

During its epicyclic motion, the star's azimuthal speed $\dot\phi$ must vary so that the angular momentum L_z remains constant:

$$\dot\phi = \frac{L_z}{R^2} = \frac{\Omega(R_g)R_g^2}{(R_g + x)^2} \approx \Omega(R_g)\left(1 - \frac{2x}{R_g} + \cdots\right). \qquad (3.73)$$

Substituting from Equation 3.70 for x and integrating, we have

$$\phi(t) = \phi_0 + \Omega(R_g)t - \frac{1}{R_g}\frac{2\Omega}{\kappa}X\sin(\kappa t + \psi), \qquad (3.74)$$

where ϕ_0 is an arbitrary constant. Here, the first two terms give the guiding center's motion. The third represents harmonic motion with the same frequency as the x oscillation in radius, but 90° out of phase, and larger by a factor of $2\Omega/\kappa$ (see Figure 3.9). The epicyclic motion is *retrograde*, namely in the opposite sense to the guiding center's motion; it speeds the star up closer to the center, slowing it down when it is further out.

In two simple cases, the epicyclic frequency κ is a multiple of the angular speed Ω of the guiding center. In the gravitational field of a point mass, $\Omega(r) \propto r^{-3/2}$ and so $\kappa = \Omega$. The star's orbit is an ellipse with the attracting mass at one focus; the epicycles are twice as long in the ϕ direction as in x, rather than circular, as assumed by Ptolemy, Copernicus, and others who used epicycles to describe planetary motions. Within a sphere of uniform density, $\Omega(R)$ is constant and $\kappa = 2\Omega$. A star moves harmonically in an ellipse which is symmetric about the center, making two excursions in and out during one circuit around, and the epicycles are circular. The potential of the Galaxy is intermediate between these two, so that $\Omega < \kappa < 2\Omega$. Near the Sun, $\kappa \approx 1.4\Omega$. The orbits of stars do not close on themselves; Figure 3.10 shows that they make about 1.4 oscillations in and out for every circuit of the Galaxy. We will see in Section 5.5 how stars with

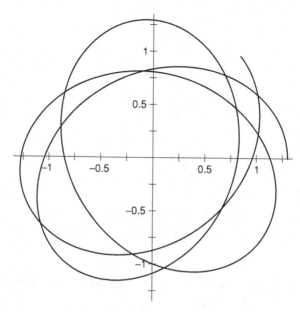

Fig. 3.10. The path of the star of Figure 3.8, viewed from above the Galactic plane; the orbit started with ($R = 1.3$, $\phi = 0$) and ($\dot{R} = 0$, $R\dot{\phi} = 0.4574$).

guiding centers at different radii R_g can be arranged on their epicycles to produce a spiral pattern in the disk.

Near the Sun, the period of the epicycles is about 170 Myr, far too long for us to watch stars complete their circuits. But we can measure the velocities of stars close to us, at $R \approx R_0$. Some of these will have guiding centers further out than the Sun, so they are on the inner parts of their epicycles, while others have their guiding centers at smaller radii. Because of its epicyclic motion, a nearby star with its guiding center at $R_g > R_0$ moves *faster* in the tangential direction than a circular orbit at our radius. Equation 3.73 gives its relative speed v_y as

$$v_y = R_0[\dot{\phi} - \Omega(R_0)] \approx R_0\left[\Omega(R_g) - 2x\frac{\Omega(R_g)}{R_g} - \Omega(R_0)\right]. \quad (3.75)$$

Recalling that $R_0 = R_g + x$, and dropping terms in x^2, we have

$$v_y \approx -x\left[2\Omega(R_0) + R_0\left(\frac{d\Omega}{dR}\right)_{R_0}\right] = -\frac{\kappa^2 x}{2\Omega} \quad \text{or} \quad 2Bx. \quad (3.76)$$

We do not know the value of x for any particular star, so we take an average over all the stars we see:

$$\langle v_y^2\rangle = \left(\frac{\kappa^2}{2\Omega}\right)^2 \langle x^2\rangle = \frac{\kappa^2}{4\Omega^2}\langle v_x^2\rangle. \quad (3.77)$$

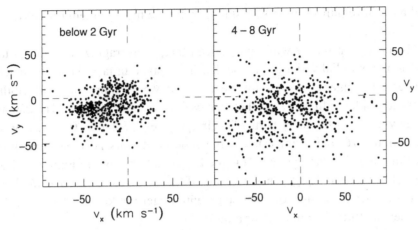

Fig. 3.11. The dispersion in velocities v_x and v_y for F and G dwarfs near the Sun increases with age. The youngest stars show a *vertex deviation*: v_x and v_y tend to have the same sign. Those stars have not yet had time to move away from the groups in which they were born. The average value of v_y is increasingly negative for older stars with larger random speeds – B. Nordström *et al.* 2004 *AAp* **418**, 98.

Since $\kappa < 2\Omega$, $\langle v_y^2 \rangle < \langle v_x^2 \rangle$; even though the epicycles are longer in the tangential y direction, the nearby stars have larger random speeds in the radial x direction. The tangential velocity dispersion is reduced because the epicycles of stars that come from further out in the Galaxy are carrying them in the same direction as the Galactic rotation, augmenting the slower motion of their guiding centers. Conversely, the epicyclic motion of stars visiting from smaller radii opposes their faster guiding-center motion. For the 'thin-disk' F and G stars of Table 2.1, we have

$$2 \lesssim \langle v_x^2 \rangle / \langle v_y^2 \rangle \lesssim 3. \tag{3.78}$$

Measuring this ratio for larger groups of nearby stars provides our best estimate of the constant B; it is about $-12\,\mathrm{km\,s^{-1}\,kpc^{-1}}$.

Problem 3.21 Use the result of Problem 2.17 to show that

$$\langle v_y^2 \rangle = -\frac{B}{A - B} \langle v_x^2 \rangle,$$

so that for a flat rotation curve we expect $\langle v_x^2 \rangle / \langle v_y^2 \rangle = 2$. Results from larger studies give this ratio as 2.2; if $A = 14.8 \pm 0.8\,\mathrm{km\,s^{-1}\,kpc^{-1}}$, what is B?

Figure 3.11 and Table 2.1 show that older stars have larger random speeds. But why should the orbits change within a few gigayears, when the relaxation time of Equation 3.55 is $\sim 10^{13}$ yr near the Sun? Clumps of stars and gas in the spiral arms have pulled on passing starts, each time tugging them further from a circular orbit.

The observant reader will have noticed that, in averaging, we did not take account of any radial variations in the density of stars. In fact, the stellar density is higher in the inner Galaxy, so that near the Sun we see more stars with guiding centers at smaller radii than stars that visit us from the outer Galaxy. The majority of stars will be on the outer parts of their epicycles, with $x > 0$; so, according to Equation 3.76, the average tangential motion of stars near the Sun should fall behind the circular velocity. This prediction is borne out in Figure 3.11 and Table 2.1; the average $\langle v_y \rangle$ is negative, an effect known as *asymmetric drift*. The drift is stronger for groups of older stars, with larger random speeds, since their orbits deviate further from circular motion.

Problem 3.22 Show from Equation 3.71 that, within a spherical galaxy of constant density, $\kappa = 2\Omega$, and the Oort constants are $A = 0$ and $B = -\Omega$. For the 'dark-halo' potential of Equation 2.19, find $\Omega(r)$ and $\kappa(r)$. Check that they agree at small radii with those for a uniform sphere of density $V_H^2/(4\pi G a_H^2)$, and that $\kappa \to \sqrt{2}\Omega$ as r becomes large. Plot Ω, κ, and $\Omega - \kappa/2$ against radius for $0 < r < 5a_H$. Show that $\Omega - \kappa/2$ approaches zero both as $r \to 0$ and as r becomes large. We will see in Section 5.5 that this is why two-armed spirals are so prominent in galaxy disks.

Problem 3.23 We saw in Section 2.3 that the Sun has $v_x \approx -10\,\mathrm{km\,s^{-1}}$ and $v_y \approx 5\,\mathrm{km\,s^{-1}}$; how do we know that its guiding center radius $R_g > R_0$? Assuming the Milky Way's rotation curve to be roughly flat, with $V(R) = R\Omega(R) = 200\,\mathrm{km\,s^{-1}}$ and $R_0 = 8\,\mathrm{kpc}$, find κ and Oort's constant B. Use Equations 3.70 and 3.76 to show that the extent of the Sun's radial excursions is $X = 0.35\,\mathrm{kpc}$, and that $R_g \approx 8.2\,\mathrm{kpc}$.

3.4 The collisionless Boltzmann equation

In the last section, we looked at the orbit of an individual star in the Galaxy's gravitational field. We can also describe the stars in a galaxy as we usually describe atoms in a gas: not by following the path of each atom, but by asking about the density of atoms in a particular region and about their average motion. For simplicity, we assume here that all the stars have the same mass m.

The *distribution function* $f(\mathbf{x}, \mathbf{v}, t)$ gives the probability density in the six-dimensional *phase space* of $(\mathbf{x}, \mathbf{v})$. The average number of particles (stars or atoms) in a cube of sides Δx, Δy, and Δz centred at $\mathbf{x}$, that have x velocity between v_x and $v_x + \Delta v_x$, y velocity between v_y and $v_y + \Delta v_y$, and z velocity between v_z and $v_z + \Delta v_z$, is

$$f(\mathbf{x}, \mathbf{v}, t)\Delta x\, \Delta y\, \Delta z\, \Delta v_x\, \Delta v_y\, \Delta v_z. \qquad (3.79)$$

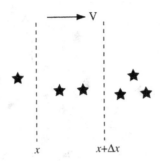

Fig. 3.12. Flow in and out of the region between x and $x + \Delta x$ is described by the equation of continuity.

The *number density* $n(\mathbf{x}, t)$ at position $\mathbf{x}$ is the integral over velocities

$$n(\mathbf{x}, t) \equiv \int_{-\infty}^{\infty} \int_{-\infty}^{\infty} \int_{-\infty}^{\infty} f(\mathbf{x}, \mathbf{v}, t) dv_x \, dv_y \, dv_z. \tag{3.80}$$

Averages such as the mean velocity $\langle \mathbf{v}(\mathbf{x}, t) \rangle$ are also given by integrals:

$$\langle \mathbf{v}(\mathbf{x}, t) \rangle \, n(\mathbf{x}, t) \equiv \int_{-\infty}^{\infty} \int_{-\infty}^{\infty} \int_{-\infty}^{\infty} \mathbf{v} f(\mathbf{x}, \mathbf{v}, t) dv_x \, dv_y \, dv_z. \tag{3.81}$$

We want to find equations to relate changes in the density and the distribution function, as stars move about in the Galaxy, to the gravitational potential $\Phi(\mathbf{x}, t)$. For simplicity, we look at stars moving only in one direction, x. At time t, the number of stars between x and $x + \Delta x$ in the 'box' of Figure 3.12 is $n(x, t)\Delta x$. Suppose that these stars move at speed $v(x) > 0$; how does $n(x)$ change with time? After a time Δt, all the stars that are now between $x - v(x)\Delta t$ and x will have entered the box, while those now within distance $v(x + \Delta x)\Delta t$ of the end will have left it. So the average number of stars in the box changes according to

$$\Delta x[n(x, t + \Delta t) - n(x, t)] \approx n(x, t)v(x)\Delta t - n(x + \Delta x, t)v(x + \Delta x, t)\Delta t. \tag{3.82}$$

Taking the limits $\Delta t \to 0$ and $\Delta x \to 0$ gives us

$$\frac{\partial n}{\partial t} + \frac{\partial (nv)}{\partial x} = 0. \tag{3.83}$$

This is the *equation of continuity*; it must hold if no stars are destroyed so that they disappear from our bookkeeping, and no extra stars are added. For example, if $v > 0$ and $\partial n/\partial x > 0$, as in Figure 3.12, the density of stars in our box must fall with time.

The *collisionless Boltzmann equation* is like the equation of continuity, but it allows for changes in velocity and relates the changes in $f(\mathbf{x}, \mathbf{v}, t)$ to the forces

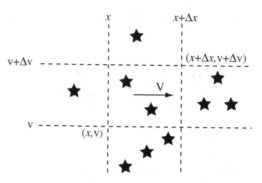

Fig. 3.13. Flow in and out of a box in the phase space (x, v) is described by the collisionless Boltzmann equation.

acting on individual stars. To derive it, we assume that two-body encounters are unimportant, so that the acceleration $d\mathbf{v}/dt$ of an individual star depends only on the smoothed potential $\Phi(\mathbf{x}, t)$. In Figure 3.13, we look at stars in the center box; these lie between x and $x + \Delta x$ and move at speeds between v and $v + \Delta v$. After a time Δt, we again find that stars now between $x - v\,\Delta t$ and x will have entered the box, while those now within distance $v\,\Delta t$ of the end have left it. Here we have specified x and v independently, so v does not depend on x. The number of stars in the box has increased by approximately

$$\Delta v\,\Delta t[vf(x, v, t) - vf(x + \Delta x, v, t)] \approx -v\,\Delta x\,\Delta v\,\Delta t\,\frac{\partial f}{\partial x}. \qquad (3.84)$$

But the number of stars in the center box also changes because the stars' speeds are altered by the applied forces. Suppose that $dv/dt > 0$, so that stars are all being accelerated in the x direction. Then, after time Δt, they will all be moving faster by approximately $\Delta t \cdot dv/dt$. Stars now moving with speeds between v and $v - \Delta t \cdot dv/dt$ will have come into the center box, because they will be moving at speeds faster than v, while those with speeds now just below $v + \Delta v$ will have left it. In total, the center box has gained a number of stars given by

$$\Delta x\,\Delta v[f(x, v, t + \Delta t) - f(x, v, t)]$$
$$\approx -v\,\Delta x\,\Delta v\,\Delta t\,\frac{\partial f}{\partial x} + \Delta x\,\Delta t\left[\frac{dv}{dt}f(x, v, t) - \frac{dv}{dt}f(x, v + \Delta v, t)\right]. \qquad (3.85)$$

In the limit that all the Δ terms are small, we have

$$\frac{\partial f}{\partial t} + v\frac{\partial f}{\partial x} + \frac{dv}{dt}(x, v, t) \cdot \frac{\partial f}{\partial v} = 0.$$

But a star's acceleration does not depend on how fast it is moving, only on its position: $dv/dt = -\partial\Phi(x, t)/\partial x$. Thus we have the one-dimensional collisionless Boltzmann equation:

$$\frac{\partial f}{\partial t} + v\frac{\partial f}{\partial x} - \frac{\partial\Phi}{\partial x}(x, t) \cdot \frac{\partial f}{\partial v} = 0. \qquad (3.86)$$

In three dimensions, the collisionless Boltzmann equation takes the form

$$\frac{\partial f(\mathbf{x}, \mathbf{v}, t)}{\partial t} + \mathbf{v} \cdot \nabla f - \nabla \Phi \cdot \frac{\partial f}{\partial \mathbf{v}} = 0. \tag{3.87}$$

Equation 3.87 holds if stars are neither created nor destroyed, and if they also change their positions and velocities smoothly. Close encounters between stars can alter their velocities much faster than their motion changes in the smoothed potential. When these are important, we include their effects as an extra 'collisional' term on the right-hand side.

Often, we do not solve the collisionless Boltzmann equation explicitly, but rather integrate to take velocity-moments. Integrating Equation 3.86 over velocity, and using the definitions 3.80 and 3.81, we find

$$\frac{\partial n(x, t)}{\partial t} + \frac{\partial}{\partial x}(n(x, t)\langle v(x, t)\rangle) - \frac{\partial \Phi}{\partial x}(x, t)[f]_{-\infty}^{\infty} = 0. \tag{3.88}$$

When $f(x, v, t)$ is well behaved, tending to zero as $|v| \to \infty$, the last term is zero. We arrive back at Equation 3.83, with the velocity $v = \langle v(x, t)\rangle$.

Multiplying Equation 3.86 by v and then integrating gives

$$\frac{\partial}{\partial t}[n(x, t)\langle v(x, t)\rangle] + \frac{\partial}{\partial x}[n(x, t)\langle v^2(x, t)\rangle] = -n(x, t)\frac{\partial \Phi}{\partial x}; \tag{3.89}$$

here the average of the squared velocity $\langle v^2 \rangle$ is defined just as for $\langle v \rangle$, and we have integrated by parts, assuming that $fv \to 0$ as $|v| \to \infty$. The velocity dispersion σ is defined by $\langle v^2(x, t)\rangle = \langle v(x, t)\rangle^2 + \sigma^2$; rearranging terms with the help of Equation 3.88 and dividing by n, we have

$$\frac{\partial \langle v \rangle}{\partial t} + \langle v \rangle \frac{\partial \langle v \rangle}{\partial x} = -\frac{\partial \Phi}{\partial x} - \frac{1}{n}\frac{\partial}{\partial x}[n\sigma^2(x, t)]. \tag{3.90}$$

This is analogous to Euler's equation of fluid mechanics, with the term in σ^2 replacing the pressure force $-\partial p/\partial x$. In a fluid, the equation of state specifies the pressure at a given density and temperature. For a stellar system there is no such relation; but sometimes we can make progress by using measured quantities, as in the next subsection.

3.4.1 Mass density in the Galactic disk

We can use the collisionless Boltzmann equation and the observed vertical motions of stars to find the mass in the Galactic disk near the Sun. We select a *tracer* population of stars (for example, the K dwarf stars) and measure its density $n(z)$ at height z above the disk's midplane. Our coordinates are now (z, v_z) instead of (x, v). We assume that the potential $\Phi(z)$ does not change with time, and that our stars are *well mixed*, so that the distribution function f and the density n are also time-steady. Looking high above the plane, $\langle v_z \rangle n(z) \to 0$; so Equation 3.88 tells us that the mean velocity $\langle v_z \rangle = 0$ everywhere. In Equation 3.90, we write σ_z for the velocity dispersion; the terms on the left-hand side vanish, giving

$$\frac{d}{dz}\left[n(z)\sigma_z^2\right] = -\frac{\partial \Phi}{\partial z} n(z). \tag{3.91}$$

So, if we measure how the density of our stars and their velocity dispersion change with z, we can find the vertical force at any height.

Poisson's equation, Equation 3.9, relates that force to the mass density $\rho(\mathbf{x})$ of the Galaxy. Assuming that the Milky Way is axisymmetric, so that ρ and Φ depend only on (R, z), we have

$$4\pi G\rho(R, z) = \nabla^2\Phi(R, z) = \frac{\partial^2 \Phi}{\partial z^2} + \frac{1}{R}\frac{\partial}{\partial R}\left(R\frac{\partial \Phi}{\partial R}\right). \tag{3.92}$$

The density $\rho(R, z)$ here includes *all* the mass in the disk: luminous stars, gas, white dwarfs, brown dwarfs, black holes, and dark matter. Writing $\partial\Phi/\partial R = V^2(R)/R$, where $V(R)$ is the rotation speed in a circular orbit at radius R, we have

$$4\pi G\rho(R, z) = \frac{d}{dz}\left\{-\frac{1}{n(z)}\frac{d}{dz}\left[n(z)\sigma_z^2\right]\right\} + \frac{1}{R}\frac{d}{dR}[V^2(R)]. \tag{3.93}$$

Near the Sun, $V(R)$ is nearly constant, so the last term is very small. The density $\rho(R_0, 0)$ in the midplane of the disk has recently been estimated from the velocities of nearby A stars, measured with the Hipparcos satellite, to be in the range $(70–100)\,\mathcal{M}_\odot$ per $1000\,\mathrm{pc}^3$.

To find the volume density ρ, the observationally determined quantity $n(z)$ has to be differentiated twice, which amplifies small errors. We can more accurately determine the surface mass density $\Sigma(<z)$ within some distance z of the midplane. Assuming that the disk is symmetric about $z = 0$, we integrate Equation 3.93 to find

$$2\pi G\Sigma(<z) \equiv 2\pi G\int_{-z}^{z}\rho(z')dz' \approx -\frac{1}{n(z)}\frac{d}{dz}\left[n(z)\sigma_z^2\right]. \tag{3.94}$$

Jan Oort in 1932 was the first to try this, measuring $n(z)$ for bright F dwarfs and K giants. He assumed that σ_z did not vary with height, and found that $\Sigma(<700\,\mathrm{pc}) \approx 90\mathcal{M}_\odot\,\mathrm{pc}^{-2}$. But, beyond a kiloparsec from the midplane, the function he derived for $\Sigma(<z)$ began to *decrease*, indicating anti-gravitating matter with $\rho(z) < 0$, or, more probably, a failure of the hypothesis that σ_z is constant.

Recent work with fainter K dwarf stars, which are more numerous and more evenly spread out in space, indicates that σ_z increases with height. At $250\,\mathrm{pc}$ from the plane, the vertical dispersion $\sigma_z \approx 20\,\mathrm{km\,s^{-1}}$, growing to $30\,\mathrm{km\,s^{-1}}$ at

$z \sim 1$ kpc; at greater heights, a larger fraction of the stars belongs to the thick disk. Taking account of the increasing dispersion, these studies find $\Sigma(< 1100\,\text{pc}) \approx$ (70–80 $\mathcal{M}_\odot\,\text{pc}^{-2}$. Some of this mass must be in the halo, so the surface density of the disk itself is probably between $50\mathcal{M}_\odot$ and $60\mathcal{M}_\odot\,\text{pc}^{-2}$.

We can compare this dynamical estimate with the mass that has been observed in gas and stars. Near the Sun, molecular gas probably amounts to $\sim 2\mathcal{M}_\odot\,\text{pc}^{-2}$. There is roughly $8\mathcal{M}_\odot\,\text{pc}^{-2}$ of neutral atomic hydrogen, and about $2\mathcal{M}_\odot\,\text{pc}^{-2}$ of ionized gas, though all these numbers are uncertain by at least 30%. Main-sequence stars more massive than M dwarfs, and the giants, are easily counted in surveys; making allowance for the hard-to-find low-mass stars and stellar remnants such as white dwarfs and neutron stars gives a total of about $(25–40)\mathcal{M}_\odot\,\text{pc}^{-2}$ in stars. Thus the mass in gas and stars is $(40–55)\mathcal{M}_\odot\,\text{pc}^{-2}$. We see that the disk does not contain much of the Galaxy's 'dark matter'.

Problem 3.24 Use the divergence theorem to show that the potential at height z above a uniform sheet of matter with surface density Σ is

$$\Phi(\mathbf{x}) = 2\pi G \Sigma |z|. \qquad (3.95)$$

Show that the vertical force does not depend on z, and check that $\nabla^2 \Phi = 0$ when $z \neq 0$. Suppose that the mass of the Galaxy was all in a flat uniform disk; use Equation 3.91 to find the density $n(z)$ of K dwarfs, assuming that they have a constant velocity dispersion σ_z. As in the Earth's atmosphere, where the acceleration of gravity is also nearly independent of height, show that $n(z)$ drops by a factor of e as $|z|$ increases by $h_z = \sigma_z^2/(2\pi G\Sigma)$. Estimate h_z near the Sun, taking $\sigma_z = 20\,\text{km s}^{-1}$.

3.4.2 Integrals of motion, and some of their uses

Often we are interested in solutions of the collisionless Boltzmann equation that describe stars moving in an unchanging gravitational potential $\Phi(\mathbf{x})$, such that their distribution function is also constant. Then, it is frequently useful to write $f(\mathbf{x}, \mathbf{v})$ in terms of *integrals of the motion*. These are functions $\mathcal{I}(\mathbf{x}, \mathbf{v})$ of a star's position $\mathbf{x}$ and velocity $\mathbf{v}$ that remain constant along its orbit. One example is the energy per unit mass, $E(\mathbf{x}, v) = \mathbf{v}^2/2 + \Phi(\mathbf{x})$, which is an integral of the motion whenever the potential $\Phi(\mathbf{x})$ does not depend on time. In an axisymmetric potential $\Phi(R, z, t)$, the z component of the angular momentum L_z is an integral; in a spherical potential, the total angular momentum $\mathbf{L}$ is an integral.

For any function $\mathcal{I}$ that is constant along the orbit, we have

$$\frac{d}{dt}\mathcal{I}(\mathbf{x}, \mathbf{v}) \equiv \frac{\partial \mathbf{x}}{\partial t} \cdot \nabla \mathcal{I} + \frac{\partial \mathbf{v}}{\partial t} \cdot \frac{\partial \mathcal{I}}{\partial \mathbf{v}} = 0, \text{ or } \mathbf{v} \cdot \nabla \mathcal{I} - \nabla \Phi \cdot \frac{\partial \mathcal{I}}{\partial \mathbf{v}} = 0. \qquad (3.96)$$

This looks suspiciously like Equation 3.87, which we can write as

$$\frac{\mathrm{d}f}{\mathrm{d}t} \equiv \frac{\partial f}{\partial t} + \mathbf{v} \cdot \frac{\partial f}{\partial \mathbf{x}} + \frac{\partial \mathbf{v}}{\partial t} \cdot \frac{\partial f}{\partial \mathbf{v}} = 0. \tag{3.97}$$

So the phase-space density $f(\mathbf{x}, \mathbf{v}, t)$ around any particular star remains constant along its orbit. Where the density $n(\mathbf{x})$ becomes higher, the dispersion in velocities of the surrounding stars must increase, and any function $f(\mathbf{x}, \mathbf{v})$ that is a time-independent solution to the collisionless Boltzmann equation is itself an integral of the motion. Conversely, *if $f(\mathcal{I}_1, \mathcal{I}_2, \ldots)$ is any function of integrals of the motion $\mathcal{I}_1, \mathcal{I}_2, \ldots,$ then f is a steady-state solution of the equations of motion.* Often it is easy to write down at least some of the integrals of motion, and this is one way to get started on constructing time-steady solutions of the collisionless Boltzmann equation.

For example, Equation 3.68 tells us that, for disk stars on nearly circular orbits, motion perpendicular to the Galactic disk is independent of that in the disk plane, so the energy of vertical motion $E_z = \Phi(R_0, z) + v_z^2/2$ is an integral of motion. If we select some tracer population of stars that are easy to find and measure, and which are also well mixed so that their distribution function $f(z, v_z)$ is not changing with time, we can write

$$f(z, v_z) = f(E_z) = f\left(\Phi(R_0, z) + \frac{1}{2}v_z^2\right). \tag{3.98}$$

If we guessed at $f(E_z)$ and the potential $\Phi(R_0, z)$, we could integrate f over velocities to find the corresponding density $n(z)$ and the velocity dispersion σ_z at any height z. Conversely, if we measured $n(z)$ and guessed at $f(E_z)$, we could find the potential $\Phi(R_0, z)$ by using Equation 3.91. For example, we could take

$$f(E_z) = \frac{n_0}{\sqrt{2\pi\sigma^2}} \exp(-E_z/\sigma^2) \qquad \text{for } E_z < 0; \tag{3.99}$$

stars with $E_z \geq 0$ would escape, so we must set $f = 0$ there. Integrating over v_z shows the density n and velocity dispersion σ_z to be

$$n(z) = n_0 \exp[-\Phi(R_0, z)/\sigma^2], \quad \text{and} \quad \sigma_z = \sigma. \tag{3.100}$$

If $n(z)$ and σ^2 are measured, we can calculate the potential $\Phi(R_0, z)$ by using Equation 3.91; if Φ and σ^2 are known, we can find the corresponding density $n(z)$.

If we also have some reason to think that the stars described by the distribution function f provide all the gravitational force, then the density $n(\mathbf{x}, t)$

found by integrating $f(\mathbf{x}, \mathbf{v}, t)$ over velocities must be equal to the density $\rho(\mathbf{x}, t)$ in Poisson's Equation 3.9. In this case, we say that f provides a *self-consistent model* for the system. Many different self-consistent models can give rise to the same gravitational potential $\Phi(\mathbf{x}, t)$; the density $n(\mathbf{x}, t)$ is the same for all of these, but the form of f, and hence the velocities of the stars, will be different.

Problem 3.25 For stars moving vertically in the Galactic disk, suppose the distribution function $f(z, v_z)$ to be given by Equation 3.99. When the disk is symmetric about the plane $z = 0$, then $\mathrm{d}\Phi(z)/\mathrm{d}z = 0$ at $z = 0$, and we can choose $\Phi(0) = 0$ too. Find the integral giving the density of stars $n(z)$: what is $n(0)$?

To construct a self-consistent model, let $\Phi(z) = \sigma^2 \phi(z)$, and let the average mass of the stars be m; show from Poisson's equation that

$$2\,\mathrm{d}^2\phi/\mathrm{d}y^2 = e^{-\phi}, \qquad \text{where } y = z/z_0 \text{ and } z_0^2 = \sigma^2/(8\pi G m n_0).$$

Integrate this once to find $\mathrm{d}\phi/\mathrm{d}y$, and then again (substituting $u = e^{-\phi/2}$) to find $\phi(y)$ and hence $\Phi(z)$. Show that the number density of stars is $n(z) = n_0 \operatorname{sech}^2[z/(2z_0)]$. What is its approximate form at large $|z|$?

In a spherically symmetric gravitational potential $\Phi(r)$, *any* function $f(E, \mathbf{L})$ of the energy E and angular momentum $\mathbf{L}$ per unit mass that does not include any unbound stars will describe one possible steady distribution of stars in that gravitational field. If we choose $f = f(E)$ then the random speeds are *isotropic*, the same in all directions.

Problem 3.26 When the distribution function for stars in a spherical system depends only on their energy, so that $f(\mathbf{x}, \mathbf{v}, t) = f(E)$, explain why the velocity dispersion is the same in all directions:

$$\langle v_x^2 \rangle \equiv \frac{1}{n(\mathbf{x})} \int f\left[\Phi(\mathbf{x}) + \frac{\mathbf{v}^2}{2}\right] v_x^2 \,\mathrm{d}v_x \,\mathrm{d}v_y \,\mathrm{d}v_z = \langle v_y^2 \rangle = \langle v_z^2 \rangle. \quad (3.101)$$

If we choose a form for $f(E)$, we can then calculate the density $n(r)$ as a function of the potential $\Phi(r)$. Then we use Poisson's equation to find which potentials $\Phi(r)$ lead to a self-consistent model, with $n(r) \propto \rho(r)$. For the Plummer sphere, we can combine Equations 3.11 and 3.12 to find

$$\rho_{\mathrm{P}}(r) = -\frac{3a_{\mathrm{p}}^2}{4\pi\, G^5 \mathcal{M}^4}\, \Phi_{\mathrm{P}}^5(r). \quad (3.102)$$

Taking the distribution function

$$f(E) = k(-E)^{N-3/2} \quad \text{for } E < 0, \tag{3.103}$$

where k and N are constants, the density is given by

$$n(r) = \int_{v=0}^{E=0} k \left[-\Phi(\mathbf{x}) - \frac{\mathbf{v}^2}{2} \right]^{N-3/2} 4\pi v^2 \, dv. \tag{3.104}$$

Defining a new variable θ by $v^2 = -2\Phi(r)\cos^2\theta$, we can integrate to show that $n(r) \propto (-\Phi)^N$. Thus the distribution function $f_P(E) = k(-E)^{7/2}$ gives a self-consistent model for the Plummer sphere. Its mass $\mathcal{M}$ is proportional to the constant k, while the depth of the central potential $\Phi(r = 0)$ sets the radius a_P.

Problem 3.27 Fill in the algebraic steps needed to derive Equations 3.102–3.104.

By analogy with expression 3.99, we might also try the *isothermal* distribution function

$$f_I(E) = \frac{n_0}{(2\pi\sigma^2)^{3/2}} \exp\left\{ -\left[\Phi(r) + \frac{v^2}{2} \right] \Big/ \sigma^2 \right\} \quad \text{for } E < 0, \tag{3.105}$$

as a first guess for representing a spherical galaxy or star cluster. As in the planar case, Equation 3.100 gives us $n(r)$. Putting this on the right-hand side of Poisson's equation, we get

$$4\pi G\rho(r) = \frac{1}{r^2} \frac{d}{dr} \left(r^2 \frac{d\Phi}{dr} \right) = 4\pi G m n_0 \exp\left[-\frac{\Phi(r)}{\sigma^2} \right]. \tag{3.106}$$

To find $\Phi(r)$, we must integrate this equation outward from $r = 0$. If the potential is smooth at the center, the radial force there must be zero, so we must start with $d\Phi/dr = 0$. But, however we choose $\Phi(r = 0)$, we find that the total mass is infinite. We should have expected this result, since, if the mass had been finite, the escape speed v_e of Equation 3.28 would drop below the average random speed σ at some radius. But that would contradict our assumption that all the stars were bound to the system.

Problem 3.28 Show that, in the potential Φ_{SIS} of Equation 3.14, the density $\rho(r)$ corresponding to $f_I(E)$ is exactly equal to ρ_{SIS}, if $\sigma^2 = 2\pi G\rho_0 r_0^2$, and all the stars have mass m so that $\rho_0 \equiv m n_0 = \rho(r_0)$. The distribution function $f_I(E)$ gives a self-consistent model for the singular isothermal sphere; show that the average random speed σ is $1/\sqrt{2}$ times the speed in a circular orbit.

Only if we reduce the number of stars with energies close to the escape energy does f give a self-consistent model with finite mass. The *King models*, sometimes called 'lowered isothermal models', provide a good description of non-rotating globular clusters and open clusters:

$$f_{\mathrm{K}}(E) = \frac{n_0}{(2\pi\sigma^2)^{3/2}} \exp\left[-\left(\Phi(r) + \frac{v^2}{2}\right)\middle/ \sigma^2 - 1\right] \quad \text{for } E < 0. \quad (3.107)$$

When we integrate this to find $n(r)$, and then solve Poisson's equation, the term '-1' acts to reduce the number of stars with high kinetic energy in the outer regions. The average random speed decreases and the density drops abruptly to zero at some *truncation radius* corresponding to the outer radius r_{t} of Table 2.3 and Figure 3.7.

Including a term involving angular momentum alters the balance between stars on nearly circular orbits and those that follow eccentric orbits. For example, the distribution function

$$f_{\mathrm{A}}(E, L) = f_{\mathrm{K}}(E)\exp\left[-\mathbf{L}^2/\left(2\sigma^2 r_{\mathrm{a}}^2\right)\right] \quad (3.108)$$

leads to a density $n(r)$ which is not very different from that associated with $f_{\mathrm{K}}(E)$, but it describes a cluster with fewer stars on the near-circular orbits that have high angular momentum. The effect is especially strong in the outer regions, where the angular momentum of a circular orbit is largest; outside the *anisotropy radius* r_{a}, stars have a decided preference for nearly radial orbits. We see evidence for this kind of velocity distribution in some globular clusters and elliptical galaxies.

If the distribution function f depends on only one component of the angular momentum, for example L_z, it can describe a system that is flattened along the z axis. For example, if

$$f(E, L_z) = \tilde{f}(E)L_z^2 \quad \text{for } E > 0, \quad (3.109)$$

for some function $\tilde{f}$, then very few stars have orbits taking them close to the z axis, with $L_z \approx 0$, but relatively many will follow near-circular orbits in the equatorial plane, with large L_z. Postulating that $f = f(E, L_z)$ seems at first sight to be useful for describing the motions of stars in the Galactic disk. Since v_R and v_z enter the expression for f in the same way, stars must have equal random motions in those two directions:

$$n(\mathbf{x})\langle v_z^2 \rangle \equiv \int f\left[\Phi(\mathbf{x}) + \frac{\mathbf{v}^2}{2}, Rv_\phi\right]v_z^2 \, dv_R \, dv_\phi \, dv_z = n(\mathbf{x})\langle v_R^2 \rangle. \quad (3.110)$$

But we saw in Section 2.2 that, near the Sun, disk stars have larger random speeds in R than they have in the vertical direction z. They cannot have a distribution $f(E, L_z)$, but f must depend on a *third integral* of motion. Curiously, it can be proven that, in a general axisymmetric potential $\Phi(R, z)$, there is *no* function of position and velocity other than E and L_z that is conserved along a star's orbit. This dynamical puzzle will resurface in Section 6.2, when we discuss stellar motions in elliptical galaxies.

Further reading: J. Binney and S. Tremaine, 1987, *Galactic Dynamics* (Princeton University Press, Princeton, New Jersey), Section 4.4.

4

Our backyard: the Local Group

The Local Group contains roughly three dozen galaxies within a sphere about a megaparsec in radius, centred between the Milky Way and our nearest large neighbor, the Andromeda galaxy M31. Figure 4.1 shows the brighter members. The three most prominent are M31, the Milky Way, and M33; according to the classification of Section 1.3, these are all spiral galaxies. M31 is about 50% more luminous than the Milky Way, while M33 is only 20% as luminous. Between them, these three galaxies emit 90% of the visible light of the Local Group. The only elliptical galaxy is M32, a satellite to M31. The remaining systems are irregular galaxies, or the even less luminous dwarf irregulars, dwarf ellipticals, and dwarf spheroidals. Many of these smaller galaxies are in orbit either around the Milky Way or around M31.

Table 4.1 lists known and probable members of the Local Group within a megaparsec of the Sun. The apparent brightness of each member is generally known to within 10%, except for the Milky Way, where our location within the disk presents special problems. Distances to Local Group galaxies are derived by picking out individual stars, measuring their apparent brightness, and estimating their true luminosity, using methods such as the period–luminosity relation for Cepheid variables. In this way, distances to the ten or so brightest galaxies can be measured to within 10%. But fewer stars are available in the less luminous galaxies, so the distances are less certain; for some dwarfs, they are known to no better than a factor of two.

Galaxies are not scattered at random within the Local Group; many small members are satellites of M31 or the Milky Way. The right-hand part of Figure 4.2 shows that most of the Milky Way's 11 known satellites lie close to a single plane – they may have formed from a single gas cloud captured into an orbit around the Milky Way. The Andromeda galaxy also has its brood of satellites; but many small systems are 'free fliers', remote from any larger galaxy. The Local Group is likely to contain still-undiscovered dwarf galaxies; in particular, galaxies can hide behind dust in the Milky Way's disk. The Local Group contains three spiral galaxies, but only one, small, elliptical. As is typical of groups, it is rich in

Table 4.1 Galaxies of the Local Group within 1 Mpc of the Sun: the Milky Way and its satellites are listed in **boldface**; M31 and its companions are listed in *italics*

Galaxy	Type	d (kpc)	L_V ($10^7 L_\odot$)	$V_r(\odot)$ (km s^{-1})	l (deg)	b (deg)	$\mathcal{M}(\text{HI})$ ($10^6 \mathcal{M}_\odot$)
M31 (NGC 224)	Sb	770	2700	−300	121	−22	5700
Milky Way	Sbc	8	1500	−10	0	0	4000
M33 (NGC 598)	Sc	850	550	−183	134	−31	1500
Large MC	SBm	50	200	274	280	−33	500
Small MC	Irr	63	55	148	303	−44	400
NGC 205	dE	830	40	−244	121	−21	0.4
M32 (NGC 221)	E2	770	40	−205	121	−22	<2.5^c
NGC 6822	dIrr	500	10	−56	25	−18	140^c
IC 10	dIrr	660	16	−344	119	−3	100
NGC 185	dE	620	13	−202	121	−15	0.1
NGC 147	dE	760	12	−193	120	−14	None
Sagittarius	dSph	30	8	170	6	−14	None
IC 1613 (DDO 8)	dIrr	715	6	−233	130	−61	60
WLM (DDO 221)	dIrr	950	5	−120	76	−74	60
Pegasus (DDO 216)	dIrr/dSph	760	1	−182	95	−44	3
Fornax	dSph	140	1.5	53	237	−66	<0.7
Sagittarius DIG	dIrr	1050	0.7	−78	21	−16	9
And I	dSph	790	0.5	−380	122	−25	None
Leo I (DDO 74)	dSph	270	0.5	285	226	49	None
And VII/Cas dSph	dSph	760	0.5	−307	110	−10	
Leo A (DDO 69)	dIrr	800	0.4	20	197	52	8
And VI/Peg dSph	dSph	775	0.3	−354	106	−36	
And II	dSph	680	0.2	−188	129	−29	
Sculptor	dSph	88	0.2	107	288	−83	≲0.1^c
LGS3 (Pisces)	dIrr/dSph	620	0.13	−286	127	−41	0.2
Aquarius (DDO 210)	dIrr/dSph	950	0.1	−137	34	−31	3
And III	dSph	760	0.1	−355	119	−26	None
Phoenix	dIrr/dSph	405	0.09	56	272	−69	∼0.2
Cetus	dSph	775	0.09		101	−73	
Leo II (DDO 93)	dSph	205	0.06	76	220	67	None
Tucana	dSph	870	0.06		323	−47	None
Sextans	dSph	85	0.05	225	244	42	None
Draco (DDO 216)	dSph	80	0.05	−293	86	35	None
Carina	dSph	95	0.04	223	260	−22	None
And V	dSph	810	0.04	−403	126	−15	
Ursa Minor	dSph	70	0.03	−247	105	45	None
And IX	dSph	790	0.02	−210	123	−20	
Ursa Major	dSph	∼100	0.004	−52	160	54	

Note: d is measured from the Sun; $V_r(\odot)$ is radial velocity with respect to the Sun; 'no' HI means $< 10^5 \mathcal{M}_\odot$.
^{c} HI is confused with M31's disk (M32), Galactic emission (NGC 6822), or the Magellanic Stream (Sculptor). Andromeda IV is an irregular galaxy in the background; Andromeda VIII is probably a stellar concentration in M31's disk. Additional dwarf members are still being discovered.

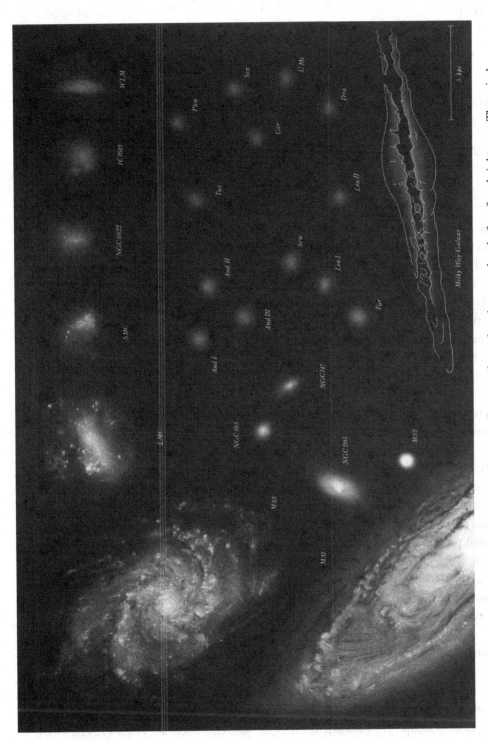

Fig. 4.1. Galaxies of the Local Group, shown to the same linear scale, and to the same level of surface brightness. The spiral and irregular galaxies stand out clearly, while the dwarf spheroidals are barely visible – B. Binggeli.

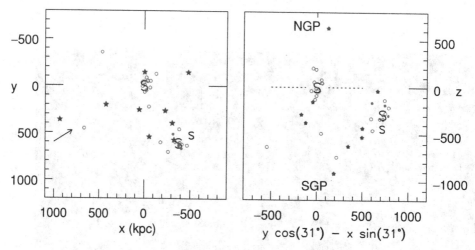

Fig. 4.2. The Local Group: our Milky Way is at the origin. Spirals are designated S; aster-isks show the Magellanic Clouds; filled stars mark irregular galaxies; circles are ellipticals or dwarf ellipticals (filled) and dwarf spheroidals (open). Left, positions projected onto the Galactic plane; axis x points to the Galactic center, y in the direction of the Sun's orbital motion. The arrow shows the direction of view in the right panel. Right, view perpendicular to the plane containing M31 and axis z toward the north Galactic pole; the dotted line marks the Galactic midplane. Many of the Milky Way's satellites, including the Magellanic Clouds, lie near a single plane.

'late-type' galaxies, spirals and irregulars, and poor in the 'early-type' giant ellip-ticals and S0 galaxies.

In the Local Group, mutual gravitational attraction is strong enough to have overcome the general expansion of the Universe. Allowing for the Sun's motion around the Galaxy, we find that the Milky Way and the Andromeda Galaxy are approaching each other instead of receding, closing at about $120 \, \mathrm{km \, s^{-1}}$. We can measure proper motions only for the Milky Way's immediate satellite galaxies (what proper motion corresponds to $V_t = 120 \, \mathrm{km \, s^{-1}}$ at a distance of $100 \, \mathrm{kpc}$?). Even these are difficult, because we must use faint distant quasars and galaxies to define our nonmoving frame of reference. But the radial velocities are easily found. They are almost all within $60 \, \mathrm{km \, s^{-1}}$ of the common motion of the Milky Way and M31; the Local Group galaxies have too little kinetic energy to escape.

Just as stars near the Sun are concentrated in the Milky Way's disk, so the galaxies within about 30 Mpc form a roughly flattened distribution. They lie near the *supergalactic plane*, approximately perpendicular to the Milky Way's disk in the direction $l = 140°$ and $l = 320°$. We discuss the supergalactic coordinate system in Section 7.1. Figure 4.3 shows the main concentrations of galaxies near the Local Group. Well-known clusters of galaxies such as Virgo (15–20 Mpc distant) and Coma (at $70h^{-1} \, \mathrm{Mpc}$) are seen to form part of larger complexes. Most of the Universe's volume is nearly empty of galaxies.

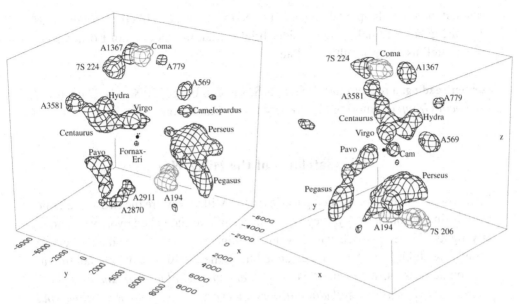

Fig. 4.3. Galaxy concentrations within $80h^{-1}$ Mpc of the Local Group: the 'mesh' encloses regions where the density is $\gtrsim 50\%$ above average. The filled dot gives our position at the origin, and axes x, y, z are as in Figure 4.2. Distance d to each galaxy is calculated from Hubble's law: axes show $H_0 d$ in km s^{-1}. Objects from Abell's catalogue of galaxy clusters are denoted A. Left, view from $(l, b) = 35°, 25°$, perpendicular to the supergalactic X–Y plane; right, view from $(l, b) = 125°, 25°$, looking nearly along that plane – M. Hudson 1993 *MNRAS* **265**, 43.

About half of all galaxies are found in clusters or groups a few megaparsecs across, and are dense enough that their gravity has by now halted the cosmological expansion. The other half lie in looser clouds and associations, within large walls and long filaments such as those in Figure 8.3; these structures are collapsing, or at least are expanding much more slowly than the Universe as a whole. Just as the Sun is a typical star, intermediate in its mass and luminosity, so the Local Group is a typical galactic environment: it is less dense than a galaxy cluster like Virgo or Coma, but contains enough mass to bind the galaxies together.

The concentration of galaxies in our Local Group presents an opportunity to study a variety of systems at close range. In particular, we can distinguish, or *resolve*, individual stars in these nearby galaxies. As with Galactic clusters, we can compare their color–magnitude diagrams with the predictions of stellar-structure theories, to determine how the *stellar population* has built up. Astronomers also take advantage of the Local Group to gather data on variable stars such as Cepheids, and to study the physical processes affecting galaxies in close proximity.

We begin this chapter with a discussion of the satellites of our Galaxy, considering some of the problems facing a small galaxy in orbit around a larger system. Section 4.2 compares the three spiral galaxies of the Local Group; in Section 4.3 we consider how these galaxies might have formed, and how elements heavier than helium are built up by their stars. In Section 4.4 we discuss the various kinds of dwarf galaxies; finally, we consider briefly what the motions of Local Group galaxies tell us about its ultimate fate.

Further reading: a recent graduate text is S. van den Bergh, 2000, *The Galaxies of the Local Group* (Cambridge University Press, Cambridge, UK).

4.1 Satellites of the Milky Way

The most prominent companion galaxies to the Milky Way are the two Magellanic Clouds; in the southern sky, they are easily visible to the naked eye, even among city lights. These gas-rich galaxies are forming new stars and star clusters in abundance. John Herschel, who extended his father William's nebula-hunting to the southern skies, noted in 1851 that *'there are nebulae in abundance, both regular and irregular; globular clusters in every state of condensation; and objects of a nebulous character quite peculiar, which have no analogue in any other region of the heavens. Such is the concentration of these objects* [star clusters] *. . . which very far exceeds the average of any other, even the most crowded part of the nebular heavens'* (*Outlines of Astronomy*, Longmans, London, p. 164). Like the Milky Way, the Magellanic Clouds have stars and star clusters with a wide range of ages. They contain variable stars, which we can compare with those of our Galaxy, and calibrate for use as 'standard candles' in estimating distances to galaxies beyond the Local Group.

By contrast, the Galaxy's dwarf spheroidal companions are so diffuse that they are almost invisible on the sky. These systems of elderly and middle-aged stars contain hardly any gaseous material from which to make fresh stars. The stars of the dwarf spheroidals contain so little mass that some of these small galaxies may be in the process of dissolving, as they are pulled apart by the Milky Way's gravitational field.

4.1.1 The Magellanic Clouds

The *Large Magellanic Cloud* (LMC) measures $15° \times 13°$ on the sky, so its long dimension is about 14 kpc; the *Small Magellanic Cloud* (SMC) covers $7° \times 4°$, extending roughly 8 kpc. The LMC has about 10% of the Milky Way's luminosity, $L \approx 2 \times 10^9 L_\odot$, and is the fourth most luminous member of the Local Group; the SMC is about ten times fainter; see Table 4.1. The LMC, the prototype of the Sm class of 'Magellanic spirals', is basically a flat disk, tilted by $20°–30°$ to the plane of the sky; the rotation speed measured from the HI gas reaches $80 \, \mathrm{km \, s^{-1}}$. It has a strong bar, with only one stubby spiral arm (Figure 4.4). The disk gas does not rotate symmetrically about the bar; instead, the orbits are centred about 0.9 kpc or $1°$ to the northwest of the brightest region. The SMC is very different; it is

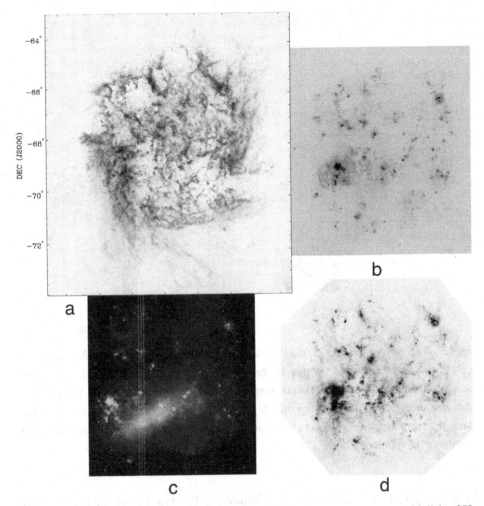

Fig. 4.4. The Large Magellanic Cloud: (a) the extended and fairly symmetric disk of Hɪ gas; (b) in Hα we see hot gas around young massive stars, with 30 Doradus the most prominent bright region; (c) an optical image shows the dense stellar bar and clumps of young stars, with 30 Doradus near the end of the bar, above and to the left; and (d) infrared light at 24 μm shows dust heated by young stars. The Hɪ map is 10° across, or ∼ 8.5 kpc; others are 7° – S. Kim and L. Staveley-Smith; K. Henize (courtesy of the Observatories of the Carnegie Institution of Washington); Spitzer.

an elongated 'cigar' structure seen roughly end-on, with a depth of about 15 kpc along the line of sight. Its stars show no organized motions.

Some astronomers would classify both the Magellanic Clouds as irregular galaxies. They have a profusion of young stars (as shown by the color–magnitude diagram of Figure 4.5), and there is less dust to block this light than in the Milky Way. Thus the Clouds are blue in visible light and very bright in the ultraviolet. Star-forming regions are spread throughout both systems, and they are rich in

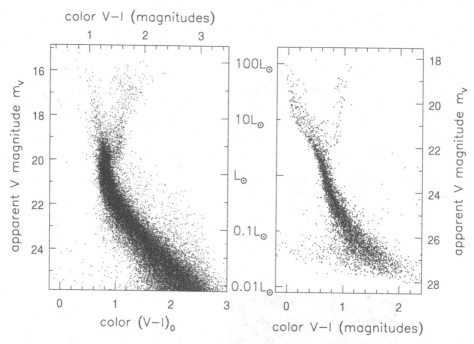

Fig. 4.5. Left, a color–magnitude diagram for stars in the bulge of the Milky Way; the lower scale shows color corrected for the reddening effects of dust. The wide main sequence indicates a range of stellar ages, but no horizontal branch is visible (cf. Figure 2.14). Right, stars in a small patch in the disk of the Large Magellanic Cloud. Note the luminous blue stars; the main sequence is bluer than that in the bulge because the stars are poorer in metals – J. Holtzman.

hydrogen gas, the raw material of star formation. The Hα map of the LMC shows holes, loops, and filaments, which are also present in the HI disk (Figure 4.4). Some of these coincide with sites of recent starbirth, where supernovae and the winds of hot stars have given the surrounding interstellar gas enough momentum to push the cooler HI gas aside, forming a large hot bubble. This morphology is typical of irregular galaxies; see Section 4.4.

Each of the Magellanic Clouds contains several hundred million solar masses of neutral hydrogen. The ratio of the HI mass to the luminosity in blue light (both being measured in solar units) is a useful measure of a galaxy's progress in converting gas into stars. For the Milky Way, $\mathcal{M}(\mathrm{HI})/L_B \approx 0.1$, for the LMC it is about 0.2, and for the SMC and irregular galaxies, $\mathcal{M}(\mathrm{HI})/L_B \sim 1$. Dwarf spheroidal galaxies contain hardly any HI gas.

A 'bridge' of gas, containing young star clusters, connects the two Clouds, while a series of large gas clouds trails beyond the SMC (Figure 4.6). This *Magellanic Stream* wraps a third of the way around the sky, approximately on a Great Circle through $l = 90°$ and $l = 270°$. It contains a further $2 \times 10^8 \mathcal{M}_\odot$ of HI gas.

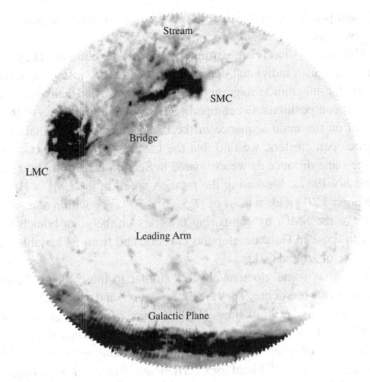

Fig. 4.6. HI in the Magellanic Clouds and the Magellanic Stream. The map is centred at the south celestial pole, extending to $\delta = -62°$; right ascension 0^h is at the top – reprinted by permission from M. Putman *et al.* 1998 *Nature* **394**, 752; © 1998, Macmillan Magazines Ltd.

Problem 4.1 Use the data of Tables 1.4–1.6 to estimate approximate spectral types for the brightest stars of the LMC, in the right-hand panel of Figure 4.5.

The Magellanic Clouds are in orbit about each other, and they also orbit the Milky Way in a plane passing almost over the Galactic pole. Just as an Earth satellite loses energy to the resistance of the upper atmosphere, so the orbit of the Clouds is slowly decaying as energy is drained into random motions of stars in the Galaxy; we will discuss this process further in Section 5.6. We can explain the present positions and velocities of the Magellanic Clouds and the Stream if the Clouds are on an eccentric plunging orbit around the Galaxy, with a period of about 2 Gyr, and made their closest approach to the Milky Way between 200 and 400 million years ago. The centers of the Large and Small Clouds are now about 25 kpc apart, but they probably came within 10 kpc of each other during their last perigalactic passage. At that time, the gravitational attraction of the LMC pulled out of the SMC the neutral hydrogen gas that we now see as the Magellanic Stream. The combined gravity of the Milky Way and the LMC has obviously distorted the SMC, and perhaps even destroyed it as a bound system; the pieces are now drifting slowly apart.

The Magellanic Clouds are extremely rich in star clusters. They are close enough to distinguish individual main-sequence stars; so, just as in Section 2.2, we can use color–magnitude diagrams of these clusters to find their ages, distances, and chemical compositions. On comparing the apparent brightness of those stars that are still on the main sequence in the LMC's clusters with that of stars in the Galactic open clusters, we find that the LMC is 50 kpc from the Sun. This is about the same distance as we estimated in Section 2.2, from observations of gas around SN 1987a. Measuring the rotation speed in the LMC's HI disk and using Equation 3.20 yields a mass of $(1.5–2) \times 10^{10} \mathcal{M}_\odot$ within about 11 kpc of its center. For the SMC, by comparing the stars on the giant branch in its old clusters with those in Galactic globular clusters, and from its variable stars, we find a distance of around 60 kpc.

The LMC has some globular clusters similar to those of our Milky Way, although somewhat less dense. They are old (≥ 10 Gyr) and poor in heavy elements: some have less than $1/100$ of the solar abundance of metals. In contrast to the Milky Way, the old metal-poor stars and clusters do not form a metal-poor halo; instead, they lie in a thickened disk. Their random motions are $\sigma_r \approx 25–35$ km s^{-1}, larger than the ~ 10 km s^{-1} found for the HI clouds. The metal-poor objects show strong asymmetric drift (see Section 2.2); they rotate at only 50 km s^{-1}, more slowly than the gas disk.

Hardly any of the LMC's clusters have ages in the range 4−10 Gyr; this galaxy may have made very few stars during that period. There are many younger clusters and associations; some of these may have formed about 50 Myr ago, when the LMC and SMC had their last close passage. Some of these are 100 times more populous than most Galactic open clusters; they may be young versions of the LMC's globular clusters. The most luminous is the cluster R136 in nebula 30 Doradus, which is the very bright peak on the left in the Hα map of Figure 4.4. The cluster is about 3.5 Myr old, and in blue light its luminosity $L_B \approx 10^7 L_\odot$. The youngest stars, and the interstellar gas, are the richest in heavy elements, with a third to a half of the solar proportion of metals.

The star clusters of the SMC cover the same age range as found in the LMC, but there is no gap in time during which few clusters were formed. The bulk of their stars may have intermediate ages, between a few gigayears and ~ 12 Gyr. The gas and youngest star clusters are poorer in metals than those of the LMC, with only 20%–30% of the solar abundance.

4.1.2 Variable stars as 'standard candles'

The RR Lyrae and Cepheid variable stars are useful for finding distances to galaxies within the Local Group and beyond. We discussed RR Lyrae stars in Section 2.2; they are low-mass stars which are burning helium in their cores, with

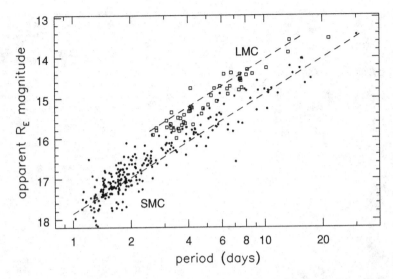

Fig. 4.7. Apparent magnitude and period in days, for Cepheids in the Large Magellanic Cloud (squares) and in the SMC (filled dots); dashed lines show mean period–luminosity relations – J.-P. Beaulieu.

$L \approx 50L_\odot$, and varying in brightness with periods of about half a day. Cepheid variables are massive helium-burning stars, with luminosity ranging up to $1000L_\odot$ and pulsation periods from one to fifty days. Both types of star are fairly easy to identify by taking several images of a galaxy at intervals suitably spaced in time, and searching for stars that have varied their brightness in the expected way.

Henrietta Leavitt found in 1912 that, of the Cepheids in the Large Magellanic Cloud, the brighter stars varied with longer periods (Figure 4.7). Since the stars are all at about the same distance from us, the apparently brighter stars are in fact more luminous; we have a *period–luminosity relation*. If we measure the period and apparent brightness of Cepheids in another galaxy and assume that the stars have the same luminosity as their LMC counterparts with the same period, we can estimate the galaxy's distance by using Equation 1.1. Care is needed, because the star's light output also depends on its composition; Cepheids in the disk of the Milky Way, where the metal abundance is high, are brighter than stars with the same period that have a smaller fraction of heavy elements. We must also correct for the effect of interstellar dust in dimming and reddening the stars. With the Hubble Space Telescope, we can find distances to galaxies within 2–3 Mpc using RR Lyrae stars; Cepheids are useful out to about 30 Mpc.

This technique of finding objects in a far-off galaxy which resemble those found closer by, and assuming that the distant objects have the same luminosity as their nearby counterparts, is called the *method of standard candles*. Often we have no other way to estimate the distance, but this method can lead us badly astray; the history of extragalactic astronomy includes many instances in which

Fig. 4.8. The Fornax dwarf spheroidal galaxy; it is far more diffuse than the Large Magellanic Cloud. Bright objects are stars in the Milky Way – D. Malin, Anglo-Australian Observatory.

the derived distances were hopelessly wrong. For example, in the 1920s Hubble observed Cepheids in the disk of M31 and derived their distance by assuming that they had the same luminosity as apparently similar stars in the Milky Way. But the distances of Cepheids in the Milky Way's disk, and thus their luminosities, had been underestimated because their light was dimmed by interstellar dust. Also, the W Virginis variable stars in Galactic globular clusters, which were thought to be as bright as the Cepheids, are in fact significantly dimmer. Because of these errors, Hubble concluded that the Cepheids in M31 were 1.5 magnitudes dimmer than in fact they are, so his distance to the galaxy was only half of what it should have been. Using Equation 1.28, he then arrived at an expansion age t_H for the Universe which was obviously less than the age of the Earth!

4.1.3 Dwarf spheroidal galaxies

The Milky Way's retinue also includes at least ten dwarf spheroidal galaxies, which are named after the constellations in which they appear. Their surface brightness is about a hundred times less than that of the Magellanic Clouds, and Figure 4.8 shows how hard it can be to spot them among the numerous Galactic stars in the foreground. The first of the Milky Way's dwarf spheroidals to be discovered were Sculptor and Fornax, in 1938; Sagittarius was found only in

Table 4.2 Dwarf galaxies, compared with the nuclear star cluster of M33, and three Milky Way globular clusters

System	L_V $(10^7 L_\odot)$	σ_r (km s^{-1})	r_c (pc)	r_t (pc)	t_{sf} (Gyr)	$\mathcal{M}/L_V$ $(\mathcal{M}_\odot/L_\odot)$	$\log_{10}(Z/Z_\odot)$ range
NGC 147 dE	12	20–30	260	1000	3–5	7 ± 3	−1.5 to −0.7
NGC 185 dE	13	20	170	2000	<0.5	5 ± 2	−1.2 to −0.8
Pegasus dIrr	1	9(HI)		500(HI)	<0.1	2–4	−2.3 to −1.7
Fornax dSph	1.5	13	400	5000	<2	~15	−2 to −0.4
M33 nucleus	0.25	24	<0.4		<1:	~1	−1.9 to −0.7
Sculptor dSph	0.2	9	200	2000	>10	~10	−2.6 to −0.8
ω Cen gc	0.1	20	4	70	>10	2.5	−1.6 to −1.2
M15 gc	0.04	12	<0.01	85	>10	2	−2.15
Carina dSph	0.04	7	200	900	2–10	~40	−2.7 to −0.3
M92 gc	0.02	5	0.5	50	>10	1.5	−2.15

Note: The velocity dispersion σ_r is highest at the center; at the core radius r_c, the surface brightness falls to half its central value, dropping to near zero at truncation radius r_t; t_{sf} the time since last significant star formation, with : indicating an uncertain value; $Z/Z_\odot$ is metal abundance compared with that of the Sun. HI denotes a measurement from HI gas, not stars; globular clusters are labelled gc.

Ursa Major system in 2005. Dwarf galaxies of low surface brightness are still being found today. In contrast to the Magellanic Clouds, the dwarf spheroidals are effectively gas-free, and they contain hardly any stars younger than 1–2 Gyr. All of them have some very old stars, such as RR Lyrae variables which require at least 10 Gyr to evolve to that stage. These systems began forming their stars as early as did 'giant' galaxies like the Milky Way.

The smallest of the dwarf spheroidal galaxies are only about as luminous as the larger globular clusters, although their radii are much larger (Table 4.2). But our satellite dwarf spheroidals are really galaxies, not just another form of star cluster. Fornax, and probably Sagittarius, have globular clusters of their own. Unlike star clusters within the Milky Way, the dwarf galaxies did not form all their stars at once; they all include stars born over several gigayears, from gas with differing proportions of heavy elements. Figure 4.9 shows the color–magnitude diagram for stars in the Carina dwarf, along with computed isochrones for metal-poor stars. Only about 2% of the stars are younger than about 2.5 Gyr, and the rest appear to have been born in three bursts, approximately 3, 7, and 15 Gyr ago. Even the most luminous of the dwarf spheroidals are only about 1/30 as rich in heavy elements as the Sun, and the less luminous systems are even more metal-poor; see Table 4.2. According to a simple model to be discussed in Section 4.3, we would expect a galaxy that had turned all its gas into stars to have roughly the solar abundance of heavy elements. Their low metallicity suggests that these galaxies lost much of their metal-enriched gas into intergalactic space.

Using the information in Table 4.2, we can estimate the masses of dwarf spheroidal galaxies from their sizes and the radial velocities of their stars. Stellar random speeds are not very different from those measured in globular clusters, but

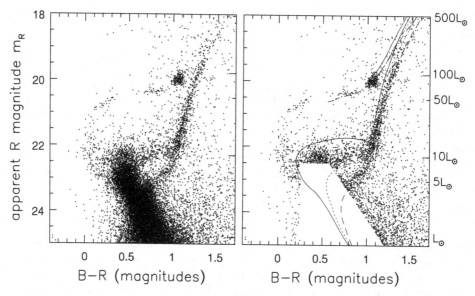

Fig. 4.9. Left, a color–magnitude diagram for the Carina dwarf spheroidal galaxy. Right, superposed isochrones give the locus of metal-poor stars ($Z = Z_\odot/50$) at ages of 3 Gyr (solid), 7 Gyr (dotted), and 15 Gyr (dashed). We see young red clump stars at $B - R, m_R =$ (1, 20), and old stars on the horizontal branch. Carina's distance modulus is taken as $(m - M)_0 = 20.03$; dust reddening is assumed to dim stars by 0.108 magnitudes in B and 0.067 magnitudes in R – T. Smecker-Hane; A. Cole, Padova stellar tracks.

random speeds are not very different from those measured in globular clusters, but the stars in dwarf galaxies are spread over distances ten or a hundred times as great. So, if we assume that these galaxies are in a steady state, and use the virial theorem, Equation 3.44, to calculate the masses, we find that the ratio of mass to light $\mathcal{M}/L$ is much greater than that for globular clusters. For the lowest-luminosity dwarf spheroidals, Ursa Minor, Carina, and Draco, $\mathcal{M}/L$ is even higher than that measured for the Milky Way (Section 2.3) or in spiral galaxies (Section 5.3). Dwarf spheroidal galaxies may consist largely of dark matter, with luminous stars as merely the 'icing on the cake'.

Problem 4.2 The Carina dwarf spheroidal galaxy has a velocity dispersion σ three times less than that at the center of the globular cluster ω Centauri, while Carina's core radius is 50 times greater. Use the virial theorem to show that Carina is about six times as massive as ω Centauri, so $\mathcal{M}/L$ must be 15 times larger.

Another possibility is that some of the dwarf spheroidal galaxies are not in equilibrium, but are being torn apart by the Milky Way's gravitational field. Sagittarius, the most recently discovered dwarf spheroidal, is almost certainly losing some of its stars. It lies nearly in the plane of the Galactic disk, only 20 kpc from the Galactic center. It is strongly distorted and spreads over $22° \times 7°$ in the sky, corresponding to the fairly large extent of 12 kpc $\times$ 4 kpc. To ask whether other galactic satellites are likely to hold themselves together, we now look at the conditions under which a star cluster or satellite galaxy could survive in the Milky Way's gravity.

4.1.4 Life in orbit: the tidal limit

As a small galaxy or a star cluster orbits a larger system, its stars feel a combined gravitational force that is changing in time: they can no longer conserve their energies according to Equation 3.27. This is the famously insoluble 'three-body problem', in which many of the possible orbits are chaotic; a small change to a star's position or velocity has a huge effect on its subsequent motion. But, if the satellite follows a circular orbit, and the gravitational potential is constant in a frame of reference rotating uniformly about the center of mass of the combined system, we can define an *effective potential* Φ_{eff} for the star's motion, and find a substitute for the no-longer-conserved energy.

If a vector $\mathbf{u}$ is constant in an inertial frame, which does not rotate, then an observer in a frame rotating with constant angular velocity $\mathbf{\Omega}$ will see it changing at the rate $d\mathbf{u}/dt' = -\mathbf{\Omega} \times \mathbf{u}$, where d/dt' denotes the derivative measured by the rotating observer. (Check this by taking Cartesian coordinates in the inertial frame, and writing $\mathbf{\Omega} = \Omega\mathbf{z}$; look at how a vector along each of the x, y, z axes changes for a rotating observer.) Suppose that a star has position $\mathbf{x}$ and velocity $\mathbf{v}$ relative to an inertial frame. Then, if the rotating observer chooses coordinates such that the star's position $\mathbf{x}'$ in that frame instantaneously coincides with $\mathbf{x}$, he or she measures its velocity as

$$\mathbf{v}' \equiv \frac{d\mathbf{x}'}{dt'} = \mathbf{v} - \mathbf{\Omega} \times \mathbf{x}. \tag{4.1}$$

For the rotating observer, the star's velocity $\mathbf{v}'$ changes at the rate

$$\frac{d\mathbf{v}'}{dt'} = \frac{d\mathbf{v}'}{dt} - \mathbf{\Omega} \times \mathbf{v}' = \frac{d\mathbf{v}}{dt} - \mathbf{\Omega} \times \mathbf{v} - \mathbf{\Omega} \times \mathbf{v}'$$
$$= -\nabla\Phi - 2\mathbf{\Omega} \times \mathbf{v}' - \mathbf{\Omega} \times (\mathbf{\Omega} \times \mathbf{x}) \tag{4.2}$$

The scalar product of $\mathbf{v}'$ with the last term is (see Table A.2)

$$-\mathbf{v}' \cdot \mathbf{\Omega} \times (\mathbf{\Omega} \times \mathbf{x}) = \Omega^2(\mathbf{x} \cdot \mathbf{v}') - (\mathbf{v}' \cdot \mathbf{\Omega})(\mathbf{\Omega} \cdot \mathbf{x}) = \frac{1}{2}\frac{d}{dt'}[(\mathbf{\Omega} \times \mathbf{x})^2].$$

Since $\mathbf{v}' \cdot (\boldsymbol{\Omega} \times \mathbf{v}') = 0$ and $\mathbf{x}' = \mathbf{x}$, taking the scalar product of $\mathbf{v}'$ with Equation 4.2 gives

$$\frac{1}{2} \frac{d}{dt'} [\mathbf{v}'^2 - (\boldsymbol{\Omega} \times \mathbf{x}')^2] = -\mathbf{v}' \cdot \nabla \Phi(\mathbf{x}'). \qquad (4.3)$$

If $\boldsymbol{\Omega}$ is chosen to follow the satellite in its orbit, then in the rotating frame the gravitational potential Φ does not depend on time; so the potential at the particle's position changes at the rate $d\Phi/dt' = \mathbf{v}' \cdot \nabla \Phi$. If we define the *Jacobi constant* E_J by

$$E_J = \frac{1}{2} \mathbf{v}'^2 + \Phi_{\text{eff}}(\mathbf{x}'), \qquad \text{where } \Phi_{\text{eff}}(\mathbf{x}') \equiv \Phi(\mathbf{x}') - \frac{1}{2}(\boldsymbol{\Omega} \times \mathbf{x}')^2, \quad (4.4)$$

then Equation 4.3 says that E_J does not change along the star's path. We can write the Jacobi constant in terms of the star's energy E and its angular momentum $\mathbf{L}$ per unit mass, as measured in the inertial frame:

$$E_J = \frac{1}{2}(\mathbf{v} - \boldsymbol{\Omega} \times \mathbf{x})^2 + \Phi_{\text{eff}} = \frac{1}{2}\mathbf{v}^2 + \Phi(\mathbf{x}, t) - \boldsymbol{\Omega} \cdot (\mathbf{x} \times \mathbf{v}) = E - \boldsymbol{\Omega} \cdot \mathbf{L}.$$
$$(4.5)$$

Problem 4.3 You can check that E_J is indeed constant by taking $\boldsymbol{\Omega}$ along the z axis, and looking at a particle moving in the x–y plane. Show from Equation 4.4 that $E_J = (v_{x'}'^2 + v_{y'}'^2)/2 + \Phi(\mathbf{x}') - \Omega^2(x'^2 + y'^2)/2$. Write the rate dE_J/dt' at which E_J changes along the particle's path, as measured by the rotating observer: you can use Equations 4.1 and 4.2 to find the derivatives dx'/dt', dv_x'/dt', etc The rate should be zero, showing that E_J is conserved along the particle's orbit. Now allow motion in the z direction, which does not contribute to $\boldsymbol{\Omega} \times \mathbf{x}$ or $\boldsymbol{\Omega} \times \mathbf{v}$, in your calculation to show that E_J is still conserved.

The simplest calculation of a tidal limit is one in which point masses m and $\mathcal{M}$, respectively, represent the satellite and the main galaxy. They are separated by distance D, while orbiting their common center of mass C with angular speed Ω. If we measure distance x from the satellite m toward $\mathcal{M}$, C lies at position $x = D\mathcal{M}/(\mathcal{M} + m)$; along the line joining the two systems,

$$\Phi_{\text{eff}}(x) = -\frac{G\mathcal{M}}{|D - x|} - \frac{Gm}{|x|} - \frac{\Omega^2}{2}\left(x - \frac{D\mathcal{M}}{\mathcal{M} + m}\right)^2. \qquad (4.6)$$

The effective potential Φ_{eff} has three maxima, at the first three *Lagrange points* (Figure 4.10). The middle point L_1 is the lowest; the next lowest point, L_2, lies behind the satellite; and L_3 is behind the main galaxy. A star for which $E_J < \Phi_{\text{eff}}(L_1)$ must remain bound to either $\mathcal{M}$ or m; it cannot wander between them.

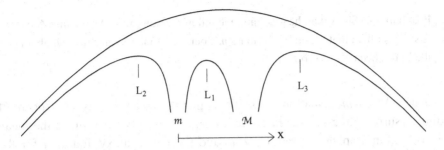

Fig. 4.10. The lower curve gives the effective potential Φ_{eff} along the line joining point masses m and $\mathcal{M}$. The Lagrange points L_1, L_2, and L_3 are extrema of $\Phi_{\text{eff}}(x)$. The upper curve shows the quadratic final term of Equation 4.6.

The Lagrange points are found by solving

$$0 = \frac{\partial \Phi_{\text{eff}}}{\partial x} = -\frac{G\mathcal{M}}{(D-x)^2} \pm \frac{Gm}{x^2} - \Omega^2 \left(x - \frac{D\mathcal{M}}{\mathcal{M}+m} \right). \qquad (4.7)$$

The acceleration $\Omega^2 D\mathcal{M}/(\mathcal{M}+m)$ of m as it circles C is due to the gravitational attraction of $\mathcal{M}$. By analogy with Equation 3.20,

$$\Omega^2 \frac{D\mathcal{M}}{\mathcal{M}+m} = \frac{G\mathcal{M}}{D^2}, \quad \text{so} \quad \Omega^2 = \frac{G(\mathcal{M}+m)}{D^3}. \qquad (4.8)$$

If the satellite's mass is much less than that of the main galaxy, L_1 and L_2 will lie close to m. We can substitute for Ω^2 in Equation 4.7, and expand in powers of x/D, to find

$$0 \approx -\frac{G\mathcal{M}}{D^2} - 2\frac{G\mathcal{M}}{D^3}x \pm \frac{Gm}{x^2} - \frac{G(\mathcal{M}+m)}{D^3}\left(x - \frac{D\mathcal{M}}{\mathcal{M}+m} \right). \qquad (4.9)$$

So at the Lagrange points L_1 and L_2, respectively,

$$x = \pm r_J, \quad \text{where } r_J = D\left(\frac{m}{3\mathcal{M}+m} \right)^{1/3}. \qquad (4.10)$$

Stars that cannot stray further from the satellite than r_J, the *Jacobi radius*, will remain bound to it: r_J is sometimes called the Roche limit. Note that L_1 is *not* the point where the gravitational forces from $\mathcal{M}$ and m are equal, but lies further from the less massive body. The Lagrange points are important for close binary stars; if the outer envelope of one star expands beyond L_1, its mass begins to spill over onto the other.

Problem 4.4 Show that the gravitational pull of the Sun (mass $\mathcal{M}$) on the Moon is stronger than that of the Earth (mass m), but the Moon remains in orbit about the Earth, because its orbital radius $r < r_J$.

When $\mathcal{M} \gg m$, Equation 4.10 tells us that the mean density in a sphere of radius r_J surrounding the satellite, $3m/(4\pi r_J^3)$, is exactly three times the mean density within a sphere of radius D around the main galaxy. Ignoring for the moment the force from the main galaxy, Equation 3.23 tells us that the period of a star orbiting the satellite at distance r_J would be roughly equal to the satellite's own orbital period. The satellite can retain those stars close enough to circle it in less time than it takes to complete its own orbit about the main galaxy, but it will lose its hold on any that are more remote.

Problem 4.5 If the mass $\mathcal{M}$ is replaced by the 'dark-halo' potential of Equation 2.19, show that the mass within radius $r \gg a_H$ of its center is $\mathcal{M}(<r) \approx r V_H^2/G$. A satellite with mass $m \ll \mathcal{M}(<D)$ orbits at radius $D \gg a_H$. By substituting the force from the dark halo for that of the point mass $\mathcal{M}$ in Equation 4.7, show that, instead of Equation 4.10, we have

$$r_J = D\left[\frac{m}{2\mathcal{M}(<D)}\right]^{1/3}. \tag{4.11}$$

In general, star clusters and satellite galaxies do not follow circular orbits. We might expect that the force of gravity at the closest approach will determine which stars remained bound. The truncation radius r_t of Section 3.2 and Table 4.2, where the density of stars drops to zero, should then be approximately equal to r_J at the pericenter of the orbit. This appears to hold for the Milky Way's globular clusters, but some of the Magellanic Cloud globulars overflow the Jacobi limit at their estimated perigalactic radius. Stars that are no longer bound to these clusters can perhaps still remain close by for a few orbits about the galaxy.

The LMC's disk is now safely stable against disruption by the Milky Way. Calculations of the orbit of the Magellanic Clouds and Stream indicate that it is now close to pericenter, and that the speed of a circular orbit about the Milky Way at the LMC's present distance of 50 kpc is about the same as that near the Sun, $\sim 200\,\mathrm{km\,s^{-1}}$. Using Equation 3.20, we estimate the mass within the LMC's orbit as about $5 \times 10^{11}\mathcal{M}_\odot$. The LMC's mass is about $10^{10}\mathcal{M}_\odot$, so, by Equation 4.11,

$$r_J \approx 50\,\mathrm{kpc} \times \left(\frac{10^{10}\mathcal{M}_\odot}{2 \times 5 \times 10^{11}\mathcal{M}_\odot}\right)^{1/3} \approx 11\,\mathrm{kpc}. \tag{4.12}$$

The LMC's disk lies safely within this radius, but we see that the SMC is too distant from the LMC to remain bound to it. The problem below shows that some dwarf galaxies are probably being torn apart by the Milky Way's gravitational field.

> **Problem 4.6** The Sagittarius dwarf spheroidal galaxy is now about 20 kpc from the Galactic center: find the mass of the Milky Way within that radius, assuming that the rotation curve remains flat with $V(R) \approx 200 \, \mathrm{km \, s^{-1}}$. Show that this dwarf galaxy would need a mass of about $6 \times 10^9 \mathcal{M}_\odot$ if stars 5 kpc from its center are to remain bound to it. Show that this requires $\mathcal{M}/L_V \sim 70$, which is much larger than the values listed in Table 4.2.

4.2 Spirals of the Local Group

The Local Group contains three spiral galaxies: our own Milky Way, the Andromeda galaxy M31, and M33. At a distance of 770 kpc, M31 is the most distant object that can easily be seen with the unaided eye; M33 is only slightly further away but is much harder to spot. By comparing these three systems with each other, we see what properties spiral galaxies have in common, and how they differ.

4.2.1 The Andromeda galaxy

M31, shown in Figure 4.11, is in all respects a bigger galaxy than our Milky Way. It is ~50% more luminous; the disk scale length h_R defined by Equation 2.8 is 6–7 kpc, twice as large as in the Milky Way; and it rotates faster, with speed $V(R)$ over most of the disk about $260 \, \mathrm{km \, s^{-1}}$, or 20%–30% higher than in our Galaxy. In addition to about 300 known globular clusters, over twice as many as in the Milky Way, M31 has its own satellite galaxies. These include the elliptical galaxy M32, three dwarf ellipticals, and several dwarf spheroidals.

The central bulge of M31 is larger in proportion than that of the Milky Way, providing 30%–40% of the measured luminosity. The apparent long axis of the bulge does not line up with the major axis of the disk further out. Either the bulge is not axisymmetric and would look somewhat oval if seen from above the disk, or its equator must be tipped relative to the plane of the disk. The bulge is faint in ultraviolet light, because it contains few young stars. As in our Galaxy, the bulge stars are all at least a few gigayears old and are generally rich in heavy elements. The bulge contains dilute ionized gas, and a few denser clouds of HI gas and dust, which are seen as dark nebulae.

At the center is a compact semi-stellar nucleus. In Hubble Space Telescope images the nucleus proves to have two separate concentrations of light about 0.5″

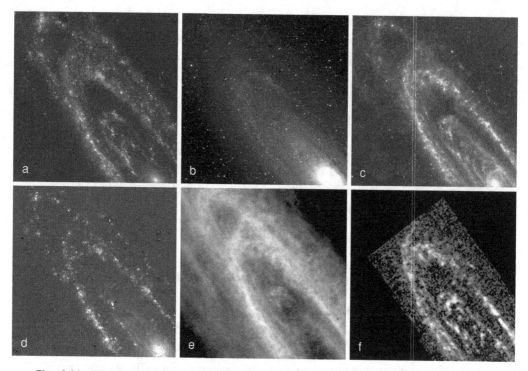

Fig. 4.11. The Andromeda galaxy, M31: (a) in ultraviolet light; (b) *B*-band image shows the prominent bulge; (c) infrared light of warm dust at 24 µm; (d) Hα image shows the 'ring of fire' and Hɪɪ regions in the spiral arms; (e) Hɪ gas; and (f) emission of CO tracing dense molecular gas – K. Gordon. (a) Galex FUV, D. Thilker *et al.* 2005 *ApJ* **619**, L67; (b) and (d) D. Thilker and R. Walterbos; (c) Spitzer; K. Gordon *et al.* 2005 *ApJ* **638**, L87; (e) R. Braun *et al.*; and (f) Nieten *et al.* 2006 *AAp* **453**, 459.

or 2 pc apart. One of these harbors a dense central object, probably a black hole of mass $\mathcal{M}_{\mathrm{BH}} \sim 3 \times 10^6 \mathcal{M}_\odot$. The other may be a star cluster which has spiralled into the center under the influence of dynamical friction (see Section 7.1). Unlike that of the Milky Way, the nucleus of M31 is impressively free of gas and dust.

Just as in the Milky Way, the metal-poor globular clusters of M31 follow deeply plunging orbits; the cluster system shows little or no ordered rotation. But the bulge also continues smoothly outward as a luminous spheroid. Most of the stars a few kiloparsecs above the disk plane are not those of a metal-poor halo; they are relatively metal-rich, and they probably form a fast-rotating system. It is as if M31's bulge has 'overflowed', largely swamping the metal-poor halo.

Like the Milky Way, M31 is a cannibal galaxy. The metal-rich halo stars are roughly 6 Gyr old; they and much of the bulge probably arrived as M31 merged with another metal-rich (and so fairly massive: see Section 4.3) galaxy. A huge stream of stars over 100 kpc long has been found, passing from northwest of the galaxy to the southeast. These stars are also more metal-rich than those of the Milky Way's halo; they may have been stripped as M31 swallowed a sizable galaxy.

Circling the bulge at a radius of about 10 kpc, the star-forming 'ring of fire' is clearly visible in Figure 4.11. Most of the young disk stars lie in this ring or just outside it; on average, M31's disk forms stars at a slower rate than does that of the Milky Way. Ionized gas in HII regions around the young massive stars glows red in Hα; in the far-infrared we see the dust heated by those stars, and the CO map shows the dense gas from which they formed. Just outside this ring, ultraviolet-bright young stars and strings of HII regions in the disk trace segments of fairly tightly wound spiral arms, where gas, dust, and stars have been compressed to a higher density. However, there is no clear large-scale spiral pattern. Because of its large bulge and moderately tightly wound spiral arms, and the relative paucity of gas and recent star formation in the inner disk, we classify M31 as an Sb galaxy, whereas our Milky Way is Sbc or Sc.

M31 has about $(4\text{--}6) \times 10^9 \mathcal{M}_\odot$ of neutral hydrogen, about 50% more than the Milky Way. Molecular gas is probably a smaller fraction of the total, so the ratio of gas mass to stellar luminosity is lower than that in our Galaxy. Figure 4.11 shows that the HI is concentrated at the ring of fire; but, as in the Milky Way, the gas extends to larger radii than the stellar disk. In the region of the spiral arms, high-resolution maps show holes in the HI disk, up to a kiloparsec across; at their edges, shells containing $(10^3\text{--}10^7)\mathcal{M}_\odot$ of dense HI gas are moving outward at $10\text{--}30\,\mathrm{km\,s^{-1}}$. At this rate, most of the holes would have taken a few megayears to reach their present sizes. Sometimes an association of massive O and B stars lies within the hole; winds from these massive stars, and recent supernova explosions, have blown away the cool gas. Holes in the inner parts of the disk tend to be smaller, perhaps because the gas is denser or the magnetic field stronger, so it is harder to push cool material out of the way.

If we measure the velocity of the HI gas at each point, we can use the fact that the clouds follow near-circular orbits to build up a three-dimensional picture of the gas disk. The outer parts are not flat but bent into an 'S' shape; the stellar disk is visibly warped in the same sense. An 'S' warp in the outer parts of spiral galaxies is quite common; the Milky Way's own disk is warped in this way, and systems with a flat disk are probably in a minority. As in our Galaxy, the HI layer flares out to become thicker at greater distances from the center. Some outlying clumps of HI gas clearly do not share the disk's rotation; lying up to 50 kpc from the center, they are the analogues of the Milky Way's high-velocity clouds (Section 2.4). The clouds are ~ 1 kpc across with $(10^5\text{--}10^6)\mathcal{M}_\odot$ of HI. Several of them lie along the same path as the giant star stream, with roughly the same velocities; perhaps they were stripped from the same satellite galaxy. Another lies close in both position and velocity to the dwarf elliptical NGC 205.

4.2.2 M33: a late-type spiral

The other spiral in the Local Group, M33, is definitely an Sc or an Scd galaxy. Its bulge is tiny; the spiral arms are more open than those in M31 and not as smooth, consisting mainly of bright blue concentrations of recently formed stars. M33 is a smaller and much less luminous galaxy than the Milky Way; the scale length is small, $h_R \approx 1.7\,$kpc, and the rotation speed $V(R)$ rises only to $120\,$km s^{-1}.

When observed in the Hα emission line, M33 displays a complex network of loops, filaments, and shells, like those in the LMC (Figure 4.4) and in the violently star-forming irregular galaxy IC 10 (see Section 4.4 below). Supernova explosions, and the winds from stars, heat the surrounding gas and drive it away, thus affecting the location and rate of future starbirth. Such *feedback* has a strong effect on the way that galaxies come into being from lumps of primordial gas, and on their subsequent development.

M33 is relatively richer in H I gas than M31 or the Milky Way; there is little CO emission, reflecting either a lack of molecular gas or a smaller ratio of CO to H_2 than in the Milky Way. The latter is more likely, since the disk is rich in young stars, which are born in the dense cores of molecular clouds. Compared with the Milky Way, relatively more of the H I gas is in the warm component and less in cold dense clouds. As in M31, the H I layer has large holes, often centred on star-forming regions in the disk. M33's neutral-gas disk is very extended. The H I continues out at least to 3 Holmberg radii or about 30 kpc, which is a substantial fraction of the $\sim 200\,$kpc separating the galaxy from M31. One rather massive cloud with $10^6 \mathcal{M}_\odot$ of gas lies about 15 kpc from M33's center; a streamer of H I gas links it to the disk. The outer disk is warped, possibly by tides from M31.

At the center of M33, we find a dense *nuclear star cluster*, with no more than a small bulge around it. This cluster is more luminous than any Galactic globular, with $L_V \approx 2.5 \times 10^6 L_\odot$; its core is tiny, so the stellar density exceeds $10^7 L_\odot\,$pc^{-3}. In contrast to the single generation of stars in a globular cluster, M33's nucleus contains old, middle-aged, and young stars. There is no sign of a black hole: if one is present, $\mathcal{M}_{BH} \lesssim 10^4 \mathcal{M}_\odot$, far less than in the Milky Way and in M31. But we do see evidence for a power source other than ordinary stars. M33's nucleus is the single brightest X-ray source of the Local Group, equivalent to several of the normal binary sources.

M33 is only two or three times more luminous than the LMC, yet it has a much more symmetric spiral pattern. Low-luminosity galaxies are in general more likely than larger systems to resemble the LMC in having a strong central bar, with the brightest parts of the galaxy off-center from the outer disk. But the morphology clearly depends on factors other than the galaxy's luminosity alone.

4.3   How did the Local Group galaxies form?

We can now sketch a picture for the formation of the Milky Way and the other galaxies of the Local Group, starting from the hot dense early Universe that we discussed in Section 1.5. Roughly 350 000 years after the brilliant beginning of the Big Bang, photons of the cosmic fireball no longer had enough energy to ionize

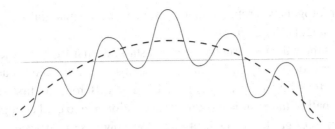

Fig. 4.12. Small galaxies form near large ones: the density of matter (wavy solid line) is a combination of small clumps within a large region that is denser than average (dashed line). Regions dense enough to collapse on themselves (above the horizontal line) tend to be clustered together.

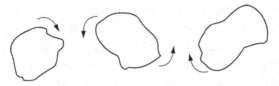

Fig. 4.13. Tidal torques: irregular lumps attract each other and begin to rotate.

hydrogen and helium. The nuclei then combined with electrons to form a gas of neutral atoms, through which light could propagate freely; the Universe became transparent. The gas that was to form the galaxies was no longer supported by the pressure of photons trapped within it. If its gravity was strong enough, a region that was denser than average would begin to collapse inward.

We will see in Section 8.5 that the denser the gas, the earlier cosmic expansion must halt and give way to contraction. These collapsing regions would not have been evenly spread in space. Where there was a general increase in density, more of the smaller surrounding clumps of matter would have been dense enough to collapse: see Figure 4.12. Clumps near the center of a large infalling region would fall toward each other, eventually merging into a single big galaxy, while those further out might become smaller satellite galaxies. We can think of all the material that is destined to come together into a single galaxy as making up a *protogalaxy*.

At their largest extent, just before they started to collapse in on themselves, the protogalaxies lay closer together than the galaxies do now, because the Universe was smaller. In general they would not have been neat spheres, but irregularly shaped lumps, tugging at each other by gravity. Mutual *tidal torques* would have pulled them into a slow rotation (Figure 4.13). There is no very definite way to calculate how much spin a galaxy would receive. In large computer simulations, which represent the forming galaxies by of many particles, each attracting the others by the force of gravity, rotation develops such that the average at any radius is about 5% of what is required for a circular orbit there. As the gas clouds within each protogalaxy collide with each other, they lose part of their energy and fall inward; the protogalaxy's rotation increases, because the material approximately conserves its angular momentum.

When did the galaxies form their earliest stars? Stellar light and the emission lines of hot gas ionized by early stars have been seen from galaxies at redshifts $z \gtrsim 6$, when the Universe was less than 1 Gyr old. Before the first stars could form, the fireball of the cosmic background radiation had to cool enough to allow star-sized lumps of gas to radiate heat away. We now observe nascent stars in the cores of molecular clouds, with temperatures $T \lesssim 20\,\mathrm{K}$; see Section 2.4. Using Equation 1.34, we see that the background radiation does not reach this temperature until redshift $z \sim 6$, hundreds of millions of years after the Big Bang.

But the very first stars were made from primeval gas, almost pure hydrogen and helium. Their atmospheres would have been much less opaque than the Sun's outer layers, and so less easily blown away by the pressure of the star's radiation. Large lumps of gas might have collapsed earlier, at higher temperatures, to form extremely massive stars with $\mathcal{M} > 100\mathcal{M}_\odot$ that could survive for long enough to allow substantial nuclear burning. When these stars exploded as supernovae, they would distribute the heavy elements that they had made to the surrounding gas. We will discuss galaxy formation and early starbirth again in Section 9.4.

4.3.1 Making the Milky Way

The first stars may have lived and died not in a galaxy-sized unit, but in smaller lumps of gas, with masses perhaps $(10^6 – 10^8)\mathcal{M}_\odot$. Here, one or two supernovae were enough to add elements such as carbon, nitrogen, and oxygen to the gas in 1/1000 or even 1/100 of the solar proportion. This is approximately what we see in the Galaxy's oldest stars, those of the metal-poor globular clusters. The stars in each cluster generally have very closely the same composition, while abundances in the Galactic gas today are far from uniform. So we think that globular clusters formed in smaller parcels of gas, where the nucleosynthetic products of earlier stars had been thoroughly mixed.

Some of the globular clusters may have been born when gas clouds ran into each other, as they fell together to form the Milky Way; the collisions would have compressed the gas, raising its density so that many stars formed in a short time. Stars, unlike gas, do not lose significant energy through collisions; so their formation halts the increase in ordered rotation. The orbits of the old metal-poor globulars and metal-poor halo stars are not circular but elongated. These orbits are oriented in random directions; the metal-poor halo has virtually no ordered rotation. This is probably because the material from which it formed did not fall far into the Galaxy before it became largely stellar.

In 1962, Olin Eggen, Donald Lynden-Bell, and Allan Sandage introduced the idea that the stars in the metal-poor halo had formed rapidly, as the proto-Milky Way collapsed under its own gravity. Equation 3.23 tells us that the time taken for a gas cloud of density ρ to fall in on itself is proportional to $1/\sqrt{\rho}$. Gas in the substructure of lumps that made the stars would have been denser than average, so material should have started to contract sooner, and turned into stars before the galaxy-sized cloud had gone far in its own collapse. The problem below shows that the whole process could have been completed within a few tenths of a gigayear.

Problem 4.7 For a galaxy like our Milky Way with a mass of $10^{11} \mathcal{M}_\odot$ and radius 10 kpc, find the average density. The virial theorem tells us that, if a galaxy of stars collapses from rest, then, after it has come to equilibrium, it will be eight times denser than at the start: see the discussion following Problem 8.31. Show that, for the proto-Milky Way, the free-fall time of Equation 3.23 was $t_{\text{ff}} \sim 300\,\text{Myr}$. This is about ten times longer than a protostar of solar mass takes to reach the main sequence. For the Sculptor dwarf with $\mathcal{M} \sim 2 \times 10^7 \mathcal{M}_\odot$ and radius 2 kpc (Table 4.2), show that the average density is only 1/40 of the Milky Way's, so the collapse time $t_{\text{ff}} \sim 2\,\text{Gyr}$.

By contrast, the material that became the Milky Way's rotating disk had to lose a considerable amount of its energy. We saw in Section 3.3 that a circle is the orbit of lowest energy for a given angular momentum. Today's thin-disk stars occupy nearly circular orbits because they were born from gas that had lost almost as much energy as possible. The thick-disk stars, and the more metal-rich globular clusters, predate most of the thin disk. They may have been born from gas clouds that had yielded up less of their energy, but still formed a somewhat flattened rotating system. By the time that the earliest thin-disk stars were born, 8–10 Gyr ago, heavy elements produced by earlier generations of stars had enriched the gas, to perhaps 10%–20% of the solar abundance.

Today, disk gas near the Sun follows nearly circular orbits with speed $V(R) \approx 200\,\text{km s}^{-1}$. If tidal torques gave this material a rotational speed only 5% of that needed for a circular orbit, the gas must subsequently have fallen inward until it reached an orbit appropriate for its angular momentum. We can use Equation 3.29 to estimate where the local gas must have been when the tidal torques were operating. If the Milky Way's gravitational potential corresponds to a flat rotation curve, with $V(R)$ constant, then this gas must have fallen in from a distance $R \sim 100\,\text{kpc}$; the gas around galaxies must have extended much further out at earlier times. The disk material had to remain gaseous as it moved inward, forming only very few stars, so that it could continue to radiate away energy. It may have been able to do this because it was much less dense than the gas that had earlier given birth to the globular clusters.

The color–magnitude diagram of Figure 4.5 shows no horizontal branch in the Galactic bulge. Even allowing for their higher metal content, very few of the bulge stars can be as old as the globular clusters. The over-whelming majority have ages less than 8–10 Gyr, and some may be much younger. We do not yet know how the bulge stars were made. They may have formed in the dense center of the protogalactic gas that was to make up the Milky Way; the bulge might have grown out of a dense inner region of the disk; or its stars may be the remains of dense clusters that fell victim to dynamical friction, and spiralled into the center; see Section 7.1. The central kiloparsec of galaxies such as M33 and the LMC is not as dense as the inner Milky Way; the low density may have prevented a bulge from developing.

Once the dense central bulge had come into being, the gravitational force of the whole Galaxy would have helped it to hold onto its gas. By trapping the hot and fast-moving debris from supernovae, the bulge formed large numbers of metal-rich stars. Both in the Local Group and beyond (see Section 6.3), the stars of more luminous galaxies are richer in heavy elements. Their stronger gravity prevents metal-bearing gas from escaping, and it is incorporated into stars.

Much of the Milky Way's dark matter is in its outskirts, beyond most of the stars of the disk. In Section 5.3 we will see that the same is true of most spiral galaxies. Why does nonluminous material lie mainly in the Galaxy's outer reaches? Since its composition remains unknown, we lack a definite answer. However, if we presume that all forms of matter were mixed evenly at early times, then the dark matter must have had less opportunity than the star-stuff to get rid of its energy. It would then be left on orbits taking it far from the Galactic center. A dark halo of the weakly interacting massive particles (WIMPs) of Section 1.5 could never radiate away energy as heat; so it is bound to remain more extended than the gaseous and stellar body. If the dark matter consists of compact objects such as brown dwarfs or black holes, we would expect that these formed very early in the Milky Way's collapse, probably predating even the globular clusters.

The Milky Way is still under construction today. As we saw in Section 2.2, stars of the Sagittarius dwarf spheroidal galaxy are being added to the Galactic halo. Near the Sun, groups of young metal-poor halo stars have been found, that may be the remnants of another partially digested dwarf galaxy. The orbit of the Magellanic Clouds has been shrinking, and in Section 7.1 we will see that the LMC will probably fall into the Milky Way within 3–5 Gyr. Like meteoric cratering in the solar system, these late additions represent the final stages of assembly.

4.3.2 The buildup of heavy elements

During its life, a galaxy turns gas into stars. Each star burns hydrogen and helium to form heavier elements, which are returned to the interstellar gas at the end of its life. We might define a 'clock' for galactic aging by the mass of stars born and of metals produced, per unit mass of gas that was present initially. Near the Sun, we see some correspondence between the time told by this 'metal-production clock' and time as measured by stellar aging; Figure 4.14 shows that older disk stars in general contain little iron, while recently formed stars have larger abundances. We saw in Section 2.2 that the open clusters of the Milky Way's thin disk are both younger and more metal-rich than the stars and globular clusters of the thick disk, while the globular clusters of the halo are the oldest and the poorest in heavy elements.

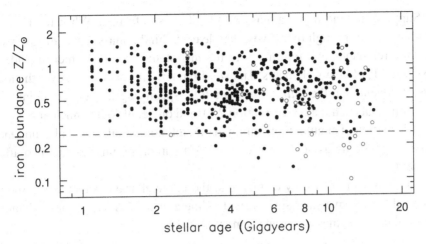

Fig. 4.14. Nearby F and G stars show a large scatter in iron abundance at any age; but younger stars tend to be richer in iron. Stars with 'thick-disk' metal abundance (below the dashed line) often move faster than at $80 \, \mathrm{km \, s^{-1}}$ relative to the local standard of rest (open circles) – B. Nordström *et al.* 2004 *AAp* **418**, 98.

At one stage it was thought that galactic contents could be divided simply into two components. Young stars and metal-rich material in the disk formed Population I, while the old metal-poor stars in the bulge and stellar halo belonged to Population II. (Astronomers sometimes refer to the first stars, made from the hydrogen and helium of the Big Bang without any heavy elements, as Population III.) We now know that this is an oversimplification. For example, the bulges of M31 and the Milky Way are several gigayears old, but they are metal-rich. Dwarf irregular galaxies, and the outer parts of normal spirals, contain young metal-poor stars born within the past 100 Myr.

Faced with this complexity, we retreat to a drastically simplified description of how the metals in a galaxy might build up over time. This is the *one-zone, instantaneous recycling* model. We assume that a galaxy's gas is well mixed, with the same composition everywhere, and that stars return the products of their nuclear fusion to the interstellar gas rapidly, much faster than the time taken to form a significant fraction of the stars. Initially, we assume that no gas escapes from the galaxy or is added to it – this is a *closed-box* model – and that all elements heavier than helium maintain the same proportion relative to each other. We define

- $\mathcal{M}_g(t)$ as the mass of gas in the galaxy at time t;
- $\mathcal{M}_\star(t)$ to be the mass in low-mass stars and the white dwarfs, neutron stars, and black holes that are the remnants of high-mass stars (the matter in these objects remains locked within them throughout the galaxy's lifetime); and, finally,
- $\mathcal{M}_h(t)$ is the total mass of elements heavier than helium in the galactic gas; the *metal abundance* in the gas is then $Z(t) = \mathcal{M}_h/\mathcal{M}_g$.

Suppose that, at time t, a mass $\Delta'\mathcal{M}_\star$ of stars is formed. When the massive stars have gone through their lives, they leave behind a mass $\Delta\mathcal{M}_\star$ of low-mass stars and remnants, and return gas to the interstellar medium which includes a mass $p\,\Delta\mathcal{M}_\star$ of heavy elements. The *yield p* represents an average over the local stars; it depends on the *initial mass function*, specifying the relative number of stars formed at each mass (see Section 2.1), and on details of the nuclear burning. The distribution of angular momentum in the stellar material, its metal abundance, stellar magnetic fields, and the fraction of stars in close binaries can also affect the yield.

The mass $\mathcal{M}_h$ of heavy elements in the interstellar gas alters as the metals produced by massive stars are returned, while a mass $Z\,\Delta\mathcal{M}_\star$ of these elements is locked into low-mass stars and remnants. We have

$$\Delta\mathcal{M}_h = p\,\Delta\mathcal{M}_\star - Z\,\Delta\mathcal{M}_\star = (p - Z)\Delta\mathcal{M}_\star; \tag{4.13}$$

so the metallicity of the gas increases by an amount

$$\Delta Z \equiv \Delta\left(\frac{\mathcal{M}_h}{\mathcal{M}_g}\right) = \frac{p\,\Delta\mathcal{M}_\star - Z[\Delta\mathcal{M}_\star + \Delta\mathcal{M}_g]}{\mathcal{M}_g}. \tag{4.14}$$

If no gas enters or leaves the system, the total in gas and stars remains constant, and $\Delta\mathcal{M}_\star + \Delta\mathcal{M}_g = 0$. When the production of an element in stars does not depend on the presence of other heavy elements in the stellar material, we call it a *primary* element. If we deal with primary elements, p is independent of Z and we can integrate Equation 4.14 to find how the metal abundance in the gas builds up. We have

$$Z(t) = Z(t = 0) + p \ln\left[\frac{\mathcal{M}_g(t = 0)}{\mathcal{M}_g(t)}\right]. \tag{4.15}$$

The metallicity of the gas grows with time, as stars are made and gas is used up. The mass of stars $\mathcal{M}_\star(t)$ formed before time t, and so with metallicity less than $Z(t)$, is just $\mathcal{M}_g(0) - \mathcal{M}_g(t)$; we have

$$\mathcal{M}_\star(<Z) = \mathcal{M}_g(0)[1 - \exp\{-[Z - Z(0)]/p\}] \tag{4.16}$$

Time does not appear explicitly here; the mass $\mathcal{M}_\star(<Z)$ of slowly evolving stars that have abundances below the given level Z depends only on the quantity of gas remaining in the galaxy when its metal abundance has reached that value. This simple model explains a basic fact: where the gas density is high in relation to the number of stars formed, the average abundance of heavy elements is low. In gas-rich regions such as in the Magellanic Clouds, or the outer disks of spiral galaxies, the stars and gas are relatively poor in metals (see Figure 4.15).

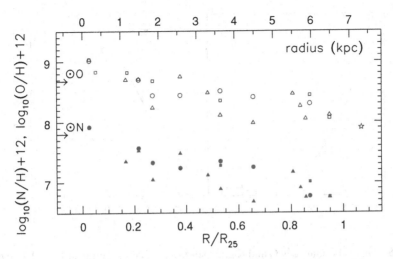

Fig. 4.15. Metal abundance of disk gas in M33, shown as logarithms of the number of atoms of oxygen (open symbols) and nitrogen (closed symbols) for each 10^{12} atoms of hydrogen, plotted against radius given as a fraction of R_{25}, where the surface brightness is 25 mag arcsec^{-2} in the B band. Horizontal arrows give the Sun's abundance – D. Garnett.

Once all the gas is gone, this model predicts that the mass of stars with metallicity between Z and $Z + \Delta Z$ should be

$$\frac{\mathrm{d}\mathcal{M}_\star(<Z)}{\mathrm{d}Z}\Delta Z \propto \exp\{-[Z(t) - Z(0)]/p\}\Delta Z. \qquad (4.17)$$

Figure 4.16 shows the number of G and K giant stars at each metallicity observed in Baade's Window, a partial clearing of the dust in the disk near $l = 1°, b = -4°$, where we have a good view of the bulge. Our simple model provides a good approximation to the observed numbers if the gas originally lacked any metals, and the yield $p \approx Z_\odot$. The bulge may have managed to retain all its gas, and turn it completely into stars.

> **Problem 4.8** Show that, if stars are made from gas that is initially free of metals, so that $Z(0) = 0$, the closed-box model predicts that, when all the gas is gone, the mean metal abundance of stars is exactly p.

In other cases, the simple model is clearly wrong. Within individual globular clusters orbiting the Milky Way, there is no gas, and the stars all have the same, metal-poor, composition. These clusters must have formed out of gas that mixed very thoroughly after its initial contamination with heavy elements; any material not used in making their single generation of stars would have been expelled promptly. Dwarf spheroidal galaxies contain very little gas, although their stars have metal abundances 30–100 times lower than that in the Galactic bulge. It is

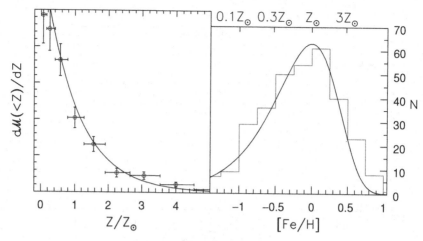

Fig. 4.16. Metal abundance in G and K giant stars of the Galactic bulge. Left, relative number in each range of metal fraction Z; right, number in each bin of $[Fe/H] = \log_{10}(Z/Z_\odot)$. Solid curves show the prediction of a closed-box model with $p = Z_\odot$ and gas that is initially free of metals: note the tail at high Z – Sadler *et al.* 1996 *AJ* **112**, 171; and Fullbright *et al.* 2006 *ApJ* **636**, 821.

possible that the dwarf spheroidals formed very few massive stars, and so produced only small amounts of metals. A more likely explanation is that, as in the globular clusters, most of the heavy elements have been lost. Interstellar gas could easily escape the weak gravitational force, and only a small fraction of the hot metal-rich material from supernovae would have mixed with cool gas, to be incorporated into a new generation of stars. To use Equation 4.15, we must redefine p as the *effective yield*, taking into account metals lost from the system. For the globular clusters, $p \approx 0$; p is always less than the true yield of metals produced.

Near the Sun, the Milky Way's disk contains $(30–40)\mathcal{M}_\odot\,pc^{-2}$ of stars, together with about $13\mathcal{M}_\odot\,pc^{-2}$ in gas, for a total of $\sim 50\mathcal{M}_\odot\,pc^{-2}$; see Section 3.4. The local disk gas has roughly the same average abundances as the Sun. If heavy elements were originally absent, and no gas had entered or left the solar vicinity, Equation 4.15 would give the yield p as

$$Z(\text{now}) \approx Z_\odot \approx p\ln(50/13), \quad \text{so} \quad p \approx 0.74 Z_\odot. \qquad (4.18)$$

But there is a difficulty with the closed-box model for the solar neighborhood: look at its prediction for metal-poor disk stars. Equation 4.16 requires

$$\frac{\mathcal{M}_\star(<Z_\odot/4)}{\mathcal{M}_\star(<Z_\odot)} = \frac{1 - \exp[-Z_\odot/(4p)]}{1 - \exp(-Z_\odot/p)} \approx 0.4; \qquad (4.19)$$

nearly half of all stars in the local disk should have less than a quarter of the Sun's metal content. In fact, of a sample of 132 G dwarf stars in the solar neighborhood, just 33 were found to have less than 25% of the solar abundance of iron, and only one below 25% of the solar fraction of oxygen. This discrepancy is known as the

G-dwarf problem, since it was first discovered in these, the most luminous stars for which we still have a sample of old objects.

A possible solution to the G-dwarf problem is that the gas from which the disk was made already had some metals before it arrived in the solar vicinity. Heavy elements produced by the earliest stars could have mixed with the gas that eventually formed the disk, to 'pre-enrich' it. In that case, we should expect all stars to have metal abundance above some minimum value. Setting $Z(0) \approx 0.15 Z_\odot$ gives approximately the observed distribution of metal abundance locally; see the following problem.

Problem 4.9 If the disk gas had $Z = 0.15 Z_\odot$ at $t = 0$ when stars first began to form, while $Z(\text{now}) \approx Z_\odot$, and $\mathcal{M}_g(t = 0)/\mathcal{M}_g(t) = 50/13$, use Equation 4.15 to show that $p \approx 0.63 Z_\odot$. From Equation 4.16, show that about 20% of low-mass stars should have $Z < Z_\odot/4$ today.

But it is also possible, and even likely, that star formation near the Sun started before the gaseous raw material had been fully assembled. In that case, the first stars would enrich only a small amount of gas to a moderately high metal abundance. Subsequent inflow of fresh metal-deficient gas would dilute that material, preventing the abundance from rising as fast as the closed-box model predicts; see the following problem. Incomplete mixing could explain the large dispersion in stellar abundance at any given age (see Figure 4.14). Since gas in the outer parts of galaxies is poorer in heavy elements, a slow inward flow, perhaps caused by energy loss in passing through shocks in the spiral arms, would dilute the metals in the local disk. Long-lived stars formed at early times should also return metal-poor gas as they age. If enough of this gas were released, the fraction of metals in newly made stars might even decline with time.

Problem 4.10 Suppose that the inflow of metal-poor gas is proportional to the rate at which new stars form, so that $\Delta \mathcal{M}_\star + \Delta \mathcal{M}_g = \nu \Delta \mathcal{M}_\star$ for some constant $\nu > 0$. Show that Equation 4.14 becomes

$$\Delta Z = \frac{(p - \nu Z)\Delta \mathcal{M}_\star}{\mathcal{M}_g} = \frac{p - \nu Z}{\nu - 1} \frac{\Delta \mathcal{M}_g}{\mathcal{M}_g}, \tag{4.20}$$

so that the metallicity in the gas is

$$Z(t) = \frac{p}{\nu} \left\{ 1 - \left[\frac{\mathcal{M}_g(t)}{\mathcal{M}_g(0)} \right]^{\nu/(1-\nu)} \right\}, \tag{4.21}$$

which can never exceed p/ν. (See Pei and Fall 1995, *ApJ* **454**, 69.) Taking $\nu < 0$ would correspond to the escape of gas: see below.

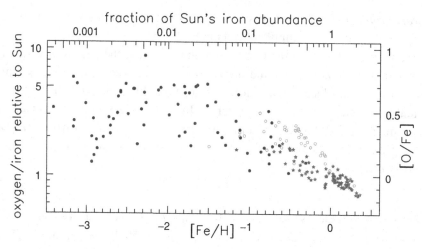

Fig. 4.17. Oxygen is more abundant relative to iron in metal-poor stars. Filled dots show stars of the Milky Way's halo, open dots represent thick-disk stars, following a slightly different relation from that for the thin-disk stars (star symbols) – T. Bensby.

The abundances of heavy elements vary relative to each other: Figure 4.17 shows that stars with low metal abundance have *more* oxygen relative to the amount of iron than do stars like the Sun. This happens because these elements are made in stars of different masses. Stars more massive than $10 \mathcal{M}_\odot$ end their lives by exploding as a Type II supernova. They release mainly lighter elements with fewer than about 30 neutrons and protons, such as oxygen, silicon, and magnesium, back into the interstellar gas. Most of the heavier nuclei such as iron, which are made in the star's core, are swallowed up into the remnant neutron star or black hole. These massive stars go through their lives within 100 Myr, whereas the local disk has made its stars fairly steadily over the last 8–10 Gyr; so the instantaneous-recycling assumption is reasonable.

Not all of the 'lighter' heavy elements are produced in very massive stars. Stars only slightly more massive than the Sun do not become supernovae, but add carbon and nitrogen to the interstellar gas. These elements are produced during helium burning, and the stars dredge them up into the envelope during helium flashes. Later, at the tip of the *asymptotic giant branch*, those outer layers are ejected; see Section 1.1. Carbon and silicates often condense as dust grains in the cool stellar atmosphere. These stars often take far longer than 100 Myr to make their contribution to the Galaxy's store of heavy elements.

The main source of iron is supernovae of Type Ia. We saw in Section 1.1 that these occur when a white dwarf in a binary system collapses under its own gravity. If matter is added to a white dwarf in a binary system, taking its mass above the *Chandrasekhar limit* of $1.4 \mathcal{M}_\odot$, it can no longer support its own weight, and it implodes. This heats the interior, triggering nuclear burning, which blows the star apart. No remnant is left; all the iron, nickel, and elements of similar atomic mass are released back to the interstellar gas. Many of the stars that explode as Type Ia supernovae do so only at ages of a gigayear or more; instantaneous recycling is a poor approximation. In stars formed during the first few gigayears of a galaxy's life, we expect the ratio of elements such as oxygen and magnesium to iron to be higher than it is in the Sun.

To predict the metal abundance of stars in a galaxy, we first need to estimate the birthrate of stars at each mass. We can then calculate how much of each element is released back to the interstellar gas over time. In the Milky Way, long-lived stars that were born soon after the disk formed are now returning metal-poor gas, so the abundance of heavy elements in the gas is growing only slowly. The chemical composition of the new stars also depends on how gas moves within the galaxy, since Figure 4.15 shows that the clock of metal enrichment runs more slowly in the outer regions. We require *chemodynamical* models, which are still under development.

Further reading: for a graduate-level treatment of this subject, see B. E. J. Pagel, 1997, *Nucleosynthesis and Chemical Evolution of Galaxies* (Cambridge University Press, Cambridge, UK) Chapters 7 and 8; and D. Arnett, 1996, *Supernovae and Nucleosynthesis* (Princeton University Press, Princeton, New Jersey).

4.4 Dwarf galaxies in the Local Group

The Local Group contains two main types of dwarf galaxies. In galaxies of the first type, the dwarf ellipticals and the much more diffuse dwarf spheroidals, almost all the stars are at least a few gigayears old. These systems contain little gas to make any new stars. By contrast, dwarf irregulars are tiny, gas-rich galaxies with active star formation, and a profusion of recently formed blue stars. Like the dwarf spheroidals of Section 4.1, the dwarf irregulars are diffuse systems. While all the Local Group's dwarf ellipticals, and most of its dwarf spheroidal galaxies, orbit either the Milky Way or M31, many of the dwarf irregulars are not satellites of larger systems, but 'free fliers'. We can use color–magnitude diagrams to chart the star-forming histories of these different varieties of dwarf, and to investigate the relationship between them.

Remarkably, all the dwarf galaxies of the Local Group contain some stars on the horizontal branch, which may include RR Lyrae variables. Horizontal branch stars are at least 10 Gyr old, so the dwarf galaxies made their first stars in the first 2–3 Gyr of cosmic history. We will see in Section 8.5 that cosmological models with *cold dark matter* predict exactly this behavior.

4.4.1 Dwarf ellipticals and dwarf spheroidals

We discussed the Milky Way's dwarf spheroidals in Section 4.1: they are not much more luminous than a globular cluster, but so diffuse as to be almost invisible on the sky. Andromeda's dwarf spheroidals are yet harder to observe; they appear to be very similar to our own. The dwarf ellipticals are more luminous versions of the dwarf spheroidals, with $L \gtrsim 3 \times 10^7 L_\odot$ or $M_V \lesssim -14$. In the Local Group, they are represented by three of M31's satellites: NGC 147, NGC 185, and NGC 205. Table 4.2 shows that their sizes are similar to those of the dwarf spheroidals, but they are more luminous, so the stellar density is higher.

Because they are close to Andromeda, the dwarf ellipticals are vulnerable to tidal damage. In NGC 205 the random speed of the stars is greater at larger radii, instead of being smaller as Equation 3.47 would imply for an isolated galaxy, because M31 is pulling at the outer stars. Both dwarf spheroidals and dwarf ellipticals appear quite oval on the sky rather than round, yet their stars show no pattern of ordered rotation. We will see in Section 6.2 that these galaxies may have no axis of symmetry; their shapes are probably triaxial.

Both NGC 205 and NGC 185 show a few patches of dust, and we can trace small amounts of cool gas by its HI and CO emission. Most of the stars in these two dwarf ellipticals date from at least 5 Gyr ago. Near the center, however, a small number (amounting to no more than $10^6 \mathcal{M}_\odot$) have ages between 100 and 500 Myr; gas lost by the old stars may have supplied the raw material for continued starbirth. By contrast, NGC 147, although otherwise similar, shows no sign of very recent star formation. Its nuclear region contains a very few stars in early middle age, born only a few gigayears ago. In the outer parts, the overwhelming majority is at least 5 Gyr old.

M32, the most luminous satellite of M31, contains virtually no cool gas and has no stars younger than a few gigayears. But Figure 4.18 shows its central brightness to be one of the highest yet measured for any galaxy. High-resolution images from the Hubble Space Telescope reveal no constant-brightness inner core; the density continues to climb, to $\gtrsim 10^6 L_\odot \, \mathrm{pc}^{-3}$ within the central parsec. A black hole of a few million solar masses may lurk at its center. Although its luminosity is within the normal range for dwarf ellipticals, its very high density suggests that M32 is a miniature version of a normal or 'giant' elliptical galaxy.

Perhaps M32 is only the remnant center of a much larger galaxy. It lacks globular clusters, whereas the less luminous dwarf elliptical satellites of M31 do have them. The stars at its center are red, and approximately as rich as the Sun in heavy elements; this is typical of more massive galaxies, that have stronger gravity to confine metal-rich gas from exploding stars; see Section 6.4. The outer regions of M32 still have an elliptical shape, but its long axis is twisted away from that of the inner regions. Tidal forces of M31 have probably affected the orbits of the outermost stars. We cannot yet measure the distances well enough to know how far M32 is from M31, or whether it could have passed close enough for M31's gravity to strip off the outer stars, as described in Section 4.1.

The motion of the stars in M32 is intermediate between that in the disk of the Milky Way and that in its metal-poor halo. The galaxy is slightly flattened, and its stars orbit in a common direction, but they also have considerable random motions. We can measure the degree of ordered rotation in a galaxy by the ratio

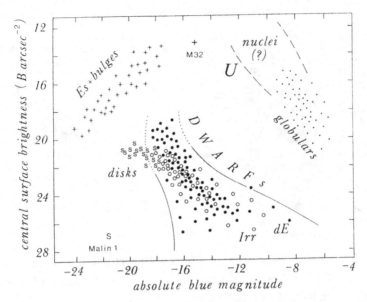

Fig. 4.18. Dwarf and giant galaxies occupy different regions in a plot of absolute V-magnitude and measured central surface brightness; because of 'seeing', the true peak brightness may be higher. On the left, luminous elliptical galaxies and the bulges of disk systems have very high surface brightness at their centers. 'U' marks ultracompact dwarf ellipticals (see Figure 6.6). The rightmost of the 'dE' points (filled circles) represent what this text calls dwarf spheroidals; open circles mark irregular and dwarf irregular galaxies. Disks of spiral galaxies are marked 'S'. Malin 1 is a low-surface-brightness galaxy; see Section 5.1 – B. Binggeli.

V/σ of the average speed V of the stars in the direction of rotation to their velocity dispersion σ. In the Milky Way's disk, stars like the Sun have $V \approx 200\,\text{km s}^{-1}$ while $\sigma \approx 30\,\text{km s}^{-1}$, so $V/\sigma \approx 7$. In M32, $V/\sigma \sim 1$, while for the dwarf spheroidals it is much less than unity. By analogy with the thermal motion of atoms in a gas, we refer to disks with high values of V/σ as 'cold'; 'hot' systems are those in which random motions are relatively more important, so that V/σ is low. The stronger the influence of ordered rotation, the more disklike an object must be. Within the solar system, the giant planets Jupiter and Saturn, with a 'day' only ten hours long, are considerably more flattened at the poles than the compact and slow-rotating Earth. We will see in Section 6.2 that not all flattened galaxies rotate fast; but strongly rotating galaxies must always be flattened.

4.4.2 Dwarf irregular galaxies

Irregular galaxies are so called because of their messy and asymmetric appearance on the sky (see NGC 4449 in Figure 1.13 and NGC 55 in Figure 5.7). Starbirth

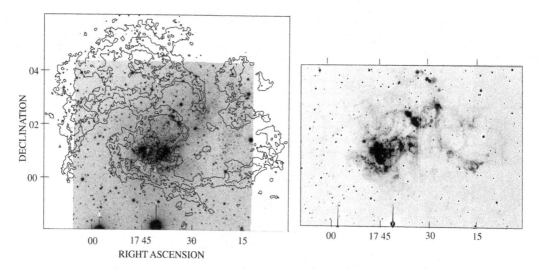

Fig. 4.19. The dwarf irregular galaxy IC 10. Left, HI contours superposed on an R-band negative image; the box measures $8'$ vertically, or 19 kpc. Right, negative image showing Hα emission from ionized gas – E. Wilcots.

occurs in disorganized patches that occupy a relatively large fraction of the disk, because the size and luminosity of star-forming regions increases only slowly with the size of the parent galaxy. Even quite small irregular galaxies can produce spectacular OB associations, as well as more normal clumps of young stars. Figure 4.18 tells us that their disks have low average surface brightness; so the bright concentrations of young stars stand out, to give the galaxy's optical image its chaotic aspect.

We draw the line between irregular galaxies and the dwarf irregulars at $L \sim 10^8 L_\odot$. Dwarf irregulars contain gas and recently formed blue stars; but in some other ways they resemble the dwarf spheroidals. Irregular galaxies are diffuse, and ordered rotational motion is much less important than in the Milky Way's disk. The stars and gas clouds have a velocity dispersion of $\sigma \sim 6$–$10\,\mathrm{km\,s^{-1}}$, but the peak rotation speed V declines at lower luminosity. In larger irregulars, $V/\sigma \sim 4$–5, falling to $V/\sigma \lesssim 1$ in the smallest dwarf irregulars. The proportion of metals in dwarf irregular galaxies is very low, generally below 10% of the solar value, and the least luminous are the most metal-poor. Oxygen in the gas of the smallest systems such as Leo A is present only at about 1/30 of the solar level, while the more massive galaxy NGC 6822 has about 1/10 of the solar abundance.

Dwarf irregulars tend to be brighter than the dwarf spheroidals, but this is only because of their populations of young stars. They contain relatively large amounts of gas, seen as neutral hydrogen, and the gas layer often extends well beyond the main stellar disk. Figure 4.19 shows the HI layer of IC 10; as in Figure 4.4, we see large 'holes' blown by the winds of supernovae and hot stars. Ionized gas

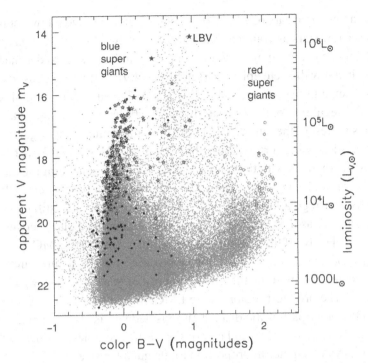

Fig. 4.20. Color–magnitude diagram for a galaxy in vigorous starbirth: the Sc spiral galaxy M33. Larger symbols show stars classified by their spectra. Open star-shapes show blue supergiants; filled stars are luminous blue variables; filled diamonds are Wolf–Rayet stars; and open circles indicate red supergiants. Most points between the vertical 'plumes' of blue and red supergiants represent fore-ground stars in our Milky Way. For an irregular galaxy like the Small Magellanic Cloud this diagram would look similar, but with fewer stars of each type – P. Massey.

shines brightly, showing that young stars have formed in the shells of denser gas surrounding the holes. This galaxy has little if any organized rotation.

The color–magnitude diagram of an irregular galaxy shows many short-lived massive stars. It looks very much like Figure 4.20 for the spiral M33, also an enthusiastic star-former. O and B main-sequence stars occupy the lower half of the diagram, below $10^4 L_{V,\odot}$. The larger symbols mark massive stars that have left the main sequence. A nearly-vertical 'plume' of blue supergiants, which burn helium or heavier elements in their cores (recall Figure 1.4), joins the rather wide main sequence from above. The luminous blue variables are among the most luminous stars known; their light output varies as they rapidly slough off their outer envelopes of hydrogen. Wolf–Rayet stars were born with $\mathcal{M} \gtrsim 40\mathcal{M}_\odot$, and live for less than 10 Myr. They have lost their outer hydrogen, exposing a hot layer of helium, carbon, and nitrogen. Red and yellow supergiants are massive stars burning hydrogen or helium in a shell.

Some dwarf galaxies, such as Phoenix and LGS3, are classified as intermediate between dwarf irregulars and dwarf spheroidals. Almost all their stars are more than a few gigayears old, but they contain a little gas and a few young stars. Fornax has a few stars as young as 500 Myr, so this dwarf spheroidal galaxy must have had some gas until quite recently. The Carina dwarf spheroidal made most of its stars in a few discrete episodes (see Figure 4.9); at times of peak starbirth it may have been a miniature version of Sextans A. Because of their similar structures, small irregulars like the Pegasus dwarf may be at an early stage, while dwarf spheroidals represent the late stages, in the life of a similar type of galaxy. In the dwarf spheroidals, which orbit close to the Galaxy or M31, gas may have been compressed by interactions with these large galaxies, perhaps encouraging more stars to form earlier on. By now these galaxies have used up or blown out all their gas, while the dwarf irregulars, perhaps benefiting from a quieter life, still retain theirs.

Galaxies such as the LMC might represent a transitional class between spiral galaxies and the dwarf irregulars. Like a spiral galaxy, the LMC is basically a rotating disk, but it lacks regular or symmetric spiral structure, and random motions account for more of the kinetic energy of stars and gas: $V/\sigma \sim 4$ for the old stars. The brightest region of the LMC, the central bar, is off center with respect to the outer part of the disk; such lopsidedness is also common in dwarf irregulars. The neutral-hydrogen layer of the LMC has 'holes' similar to those in IC 10, but they are smaller in proportion to the galaxy's size.

In the Local Group and beyond, dwarf galaxies are not simply smaller or less luminous versions of bigger and brighter galaxies. Table 4.2 shows that dwarf elliptical galaxies all have about the same physical size; the core radius is always $r_c \sim 200$ pc. Thus Figure 4.18 implies that the more luminous dwarfs have higher surface brightness. But, among the normal or 'giant' ellipticals, the most luminous galaxies are also the most diffuse. We will see in Section 6.1 that the core radius is so much larger at higher luminosity that among these the central surface brightness is *lower* in the most luminous systems. Because of these contrasting trends, we think that dwarf galaxies probably developed by processes different from those that produced the giant spiral and elliptical galaxies.

4.5 The past and future of the Local Group

The galaxies of the Local Group are no longer expanding away from each other according to Hubble's law. Their mutual gravitational attraction, and that of any matter present between the galaxies, has been strong enough to pull the group members back toward each other. The Milky Way and M31 are now approaching each other; these two galaxies will probably come near to a head-on collision within a few gigayears. We can use the orbits to make an estimate of the total mass within the Local Group. Readers will not be surprised that this analysis reveals yet more 'dark matter'.

We start by assuming that all the mass of the Local Group lies in or very close to the Milky Way or M31, and we treat these two galaxies as point masses m and

$\mathcal{M}$. They are now separated by $r \approx 770\,\text{kpc}$, and they are closing on each other with $dr/dt \approx -120\,\text{km s}^{-1}$.

Problem 4.11 Show that the separation $\mathbf{x}_{\mathcal{M}} - \mathbf{x}_m$ of two point masses m and $\mathcal{M}$ moving under their mutual gravitational attraction obeys

$$\frac{d^2}{dt^2}(\mathbf{x}_{\mathcal{M}} - \mathbf{x}_m) = -\frac{G(m+\mathcal{M})(\mathbf{x}_{\mathcal{M}} - \mathbf{x}_m)}{|\mathbf{x}_{\mathcal{M}} - \mathbf{x}_m|^3}; \tag{4.22}$$

the separation between the two objects obeys the same equation as a star of small mass, attracted by a mass $m + \mathcal{M}$.

Problem 4.12 For a star orbiting in the plane $z = 0$ around a much larger mass $\mathcal{M}$, its distance r from $\mathcal{M}$ changes according to Equation 3.65:

$$\frac{d^2 r}{dt^2} - \frac{L_z^2}{r^3} = -\frac{G\mathcal{M}}{r^2}, \tag{4.23}$$

where L_z is the conserved z angular momentum. By substituting into this equation, show that its path can be written in terms of the parameter η as

$$r = a(1 - e\cos\eta), \qquad t = \sqrt{\frac{a^3}{G\mathcal{M}}}(\eta - e\sin\eta) \text{ for } a = \frac{L_z^2}{G\mathcal{M}(1-e^2)}. \tag{4.24}$$

This orbit is an ellipse of eccentricity e, with semi-major axis a; the time t is measured from one of the pericenter passages, where $\eta = 0$.

The Milky Way and M31 began moving apart at the Big Bang. So combining Equations 4.22 and 4.24 tells us that, at the present time $t = t_0$, their relative distance r is changing at the rate

$$\frac{dr}{dt} = \frac{dr/d\eta}{dt/d\eta} = \sqrt{\frac{G(m+\mathcal{M})}{a}}\frac{e\sin\eta}{1-e\cos\eta} = \frac{r}{t_0}\frac{e\sin\eta(\eta - e\sin\eta)}{(1-e\cos\eta)^2}. \tag{4.25}$$

Since $dr/dt < 0$, the galaxies are approaching pericenter; $\sin\eta < 0$. In a nearly circular orbit, with $e \approx 0$, the speed of approach is a very small fraction of the orbital speed, which would imply a large total mass. The smallest combined mass that M31 and the Milky Way could have is given by assuming that the orbit is almost a straight line, with $e \approx 1$, and that the galaxies are falling together for the first time, so that $\pi < \eta < 2\pi$. Using the measured values of r and dr/dt, and setting $12\,\text{Gyr} \lesssim t_0 \lesssim 15\,\text{Gyr}$, we can find η by equating the leftmost and rightmost terms of Equation 4.25. Substituting η into the third term gives $m + \mathcal{M} \approx (4\text{--}5) \times 10^{12}\mathcal{M}_\odot$; the larger mass corresponds to the smaller age.

This is more than ten times the mass that we found for the Milky Way in Problem 2.18; even beyond $2.5R_0$, there is yet more dark matter. We have only 3–4 Gyr until $\eta = 2\pi$, at the next close passage. Since there are no large concentrations of massive galaxies near the Local Group, that could have pulled on M31 and the Milky Way to give them an orbit of high angular momentum, it is quite likely that $e \approx 1$. In that case, we will come close to a direct collision: M31 and the Milky Way could merge to form a single larger system.

Problem 4.13 Taking $e = 1$ in Equation 4.24, and giving r and dr/dt their current measured values, use that and Equation 4.25 to show that $\eta = 4.2$ corresponds to $t_0 = 12.8\,\mathrm{Gyr}$, and $a = 520\,\mathrm{kpc}$. Use Equation 4.24 to show that the combined mass $m + \mathcal{M} \approx 4.8 \times 10^{12}\mathcal{M}_\odot$. Show that the Milky Way and M31 will again come close to each other in about 3 Gyr. Use the data of Table 4.1 to estimate L_V for the Local Group as a whole, and show that the overall mass-to-light ratio $\mathcal{M}/L \gtrsim 80$ in solar units.

By repeating your calculation for $\eta = 4.25$, show that $t_0 = 14.1\,\mathrm{Gyr}$ and $m + \mathcal{M} \approx 4.4 \times 10^{12}\mathcal{M}_\odot$: a greater cosmic age corresponds to a smaller mass for the Local Group.

The following chapter is devoted to spiral and S0 galaxies. It is often useful to think of M31, the Milky Way, and M33 as typical spiral galaxies. Like them, most spirals live in groups. In Section 7.1 we will see that collisions between group galaxies are fairly common; at earlier times, when the Universe was denser, they would have been even more frequent. As the disks crash into each other, their gas is compressed, swiftly converting much of it into stars. Material from the outer disks will be stripped off as 'tidal tails'. A few gigayears later a red galaxy might remain, largely free of gas or young stars. Some astronomers believe that many of the giant elliptical galaxies that we see today are the remnants of such galactic traffic accidents.

5

Spiral and S0 galaxies

The main feature of a spiral or S0 galaxy is its conspicuous extended stellar disk. Stars in the disk of a large spiral galaxy, like our Milky Way, follow nearly circular orbits with very little random motion. Ordered rotation accounts for almost all the energy of motion, with random speeds contributing less than $\sim 5\%$: the disk is dynamically 'cold'. In smaller galaxies, random motions are proportionally larger, but most of the disk's kinetic energy is still in rotation. Because the stars have little vertical motion perpendicular to the disk plane, the disk can be quite thin.

Spiral galaxies are distinguished from S0 systems by the multi-armed spiral pattern in the disk. The disks of spiral galaxies still retain some gas, whereas S0 systems have lost their disk gas, or converted it into stars. Both S0 and spiral galaxies can show a central linear bar; in Figure 1.11, the sequence of barred galaxies SB0, SBa, . . . , SBm runs parallel to the 'unbarred' sequence S0, Sa, . . . Apart from the bar and spiral arms, the stellar disks of large galaxies are usually fairly round; but many smaller systems are quite asymmetric.

Most giant disk galaxies – those with $M_B \lesssim -19$ or $L_B \gtrsim 6 \times 10^9 L_\odot$ – are composite systems. Many of them probably have a metal-poor *stellar halo* like that of the Milky Way (see Figure 1.8). But the halo accounts for only a few percent of the galaxy's light, and is spread over an enormous volume; so the surface brightness is low, making it difficult to study. The dense inner *bulge* is prominent in the Sa and S0 systems, less important in Sb and Sc galaxies, and absent in the Sd and Sm classes. Bulge stars have considerable random motions, and they are much more tightly packed than in the disk: near the Sun, the density of stars is $n \sim 0.1 \, \mathrm{pc}^{-3}$, whereas in bulges it is often 10 000 times higher. Bulges are generally rounder than the very flattened disks. They tend to be gas-poor, except for their innermost regions; in some respects, they are small elliptical galaxies placed inside a disk. The central hundred parsecs of the bulge may accumulate enough gas to fuel violent bursts of star formation. As in our Milky Way, the centers of many bulges host *nuclear star clusters*, the densest stellar systems. In some nuclei, we find massive compact central objects, which are probably black holes.

Spirals are the most common of the giant galaxies, and produce most of the visible light in the local Universe. In the opening section we investigate the stellar content of the disks of spiral and S0 galaxies; Section 5.2 considers the gaseous component, and its relationship to the stars. In Section 5.3 we discuss the rotation curves of spiral galaxies, and what these reveal about the gravitational forces. In most spirals, the force required to maintain the outermost disk material in its orbit cannot be accounted for by the visible portion of the galaxy, its stars and gas. The difference is attributed to 'dark' material, which we detect only by its gravity. We then pause, in Section 5.4, to consider how much the scheme of Figure 1.11, classifying galaxies according to their appearance in visible light, can tell us about their other properties. Spiral arms and galactic bars form the topic of Section 5.5; these common and prominent features prove surprisingly difficult to understand. In Section 5.6 we discuss bulges and nuclei, and speculate on how they are related to the rest of the galaxy.

5.1   The distribution of starlight

The light coming to us directly from the stars of present-day disk galaxies is mainly in the near-infrared region of the spectrum: old stars, such as K giants, give out most of their light at wavelengths close to $1\,\mu$m. Much of the blue light of hot massive stars is intercepted by surrounding dust, and re-radiated in the far-infrared, beyond $\sim 10\ \mu$m. Visible light is next in importance; ultraviolet photons do not carry much of the energy, except in irregular galaxies where dust is less plentiful, and the light of young hot stars more easily escapes absorption. Historically, galaxies have been studied mainly by optical photography. Although infrared detectors are rapidly improving, optical images still yield most of our information about galactic structure.

5.1.1   Astronomical array detectors

The standard detector for optical extragalactic astronomy is now the *charge-coupled device*, or CCD; the same devices are used to record images in modern video and electronic cameras. A CCD consists of a thin silicon wafer that will absorb light (Figure 5.1). A photon's energy sets free one or more electron–hole pairs; the electrons are collected and amplified, to produce an output signal that should be linearly proportional to the number of photons absorbed. Not every incoming photon produces an electron–hole pair; we define the *quantum efficiency* as the ratio of detected photoelectrons to incident photons. In an excellent CCD, quantum efficiency can peak above 90%, for red light with $5000\,\text{Å} \lesssim \lambda \lesssim 7000\,\text{Å}$. CCDs are analogue detectors: photons are converted into a current which is then amplified, rather than measured individually as they are in *photon-counting* systems such as photomultiplier tubes.

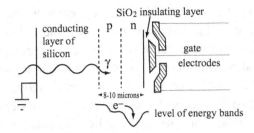

Fig. 5.1. A section through a back-illuminated CCD chip turned on its side: photons (wavy line) enter through the conducting silicon layer, liberating electrons which are attracted toward the gate electrodes. The lower curve shows how the energy of an electron varies through the p-doped and n-doped layers; gate voltages are adjusted to trap electrons just beneath the surface.

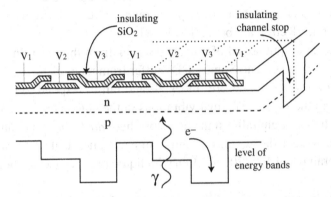

Fig. 5.2. A back-illuminated CCD chip: photons (wavy line) enter from below, producing photoelectrons which migrate toward the low point in the nearest of the potential wells. The lower curve shows the energy level of an electron just below the surface, when the voltages are set so that $V_1 < V_2 < V_3$. Insulating *channel stops* divide each column from its neighbors; dotted lines show how a column is divided into *pixels*.

Figure 5.2 shows how the surface of the CCD detector is divided into individual picture elements or *pixels*. Thin insulating layers separate the light-collecting layer of silicon into long 'channels'; along each of these *columns*, insulated electrodes are constructed in a regular pattern on the top surface of the wafer. Together with doped zones inside the detector, the electrodes produce a localized potential well that attracts and holds the photoelectrons, dividing the columns into *rows* of pixels. In typical astronomical CCDs, each pixel is a square 10–30 μm on a side; for other applications, such as video recorders, the pixels are often smaller.

The image recorded on a CCD can be that of a portion of the sky, or the spectrum of light that has been dispersed into its component colors (for example, by reflection from a grating in a spectrograph). When the exposure is complete, the image is 'read out' by using the electrode potentials like a bucket brigade. The CCD control system varies or 'clocks' their voltages in the sequence

$V_1 \rightarrow V_3, V_2 \rightarrow V_1, V_3 \rightarrow V_2$: with each step, electrons are 'tipped' into the deepest part of the potential, which moves down the column to the right in Figure 5.2. At the base of the column is a similar arrangement, collecting each pixel's electrons and moving them sequentially through amplifiers, and into an analogue-to-digital converter that gives the number of *counts* for each pixel. We can then produce a two-dimensional image by encoding the signal level from each pixel as a gray scale, where the tone varies, or by false color, with color variations indicating the intensity.

Although CCDs are excellent detectors, they are not perfect. The counts, which are stored in a computer file, should be proportional to the number of electrons collected, or the number of photons falling on the pixel. Unfortunately, the amplifiers introduce *read noise*: a random fluctuation in the count rate, expressed in units of electrons. Astronomical CCDs of mid-2000s vintage normally have read noise equivalent to 2–5 electrons per pixel per readout, and devices with read noise below one electron are being developed.

Usually, the output signal of a CCD deviates from the ideal linear behavior by a few tenths of a percent. Some CCDs suffer from manufacturing errors; most often, parts of a few columns are 'blocked' and fail to respond to light, making dark streaks in the image. Thermal vibrations in the silicon create electron–hole pairs even when no light falls on the detector; these *dark counts* are an unwanted noise source. To keep dark counts to a minimum, astronomical CCDs are operated at low temperatures of 100–200 K, often with liquid nitrogen (which boils at 77 K) as a coolant.

Energetic particles are readily detected by CCDs, giving rise to a *cosmic-ray* background. This limits the exposure times to about an hour on the ground, and less for telescopes in space, since the atmosphere screens out some cosmic rays. Often, we take two or more images of the same piece of sky, so that these 'false stars' can be removed. The potential well of each pixel can typically hold at most a few hundred thousand electrons; so, if the CCD is overexposed (e.g., by a bright star), electrons spill down the column, producing a bright streak in the image; see Figure 5.16 below. When the problem is severe, the streaks also radiate along rows; a residual image of the bright source may be seen in subsequent exposures.

CCDs are not useful at all wavelengths: even with anti-reflection coatings, ultraviolet light shortward of about 2000 Å (200 nm) can hardly penetrate into the silicon. At wavelengths $\lambda \gtrsim 11\,000$ Å (1.1 μm), infrared photons have too little energy to liberate an electron–hole pair; they travel easily through the silicon layer without being absorbed. Photons between these wavelengths give up their energy to produce a single electron–hole pair. Red light could travel through the electrodes on top of the CCD wafer and into the silicon below, but bluer photons, with $\lambda \lesssim 4500$ Å, cannot. So blue-sensitive CCDs are made with a thinned lower layer, only 15–20 μm thick, and are *back-illuminated* so that the light shines in from below, as in Figures 5.1 and 5.2. CCDs are also efficient X-ray detectors; at wavelengths below 1000 Å, each photon produces many electron–hole pairs. The number of electrons is proportional to the photon energy, so the CCD also provides spectral information. In this respect, energetic particles like cosmic rays behave similarly to X-rays, and CCDs are sometimes used to detect them.

When measuring light from an object using a CCD, the astronomer must compensate for the fact that the pixels are not exactly uniform: some are more sensitive than others. *Flat fields*, images of blank twilight sky, or a diffuse screen (often called the 'white spot') on the interior of the telescope dome, are taken to measure and to correct for this effect. The CCD must be *calibrated*, by observing stars of known brightness to determine how many counts correspond to a given flux or magnitude. To produce an image of a galaxy, we must subtract out the contribution of the night sky, since only the brightest parts of galaxies are brighter than the sky: see Section 1.3. We routinely measure surface brightnesses down to $\lesssim 1\%$ of the sky level by using dome or sky flat fields. Special techniques are required to do much better, because the CCD's sensitivity changes slowly during the night, limiting the accuracy of the flat fields.

We can estimate the quality of measurements with a CCD by calculating a *signal-to-noise ratio*, S/N. The simplest case is when statistical noise in the number of detected photons is the main source of error. We say that the CCD has a *gain g* if a signal with C counts in some pixel corresponds to S detected photons (or equivalently, S electrons captured), where $S = gC$. The number of photons arriving at that pixel from the galaxy under study has a random fluctuation of approximately $\sqrt{S}$; so the *noise* or 1σ error in S is $N = \sqrt{gC}$. When we are *photon-noise-limited*, then $S/N = \sqrt{gC}$; this is the *lowest* possible level of noise.

The read noise R of a CCD is given in units of photoelectrons; including this, we can write $S/N = gC/\sqrt{gC + R^2}$. At low signal levels, R is the dominant noise source; we are *read-noise-limited*. In this case $R^2 \gg gC$ and the signal-to-noise ratio S/N grows linearly with exposure time t. When photon noise dominates, $S/N \propto \sqrt{gC}$, which grows only as $\sqrt{t}$, and doubling the signal-to-noise ratio of a measurement requires that we observe four times longer. Extending this approach to include other noise sources, such as flat field errors, gives us a basis to predict the performance of real CCDs and other similar detectors.

We may wish to measure the *surface brightness* $I(\mathbf{x})$, the amount of light per square arcsecond at position $\mathbf{x}$ in the image of a galaxy, or the *flux* F_λ in a spectrum. A CCD records only the total amount of light falling on each pixel during an exposure; it gives us no information on where a photon has landed within the pixel. So we can measure the angular size of a feature in the image only if its light is spread over at least two pixels on the CCD. To produce an image of the sky showing details $1''$ across, the pixels must correspond to a size $\lesssim 0.5''$ on the image. Similarly, to obtain a spectrum with wavelength resolution $\Delta\lambda$, our spectrograph must disperse light in this wavelength interval over at least two pixels on the CCD. The physical size of CCDs is limited, currently to about 10 cm × 10 cm; so we must compromise between high resolution and a large field of view. Images of objects extending over a large region on the sky are sometimes built up from a *mosaic* of several exposures, each covering only part of the object. In using a CCD, we must match its properties to those of the instrument for which it serves as the detector and those of the telescope; see the following problem.

Problem 5.1 You have a CCD camera with 2048×2048 pixels, each 24 μm square. The angle that a pixel covers on the sky is the angle that it subtends at a distance FL, the *focal length* of the telescope. This in turn depends on the *focal ratio*, f, and the diameter D of the mirror: $FL = fD$. Show that each pixel corresponds to $5''/(fD)$, where D is in meters, and the whole CCD covers a region $168'/(fD)$ across.

You can use this camera on a 0.6 m telescope with $f = 7.5$ (usually written $f/7.5$), on a 4 m telescope at $f/7.5$, or at the prime focus of the 4 m telescope, which has $f/2.3$. What is its pixel scale and field of view at each location? If atmospheric seeing blurs the image to $0.8''$ across, where would you want to use the camera in order to get the sharpest images? If you take images of a nearby galaxy $10'$ across, which option would you use to obtain a brightness profile $I(\mathbf{x})$ for the entire galaxy?

Array detectors with large numbers of pixels are used for astronomical observations from X-rays into the infrared. The physical mechanisms that convert photons into an electrical signal depend on the spectral region, but the basic principles are similar to those for CCDs, and the signal-to-noise ratio S/N for an observation depends similarly on the number of photons detected and the sources of noise within the instrument. Unlike CCDs, most of the devices used for X-ray and ultraviolet observations count photons; each detected photon produces exactly one output count, so read noise is absent. But the speed of the electronics then limits both the maximum rate at which photons arriving at any one pixel can be counted and the counting rate across the entire array. A photon-counting device may saturate when observing a bright source.

Infrared detectors, like CCDs, are analogue devices; they rely on a variety of physical processes. In the *thermal infrared*, at wavelengths $\gtrsim 2$ μm, telescopes and the Earth's atmosphere, at temperatures ~ 300 K, radiate strongly. This extra light must be taken into account when estimating the expected signal-to-noise ratio. Cooled telescopes in space can observe in the infrared without this unwanted background, but must still contend with the emission of warm dust in the solar system. At radio frequencies, instead of detecting photons, we take advantage of the wave nature of light; see Section 5.2. Our current perspective on galaxies has been heavily influenced by our ability to observe at many wavelengths, spanning the electromagnetic spectrum.

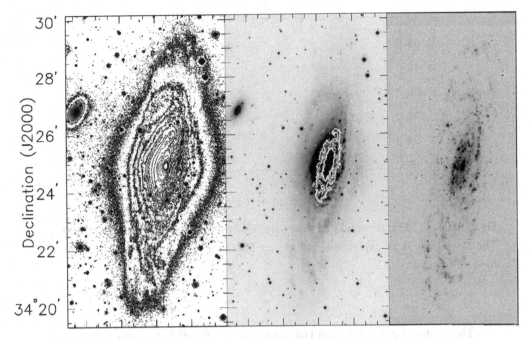

Fig. 5.3. Sb spiral galaxy NGC 7331. Left, isophotes in the R band; center, negative image in the R band, including a background galaxy, with contours of CO emission overlaid. Right, negative image in Hα, showing HII regions in the spiral arms – A. Ferguson, M. Thornley, and the BIMA survey of nearby galaxies.

Further reading: G. H. Rieke, 1994, *Detection of Light: From the Ultraviolet to the Submillimeter* (Cambridge University Press, Cambridge, UK). For a wider wavelength range, see P. Léna, F. Lebrun, and F. Mignard, *Observational Astrophysics*, 2nd edition (English translation, 1998; Springer, Berlin). For spectrograph design, see D. F. Gray, 1992, *The Observation and Analysis of Stellar Photospheres*, 2nd edition (Cambridge University Press, Cambridge, UK) – the last two are advanced texts. On statistics and observational uncertainties, see P. R. Bevington and D. K. Robinson, 1992, *Data Reduction and Error Analysis for the Physical Sciences*, 2nd edition (McGraw-Hill, New York).

5.1.2 Surface photometry of disk galaxies

The center panel of Figure 5.3 shows the surface brightness $I(\mathbf{x})$ of the Sb spiral NGC 7331 from a CCD image in the R band, around 6400Å. Here, the galaxy resembles M31 in having a bright center, a large central bulge, and tightly wrapped spiral arms in the disk. At optical wavelengths, atmospheric turbulence, or *seeing*, sets a limit to the smallest structures that can be distinguished by a conventional ground-based telescope. In this image, even the foreground stars are spread out into a disk about $1''$ across.

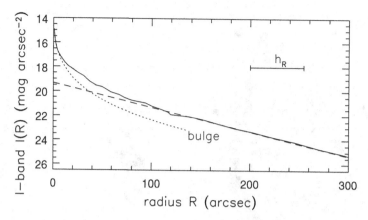

Fig. 5.4. NGC 7331: the solid line shows surface brightness in the I band, near 8000 Å. The dashed line is an exponential with $h_R = 55''$; the dotted line represents additional light – R. Peletier.

The contours in the left panel of Figure 5.3 are the R-band *isophotes*, lines of constant surface brightness. The isophotes are fairly circular in the bulge region, becoming elliptical in the disk, apart from where they are affected by the spiral arms; they are ragged at the outer edge where the signal-to-noise ratio is low. If we assume that the disk is circular and very thin, it will appear as an ellipse with axis ratio $\cos i$ when we view it at an angle i from face-on. Real disks have some thickness, so they appear somewhat rounder at any particular viewing angle; see Section 6.1. For NGC 7331 the diameter along the minor axis of the disk isophotes is only 0.35 of that measured along the major axis; the galaxy is inclined at about 75° from face-on.

Each square arcsecond at the center of the galaxy emits about ten thousand times as much light as the same area at $R = 300''$; the center is 100 times brighter than the sky, while the outer regions fade to about 1% of the sky brightness. Our printed image cannot show so much contrast, so we use a graph like Figure 5.4. If we can ignore the absorbing dust within the disk, the surface brightness is larger by a factor $1/\cos i$ than if we had seen the disk face-on. Using this or another method, we can correct to what we would observe if NGC 7331 had been face-on, and then find the average surface brightness $I(R)$ at distance R from the center.

Surface brightness is generally given in units of mag arcsec^{-2}: the flux coming from each square arcsecond of the galaxy, expressed as an apparent magnitude. At the center of NGC 7331, the I-band surface brightness $I_I(0) = 15$ mag arcsec^{-2}. Galaxies do not have sharp outer edges; for historical and technical reasons, we usually measure the size within a given isophote in the B band, centred near 4400 Å. For NGC 7331, the radius at the isophote $I_B = 25$ mag arcsec^{-2} is $R_{25} = 315''$.

Problem 5.2 Show that a central surface brightness of $15 \, \mathrm{mag \, arcsec^{-2}}$ in the I band corresponds to $18\,000 L_\odot \, \mathrm{pc^{-2}}$; explain why this does not depend on the galaxy's distance. About how far does $I_B = 25 \, \mathrm{mag \, arcsec^{-2}}$ fall below the brightness of the night sky, given in Table 1.9?

Integrating the surface brightness over the whole of an image such as the one in Figure 5.3, and extrapolating to allow for those parts of the galaxy too faint to measure reliably, gives the *total apparent magnitude*. Catalogues usually give these in the B band, or the V band at 5500 Å. A face-on galaxy would be brighter than the same galaxy seen at an angle, since starlight leaving the disk at a slant must travel further through the absorbing layer of interstellar dust in the disk. So, to compare the true brightnesses of galaxies that we observe at different inclinations, we must try to compensate for dust dimming. Finally, we must correct for the effect of foreground dust, within our own Milky Way. Catalogues often quote total magnitudes corrected to face-on viewing, denoted by ^{0}. For NGC 7331, the corrected total B-magnitude is $B_T^0 = 9.37$, and in V it is $V_T^0 = 8.75$, so $(B - V)^0 \approx 0.6$; taken as a whole, the galaxy is about the same color as the Sun.

In the stellar disk, when we average over features like spiral arms, the surface brightness $I(R)$ often follows approximately an exponential form,

$$I(R) = I(0) \exp(-R/h_R), \tag{5.1}$$

as we expect if the density of stars in the disk varies according to the double-exponential formula of Equation 2.8. The long-dashed line in Figure 5.4 shows the exponential slope with *scale length* $h_R = 55''$. For most disk galaxies, $1 \, \mathrm{kpc} \lesssim h_R \lesssim 10 \, \mathrm{kpc}$. When measured in the B band rather than in I, h_R is typically about 20% longer because the disks become redder toward the center. The inner part of the disk is typically richer in heavy elements (as Figure 4.15 shows for M33), and metal-rich stars are redder (Figure 1.5). The outer disk may have a larger fraction of young blue stars, or it may be less heavily reddened by dust; frequently it is hard to separate these effects. The exponential part of the stellar disk in many, though not all, disk galaxies appears to end at some radius R_{max}, usually in the range 10–30 kpc or $(3–5)h_R$. Beyond R_{max} the surface brightness decreases more sharply; but this is not the edge of the galaxy, since HI gas and some disk stars may be found still further out.

Problem 5.3 For NGC 7331, the radial velocity $V_r = 820 \, \mathrm{km \, s^{-1}}$; if $H_0 = 60 \, \mathrm{km \, s^{-1} \, Mpc^{-1}}$, find its distance by using Hubble's law, Equation 1.27. Show that its V-band luminosity $L_V \approx 5 \times 10^{10} L_{V,\odot}$. Is NGC 7331 more or less luminous than the Milky Way and M31? Show that $h_R \approx 3.6 \, \mathrm{kpc}$. What is the disk

Fig. 5.5. A V-band image of Sa galaxy M104, the 'Sombrero' (NGC 4594): this is a luminous galaxy with $L_V \approx 8 \times 10^{10} L_\odot$, about 10 Mpc away. Note the large bulge and numerous globular clusters – A. Cole, WIYN telescope.

radius R_{25} in kiloparsecs? How does it compare with the size of the Milky Way's disk?

From Figure 5.4 and Problem 5.2, show that the disk's extrapolated central surface brightness in the I band, is $I(0) \approx 325 L_\odot \, pc^{-2}$. Using Equation 5.1 for the disk's luminosity, show that $L_D = 2\pi I(0) h_R^2 \approx 3 \times 10^{10} L_{I,\odot}$: the disk gives roughly 60% of the total light. What is the surface density of starlight at $R = 8$ kpc? How does this compare with what we found near the Sun, in Problem 2.8?

At small radii, the surface brightness $I(R)$ rises above the level which Equation 5.1 predicts for an exponential disk; the additional light comes from the central *bulge*. In Figure 5.3, the rounder inner isophotes tell us that the bulge of NGC 7331 is an ellipsoid, not a flat disk. The bulge is generally more important in S0 and Sa galaxies: in Figure 5.5, most of the light of the Sa galaxy M104 comes from the very large bulge. Some bulges are clearly round like these, while others are almost as flat as the disk. Since it contains mainly older, redder, stars, the bulge's contribution is relatively larger in red light and smaller in the blue.

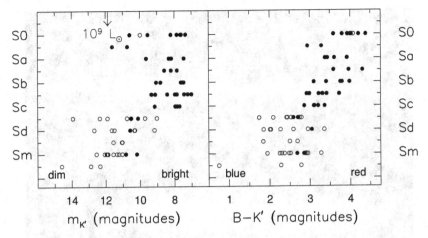

Fig. 5.6. Apparent magnitude $m_{K'}$ and $B - K'$ color of galaxies in the Ursa Major group, plotted by galaxy type. Galaxies to the right of the arrow have $L > 10^9 L_\odot$ at the group's distance of 15.5 Mpc. Open circles show galaxies for which the disk has lower central brightness: $I_{K'}(0) > 19.5$. On average, S0 galaxies are luminous and red, while the Sd and Sm systems are fainter and bluer – M. Verheijen.

In general, galaxies become bluer and fainter along the sequence from S0 to Sd and Sm. Figure 5.6 shows the apparent magnitude $m_{K'}$, in the K' bandpass at 2.2 μm, and the $B - K'$ color, for galaxies in the Ursa Major group. This is a moderately dense nearby group of galaxies, about 15 Mpc from us, roughly in the direction of the Virgo cluster. Because all the galaxies have nearly the same distance from us, those that appear brightest, with small values of $m_{K'}$, are also the most luminous. But one disadvantage of observing a group is that we do not obtain a complete sample of all types of galaxy, and may not even have a fair sample of 'average' galaxies.

Figure 5.6 shows that the S0, Sa, Sb, and Sc galaxies in the Ursa Major group are all more luminous than $10^9 L_\odot$; the dimmer spirals are all of types Scd, Sd, and Sm. The color of the S0 galaxies is approximately that of a K giant; since young blue stars are absent, most of the light comes from red stars which have evolved past the main sequence. Sd and Sm galaxies are bluer because they have a larger fraction of young stars; their optical-band colors are similar to those of late F and G stars.

In the optical image of an edge-on disk galaxy, such as the right panel of Figure 5.16 below, we see a thin dark dust lane cutting across the middle of the disk. As in the Milky Way, the light-scattering and light-absorbing dust lies in a much thinner layer than the majority of the stars. In an edge-on disk, if we measure R along the major axis of the image, and z is distance from the midplane, then, above and below the dust lane, the surface brightness often follows approximately

$$I(R, z) = I(R) \exp(-|z|/h_z). \tag{5.2}$$

Fig. 5.7. B-band images of two very different late-type disk galaxies. Top: 'superthin' Sd UGC 7321, viewed $\lesssim 2°$ from edge-on. This is a small galaxy: $L_B \approx 10^9 L_\odot$ at $d = 10\,$Mpc – L. Matthews, WIYN telescope. Below: nearby barred Magellanic or irregular NGC 55, about $10°$ from edge-on. Its linear size is about half that of UGC 7321, and $L_B \approx 2 \times 10^9 L_\odot$ at $d = 1.5\,$Mpc. Note the fluffy disk, and off-center concentration of light in the bar – A. Ferguson.

Typically, most of the starlight comes from a disk about 10% as thick as it is wide: $h_z \approx 0.1 h_R$. Some Sc and Sd galaxies are 'superthin', with even more extreme flattening, while in Sm and irregular galaxies the disk is quite thick and fluffy (Figure 5.7). In other systems, the surface brightness drops at high z less rapidly than Equation 5.2 would predict: like the Milky Way, they have a *thick disk*.

When we use Equation 5.1 to extrapolate inward, we find that in luminous spirals the central surface brightness $I(0)$ of the disk lies below some upper bound. A recent study found almost no galaxies with $I(0)$ much brighter than the average values $I_B(0) \approx 22\,$mag arcsec^{-2} and $I_K(0) \approx 18\,$mag arcsec^{-2}. Deeper searches reveal many *low-surface-brightness* disk galaxies. Most of these are faint, but a few are more luminous than the Milky Way. An example is UGC 6614 with $I_B(0) \approx 24.5\,$mag arcsec^{-2}, 10 times less than the average for 'normal' galaxies, and well below the sky brightness given in Table 1.9. The scale length is large, $h_R = 14\,$kpc, so the galaxy is luminous, with $L \approx 2 \times 10^{10} L_\odot$. Its central bulge is apparently normal. But most low-surface-brightness galaxies are much less luminous than 'normal' galaxies of high surface brightness, and lack bright centers; their scale lengths h_R are not particularly large.

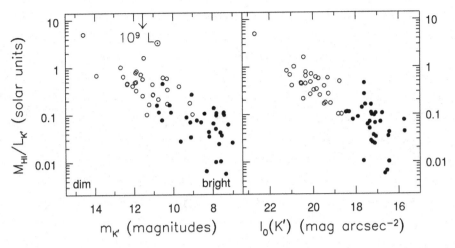

Fig. 5.8. Hɪ in galaxies of the Ursa Major group. Left, the ratio of Hɪ gas mass to luminosity in the K' band, in units of $\mathcal{M}_\odot/L_{K',\odot}$. Right, fainter galaxies have proportionately more Hɪ gas and the disk has lower extrapolated central surface brightness $I_{K'}(0)$. Open circles show low-surface-brightness galaxies with $I_{K'}(0) > 19.5$ – M. Verheijen.

Figure 5.8 shows data for galaxies in the Ursa Major group: the low-surface-brightness galaxies are the least luminous, and proportionately the richest in Hɪ gas. Like dwarf irregular galaxies, they have not been efficient at turning their gas into stars. On combining the information in this figure with that in Figure 5.6, we see that Sa and Sb galaxies on average have both higher luminosity and disks of higher central brightness than do the Sd systems. Irregular galaxies have even lower values for both quantities; see Section 4.1.

Problem 5.4 According to a study of high-surface-brightness spirals (*ApJ* **160**, 811; 1970), the disks all reach $I_B(0) \approx 21.7\,\text{mag arcsec}^{-2}$; this is *Freeman's law*. How many $L_\odot$ does the central square parsec radiate (see Problem 1.14)? If its absolute magnitude $M_B = -20.5$, how many $L_\odot$ does the galaxy emit in the B band? If we ignore light from the bulge, show that the exponential disk must have $h_R \approx 5.4\,\text{kpc}$, while $R_{25} \approx 3h_R$, and 80% of the light falls within this radius. For a low-surface-brightness galaxy with the same total luminosity, but $I_B(0) = 24.5\,\text{mag arcsec}^{-2}$, show that $< 10\%$ of the light comes from $R < R_{25}$.

Now consider many spiral disks with $L_B = 2.5 \times 10^{10} L_{B,\odot}$; the larger the length h_R, the smaller $I(0)$ must be. For $1\,\text{kpc} < h_R < 30\,\text{kpc}$, plot R_{25} (in kpc) against h_R, and R_{25} against $I(0)$. Show that R_{25} is small when h_R is small, rises to a maximum, and declines to zero at $h_R \approx 24\,\text{kpc}$. Explain why galaxies with $I(0)$ more than ten times lower than Freeman's value might have been missed from his 1970 sample. (Very small galaxies are also difficult to study: those with $R_{25} < 30''$, or 6 kpc at $d \approx 40\,\text{Mpc}$, are likely to be omitted from catalogues.)

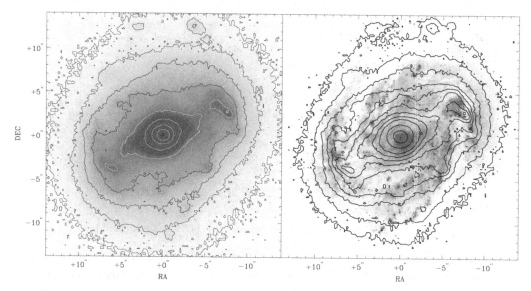

Fig. 5.9. A negative image of inner parts of Sbc galaxy M100 (NGC 4321): $26'' = 2$ kpc. Left, K-band image and isophotes at 2.2 μm, showing a central bar. Right, Hα (visible light) emission from gas around young massive stars, with K-band isophotes superposed; the bar is hidden by dust – J. Knapen 1995 *ApJ* **443**, L73.

We can also use solid-state devices to detect infrared photons; large arrays are available for wavelengths $\lambda \leq 40$ μm. These cannot be made from silicon, but must use semiconductors with lower bandgap energies, such as indium antimonide. To reduce dark counts, they must be operated very cold, at cryogenic temperatures. Dust grains affect longer-wavelength light less strongly, so infrared images of the central parts of galaxies can reveal stars which at optical wavelengths are hidden behind dust.

An infrared image in the 2.2 μm K band of the central few arcseconds of the spiral galaxy M100 reveals a central bar, with little sign of the four spiral arms so clearly outlined in visible light (Figure 5.9). Because hot young O and B stars emit relatively little of a galaxy's near-infrared light, images at these wavelengths are used to investigate the old stellar populations. In the infrared, spiral arms generally appear smoother and less prominent, showing that much of their light comes from young massive stars, and that their appearance at visible wavelengths is heavily influenced by dust lanes.

Problem 5.5 At what wavelength does a blackbody at 300 K emit most of its radiation? By how much is the flux F_λ reduced at $\lambda = 10$ μm if $T = 100$ K? For infrared work, it is best to cool those parts of a telescope from which light can fall onto the detector (e.g., with liquid nitrogen).

Fig. 5.10. SBb barred spiral galaxy NGC 3351 (M95). The left image combines ultraviolet light at 1530 Å and 2300 Å. We do not see the bar, since it lacks young blue stars; star-forming knots give the spiral arms a fragmented appearance. Right, in visible light we see a strong central bar, surrounded by a ring and smooth spiral arms – GALEX.

At ultraviolet wavelengths 1000 Å $\lesssim \lambda \lesssim$ 3000 Å, almost all the light of a normal galaxy comes from its hottest stars. An ultraviolet image of a spiral galaxy gives us a snapshot of the star formation that is renewing the disk; we do not see the more smoothly distributed older stars. In the left panel of Figure 5.10 the central bar, which is made largely of old and middle-aged stars, has disappeared. We see bright islands of light along the spiral arms, where clumps of short-lived massive stars have been born in giant molecular clouds. Their brilliance depends both on how many hot stars are concentrated there and on gaps in the obscuring dust that allow light to shine through. Figure 5.10 confirms that the spiral pattern stands out in photographs because of its luminous young stars. When we look in Section 9.4 at the optical images of galaxies at redshifts $z \gtrsim 1$, we are seeing the light which they emitted as ultraviolet radiation, redshifted by cosmic expansion. We must be careful to compare these with ultraviolet images of nearby galaxies, rather than with their optical appearance.

Since dust is very opaque at these shorter wavelengths, it can absorb much of a galaxy's ultraviolet light. In a galaxy like NGC 7331, 20%–30% of the ultraviolet photons are absorbed, but only a few percent escape from the most vigorously star-forming galaxies. The dust re-emits that energy at infrared wavelengths. Figure 2.24 shows the infrared spectrum of the star-forming inner ring of NGC 7331. Most of the energy is radiated near 100 μm, by dust with $T \sim$ 20–30 K. At shorter wavelengths we see emission from heated molecular hydrogen and from very large molecules, namely the polycyclic aromatic hydrocarbons.

The disks of spiral galaxies are not bright in X-rays. The interstellar gas is cool and supernova remnants expand supersonically; gas heated by their strong shocks, the hot winds of young stars, and binary stars are the main sources of X-rays. In a starburst (see Section 5.6 below), the stellar winds and supernova remnants overlap to heat a bubble of gas that can break through surrounding dense gas to leave the galaxy as a *superwind*.

5.2 Observing the gas

Most of the gas of a spiral galaxy is found in the disk; the cool atomic (HI) and molecular (H$_2$) hydrogen form the raw material out of which the galaxy makes new stars. In Section 5.1, we saw how to locate concentrations of gas indirectly, by looking in optical images for the dark obscuring dust lanes. Cool gas can be seen directly by its emission in spectral lines at radio wavelengths, which propagate unhindered through the dusty disk. If it is ionized by the ultraviolet radiation of hot stars, or by a shock wave, we see it in optical emission lines such as Hα. Because we observe the gas in its spectral lines, we can also measure its velocity at each point within the galaxy, and so explore its motion.

5.2.1 Radio-telescope arrays

We observe cool gas mainly at radio wavelengths, longer than about a millimeter. Because of the longer wavelengths, radio telescopes have fuzzier vision than optical telescopes. When observing at wavelength λ with a telescope of diameter D, sources that lie within an angle λ/D of each other on the sky will appear blended together: this is the *diffraction limit*, which no telescope can escape. At optical wavelengths of $\lambda \approx 5000$ Å, the diffraction limit would allow images as sharp as $1''$ from a small telescope of $D = 10$ cm, while at 1 mm we would need a telescope 200 m across to achieve the same resolution. The largest single-dish radio telescopes are about 100 m in diameter, and most are a few tens of meters across; so in general their images are not as sharp as those from optical telescopes. To make an image with $1''$ resolution at $\lambda = 20$ cm would require a telescope 40 km in diameter; a single dish of this size would collapse under its own weight. At centimeter wavelengths, and increasingly in the millimeter region as well, we use an array of smaller telescopes for *aperture synthesis*.

To see how aperture synthesis works, we can think of radio waves as oscillating electric and magnetic fields. If the wavelength is λ, the electric field $E \propto \cos(2\pi ct/\lambda)$; the voltage V induced at the focus of a radio telescope is proportional to the electric field E at that instant. Suppose that we observe with two telescopes, as in Figure 5.11. Waves from a source at elevation θ must travel an extra distance $d \cos \theta$ to arrive at telescope 2, so the wavecrests, and the peaks in voltage, are delayed by a time $d \cos \theta/c$ relative to those at telescope 1. The

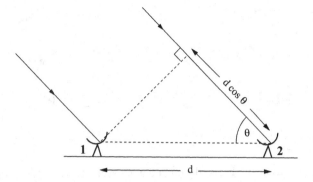

Fig. 5.11. A two-element interferometer.

voltages V_1 and V_2 at the feeds of the two telescopes are

$$V_1 \propto \cos\left(\frac{2\pi ct}{\lambda}\right) \quad \text{and} \quad V_2 \propto \cos\left(\frac{2\pi(ct - d\cos\theta)}{\lambda}\right). \tag{5.3}$$

We multiply the signals V_1 and V_2 in a *correlator*, and then filter the output to remove rapid oscillations. The result is the *fringe pattern*: a signal $S \propto \cos(2\pi d\cos\theta/\lambda)$, which varies slowly as the Earth rotates, changing the angle θ. Two sources that are nearby on the sky will produce completely different fringe patterns if the elevation $\cos\theta$ differs by an amount $\lambda/(2d)$. So this simple *interferometer* can distinguish sources separated only by an angle $\lambda/(d\sin\theta)$ – it has the resolving power of a large single dish of diameter $d\sin\theta$. The vector separating the two dishes is called the *baseline*. The interferometer's resolution is set by the component of the baseline that lies perpendicular to a vector pointing at the source.

Our two-element interferometer cannot measure the source's position: it can only tell us the value of $\cos\theta$ to within a multiple of λ/d. To get more information, we add more pairs of telescopes, with different baselines. If we could cover all the area of a circle of radius d with small dishes, and examine the signals from each of them in conjunction with every other, we would effectively build up, or synthesize, the whole picture that we would be able to see with a giant dish of diameter d.

If the source is not varying with time, then we do not need to have all the small telescopes in place at once. We can space them in an east–west line, as in the synthesis telescope at Westerbork, in the Netherlands. As seen from the source, the dishes move around each other in elliptical tracks as the Earth rotates (Figure 5.12). The Earth's rotation gives us a large number of baselines at a range of angles, even with only a few dishes. In twelve hours, we could synthesize a map of our source as it would be seen by a large elliptical telescope. When the angle of the source from the celestial equator, its *declination*, is δ, our elliptical

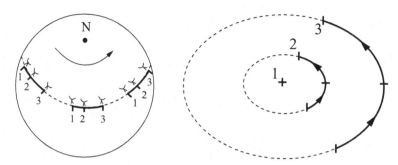

Fig. 5.12. Earth-rotation synthesis: during a day's observation, as seen from the source, the baseline vectors from dishes 2 and 3 to dish 1 sweep out an ellipse.

'telescope' is smaller in the north–south direction than in the east–west direction, by a factor of $|\sin \delta|$. To speed up the observations, radio telescopes like the Very Large Array (VLA) in New Mexico place the dishes in a large 'Y'. A four-hour observation then includes baselines at a full range of angles, and even a brief 'snapshot' contains enough information to construct a crude map.

The correlated signal S from each combination of two small telescopes samples the *Fourier transform* of the brightness distribution on the sky. We measure the transform at a wave vector that is proportional to the component of the baseline separating the two dishes, in the direction perpendicular to the line of sight to the source. To map a source that is more complicated than just a single point, we must use a computer to invert the Fourier transform, and construct the image seen by the large 'telescope'. A telescope 'dish' synthesized in this way has big holes in it, because the small dishes fill only a fraction of its area: we can sample the Fourier transform only at a limited set of points. Depending on how the telescopes are placed, this can make pointlike sources appear elongated or star-shaped instead of round; computer processing is used to correct the images.

The synthesized telescope has much less light-collecting area than a filled dish of the same diameter would have; thus, mapping faint sources can require many hours. The radio sky is so much darker than at visible wavelengths that we can ignore it; we do not need to subtract its contribution or worry about how it varies during a long observation. However, radio telescopes can suffer interference of terrestrial origin and from strong radio sources in other parts of the sky.

Each dish of diameter D receives radiation only from a patch on the sky which is λ/D across; this limits the region we can map with each pointing of the synthesis array. If the shortest distance between any pair of small telescopes is d_{min}, then sources that cover an angle on the sky larger than λ/d_{min} will hardly be detected at all. Since the dishes cannot be closer together than their physical diameter D, maps from synthesis telescopes always make the gas look clumpier than it really is. Synthesis maps should be completed by adding in the information from a single telescope of diameter at least $2D$; but often this is not practical.

Problem 5.6 In the VLA's C-array configuration, the most widely separated dishes are 3.4 km apart, and the closest are 73 m from each other. Show that the resolution in the 21 cm line of H I is roughly $13''$; explain why structures larger than $\sim 6'$ are missing from the maps.

In synthesis telescopes with diameters up to 100 km, the dishes are connected by cabling, and their signals are combined online. With widely separated dishes, we can use an atomic clock at each one for synchronization, and combine the recorded streams of data after the observations are finished. In *very long baseline interferometry* (VLBI), the combined 'telescope' can be as big as the whole Earth, so we can make images of sources less than $0.001''$ across. Putting one of the dishes into space allows even wider separations and finer resolution.

Further reading: B. Burke and F. Graham-Smith, 1996, *Introduction to Radio Astronomy* (Cambridge University Press, Cambridge, UK); and G. L. Verschuur and K. I. Kellermann, eds., 1974, *Galactic and Extragalactic Radio Astronomy* (Springer, New York); Chapters 10 and 11 of this *first* edition give a clear explanation of aperture synthesis.

5.2.2 Cool gas in the disk

Figure 5.13 shows neutral hydrogen (H I) gas in the galaxy NGC 7331, observed in the 21 cm emission line. Since the gas is moving within the galaxy, its line emission is Doppler shifted according to the radial velocity V_r. So we set the telescope to observe simultaneously in a number of closely spaced frequency channels; typically, each covers a few kilometers per second in velocity. For the most part, H I in galaxy disks is *optically thin*; the 21 cm line suffers little absorption, so the mass of gas is just proportional to the intensity of its emission. Just as for visible light, the radio power that we receive from a given cloud of H I gas decreases with its distance d as $1/d^2$. We can use the result of Problem 1.9 to find its mass, integrating the flux F_ν over the frequencies corresponding to the galaxy's gas. When we measure d in Mpc, F_ν in janskys, and V_r in km s^{-1},

$$\mathcal{M}(\text{H I}) = 2.36 \times 10^5 \mathcal{M}_\odot \times d^2 \int F_\nu \left[1421\,\text{MHz} \times \left(1 - \frac{V_r}{c} \right) \right] dV_r. \quad (5.4)$$

For NGC 7331, assuming $d = 14$ Mpc, our measured flux corresponds to $1.1 \times 10^{10} \mathcal{M}_\odot$ of atomic hydrogen; this is twice as much as in M31.

A deep H I map traces gas to $\sim 10^{19}$ H atoms cm^{-2}, corresponding to $0.1\mathcal{M}_\odot$ pc^{-2}. Generally, H I maps are not as sharp as optical images; because the signal is weak, we use relatively coarse resolution to improve the signal-to-noise ratio. Adding the emission from all channels gives the overall distribution of

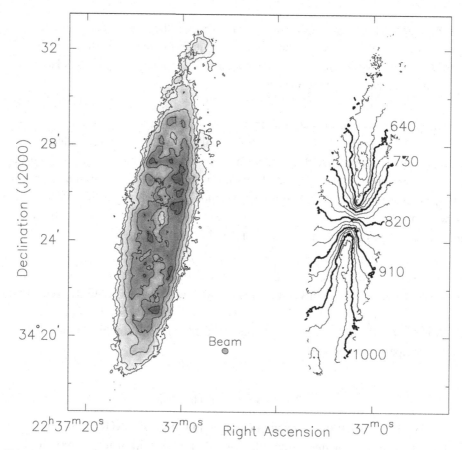

Fig. 5.13. HI gas in NGC 7331, observed with the VLA. Left, gas surface density; at $d = 14\,\text{Mpc}$, we see $11.3 \times 10^9 \mathcal{M}_\odot$ of HI, and $1' = 4\,\text{kpc}$. The outer contour shows diffuse gas, at $N_\text{H} = 2.8 \times 10^{19}\,\text{cm}^{-2}$; higher levels are at 1.2, 3.3, 6.4, and $9.5 \times 10^{20}\,\text{cm}^{-2}$. The small oval $15.7'' \times 13.7''$ shows the half-power width of the telescope *beam*: a pointlike source would appear with roughly this size and shape. Right, contours of gas velocity V_r, spaced $30\,\text{km}\,\text{s}^{-1}$ apart – M. Thornley and D. Bambic.

gas. We can also make a contour plot like Figure 5.13 showing the average radial velocity of gas at each position.

The gas clearly lies in the galaxy's disk. As in the Milky Way and M31, the center is largely gas-poor, while HI is piled up in a ring several kiloparsecs in radius. Figure 5.14 shows how the surface density of HI varies with radius. The gas is spread out much more uniformly than the stellar light; the peak density in the ring is only a few times larger than average, much less than the 10 000-fold variation in surface brightness that we saw in Figure 5.4. The HI disk is larger than that of the stars; on measuring its size at the radius where the density has fallen to $1\mathcal{M}_\odot\,\text{pc}^{-2}$, we find that it extends to about twice the optical size R_{25}. As Figure 5.15 shows, this is typical for spiral galaxies. The HI layer extends significantly beyond $2R_{25}$ in only 10%–20% of spirals; NGC 3351, in Figure 5.10, has an enormous HI disk, stretching out to $4R_{25}$.

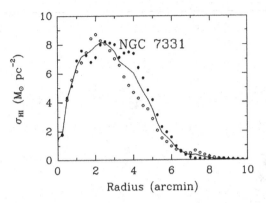

Fig. 5.14. In NGC 7331, the average surface density of HI gas at each radius, calculated separately for northern (filled dots) and southern (open circles) halves of the galaxy; the solid curve shows the average – K. Begeman.

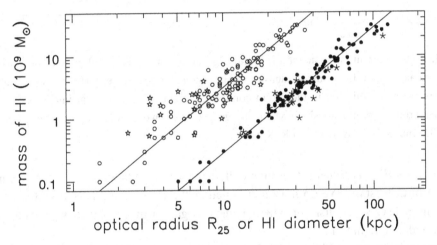

Fig. 5.15. The mass of HI gas in disk galaxies increases as the square of the optical radius R_{25} (open symbols) and the diameter at which the surface density of HI drops to $1\mathcal{M}_\odot\,\mathrm{pc}^{-2}$ (filled symbols): sloping lines show $10\mathcal{M}_\odot\,\mathrm{pc}^{-2}$ of HI within R_{25} and $3.6\mathcal{M}_\odot\,\mathrm{pc}^{-2}$ within the HI diameter. Five-pointed symbols represent low-surface-brightness galaxies – A. Broeils and E. de Blok.

The average column density of neutral hydrogen is about the same in all spiral galaxies, even those with low-surface-brightness disks. This may be due to *self-shielding*. As we discussed in Section 2.4, a thick-enough layer of HI gas can absorb virtually all the ultraviolet photons that would have enough energy to break up hydrogen molecules. If the surface density of gas is higher than about $4\mathcal{M}_\odot\,\mathrm{pc}^{-2}$, those atoms deepest in the layer will combine into H_2 molecules, protected by the atomic gas above and below.

In the outermost reaches of the galaxy, we find gas but very few stars. It is not completely clear why the outer disk displays so little enthusiasm for star formation. Perhaps most of the gas is too diffuse to pull itself together by gravity;

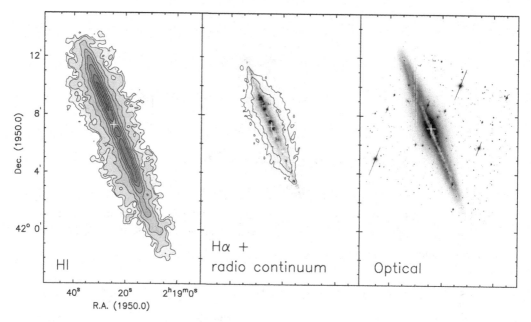

Fig. 5.16. Stars and interstellar gas in the edge-on Sc galaxy NGC 891: the cross marks
the galaxy center. Left, surface density of HI gas; center, an image in Hα, with contours
showing radio emission in the 20 cm continuum; right, *R*-band optical image; sloping
lines through bright stars show where charge has overflowed into adjoining pixels of the
CCD column – R. Swaters and R. Rand.

the disk's *differential rotation* tugs collapsing gas clouds apart before they can
give birth to stars. We do not know whether the edge of the HI marks the end
of the galaxy's gas disk, or whether there is gas further out that is ionized by
intergalactic radiation.

    The left panel of Figure 5.16 shows the distribution of HI in the unusual edge-
on Sc galaxy NGC 891. The dense gas layer in the disk is surrounded by more
diffuse HI, continuing out to ∼ 5 kpc above and below the midplane. The gas may
have been pushed out of the disk by vigorous star formation, perhaps forming
'holes' like those in the HI disks of star-forming galaxies in the Local Group
(Figures 4.4 and 4.19). NGC 891 has a remarkably large amount of gas above
the disk, but all spiral galaxies show some. The gas is 'effervescent', bubbling up
from the disk; much of it probably cools and then falls back in. There is much
more HI gas on the southern side of this galaxy than to the north; the outer parts
of galactic disks are quite commonly lopsided.

    The right panel of Figure 5.16 shows NGC 891 in the red *R* band; a thin
absorbing lane of dust bisects the galaxy. The middle panel shows that Hα emission, from gas ionized by the ultraviolet radiation of hot stars, is brightest at small
radii and close to the disk midplane, but diffuse emission persists even up to
heights of 5 kpc. The 20 cm radio continuum emission comes from roughly the
same region as the Hα. Part of this is *free–free* radiation from the hot ionized gas

and part is *synchrotron* radiation from fast-moving electrons accelerated in supernova remnants; see Section 1.2. The diffuse emission shows that both ultraviolet light and fast electrons can escape from the galactic disk, where they originate. The gas of the disk must be lumpy, so that both photons and fast particles can stream out between the clumps or clouds. High above the disk, this galaxy probably has a halo of hot gas like that of the Milky Way, at temperatures close to a million degrees.

We cannot search directly for emission lines from cool molecular hydrogen; see Section 1.2. Instead, we rely on detecting the spectral lines of molecules such as carbon monoxide, generally at millimeter wavelengths. Radio receivers at centimeter wavelengths work better than those in the millimeter region; and, locally, the ratio of CO molecules to H_2 is only about 10^{-4}. Hence observations of molecular gas are less sensitive than those for HI; it is usually easier to detect a given mass of atomic gas than the same amount of molecular material. But, because of the shorter wavelength, maps of molecular gas often have better spatial resolution than those in HI. In Figure 5.3 we see CO emission from a ring within $2'$, or $2.2h_R$, of the center of NGC 7331, corresponding to about $3 \times 10^9 \mathcal{M}_\odot$ of H_2. CO emission in spiral galaxies is generally most intense in the inner regions; there, most of the gas is molecular. Some spirals, like the Milky Way and NGC 7331, have an inner ring of dense molecular gas, while in others the CO emission peaks at the center. Unlike HI, molecular gas is not usually detected beyond the stellar disk, and the majority of gas in the disk is atomic.

Often, we use the ratio of the HI mass $\mathcal{M}(\text{HI})$ to the blue luminosity to measure how gas-rich a galaxy is. This ratio does not depend on the galaxy's distance d, because the apparent brightness in visible light, and the radio power received, both diminish as $1/d^2$. In S0 and Sa galaxies, $\mathcal{M}(\text{HI})/L_B \sim (0.05\text{--}0.1)\mathcal{M}_\odot/L_{B,\odot}$; it is about ten times larger in the gas-rich Sc and Sd systems. The Sc, Sd, and Sm galaxies have been consuming their gas supply fairly slowly; a smaller fraction of their light comes from older, redder, stars and relatively more from hot young stars. From Figure 5.6, we saw that these galaxies have the bluer colors expected when young stars predominate. In them, young associations and their HII regions stand out more clearly, and the individual HII regions around these associations are also larger than those in Sa and Sb galaxies.

Problem 5.7 For a distance $d = 14\,\text{Mpc}$, find the blue luminosity of NGC 7331 from its apparent total magnitude. Show that $\mathcal{M}(\text{HI})/L_B \approx 0.2$ in solar units. Assuming the galaxies to have the same $B - V$ color as the Sun, use Table 4.1 to show that M31 has about the same $\mathcal{M}(\text{HI})/L_B$ as NGC 7331, and compute this ratio for the Magellanic Clouds. On the basis of Problem 2.8, assume that $\mathcal{M}/L \gtrsim 2$ for the stars in NGC 7331; show that, even if we include the molecular gas, the mass in its stars is at least five times larger than that in cool gas. This galaxy has already converted most of its gas into stars.

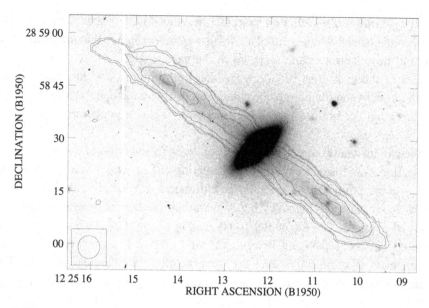

Fig. 5.17. A negative image of the S0 galaxy UGC 7576 in the V band: a thin ring of dust, gas, and stars orbits over the galaxy's pole. Contours show $5 \times 10^9 \mathcal{M}_\odot$ of HI gas in the polar ring; the disk of the S0 galaxy has hardly any cool gas. The circle on the left shows the radio telescope's beam – A. Cox.

The gas content of S0 galaxies is very different from that of spirals. The stellar disks of most S0s have so little gas that it is hardly measurable. Star formation or some other process has depleted the gas, leaving behind 'stellar fossils': disks where no significant numbers of new stars have been made over the past few gigayears. But a few S0 and elliptical galaxies have $\gtrsim 10^{10} \mathcal{M}_\odot$ of HI, as much as the most gas-rich spirals. Often this gas does not lie in the galaxy disk, or share the orbital motion of the disk stars, but forms a tilted ring encircling the galaxy. Occasionally, the rotation of the gas is *retrograde*, opposite to that of the disk stars; in other cases – see Figure 5.17 – it orbits as a polar ring, perpendicular to the galaxy's disk. Because the angular momentum of the ring and that of the central galaxy are so different, we think that the gas was captured at a late stage, after the galaxy's central body had formed. In a few S0 galaxies, such as NGC 4550, a substantial minority of the disk stars rotate in the opposite sense to the majority: they presumably formed from late-arriving gas, which was caught into a retrograde orbit. Counter-rotating stars are not common; they are detected in fewer than 1% of all disk galaxies.

5.3 Gas motions and the masses of disk galaxies

We saw in Section 2.3 that the stars and gas of the Milky Way account for only a fraction of its mass; most of it is 'dark', undetectable except through its gravitational attraction. The same is true for most spiral galaxies. If we measure the

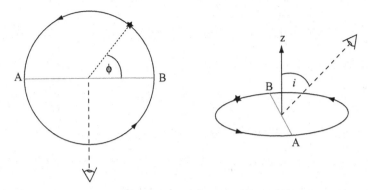

Fig. 5.18. Left, a rotating disk viewed from above. Azimuth ϕ, measured in the disk plane, gives a star's position in its orbit; an observer looks from above the disk, perpendicular to diameter AB. Right, the observer's line of sight makes angle i with the disk's rotation axis z.

speed $V(R)$ of a circular orbit at radius R, even in a flattened galaxy we can use the radial-force equation

$$\frac{V^2(R)}{R} = \frac{G\mathcal{M}(<R)}{R^2} \tag{3.20}$$

for a rough estimate of the mass $\mathcal{M}(<R)$ within that radius. We saw in Section 5.2 that the HI layer of a spiral galaxy generally extends out about twice as far as the stellar disk, as measured by the isophotal radius R_{25}. Usually, the gas speeds are approximately constant to the edge of the HI disk, implying that $\mathcal{M}(<R)$ continues to rise; the outer parts of the galaxy contain much mass but emit little light. This section discusses how we determine a spiral galaxy's *rotation curve* $V(R)$, and what we know about gravitational forces and dark matter in disk galaxies.

5.3.1 The rotation curve

The predominant motion of gas in a spiral galaxy is rotation; random speeds in the HI gas are typically only 8–10 $\mathrm{km\,s^{-1}}$, even less than for the stars. So the *asymmetric drift*, which we discussed in Section 2.2, is small; we can assume that, at radius R, a gas cloud follows a near-circular path with speed $V(R)$. All we can detect of this motion is the radial velocity V_r toward or away from us; its value at the galaxy's center, V_{sys}, is the *systemic velocity*. Suppose that we observe a disk in pure circular rotation, tilted at an angle i to face-on, as in Figure 5.18. We can specify the position of a star or gas cloud by its radius R and azimuth ϕ, measured in the disk from the diameter AB lying perpendicular to our viewing direction. There, the radial velocity is

$$V_r(R, i) = V_{\text{sys}} + V(R)\sin i \cos \phi. \tag{5.5}$$

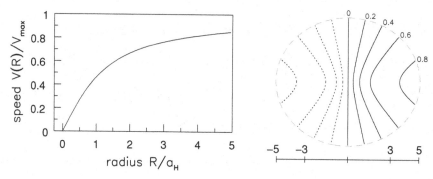

Fig. 5.19. Left, the rotation curve $V(R)$ in the 'dark-halo' potential of Equation 2.19, in units of $V_H = V_{max}$. Right, the spider diagram of $V_r - V_{sys}$ for a disk observed 30° from face-on; contours are marked in units of $V_H \sin 30°$, with negative velocities shown dotted.

Contours of constant V_r connect points with the same value of $V(R)\cos\phi$, forming a 'spider diagram' like that in Figure 5.19. The line AB is the *kinematic major axis*, the azimuth where V_r deviates furthest from V_{sys}. In the central regions, where approximately $V(R) \propto R$, the contours are parallel to the minor axis; further out, where the rotation speed is nearly constant, they run radially away from the center. If $V(R)$ begins to fall, the extreme velocity contours close back on themselves. Taking V_{max} as the peak of the galaxy's rotation curve, the spread between the largest and the smallest values of the measured velocity is $W = 2V_{max} \sin i$.

> **Problem 5.8** In a galaxy where the potential follows the Plummer model of Equation 3.11, use Equation 3.20 to find the rotation curve $V(R)$; show that $V_{max}^2 = V^2(\sqrt{2}a_P) = 2G\mathcal{M}/(3\sqrt{3}a_P)$. Sketch $V(R)$ for $R \leq 4a_P$. For inclination $i = 30°$, draw a spider diagram with contours of $V_r - V_{sys}$ at 0.2, 0.4, 0.6, and 0.8 times $V_{max} \sin i$; show that the last of these forms a closed loop.

In many galaxies, like our Milky Way, the inclination i changes with radius; the gas disk is warped. Then, the outer part of the spider is generally rotated with respect to the inner region; we can use that twist in the kinematic major axis to deduce the amount of warping. If $V_r \neq 0$ along the apparent minor axis, the gas must have some motion toward or away from the galaxy's center. We will see in Section 5.5 that radial motion is characteristic of barred galaxies, where the gas follows elongated oval orbits. Random motions of the gas clouds, and streaming velocities induced by the spiral arms (see Section 5.5 below), also distort the contours. But the characteristic spider pattern is usually visible: we see it in the velocity field in the right panel of Figure 5.13.

We can find $V(R)$ and the inclination i by choosing values so that the computed velocity contours are close to those measured in H I. Figure 5.20 shows the rotation curve $V(R)$ derived from H I and CO observations of NGC 7331. It climbs steeply

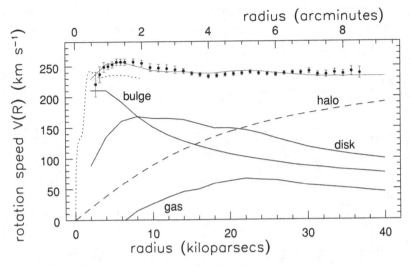

Fig. 5.20. Points give the rotation curve of NGC 7331, as found from the HI map of Figure 5.13; vertical bars show uncertainty. CO gas (dotted), observed with a finer spatial resolution, traces a faster rise. The lower solid curves show contributions to $V(R)$ from the gas disk, the bulge, and the stellar disk. A dark halo (dashes) must be added before the combined rotation speed (uppermost curve) matches the measured velocities – K. Begeman and Y. Sofue.

over the first 1–2 kpc, then remains approximately flat out to the last measured points, at around 37 kpc. As in many giant spirals, the rising part of the rotation curve is very steep; often, as here, HI observations lack the spatial resolution to follow the rapid climb. At all radii, the angular speed $V(R)/R$ is decreasing; gas further out takes longer to complete an orbit about the galactic center. This *differential rotation* is typical of spiral galaxies.

5.3.2 Dark matter in disk galaxies

We can compare the rotation curve of Figure 5.20 with what we would expect if the mass of the galaxy had been concentrated entirely in its stars and gas. For the stellar disk and the bulge, we assume that the density of stars is proportional to the R-band light, and guess at the mass-to-light ratio $\mathcal{M}/L$. For the gas disk, the surface density is approximately 1.4 times that measured in HI, since helium contributes a mass about 40% of that in hydrogen; see Section 1.5. We calculate the contributions to the radial force from each component separately, and we add them to find the total. Thus $V^2(R)$ for the galaxy is the sum of contributions from the various parts.

Taking the bulge to be nearly spherical, we can find its inward force from Equation 3.20. Because the stellar and gas disks are flattened, their force can point either inward or outward. At $R \lesssim 6$ kpc, the force from the gas disk is outward,

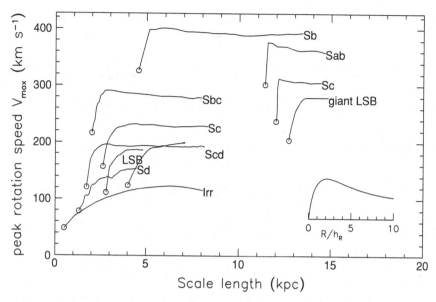

Fig. 5.21. Rotation curves for disk galaxies of various types. Open circles show the scale length h_R of the stellar disk and the peak rotation speed V_{max} for each galaxy. Curves are plotted in units of R/h_R, to the same horizontal scale as for the inset, showing $V(R)$ for the exponential disk (Equation 5.1). LSB denotes a low-surface-brightness galaxy; the 'giant LSB' with $V_{max} \approx 200$ km s^{-1} is UGC 6614. The measured rotation does not fall as it should if the stellar disk contained most of the mass – A. Broeils and E. de Blok.

making a negative contribution to $V^2(R)$. In Figure 5.20, the ratio M/L has been adjusted so that gas and luminous stars account for as much of the galaxy's rotation as possible: this is the 'maximum-disk' model. If no other matter had been present, we see from Figure 5.20 that the rotation speed should have begun to fall at around 20 kpc from the center. Like the Milky Way, this galaxy contains substantial mass in regions beyond the visible stellar disk. The curve labelled 'halo' shows how a spherical halo of dark matter could provide enough inward force to account for the measured rotation speed; at least 75% of the total mass appears to be dark. The outer reaches of this galaxy contain almost exclusively HI gas and unseen matter.

Problem 5.9 Using Equation 3.20, find the total mass of NGC 7331 within $R = 37$ kpc, and show that the ratio $M/L \approx 10$ in solar units. (This is considerably higher than the value $M/L \approx 3$ that we found for the gas and stars of the Galactic disk near the Sun in Problem 2.8; the visible parts of the galaxy contribute only a small fraction of its mass.)

Figure 5.21 shows the rotation curves for a number of disk galaxies of various types, found by observing the HI gas. These provide our best information about rotation in the outer parts of the galaxy. The circle at the starting point of each curve shows the scale length h_R of the stellar disk and the peak rotation speed

V_{max}, while the horizontal extent shows the number of scale lengths out to which the rotation was measured. Beyond its peak, $V(R)$ remains fairly steady; it does not drop as it should if the mass of the galaxy had been in an exponential disk. As in NGC 7331, the higher-than-expected rotation requires an additional inward force, which we ascribe to a dark halo.

Peak rotation speeds in spiral galaxies are usually 150–300 km s^{-1}. They rarely rise above 400 km s^{-1}, and the fastest measured rotation is about 500 km s^{-1}, in the S0/Sa galaxy UGC 12591. Larger galaxies, with longer scale lengths h_R, generally rotate more rapidly; these tend to be the Sa and Sb galaxies, rather than the Sc, Sd, and Sm systems. The rotation curves of Sa and Sb galaxies initially climb steeply, showing that relatively more of their mass is closer to the center. In these systems, the luminous matter in the disk and the bulge is concentrated at small radii, and the dark matter in the halo also becomes very dense there. In Sd and Sm galaxies, the rotation speed increases more gradually. These galaxies do not have large bulges, and Figure 5.8 showed us that their luminous disks have low central surface brightness. The rotation curves show that the dark halo also lacks central concentration; its core, where the density is nearly constant, must be larger in relation to the galaxy's stellar disk. Most low-surface-brightness galaxies rotate slowly, with gently ascending rotation curves like those of Sd or Sm galaxies; but there are some with higher speeds and faster-rising rotation curves.

The proportion of dark matter required to explain these rotation curves varies from about 50% in Sa and Sb galaxies to 80%–90% in Sd and Sm galaxies. There may be yet more dark material out beyond the last point where we have observed H I gas; the 'total mass' of a spiral galaxy measured in this way is only a lower bound. We must turn to the orbital motion of galaxies in binary pairs or groups to study the dark matter at larger radii: look back at Section 4.5, or forward to Chapter 7.

Galaxies and their dark matter do not provide enough mass to halt the cosmic expansion. For most galaxies, we must use Hubble's law and the recession speed V_{sys} to give a rough estimate of the distance d. Equation 1.27 tells us that

$$d \approx h^{-1}[V_{sys}\,(\text{km s}^{-1})/100]\,\text{Mpc}, \tag{1.27}$$

where $h = H_0/100\,\text{km s}^{-1}\,\text{Mpc}^{-1}$. So when we estimate the luminosity L of a galaxy from its apparent brightness by using Equation 1.1, $L \propto h^{-2}$. The mass inferred from Equation 3.20 depends on the Hubble constant as $\mathcal{M} \propto h^{-1}$, so the mass-to-light ratios derived from the rotation curves of spiral galaxies follow $\mathcal{M}/L \propto h$. For most spirals, we find $5h \lesssim \mathcal{M}/L \lesssim 25h$, in units of the Sun's mass and blue luminosity.

From Equation 1.25, the blue light of all the galaxies averages out to $2 \times 10^8 h L_\odot\,\text{Mpc}^{-3}$. So if the mass-to-light ratio for each were roughly the same as that for the gas-rich disk systems where we observe H I gas, they would contribute a density $\rho_{gal} \sim (1-5) \times 10^9 h^2 \mathcal{M}_\odot\,\text{Mpc}^{-3}$, less than 0.02 of the critical density ρ_{crit} given by Equation 1.30. The dark matter in galaxies is far short of what would be

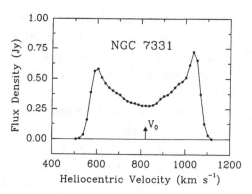

Fig. 5.22. The HI global profile for NGC 7331: radio power F_ν (in janskys) received from gas moving at each velocity, measured with respect to the Sun; $V_0 = V_{sys}$, the recession speed at the galaxy center – K. Begeman.

needed to save us from an ever-expanding Universe. What is more, Equation 1.40 tells us that the Universe contains baryons equivalent to $0.03\rho_{crit} \lesssim \rho_B \lesssim 0.07\rho_{crit}$. Most of those neutrons and protons are not in the galaxies. We will see below, in Chapter 7, that groups and clusters of galaxies contain hot diffuse intergalactic gas, which could be a storehouse for the 'hidden' baryons.

5.3.3 The Tully–Fisher relation

If we want to know only the peak rotation speed V_{max} in a galaxy, we can use a single-dish radio telescope with a large enough beam to include all the HI gas, and measure how much there is at each velocity. Figure 5.22 shows this *global profile* for NGC 7331. Because much of the gas lies at radii where $V(R)$ is nearly constant, most of the emission is crowded into two peaks near the extreme velocities. This double-horn profile is characteristic of galaxies where the rotation curve first rises, then remains roughly flat; the separation of the peaks is $W \approx 2V_{max} \sin i$. If instead we had observed a galaxy in which the rotation curve was rising at all radii, or if we looked only at the inner part of the disk in Figure 5.19, we would see a flat-topped or centrally peaked profile instead. Brighter galaxies rotate faster on average, which tells us that they are more massive. Brent Tully and J. Richard Fisher showed that the rotation speed of a galaxy increases with its luminosity, roughly as $L \propto V_{max}^\alpha$, with $\alpha \sim 4$: this is the *Tully–Fisher relation*. The observed values fall closer to a single curve when L is measured in the red or near-infrared. The blue luminosity is more likely to fluctuate over time, since young massive stars contribute much of the light. In the blue, a galaxy that has recently had a burst of star formation will temporarily be much brighter than it usually is, while V_{max} remains unchanged; so the observed luminosities will scatter widely about their mean at any given rotation speed.

Figure 5.23 plots the width of the global profile against the apparent magnitude measured at $K' \approx 2.2$ μm for galaxies in the Ursa Major group; the luminosity

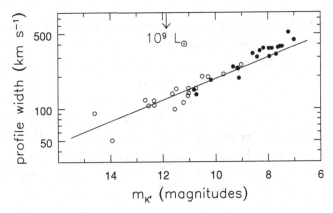

Fig. 5.23. For galaxies in the Ursa Major group: from the HI global profile, width $W/\sin i \approx 2V_{max}$ plotted against apparent K'-magnitude. Low-surface-brightness galaxies (open circles) follow the same relationship as do those of high surface brightness (filled circles). The solid line passing through $L = 3 \times 10^{10}L_\odot$, $V_{max} = 205\,km\,s^{-1}$ has slope $L \propto V_{max}^4$ – M. Verheijen.

increases slightly slower than the fourth power of V_{max}. Another recent study, measuring the galaxy light in the I band at 0.8 μm, found

$$\frac{L_I}{4 \times 10^{10}L_{I,\odot}} \approx \left(\frac{V_{max}}{200\,km\,s^{-1}}\right)^4. \qquad (5.6)$$

When the luminosity L is measured in blue light, the exponent of V_{max} is close to three. Figure 5.6 showed us that more luminous galaxies tend to be redder, so a fast-rotating luminous galaxy will be even brighter relative to a faint and slowly rotating one at red or infrared wavelengths than when both are measured in blue light. This is another of the confusing consequences of the way that the average luminosity and color of galaxies are related to their other properties.

If galaxies contained no dark matter, we could understand the Tully–Fisher relation fairly easily; see the following problem. But, since the rotation speed V_{max} is set largely by the unseen material, while the luminosity comes from the stellar disk, the link between them is puzzling. Somehow, the amount of dark matter is coordinated with the luminous mass.

The Tully–Fisher relation can be used to estimate distances to galaxies and galaxy groups; it gives us an important step on the *cosmic distance ladder*. We first calibrate the relation, using galaxies close enough that we can estimate their distances by using Cepheid variables. From the HI profile of a more distant system observed with a radio telescope, we measure V_{max}, and then use the Tully–Fisher relation to infer the galaxy's intrinsic luminosity in visible or infrared light. Comparing this with the observed apparent magnitude yields the distance. If we know the Hubble constant H_0, we can find the *peculiar velocities* of our galaxies – the amount by which their motion differs from the homogeneous and isotropic cosmic expansion. We will discuss these large-scale motions in Section 8.4.

Problem 5.10 Ignoring the bulge, use Equation 3.20 to explain why we might expect the mass $\mathcal{M}$ of a spiral galaxy to follow approximately

$$\mathcal{M} \propto V_{max}^2 h_R.$$

Show from Equation 5.1 that $L = 2\pi I(0)h_R^2$, and hence that, if the ratio $\mathcal{M}/L$ and the central surface brightness $I(0)$ are constant, then $L \propto V_{max}^4$. In fact $I(0)$ is lower in low-surface-brightness galaxies: show that, if these objects follow the same Tully–Fisher relation, they must have higher mass-to-light ratios, with approximately $\mathcal{M}/L \propto 1/\sqrt{I(0)}$.

Problem 5.11 NGC 7331 has apparent magnitude $m_I = 7.92$. Using Table 1.4 to find M_I for the Sun, and estimating V_{max} from the flat part of the rotation curve in Figure 5.20, show that Equation 5.6 gives its distance as $d \approx 16$ Mpc. If $V_{max} = V_0 = 200\,\mathrm{km\,s^{-1}}$ for the Milky Way, what does Equation 5.6 predict for its luminosity L_I? This differs from L_V given in Table 4.1; what might cause the discrepancy? If a galaxy has a peak rotation speed $V_{max} = 200\,\mathrm{km\,s^{-1}}$, and its apparent magnitude is $m_{K'} = 13$, use Figure 5.23 to show that it is about 20 times more distant than the Ursa Major galaxies.

5.4 Interlude: the sequence of disk galaxies

Edwin Hubble defined the progression from S0 galaxies through Sa to Sc spirals almost completely by the appearance of the spiral arms in visible light. Spiral arms are absent from the disks of S0 galaxies, while '*as the sequence progresses, the arms increase in bulk at the expense of the nuclear region, unwinding as they grow, until in the end they are widely open and the nucleus is inconspicuous*': the 'nucleus' here is what we would today call the galaxy's bulge. Hubble's original scheme has since been extended to include the Sd galaxies, which almost completely lack a bulge, and the Magellanic Sm systems. Nearby galaxies are still classified by a human expert, comparing optical images of the systems under study with those of galaxies that have already been assigned a type.

The modified Hubble classification is useful because other characteristics of galaxies are linked with their position along the sequence. Table 5.1 lists these, beginning with the spiral structure. Some are linked fairly directly with the spiral properties: for example, S0 galaxies, which lack the hot young stars that outline spiral arms, are redder than Sc and Sd galaxies. Figure 5.24 shows spectra of an S0 galaxy, an Sb, and an Sc. Most of the light of the S0 galaxy emerges at the longest wavelengths, where we see absorption lines characteristic of cool K stars. In the blue, we see the H and K lines of calcium, and the G band, features characteristic of the hotter solar-type stars that produce most of the light at these wavelengths; see Section 1.1. There is little light at wavelengths shorter than 4000 Å and no

Table 5.1 The sequence of luminous disk galaxies

Characteristic	S0–Sa	Sb–Sc	Sd–Sm
Spiral arms	Absent or tight		Open spiral
Color	Red: late G star	Early G star	Blue: late F star
[a]$B - V$	0.7–0.9	0.6–0.9	0.4–0.8
[a]$u - r$	2.5–3	1–3	1.5–2.5
Young stars	Few		Relatively many
HII regions	Few, small		More, brighter
Gas	Little gas		Much gas
$\mathcal{M}(\mathrm{HI})/L_B$	≲0.05–0.1		~0.25 to >1
	Luminous		Less luminous
L_B	$(1–4) \times 10^{10} L_\odot$		$(<0.1–2) \times 10^{10} L_\odot$
$I(0)$	High central brightness		Low central brightness
	Massive		Less massive
$\mathcal{M}(<R)$	$(0.5–3) \times 10^{11} \mathcal{M}_\odot$		$(<0.2–1) \times 10^{11} \mathcal{M}_\odot$
Rotation	Fast-rising $V(R)$		Slowly rising $V(R)$

[a] See Tables 1.2 and 1.3 for definitions of these wavelength bands.

prominent emission lines. By contrast, the Sc galaxy emits most of its light in the blue and near-ultraviolet part of the spectrum; the light comes mainly from hot young stars, which also heat and ionize the gas responsible for the prominent emission lines. The uppermost spectrum is for a *starburst* galaxy, where many stars have recently been born: see Section 5.6. We saw from Figure 5.8 that Sc and Sd galaxies contain a higher proportion of gas than do the Sa and Sb systems. So perhaps it is not surprising that Sc and Sd galaxies have made a larger fraction of their stars in the past gigayear.

Galactic bulges are much brighter at their centers than the disks; see Figure 5.4. Thus galaxies early in the sequence have higher central brightness $I(0)$. When galaxies are too distant to allow a clear view of the spiral arms, Figure 5.25 shows that we can base a classification on the degree to which their light is concentrated toward the center. We can also calculate the coarseness or asymmetry, both of which measure how far the galaxy's appearance differs from an axially symmetric disk. Both spiral arms and patches of vigorous star formation are absent in S0 galaxies; Figure 5.25 shows how they become more prominent along the sequence from Sb to Sm. Distant galaxies can also be classified on the basis of their spectra: Figures 8.5 and 9.15 make use of such a scheme.

Figure 5.6 showed us that S0 and Sa galaxies are on average more luminous than Sd and Sm systems, and we saw in the previous section that they also are generally more massive. Somehow, more luminous and more massive galaxies know that they must develop into one of the former types, while smaller systems become Sd and Sm galaxies. But all the relations we have described in this section hold true only on average. Although the S0, Sa, and Sb galaxies tend to have higher central surface brightness than do Sd systems, low-surface-brightness S0 galaxies are also found. Some S0 galaxies are less luminous than some Sm galaxies; the S0 systems NGC 404 and NGC 5102 both have $L_B \lesssim 10^9 L_\odot$. Sc galaxies

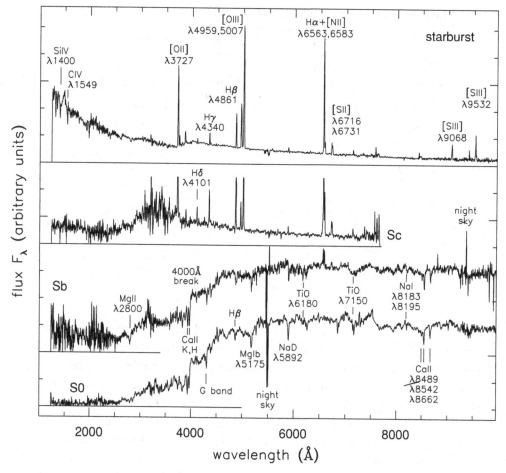

Fig. 5.24. Spectra of galaxies from ultraviolet to near-infrared wavelengths; incompletely removed emission lines from the night sky are marked. From below: a red S0 spectrum; a bluer Sb galaxy; an Sc spectrum showing blue and near-ultraviolet light from hot young stars, and gas emission lines; and a blue starburst galaxy, that has made many of its stars in the past 100 Myr – A. Kinney.

encompass a huge range in luminosity: the giant UGC 2885, with $L_B \gtrsim 10^{11} L_\odot$, is two hundred times more luminous than the Local Group's M33. From the image alone, it is almost impossible to distinguish between a small nearby spiral and a distant luminous example.

Such partial linkage of the various properties of galaxies is infuriating to theoreticians, who do not yet have much understanding of it. Because so many characteristics of galaxies are related, we must be careful in studying any particular property. Early work on rotation curves provides a cautionary tale. In local samples of galaxies within ∼20 Mpc, rotation speeds of early-type spirals were found to be higher than those of late-type Sc galaxies, inviting the conclusion that early-type spirals are in general faster rotators. But the most luminous Sc galaxies are rare,

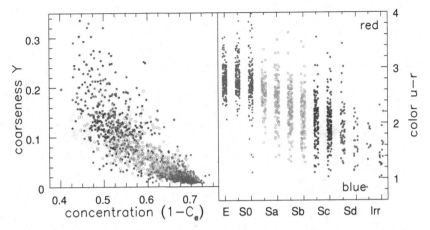

Fig. 5.25. For 1421 galaxies of the Sloan Digital Sky Survey, the left panel shows how far the r-band light is concentrated to the center, and the coarseness or deviation from a smooth image. Elliptical and S0 galaxies (filled dots) are the most concentrated, while Sc, Sd, and irregular galaxies (stars) are lumpiest, with Sa and Sb galaxies (open dots) between them. Right, average color becomes bluer along the sequence from S0 to Sd – C. Yamauchi 2005 *AJ* **130**, 1545.

and there is none close to the Milky Way. Later surveys further afield revealed very luminous and rapidly rotating Sc galaxies: V_{max} depends mainly on luminosity, through the Tully–Fisher relation.

5.5 Spiral arms and galactic bars

The photogenic arms of spiral galaxies are the most striking luminous structures in any galaxy atlas, yet spiral structure still has its puzzling aspects. Almost all giant galaxies with gas in their disks have some kind of spiral, although simple arguments imply that spiral arms should rapidly disappear. Two properties of the disk seem to be essential: differential rotation, which tends to shear any feature into a trailing arm-segment, and self-gravity, which allows the spiral to grow, fed by the energy of the galaxy's rotation. Gas appears to be required for a spiral, although not for a bar: the disks of S0 galaxies lack both gas and spiral arms, but they are as likely to be barred as are the gas-rich spiral galaxies. Beyond these simple statements, the subject becomes confusingly complex.

5.5.1 Observed spiral patterns

The arms of a spiral galaxy are bluer than the rest of the disk, and Hα emission betrays hot ionized gas around young massive stars. Since stars hot enough to emit the ultraviolet photons that ionize hydrogen atoms live only about 10 Myr, spiral arms must be sites of active star formation. Figure 5.26 compares the spiral

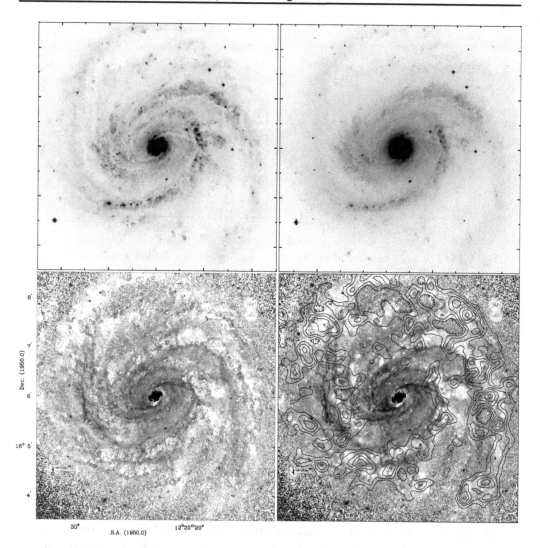

Fig. 5.26. Sbc galaxy M100: $26'' = 2$ kpc. Top, B band (left) and I band (right); in these negative images, dark dust lanes just inside the bright spiral arms appear as thin light filaments. Lower panels show $B - K$ color. Light areas are blue with young massive stars, and dark regions show red regions where dust lanes spiral into the galaxy center; the overexposed core appears as a dark hole. Contours show Hα emission (left) and HI gas (right), which are both concentrated in spiral arms; the center is largely empty of HI – J. Knapen 1996 *MNRAS* **283**, 251.

structure of the Sc galaxy M100 in blue light with that seen in the I band at about 1 μm: the arms are sharper in the blue, since more of their light comes from young stars. The lower panels show that both the ionized gas of HII regions and the cool atomic HI gas are concentrated in the spiral arms. Spiral arms stand out most clearly when they are organized into a *grand design* that can be traced over many radians in angle and a substantial range in radius. Using galactocentric polar coordinates (R, ϕ), we can describe the shape of an m-armed spiral by

Fig. 5.27. NGC 3949, a fairly luminous Sbc galaxy with $L_V \sim 7 \times 10^9 L_\odot$, shows a flocculent spiral pattern, without long continuous arm-segments – Hubble Space Telescope.

the equation

$$\cos\{m[\phi + f(R, t)]\} = 1. \tag{5.7}$$

The function $f(R, t)$ describes how tightly the spiral is wound; if $|\partial f/\partial R|$ is large, the arms are closely wrapped, whereas if it is small, they are open. The *pitch angle* i, the angle between the arm and the tangent to the circle at radius R, is given by

$$\frac{1}{\tan i} = \left| R \frac{\partial \phi}{\partial R} \right| = \left| R \frac{\partial f}{\partial R} \right|. \tag{5.8}$$

In Sa spirals, i averages about $5°$, while in Sc galaxies it is generally in the range $10° < i < 30°$.

Problem 5.12 If the pitch angle i remains constant, show that we have a *logarithmic spiral*, with $f(R, t) \tan i = \ln R + k$ for some constant k. Starting from a point on an arm, and moving outward at fixed angle ϕ, explain why we cross the next arm at a radius $\exp(2\pi \tan i /m)$ times larger.

While we often think of a typical spiral as two-armed, many have three or four arms. Some galaxies have a *flocculent* pattern, with many short arm-segments instead of a continuous spiral; see Figure 5.27, or M33 in Figure 4.1. In Magellanic SBm spirals the bar is often off-center and the spiral is asymmetric, with one arm being much the strongest. In barred spiral galaxies, the spiral arms usually appear to grow out from the ends of the bar.

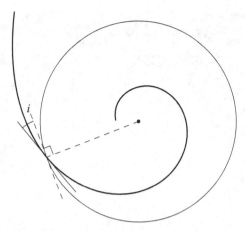

Fig. 5.28. In a disk rotating anticlockwise, where the rotation rate $\Omega(R)$ falls with radius R, stars that initially lie along a radial line are wound into a trailing spiral; the angle i is the *pitch angle* of the spiral.

We can characterize a spiral as *leading*, with the tips of the arms pointing forward in the direction of the galaxy's rotation, or *trailing*, with tips aimed in the direction opposite to the rotation. To know which applies, we must first find which side of the galaxy is nearest to us, usually by looking for differences in the obscuring effects of dust. This is not easy; but where we can determine the sense, the spiral arms *almost always trail*. In Figure 5.26, the narrow dust lanes on the concave inner sides of the spiral arms show that dust-bearing gas is being compressed there. We will see this as a sign that the arms do not contain a fixed population of stars and gas, but form a *density wave*, a 'stellar traffic jam' where stars are packed more densely and move more slowly along their orbits.

One reason why we believe many spiral arms to be density waves is that the galaxy's *differential rotation* would otherwise wind them rapidly into a very tight curl. To see why, suppose that stars are initally spread along a straight line through the galaxy's center, given by $\phi = \phi_0$ (Figure 5.28). Each star moves in its orbit with speed $V(R)$, at the angular rate $\Omega(R) = V(R)/R$, so after time t they lie on a spiral given by the curve $\phi = \phi_0 + \Omega(R)t$. In the language of Equation 5.7, we have $f(R, t) = -\phi_0 - \Omega(R)t$. Since the angular speed $\Omega(R)$ generally drops with radius, then, if we take $\Omega(R) > 0$, $f(R, t)$ increases on moving out along the arm to larger R, so ϕ must diminish. This is a trailing spiral, since the tip of the arm points in the opposite sense to the galaxy's rotation.

As time goes on, this spiral becomes ever more closely wound. Near the Sun's position in the Milky Way, where $V(R) \approx 200\,\text{km}\,\text{s}^{-1}$ is nearly constant and $R \approx 8\,\text{kpc}$, the pitch angle i of Equation 5.8 tightens according to

$$\cot i = R\left|\frac{\text{d}\Omega(R)}{\text{d}R}\right| t \approx \frac{200}{8}\left(\frac{t}{1\,\text{Gyr}}\right), \qquad \text{or } i \approx 2° \times \left(\frac{1\,\text{Gyr}}{t}\right). \qquad (5.9)$$

After only a gigayear, this spiral should be much tighter than those observed in Sc galaxies like our own. Any initial spiral pattern would suffer a similar winding-up; the stars of the spiral arms must perpetually be renewed.

5.5.2 Theories for spiral structure

Spiral structure is a complex phenomenon, and it is likely that no single process is responsible for the whole range of observed structures. Some galaxies, particularly those with flocculent spiral arms, may simply recreate their spiral patterns every few rotation periods. Once a cloud of gas has made its first stars, blast waves from the supernova explosions of short-lived massive stars will compress the surrounding gas. This may trigger the birth of yet more stars, so that star formation propagates through the gas. Differential rotation then draws the cloud out into a segment of a trailing spiral arm. By the time this fragment is strongly stretched, the gas will have been used up, the hot stars will have faded, and the region will have blended back into the disk. This model of *self-propagating star formation* will work only if the pace of starbirth can be regulated so that it neither dies out completely nor sets the whole disk afire, exhausting it of gas. It may be able to explain the fragmentary spiral arms of a galaxy like M33, but is less likely to apply to a system like M100: in Figure 5.26, the spiral arms can be traced over more than $180°$.

A spiral pattern can last longer if the stars that make it up are not on circular orbits, but are arranged in a particular order in their slightly eccentric paths: this is called a *kinematic spiral*. In Section 3.3, we saw that a star's path on a nearly circular orbit can be described as the sum of circular motion of a guiding center, at the rate $\Omega(R_g)$ appropriate to its radius R_g, and an epicyclic oscillation moving the star in and out. The azimuth of the guiding center is $\phi_{gc} = \Omega(R_g)t$, while epicyclic motion causes the star's radius to vary as

$$R = R_g + x = R_g + X\cos(\kappa t + \psi). \tag{3.70}$$

Here X is the amplitude of the radial motion, κ is the epicyclic frequency, and the constant ψ prescribes the initial radius. If we start by placing stars with their guiding centers spread around the circle at R_g, and set $\psi = 2\phi_{gc}(0)$ for each of them, they will lie on an oval, with its long axis pointing along $\phi = 0$.

At a later time t, the guiding centers move, so that $\phi_{gc}(t) = \phi_{gc}(0) + \Omega t$. The stars advance on their epicycles, to lie at radius $R = R_g + x$, where

$$x = X\cos\{\kappa t + 2[\phi_{gc}(t) - \Omega t]\} = X\cos[(2\Omega - \kappa)t - 2\phi_{gc}(t)]. \tag{5.10}$$

The long axis of the oval now points along the direction where

$$(2\Omega - \kappa)t - 2\phi = 0, \qquad \text{or} \qquad \phi = (\Omega - \kappa/2)t \equiv \Omega_p t. \tag{5.11}$$

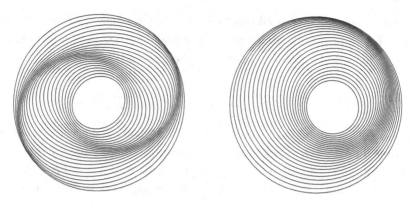

Fig. 5.29. Left, oval orbits nested to form a two-armed spiral; the equation of the pattern is $R = R_g\{1 + 0.075\cos[2(5 - 5R_g + \phi)]\}^{-1}$, and $0.3 < R_g < 1$. Right, a one-armed spiral, with $R = R_g[1 + 0.15\cos(5 - 5R_g + \phi)]^{-1}$.

We have defined the *pattern speed* Ω_p so that the pattern made up by stars with guiding center R_g will return to its original state after a time $2\pi/\Omega_p$, even though individual stars complete their orbits about the center in the shorter time $2\pi/\Omega$. A two-armed spiral can be made up from a set of nested ovals of stars with guiding centers at different radii R_g, as shown on the left of Figure 5.29. Because the pattern speed Ω_p varies with R_g, this spiral will in time also wind itself into a tight trailing pattern, but it will do so more slowly by the factor Ω_p/Ω, which is about 0.3 when the rotation curve is flat. To describe an m-armed spiral, we would set $\psi = m\phi_{gc}(0)$ in Equation 5.10; stars with a given guiding center then lie on m-fingered shapes that turn with a pattern speed $\Omega_p = \Omega - \kappa/m$. The right panel of Figure 5.29 shows a one-armed kinematic spiral.

The *density-wave theory* of spiral structure is based on the premise that mutual gravitational attraction of stars and gas clouds at different radii can offset the kinematic spiral's tendency to wind up, and will cause the growth of a pattern which rotates rigidly with a single pattern speed Ω_p. One way to test whether a spiral pattern can develop spontaneously is to examine how it would affect the orbits of disk stars: the spiral will grow only if the stars respond to its gravity by moving so as to reinforce the pattern. A star orbiting at radius R passes through an arm of an m-fold spiral pattern with frequency $m[\Omega_p - \Omega(R)]$. To see how this periodic tugging affects the stellar motions, we add a forcing term to the epicyclic equations. The calculation is lengthy but not difficult: we refer readers to Section 3.3 of Binney and Tremaine's book.

Finding how the forced motions of all the stars in turn affects the gravitational potential of the galaxy's disk is much more difficult. In general the calculation can be done only for a tightly wound spiral; it shows that stars respond so as to strengthen the spiral only if the perturbing frequency $m|\Omega_p - \Omega(R)|$ is slower than the epicyclic frequency $\kappa(R)$ at that radius. Hence, a continuous spiral *wave* can propagate only in the region between the *inner Lindblad resonance*, where $\Omega_p = \Omega - \kappa/m$, and the *outer Lindblad resonance*, where $\Omega_p = \Omega + \kappa/m$.

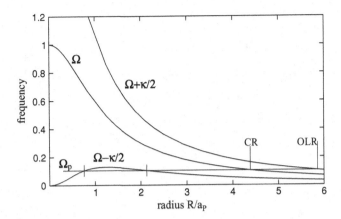

Fig. 5.30. Frequencies $\Omega(R)$ and $\Omega \pm \kappa/2$ in the Plummer potential of Equation 3.11. For pattern speed Ω_p, the $m = 2$ inner Lindblad resonances are marked by vertical ticks, the corotation radius is labelled 'CR', and the outer Lindblad resonance 'OLR'. If the pattern speed were twice as large, the inner Lindblad resonances would be absent.

Figure 5.30 shows positions of the $m = 2$ Lindblad resonances for the Plummer potential of Equation 3.11. Spirals with two or more arms always have an outer Lindblad resonance, but, if the pattern speed is high, there may be no inner Lindblad resonance. Stars beyond the outer resonance, or between the two inner resonances, find that the periodic pull of the spiral is faster than their epicyclic frequency κ; they cannot respond to reinforce the spiral, and the wave dies out. Thus we would expect to see a continuous spiral only in the region between the inner and outer Lindblad resonances. Since for $m > 2$ the spiral is restricted to just a narrow annulus of the disk, we would expect two-armed spiral patterns to be more prominent than those with three or four arms.

Problem 5.13 Show that, if the rotation curve of the Milky Way is flat near the Sun, then $\kappa = \sqrt{2}\Omega(R)$, so that locally $\kappa \approx 36\,\mathrm{km\,s^{-1}\,kpc^{-1}}$. Sketch the curves of Ω, $\Omega \pm \kappa/2$, and $\Omega \pm \kappa/4$ in a disk where $V(R)$ is constant everywhere, and show that the zone where two-armed spiral waves can persist is almost four times larger than that for four-armed spirals.

But disk stars can reinforce a spiral wave, and help it to grow, only if their random motion is small enough so as not to take them outside the spiral arms. Alar Toomre showed in 1964 that axisymmetric ($m = 0$, so nonspiral) waves can grow in a thin rotating disk of stars only if the disk is 'cold'. Stellar speeds in the radial direction, measured by the dispersion σ_R, must be low in relation to the surface density Σ of mass in the disk: we need

$$Q \equiv \frac{\kappa \sigma_R}{3.36 G \Sigma} \lesssim 1. \tag{5.12}$$

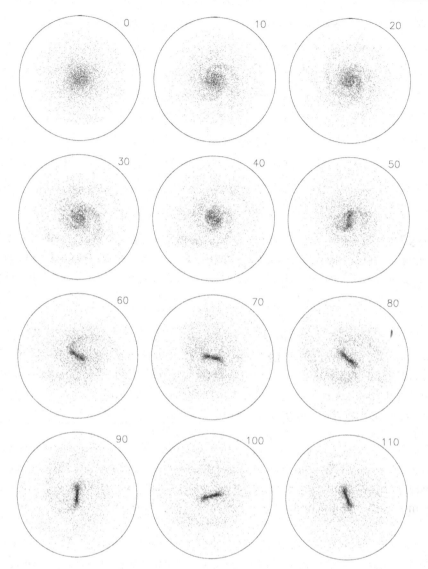

Fig. 5.31. Gravitational *N*-body simulation showing how a disk of 50 000 particles that attract each other by gravity develops first a two-armed spiral pattern, then a bar. The galactic bulge and dark halo are represented by a fixed inward force. The disk begins with $Q = 1$, as defined in Equation 5.12. For a galaxy of mass $2 \times 10^{11} \mathcal{M}_\odot$ within the disk radius of 16 kpc, the run time corresponds to 2.65 Gyr – J. Sellwood.

Computer simulations like that of Figure 5.31, which began with $Q = 1$, show that a spiral pattern generally grows if $Q \lesssim 1.2$. As it does so, the stars develop larger epicyclic motions, and Q rises; so we never expect to see a stellar disk with $Q \lesssim 1$. We can test this hypothesis in the solar neighborhood. Here, the stars making up most of the disk mass, those at least as old as the Sun, have $\sigma_R \approx 30 \, \text{km s}^{-1}$ (Section 2.2). In Section 3.4 we saw that the density in the disk $\Sigma \sim 50 \mathcal{M}_\odot \, \text{pc}^{-2}$, while $\kappa \approx 36\,\mathrm{km\,s^{-1}\,kpc^{-1}}$. Hence $Q \sim 1.4$ near the Sun, which is safely greater than unity.

Why does the disk develop trailing spiral arms, instead of leading arms? It turns out that, in a trailing spiral, the inner parts of the disk exert a torque on the outer disk, which transfers angular momentum outward and allows material at small radii to move inward. As we saw in Problem 3.8, this decreases the energy of the disk's rotational motion. By contrast, a disk could develop a leading spiral only if energy were supplied from outside; for example, by the close passage of another galaxy. The energy released by the spiral torques goes to increase the epicyclic motion of the stars.

Figure 5.31 shows a gravitational N-body simulation, following what happens to a disk of 'stars' attracting each other by gravity. It is initially axisymmetric, with the stars on nearly circular orbits. The growing spiral pattern is two-armed; a straight central bar also forms. As the stellar random speeds grow, the disk 'heats up' and the spiral eventually disappears. The disk of an S0 galaxy, where no gas is present, would behave in the same way: any spiral pattern would be short-lived. But, as we saw in Section 2.2, stars freshly born from the disk gas have very small random speeds. Continued addition of these new stars may be important in prolonging the life of a spiral pattern, or in re-creating it periodically.

The density-wave theory has not been a complete success. Spiral waves, like water waves, travel: the individual stars, or water molecules, oscillate about a fixed radius or position, but the waves move off to leave the water calm, and the disk without a spiral. Energy taken from the disk's rotational motion would be sufficient to regenerate a spiral wave, but we do not understand exactly how it could be transferred to the wave. Simulations like that in Figure 5.31 suggest that this happens only if the spiral pattern speed Ω_p is high enough for the inner Lindblad resonances to be absent. Then, waves can travel through the center of the disk, changing from a trailing to a leading spiral as they do so. The galaxy's differential rotation pulls the leading spiral pattern into a trailing one, amplifying it in the process; the new trailing waves restart the cycle.

Another possibility is that the spiral is driven by the gravitational pull of a companion galaxy, or by internal forces from a central rotating bar (see below). Many of the galaxies with the best grand-design spiral patterns have either close companions or pronounced central bars. Close to M100, we find the dwarf elliptical galaxy NGC 4322. In Figure 7.2 below, the galaxy M81 has a two-armed spiral; its two companions, the starburst galaxy M82 and NGC 3077, lie within 50 kpc. We will see in Section 7.1 that close passage of a neighboring galaxy can create at least a temporary two-armed spiral. An orbiting companion might trigger spiral arms as it came closest to the disk; if Q were close to unity, the disk stars could cooperate to strengthen the induced spiral. Similarly, a strong bar might encourage a spiral in a disk that was on the verge of producing one for itself, and would organize that pattern into a two-armed form.

Fig. 5.32. The barred galaxy NGC 1300, classified as SBb or SBbc. The spiral arms trail; note the dust lanes on the leading edge of the bar – WIYN telescope.

The gravitational pull of the spiral arms affects gas even more strongly than the stellar disk, because the random speeds of gas clouds are only 5–$10\,\mathrm{km\,s^{-1}}$, much lower than for the stars. Except immediately around the *corotation* radius, where $\Omega(R) = \Omega_p$, the linear speed $R[\Omega(R) - \Omega_p]$, with which gas and stars move into the spiral arm, is *supersonic*. Shocks develop as the gas flows into the arms, compressing it enormously. The fact that the dust lanes of Figure 5.26 lie on the concave side of the arm tells us that gas is entering the spiral from this side. So the local rotation rate $\Omega(R)$ must exceed the pattern speed Ω_p; this part of the arm lies inside the corotation radius. It takes about 10 Myr for young stars to be born in the compressed gas and begin to shine; in the lower left panel of Figure 5.26, the peak of the Hα emission from gas ionized by these stars is not on top of the dust lane, but 'downstream' of it. Radiation from the hottest of these stars also splits some of the H_2 molecules apart, raising the density of atomic hydrogen, HI, in the spiral arms (lower right).

5.5.3   Barred disks

About half of all disk galaxies show a central linear bar, containing up to a third of the total light. The ratio of the long to the short axis of the bar can be as extreme as $1 : 5$, as in Figure 5.32. If the bars were much thicker than the disks, we would expect to see anomalous central thickening in roughly half of all edge-on spiral and S0 galaxies; this is not observed, so we think that bars must be almost as flat

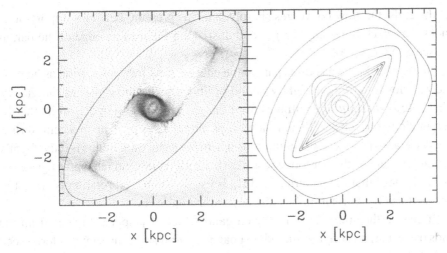

Fig. 5.33. Left, gas density from a computer simulation of flow within a bar; the solid curve outlines the bar, rotating clockwise. Right, particle orbits that close on themselves in a frame rotating with the bar. The gas flow is compressed in shocks along the leading edge of the bar, where the aligned orbits have their greatest curvature – P. Englmaier, after *MNRAS* **287**, 57 (1997).

as disks. S0 galaxies are about as likely to be barred as Sa or Sb spirals; in contrast to spiral arms, a bar can persist even if the disk is empty of gas. Barred galaxies still pose a riddle: we do not yet understand why some galaxies are barred while other, apparently similar, systems are not.

Like spirals, the figure of the bar is not static, but rotates with some pattern speed Ω_p. But, unlike spiral arms, bars are *not* density waves; most of their stars always remain within the bar. Inside a rotating bar, stars and gas no longer follow near-circular paths, but stay close to orbits that close on themselves as seen by an observer moving with the bar's rotation. We can investigate these orbits by adding a term to the epicyclic equations of Section 3.3 to represent the force of the bar. Within corotation, where $\Omega_p < \Omega(R)$, we find a family of closed orbits that line up with the long axis of the bar; see Figure 5.33. In a stable bar, most of the stars would remain fairly near to these orbits, but outside the corotation radius, all the closed orbits lie perpendicular to the bar; there are no aligned orbits for stars to follow. For this reason, we think that bars probably rotate slowly enough that corotation falls beyond the end of the bar.

Gas in the disk of a spiral galaxy can flow in toward the center only if it can get rid of its angular momentum. The strongly asymmetric gravitational forces of a bar help it to do just that. In many barred galaxies, dark dust lanes run along the leading edge of the bar; in Figure 5.32 they are visible as the light lines on the side of the bar opposite to the (trailing) spiral arms. The left panel of Figure 5.33 shows a computer simulation of the gas flow within a barred galactic disk. Within the bar, gas stays close to the aligned orbits, shown on the right of the figure.

At the ends, the flow converges sharply, and gas pressure becomes important. A shock forms, compressing the gas and dust along the leading edge of the bar, as we see in NGC 1300.

In the shock, the gas loses part of the energy of its forward motion as heat, so it drops down onto more tightly bound orbits closer to the center of the galaxy. This *energy dissipation* continues until the gas meets the rounder orbits that lie perpendicular to the bar; then, inflow stops, and the gas piles up in a central ring. The dust lanes produced in simulations where the corotation radius is only slightly beyond the end of the bar look most like those we observe in real barred galaxies. So we believe that the pattern speed Ω_p is usually high, and rotation is almost as fast as it can be; corotation lies just beyond the bar's end.

Because the spiral arms in barred galaxies usually appear to start from the ends of the bar, it is often assumed that bar and spiral share the same pattern speed and always maintain the same relative alignment. This need not be the case: in Figure 5.31, the outer spiral has a lower pattern speed than the bar. But the eye joins the spiral pattern to the bar, so that the arms always appear to start near the end of the bar. There is nothing to prevent a galaxy from developing a bar that rotates faster (or slower) than the pattern of a surrounding spiral, or from having two or more independent spirals with discrepant pattern speeds or different numbers of arms.

Further reading: D. M. Elmegreen, 1998, *Galaxies and Galactic Structure* (Prentice-Hall, Englewood Cliffs, New Jersey), unfortunately now out of print, treats spiral structure on an undergraduate level; and Chapter 6 of J. Binney and S. Tremaine, 1987, *Galactic Dynamics* (Princeton University Press, Princeton, New Jersey) gives a graduate-level discussion of the theory.

5.6   Bulges and centers of disk galaxies

Bulges, along with the centers of small elliptical galaxies, are some of the densest known stellar systems (see Figure 4.18). The Milky Way's bulge is largely hidden from us by dust; but when we look at M31 through a small telescope or binoculars, or at an 'underexposed' optical image of a disk galaxy, only the bright bulge is visible. In Figure 5.4, the surface brightness at the center of NGC 7331 rises to fifty times that of the inward-extrapolated exponential disk. The bulge is the 'big city' of a disk galaxy: a centrally placed dense region, where old and new stars are crammed tightly together. Within the central few parsecs we can find a hundred million stars, packed into a nuclear cluster; these in turn often harbor a large black hole. In these tiny central regions, densities are enormous, and timescales short.

5.6.1   Bulges

In the images of two edge-on galaxies in Figures 5.5 and 5.16, the bulges appear as round ellipsoids. Other bulges are quite flattened, almost like bright central disks.

Some, including those of the Milky Way (Section 2.2) and M31 (Section 4.2) appear barlike; they may be *triaxial* ellipsoids, with three unequal axes. SB0, SBa, and SBb galaxies have a substantial bulge as well as a stellar bar; Figure 1.13 shows the SB0 galaxy NGC 936. When seen edge-on, most bulges appear roughly elliptical, but about 20% of them have the appearance of a peanut: the isophotes dip down toward the disk's midplane at the center. Studies of gas motions within them suggest that peanut-bulges may in fact be fast-tumbling bars.

The Milky Way's bulge and that of M31 are fairly typical in being more metal-rich than the disk, and gas-poor except at the very center. On average, bulges are only slightly redder than the inner part of the disk; they continue the pattern seen in the disks, namely that galaxies become redder closer to the center. Like the colors of disks, bulge colors show a broad spread, varying from $B - R \approx 1.2$, the color of a late F star, to $B - R \approx 1.8$, corresponding to a late G star. Almost all the stars in the bulges of the Milky Way and M31 are at least several gigayears old. Beyond the Local Group we cannot resolve individual bulge stars, so we have no good estimates for their ages.

Bulge stars share a common sense of rotation about the center, but have larger random motions than the disk stars: usually the ratio of rotational to random speeds $V/\sigma \sim 1$. We would not expect round bulges to rotate very rapidly; if they did, they should be thin and disklike. In Section 6.2 we will discuss the minimum flattening that should be associated with a given amount of rotation, and we will find that large elliptical galaxies rotate less rapidly than would be possible for their flattening. Bulges and small ellipticals, by contrast, have about as much rotation as is permitted by their shapes; in these terms they are fast-rotating.

The surface brightness of a bulge is often approximated by *Sérsic's* formula

$$I(R) = I(0)\exp[-(R/R_0)^{1/n}]. \tag{5.13}$$

If the parameter $n = 1$, this is the same exponential law as Equation 5.1, whereas for $n = 4$ it is the *de Vaucouleurs* formula, which was developed to describe the light distribution in elliptical galaxies: we will meet it again in Section 6.1. According to this formula, the surface brightness should continue to increase all the way to the center; the three-dimensional density of stars would then grow without limit. Because our telescopes have a limited angular resolution, we do not know how far the stellar density increases; observed values reach thousands of stars per cubic parsec.

A good measure of a bulge's extent is the *effective radius* R_e, the radius of a circle drawn on the sky that includes half of the bulge light. A galaxy in which the disk has a large scale length h_R generally has a bulge with a large effective radius; one recent study found the ratio $R_e/h_R \approx 0.1$. R_e ranges from about 100 pc up to a few kiloparsecs in galaxies with the largest dimensions; for the bulge of the Sombrero galaxy in Figure 5.5, $R_e \approx 4$ kpc.

The prominence of the bulge, along with the appearance of the spiral arms, is used to order the disk galaxies along the sequences of Figure 1.11. In S0 and Sa galaxies the bulge produces a larger fraction of the light than it does in Sb and Sc systems, while in the Sd class it is usually absent. The bulges of the early-type S0 and Sa systems have higher surface brightness $I(0)$ at the center, but not necessarily larger sizes as measured by R_e, compared with those in Sb and Sc galaxies. Faint low-surface-brightness galaxies lack a substantial bulge; but in more luminous and more massive examples the bulge is similar to those of high-surface-brightness galaxies. On the basis of its bulge, UGC 6614 was classified as an Sa galaxy. But the spiral arms have the open character of Sc systems, although with much lower surface brightness than normal, and the disk is gas-rich with $\mathcal{M}(\text{HI})/L_B \sim 1$.

Because bulges are so much denser than the disks surrounding them, they may have their origin earlier in the history of the Universe, when the average cosmic density was higher. Alternatively, they could have been formed later, out of gas which had been able to lose a great deal of its angular momentum, so it could sink toward the center of the galaxy. Or gas may have flowed inward, causing the disk to form a central bar, which in turn became unstable, thickening to become a peanut-bulge. None of the current theories gives a plausible account of the entire range of observed bulges.

A bulge like that of the Milky Way probably formed most of its stars within the first few gigayears after the Big Bang. The Hubble Space Telescope has revealed a number of compact galaxies at redshifts $z \gtrsim 3$ which may be building their bulges. These objects seen in light that was emitted at $1400 \text{ Å} \lesssim \lambda \lesssim 1900 \text{ Å}$ have sizes comparable to those of present-day galactic bulges; their luminosities correspond to the birth of $(10–100) \, \mathcal{M}_\odot$ of new stars per year. At this rate, it would take only about 1 Gyr to make the entire bulge of a galaxy like our own. But not all bulges are so old. In the spiral galaxy NGC 7331, some of the bulge stars orbit the center in the opposite sense to the disk's rotation! The bulge of this normal-looking galaxy probably contains material that fell into the system later, from outside. NGC 7331 is the only galaxy known where bulge stars orbit counter to the disk; such a major late addition cannot be common.

5.6.2 Nuclei and central black holes

Just as water runs downhill, gas tends to flow into the center of a galaxy. In the bulge, clouds of gas shed from dying stars will collide with each other, lose energy, and sink to form a fast-rotating inner disk. As we saw in the last section, the action of a bar can also bring disk gas inward. In the central region, where the galaxy's rotation curve $V(R)$ rises linearly, the angular speed $V(R)/R$ is nearly constant: shear is absent. In contrast to the main disk, gas clouds here would not be pulled apart by differential rotation; they might collapse easily under their own gravity, becoming dense enough to make stars. Just as in the Milky Way (see Section 2.2), in most spiral galaxies we see abundant gas within the central 100 pc, and some newly formed stars.

In some galaxies, a central *starburst* is taking place: stars are being made so rapidly as to exhaust the gas supply within $\sim$100 Myr. This exuberant starbirth cannot be sustained; it will die away, perhaps leaving a compact inner disk of stars. The uppermost spectrum in Figure 5.24 is from a blue starburst galaxy. Its massive young stars are bright at short wavelengths; they ionize the gas around them to produce the strong emission lines. If the young stars are shrouded in dust, that will intercept most of their light, re-radiating the energy at infrared wavelengths. Starbursts are often triggered when gas falls into the center of a galaxy, as it is tugged by another galaxy passing nearby: we discuss this further in Section 7.1. They sometimes occur in cycles: supernovae from a starburst heat the surrounding gas and blow it out of the bulge, after which gas must accumulate before the next burst. This process might build up a dense central cluster of stars.

At the center of the Milky Way, about $10^7 \mathcal{M}_\odot$ of stars are packed into a tight cluster only $\sim$3 pc in radius. Such *nuclear star clusters* are often seen in spiral and dwarf elliptical galaxies. They differ from the open and globular star clusters of the Milky Way, where all stars share a common birth, and any leftover gas was swiftly dispersed. Gas flowing into a galaxy center has nowhere else to go than into the nuclear cluster; it provides the raw material for multiple episodes of starbirth. In a single nucleus, we might see the spectral signatures of short-lived main-sequence B stars, along with much older red giants. The cluster grows as more stars form; while the radii of nuclear star clusters are no larger than for the Milky Way's globulars, their mass, and thus the velocity dispersion, can be many times higher.

> **Problem 5.14** The nuclear star cluster of M33 has core radius $r_c \lesssim 0.4$ pc and measured velocity dispersion $\sigma = 24$ km s^{-1} (Table 4.2). The luminosity $L_V \approx 2.5 \times 10^6 L_\odot$. Approximating it crudely as a Plummer model, use Equation 3.13 and the virial theorem (see Problem 3.13) to estimate its mass $\mathcal{M}$. Show that the mass-to-light ratio $\mathcal{M}/L_V \sim 0.2$ – much lower than for a globular cluster, because young massive stars are present. (Our $\mathcal{M}/L$ is less than in Table 4.2 because the newest stars are concentrated to the cluster center; r_c reflects their distribution, rather than that of the older stars that carry most of the mass.)

As in the Milky Way, nuclear clusters can hide objects so small and so massive that they are almost certainly black holes. As material spirals around and into a black hole, its energy can be liberated more efficiently than by any other known process. Turning a mass $\mathcal{M}$ of hydrogen into helium releases merely $0.007 \mathcal{M}c^2$ as energy; throwing it into a black hole can yield $\sim 0.1 \mathcal{M}c^2$. Some large black holes, with $\mathcal{M}_{BH} \gtrsim 10^7 \mathcal{M}_\odot$, are believed to be power sources for *active galactic nuclei*: we will discuss these later in Section 9.1. Others, like that in the center of our Milky Way, are quiescent, and can be 'seen' only by their gravitational effects.

In disk galaxies with an active nucleus, we see light from the very center of the galaxy, which does not come from its stars, or gas ionized by their radiation. The nucleus shines at all wavelengths from radio to γ-rays; in visible light it is extremely luminous, sometimes as bright as the entire galaxy around it. The optical and ultraviolet spectrum shows broad emission lines, with widths corresponding to velocities $V_r > 5000 \, \text{km s}^{-1}$. Some of these lines come from atoms that are much more highly ionized even than the gas around hot stars in a starburst. Our Milky Way has only a very-low-level active nucleus, with a weak central radio source.

A striking example of an active nucleus is that in the barred spiral galaxy NGC 4258; see Figure 9.2. Within a few parsecs of its center is a rotating disk of dense gas. Radiation from the nucleus excites water molecules in the disk, so that they emit maser radiation in a line at 22.2 GHz. The bright masing spots can be located very precisely by very long baseline interferometry, and we can use the Doppler effect to measure the radial velocity V_r. The gas lies in a disk 0.015" across, which corresponds to only 0.5 pc, and orbits the center at over $1000 \, \text{km s}^{-1}$. Within the disk must lie a compact object with a mass $\mathcal{M} > 3 \times 10^7 \mathcal{M}_\odot$; see the following problem. The density exceeds $10^9 \mathcal{M}_\odot \, \text{pc}^{-3}$ or 100 000 times higher than at the center of a globular cluster: this is far too concentrated to be a cluster of normal stars.

Problem 5.15 The inner edge of the masing gas disk in NGC 4258 is about 0.004" from the center. By measuring the motion of the masing spots across the sky, Herrnstein *et al.* (1999 *Nature* **400**, 539) found its distance to be 7.2 ± 0.3 Mpc: what is the radius of the inner edge of the disk in parsecs? Gas there orbits at $1100 \, \text{km s}^{-1}$: use Equation 3.20 to show that the mass inside the disk is $\sim 4 \times 10^7 \mathcal{M}_\odot$. Supposing that the central object consists of solar-type stars, find the density n per cubic parsec, and the cross-sectional area $\sigma \sim \pi R_\odot^2$ of each star. The mean time t between collisions for any star is given by $t \approx 1/(n\sigma V)$; show that each star would collide with another about every 100 Myr, so that the cluster could not survive for long.

6

Elliptical galaxies

Elliptical galaxies look like simple objects; but they are not. As their name implies, they appear round on the sky; the light is smoothly distributed, and they lack the bright clumps of young blue stars and patches of obscuring dust which are such obvious features of spiral galaxies. Ellipticals are almost devoid of cool gas, except at the very center; in contrast to S0 systems, they have no prominent disk. Their smooth appearance suggests that, like the molecules of air in a room, their stars have had time to reach a well-mixed equilibrium state. As with stars on the main sequence, we would expect the properties of elliptical galaxies to reflect the most probable state of a fairly simple system, with 'no surprises'.

Instead, detailed studies reveal a bewildering complexity. Elliptical galaxies cover a huge range of luminosity and of light concentration. Some ellipticals rotate fast, others hardly at all. Some appear to be oblate (grapefruit shaped), while others have a *triaxial* shape with three unequal axes, like a squashed (American or rugby) football. These properties are interlinked: luminous ellipticals are more likely to be triaxial, slowly rotating, and also strong X-ray sources, while the less luminous systems are oblate and relatively rapidly rotating, and have dense stellar cusps at their centers.

It was a mistake to think that elliptical galaxies might be close to an equilibrium state, because stellar systems have a very long memory. Most of a galaxy's stars have made fewer than 100 orbits about the center; we saw in Section 3.2 that the *relaxation time* required to randomize their motions is far greater than the age of the Universe. If a galaxy was assembled in a triaxial shape, or with a dense central cusp, these characteristics would not yet have been erased. The variety among elliptical galaxies suggests that they originated by a number of different pathways. Present-day elliptical galaxies are 'fossils' of the earlier Universe; our task is to reconstruct their birth and youthful star-forming lives from the old low-mass stars that remain.

We begin this chapter with a section on photometry: how the images of elliptical galaxies appear in visible light, and what this tells us about the distribution of stars within them. Section 6.2 discusses stellar motions, and how the rotation of an elliptical galaxy is linked to its other properties. We consider what stellar orbits would be possible in a *triaxial* galaxy, with three unequal axes. In Section 6.3 we look at the stellar populations of elliptical galaxies and at their gaseous content. Elliptical galaxies are quite rich in interstellar gas, but that gas is much hotter than the gas in disk systems; it can be studied only in X-rays. Section 6.4 discusses the dark matter in elliptical galaxies and the black holes at their centers.

6.1 Photometry

The brightest galaxies in the Universe are ellipticals, but so also are some of the dimmest: these systems share little more than their general shape, and their lack of cool gas and young stars. It is useful to divide elliptical galaxies into three classes. The *luminous giant ellipticals* have $L \gtrsim L_\star$, the characteristic luminosity of a large galaxy that we defined in Equation 1.24: $L_\star \approx 2 \times 10^{10} L_\odot$, equivalent to blue magnitude $M_B \approx -20$. *Midsized* ellipticals are less luminous, but still have $L \gtrsim 3 \times 10^9 L_\odot$, or $M_B \lesssim -18$. Finally, *dwarf* elliptical galaxies are those with luminosities below $3 \times 10^9 L_\odot$. We saw in Chapter 5 that many attributes of a disk galaxy are determined by its position in the classification scheme of Figure 1.11, with galaxies in each class covering a wide range in size and luminosity. By contrast, when we measure the luminosity of an elliptical galaxy, we have largely fixed the other properties. Elliptical galaxies come in 'one size (sequence) only'.

Figure 6.1 shows the *isophotes*, contours of constant surface brightness, of four elliptical galaxies. In Figure 6.1(a), the isophotes are remarkably close to being true ellipses. The ratio of the semi-major axis a and the semi-minor axis b quantifies how far the isophote differs from a circle: the *ellipticity* ϵ is defined as $\epsilon = 1 - b/a$. Often the ellipticity is fairly constant, and the position of the center and the direction of the long axis remain stable, from the bright inner isophotes to the faint outer contours. This allows us to label elliptical galaxies by the Hubble type En, where $n = 10(1 - b/a)$; E0 galaxies appear circular on the sky, while for an E5 galaxy the short diameter of the image is half the size of the long diameter. The index n is usually rounded to the nearest whole number, roughly the accuracy to which we can classify galaxies by eye from photographic plates. Unlike the classification of a disk galaxy, the Hubble type of an elliptical galaxy depends on our viewing direction.

The light in elliptical galaxies is much more concentrated toward the center than it is in the disks of spirals. As for spirals, we can plot the surface brightness $I(R)$ on the major axis of the image against radius R, as in Figure 6.3. There, we see that the surface brightness in NGC 1399 falls by more than 10 000, and the corresponding volume density more than a millionfold, between the center and the outskirts where the galaxy disappears into the sky.

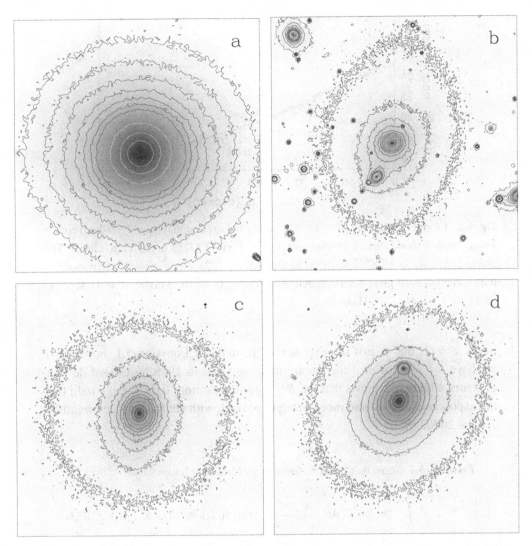

Fig. 6.1. Isophotes in the R band of four giant elliptical galaxies: (a) isophotes are elliptical (NGC 5846); (b) the long axis of the inner isophotes is roughly horizontal, twisting to near-vertical at the outer contour (EFAR J16WG); (c) diamond-shaped 'disky' isophotes, with $a_4 \approx 0.03$ (Zw 159-89 in Coma); (d) rectangular 'boxy' isophotes, with $a_4 \approx -0.01$ (NGC 4478). The compact objects, especially prominent in (b), are mainly foreground stars – R. de Jong.

Just as for the bulges of disk galaxies, we use Sérsic's empirical formula, Equation 5.13, to describe the light distribution. We can rewrite it as

$$I(R) = I(R_e)\exp\{-b[(R/R_e)^{1/n} - 1]\}, \qquad (6.1)$$

where the constant b is chosen so that a circle of radius R_e, the *effective radius*, includes half the light of the image. For $n > 1$, $b \approx 1.999n - 0.327$. When the

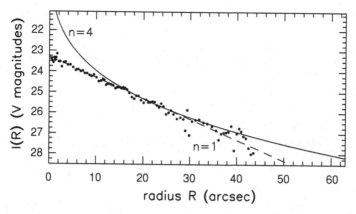

Fig. 6.2. From Equation 6.1, $R^{1/4}$ ($n = 4$: solid) and exponential ($n = 1$: dashed) curves. Points show V-band surface brightness for dE galaxy VCC753 in the Virgo cluster. This elliptical has $R_e = 15.8''$ in the B band and $I_B(R_e) = 24.4 \,\mathrm{mag\,arcsec^{-2}}$. Extrapolating the profile outward gives total apparent magnitude $B_T^0 = 16.4$; taking $d = 16\,\mathrm{Mpc}$, we find $L \approx 1.1 \times 10^8 L_\odot$ – H. Jerjen.

index $n = 1$, this is just the exponential formula of Equation 5.1. For $n = 4$, we have the $R^{1/4}$ or *de Vaucouleurs* law, proposed in 1948 by Gérard de Vaucouleurs. Figure 6.2 shows that the $R^{1/4}$ profile has more light at large radii than the exponential, but is also more strongly peaked, with central surface brightness $I(0) > 2000 I(R_e)$.

Problem 6.1 Show that the $R^{1/4}$ formula yields a total luminosity

$$L = \int_0^\infty 2\pi R I(R)\,\mathrm{d}R = 8! \frac{e^{7.67}}{(7.67)^8} \pi R_e^2 I(R_e) \approx 7.22\pi R_e^2 I(R_e). \quad (6.2)$$

(Remember that $\int_0^\infty e^{-t} t^7 \,\mathrm{d}t = \Gamma(8) = 7!$) Use a table of incomplete Γ functions to show that half of this light comes from within radius R_e.

Problem 6.2 Use Table 1.9 to show that, even at its center, the surface brightness of the galaxy of Figure 6.2 is less than half that of the night sky. Show from Figure 6.3 that, for the galaxy G675, $I(R)$ drops below sky level at $R \approx R_e$; almost half of its light comes from regions further out.

Outside the very center, the $R^{1/4}$ formula provides a fairly good description for the surface brightness of luminous and midsized elliptical galaxies, those brighter than about $3 \times 10^9 L_\odot$. Figure 6.3 shows the observed profile of the elliptical G675; it is close to a straight line except in the very bright core, where atmospheric turbulence, or *seeing*, blurs the image. The horizontal bar indicates the radius of a typical image of a pointlike star measured at half its peak intensity;

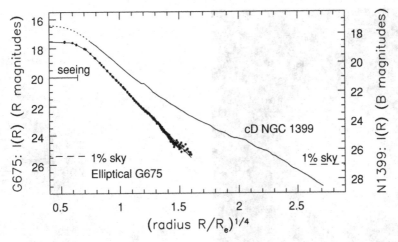

Fig. 6.3. Surface brightness of two luminous ellipticals: an $R^{1/4}$ law corresponds to a straight line. Dots show the measured R-band surface brightness for galaxy G675, in cluster Abell 2572. It has $L_V \approx 2 \times 10^{10} L_\odot$, and $R_e = 4.95''$ or 3.8 kpc. The curve gives an $R^{1/4}$ profile, smoothed by atmospheric seeing: the horizontal bar shows $1.67''$, the half-width of a stellar image. The upper curve shows the measured B-band profile of the cD galaxy NGC 1399, which is about twice as luminous as G675. For it, $R_e = 15.7'' \approx 1.4$ kpc, so measurements cover $R \lesssim 850''$ or 75 kpc. Between the dotted region, where seeing has affected the measurement, and $R \sim 2^4 R_e$, $I(R)$ follows the $R^{1/4}$ profile closely – R. Saglia and N. Caon.

this is a common way to quantify the seeing. The overplotted solid curve shows what we would expect to find if the galaxy followed exactly the $R^{1/4}$ law, but the seeing was such as to blur stellar images to their measured extent. The biggest and brightest systems are generally best fit with larger values of n, while the surface brightness of dwarf elliptical galaxies is often close to an exponential profile with $n \approx 1$, as in Figure 6.2.

The most luminous of all galaxies are the *cD galaxies*. These enormous ellipticals contain the largest stellar mass of any galaxies; they are significantly brighter than $L_\star$. In Figure 6.3, the profile of the cD galaxy NGC 1399 (with $L \approx 2L_\star$ or $4 \times 10^{10} L_\odot$) follows the $R^{1/4}$ formula fairly well, out to $R \sim 20R_e$. Beyond that, the surface brightness is higher than the formula would predict. This outer envelope of 'extra' light is characteristic of cD galaxies, which are found only at the centers of galaxy groups and clusters: NGC 1399 is in the Fornax cluster. Although it is not classified as a cD galaxy, the outer envelope of the giant elliptical galaxy M87, at the center of the Virgo cluster, is also very extended. The starlight can be traced for nearly 100 kpc on a deep photograph (Figure 6.4). These stars may belong to the cluster rather than to individual galaxies, or they might be the shredded remains of smaller galaxies that came too close to the central monster system.

Recently, arclike 'shells' and other asymmetric structures have been found in the faint outer regions of elliptical galaxies (Figure 6.5). The shells are probably

Fig. 6.4. E0 elliptical galaxy M87 (NGC 4486). Left, the round inner region – J.-C. Cuillandre, Canada–France–Hawaii Telescope. Right, the negative from a deep exposure; the three bright stars of the left image are nearly lost in the galaxy's extended and asymmetric outer parts. Starlight can be traced to the SE (lower left) for $\gtrsim 15'$ or 70 kpc – D. Malin, Anglo-Australian Observatory.

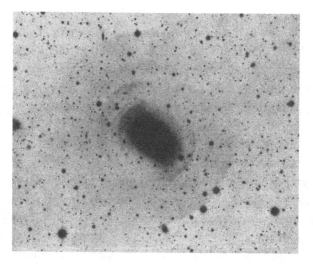

Fig. 6.5. A negative image shows faint arclike shells around elliptical galaxy NGC 3923; an out-of-focus copy was subtracted from the original photograph, allowing faint but sharp features to stand out. The picture is $18'$ across, or 110 kpc at $d \approx 21$ Mpc. This is a luminous galaxy ($L_B \approx 4 \times 10^{10} L_\odot$) in a loose group; it appears normal apart from the shells – D. Malin 1983 *ApJ* **274**, 534.

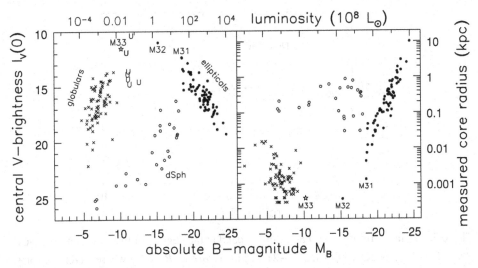

Fig. 6.6. Central surface brightness $I_V(0)$, in mag arcsec^{-2} in the V band, and core radius r_c plotted against B-band luminosity M_B. Filled circles are elliptical galaxies and bulges of spirals (including the Andromeda galaxy M31); open circles are dwarf spheroidals; crosses are globular clusters; the star is the nucleus of Sc galaxy M33. 'U' denotes an ultracompact dwarf elliptical in the Fornax or Virgo cluster – J. Kormendy and S. Phillipps.

remnants of a small galaxy which has been torn apart by the gravitational forces of the larger one, then swallowed up into it. A few of the galaxies with shells have clearly suffered a recent merger; they show faint tidal tails such as those seen in Figure 7.5. After a few gigayears, the shells and tails will be gone, leaving galaxies that appear in optical images as fairly normal ellipticals.

Unlike for disk galaxies, in an elliptical the central brightness is tightly linked to luminosity. Figure 6.6 shows the central brightness $I_V(0)$ and the *core radius* r_c, where the surface brightness drops to half its measured central value, for ellipticals, the central bulges of spirals, and star clusters. Just as with stars in a color–magnitude diagram, we see that the points are confined to certain regions of this plane.

Among the luminous and midsized ellipticals, namely those brighter than $L \sim 3 \times 10^9 L_\odot$, we find that the more luminous the galaxy, the lower its central surface brightness and the larger its core. The surface brightness in some of the most luminous galaxies, such as the cD systems, is almost as low as that in the disks of spirals. We will see in Section 7.1 that collisions between galaxies increase the internal motion of the stars within them, causing them to expand and become less tightly bound. If the most luminous ellipticals were produced by the merger of two smaller systems, that could explain why their centers are more diffuse.

The Local Group galaxy M32 lies at the other extreme, with the highest measured surface brightness: even with $L_V \approx 3 \times 10^8 L_\odot$, it is the 'smallest giant elliptical galaxy'. The extremely compact dwarf ellipticals recently discovered

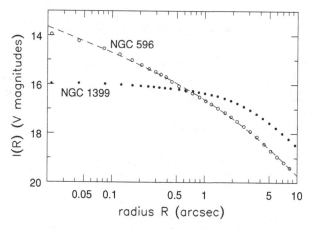

Fig. 6.7. Surface brightness $I_V(R)$ in the V band at the centers of two elliptical galaxies. The cD galaxy NGC 1399 ($M_V = -21.7$) has a *core* at $R \lesssim 1''$, where $I(R)$ is nearly constant. NGC 596 ($M_V = -20.9$) is half as luminous; the surface brightness continues to rise as a *cusp*. The dashed line shows $I(R) \propto R^{-0.55}$ – T. Lauer.

in the Fornax and Virgo clusters of galaxies may be similar. Beyond the Local Group, our observations often lack the spatial resolution to measure such very small cores.

The dimmest of the ellipticals, with $L \lesssim 3 \times 10^9 L_\odot$, or $M_B \gtrsim -18$, fall into two types: rare compact ellipticals like M32 (see Section 4.4) are distinct from the dwarf elliptical (dE) and dwarf spheroidal (dSph) galaxies described in Sections 4.1 and 4.4. Compact ellipticals have appreciable rotation, whereas the dwarf ellipticals and dwarf spheroidals do not rotate significantly. Dwarf spheroidals are less luminous versions of the dEs, with $L \lesssim 3 \times 10^7 L_\odot$ or $M_V \gtrsim -14$. Most are so diffuse as to be barely visible in images of the sky. Figure 6.6 shows that the dE and dSph galaxies have larger cores than the midsized ellipticals. The trend in surface brightness reverses itself: the central brightness is lowest in the least-luminous dE and dSph galaxies.

We understand why stars populate only certain regions of a color–magnitude diagram; both luminosity and temperature are controlled largely by the star's mass, and the zones where it burns nuclear fuel. Explaining the distribution in Figure 6.6 is harder, since it almost certainly reflects the conditions under which the galaxies formed, rather than their internal workings at the present day. The pattern tells us that galaxy formation had some regularity; the way that a galaxy came into being must be related to its mass. But this clue requires interpretation, which so far is lacking.

We gain a sharper view of galaxy centers from above the Earth's atmosphere; the Hubble Space Telescope can resolve details as fine as 0.05″ across. Luminous galaxies, like NGC 1399, generally have central cores within which the surface brightness is nearly constant. But Figure 6.7 shows that, in the less-bright elliptical

NGC 596, the surface brightness continues to rise as far in as our observations can follow; M32 shows a similar structure. Midsized galaxies, with $L \lesssim L_*$, typically have central *cusps*, not cores. If we assume that the luminosity density is proportional to the number density $n(r)$ of stars, then $n(r)$ must rise more steeply toward the center than r^{-1}; see the following problem. Even in galaxies with cores, a central plateau in surface brightness may conceal a peak in stellar density, so long as this is shallower than $n(r) \propto r^{-1}$.

These results from space observations show that we must be careful when interpreting diagrams such as Figure 6.6. The measured central surface brightness observed through the Earth's turbulent atmosphere is only a lower bound to the true value. The measured 'core radius' may in fact mark a point where the intensity profile changes its slope, rather than the outer limit of a region where the stellar density is constant.

Problem 6.3 From Figures 6.6 and 6.7, show that the measured surface brightness at the center of NGC 1399 is $I_V(0) \sim 14\,000 L_\odot\,\mathrm{pc}^{-2}$, while that of M32 is at least $10^6 L_\odot\,\mathrm{pc}^{-2}$. (Recall from Problem 5.3 that the disk of the spiral galaxy NGC 7331 reaches only $350 L_\odot\,\mathrm{pc}^{-2}$.)

Problem 6.4 When a spherical galaxy with stellar density $n(r)$ is viewed from a great distance along the axis z, show that the surface density at distance R from the center is

$$\Sigma(R) = 2 \int_0^\infty n(r)\mathrm{d}z = 2 \int_R^\infty \frac{n(r)r\,\mathrm{d}r}{\sqrt{r^2 - R^2}}. \tag{6.3}$$

If $n(r) = n_0(r_0/r)^\alpha$, show that as long as $\alpha > 1$ we have

$$\Sigma(R) = 2n_0 r_0 (r_0/R)^{\alpha-1} \int_1^\infty \frac{x^{1-\alpha}\,\mathrm{d}x}{\sqrt{x^2-1}} = \Sigma(R = r_0)(r_0/R)^{\alpha-1}. \tag{6.4}$$

(What happens if $\alpha < 1$?) The surface density $\Sigma(R)$ remains finite as $R \to 0$ if the volume density rises less steeply than $n \propto r^{-1}$.

6.1.1 The shapes of elliptical galaxies

The appearance of an elliptical galaxy depends on the direction from which we observe it. If the galaxy is symmetric about some axis, then an observer looking along that direction will always see a circular image. But, since we view galaxies with random orientation, we can use the distribution of apparent shapes to infer the average true three-dimensional figure. The isophotes may also reveal internal complexity: for example, a small disk hidden within the body of the galaxy.

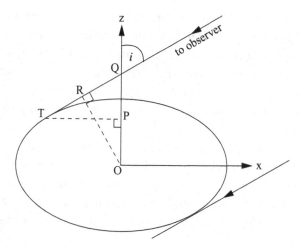

Fig. 6.8. Viewing angles for an oblate galaxy.

If a galaxy is an oblate spheroid, symmetric about the shortest axis z, then in Cartesian coordinates x, y, z, we can write the density of stars $\rho(\mathbf{x})$ as

$$\rho(\mathbf{x}) = \rho(m^2), \quad \text{where } m^2 = \frac{x^2 + y^2}{A^2} + \frac{z^2}{B^2}, \tag{6.5}$$

and $A \geq B > 0$. The contours of constant density $\rho(\mathbf{x})$ are ellipsoids $m^2 = $ constant. An observer looking directly along the z axis will see a round E0 galaxy; when viewed at an angle, the system appears elliptical. To calculate the apparent axis ratio $q = b/a$, we place the observer in the x–z plane, as in Figure 6.8, looking at an angle $0° < i < 90°$ to the axis z. The line of sight grazes the constant-density surface $m^2 = x^2/A^2 + z^2/B^2$ at the point T where it is tangent to this ellipse; so we have

$$\tan i = \mathrm{d}x/\mathrm{d}z = -(z/x)(A^2/B^2). \tag{6.6}$$

So the elliptical image has semi-major axis $a = mA$, while its semi-minor axis $b = \mathrm{OR} = \mathrm{OQ} \sin i$. Using Equations 6.5 and 6.6, we have

$$\mathrm{OQ} = \mathrm{OP} + \mathrm{PQ} = z + (-x)\cot i = B^2 m^2/z; \tag{6.7}$$

the ratio q of apparent minor to major axis is

$$q_{\mathrm{obl}} \equiv \frac{b}{a} = \frac{\mathrm{OQ} \sin i}{mA} = \frac{B^2 m}{zA} \sin i = \left(\frac{B^2}{A^2} + \cot^2 i \right)^{1/2} \sin i, \tag{6.8}$$

where we have used the definition 6.5 in the last step. Thus

$$q_{\mathrm{obl}}^2 = (b/a)^2 = (B/A)^2 \sin^2 i + \cos^2 i; \tag{6.9}$$

an oblate spheroidal galaxy never appears more flattened than its true axis ratio B/A. A prolate galaxy can be described by the same function $\rho(\mathbf{x})$ of Equation 6.5, with $A < B$. When viewed at angle i to its symmetry axis, it appears as an ellipse with the ratio q_{prol} of minor to major axis given by

$$q_{\text{prol}}^2 = [(B/A)^2 \sin^2 i + \cos^2 i]^{-1}. \tag{6.10}$$

Thus $q_{\text{prol}} \geq A/B$; again, the apparent flattening cannot be more than the true axis ratio.

Looking from a random direction, the fraction of galaxies that we see at an angle between i and $i + \Delta i$ to the polar axis is just $\sin i \cdot \Delta i$, the fraction of a sphere around each galaxy corresponding to these viewing directions. If they are all oblate with axis ratio B/A, then from Equation 6.9 the fraction $f_{\text{obl}} \Delta q$ with apparent axis ratios between q and $q + \Delta q$ is

$$f_{\text{obl}}(q)\Delta q = \frac{\sin i \cdot \Delta q}{|dq/di|} = \frac{q\,\Delta q}{\sqrt{1 - (B/A)^2}\sqrt{q^2 - (B/A)^2}}. \tag{6.11}$$

For very flattened systems with $B \ll A$, this is almost a uniform distribution; all values $q > B/A$ are about equally probable. For the disks of spiral and S0 galaxies, apparent shapes rounder than $q \approx 0.2$ are found with roughly equal frequency; so we conclude that most of their disks have $B/A \lesssim 0.2$. Very few disk galaxies show outer isophotes as elliptical as $q \lesssim 0.1$; so few disks can be as flattened as $B/A \approx 0.1$. This is consistent with what we learned in Section 5.1: the ratio of the vertical and horizontal scale lengths $h_z/h_R \approx 0.1-0.2$.

There are no elliptical galaxies in the sky more flattened than E7, or $q = 0.3$. Astronomers suspect that such a galaxy would be dynamically unstable. A system with high angular momentum would probably separate into a thin fast-rotating disk and a central bulge, rather than becoming a very flattened oblate elliptical. We know that a needle-like prolate galaxy, or a non-rotating oblate 'pancake', would buckle and thicken to a rounder shape.

Figure 6.9 shows that the apparent shapes of small elliptical galaxies are generally more elongated than those of large luminous systems. On average the midsized galaxies, with $M_B \gtrsim -20$, have apparent axis ratio $q \approx 0.75$; if these are oblate, the most common true flattening is in the range $0.55 \lesssim B/A \lesssim 0.7$. Luminous ellipticals, with $L \gtrsim L_\star$ or $M_B \lesssim -20$, have on average $q \approx 0.85$; but so few of them appear almost circular on the sky that *no* selection of oblate ellipsoidal shapes can give the observed distribution of q values. Some of these bright galaxies are probably *triaxial*, with the stellar density taking a form

$$\rho(\mathbf{x}) = \rho(m^2), \quad \text{where } m^2 = \frac{x^2}{A^2} + \frac{y^2}{B^2} + \frac{z^2}{C^2}. \tag{6.12}$$

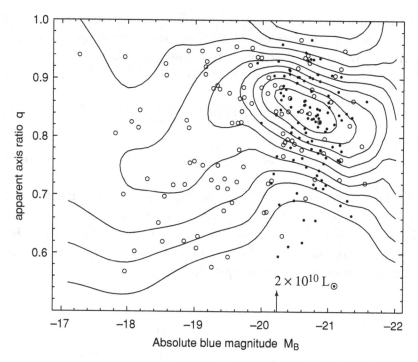

Fig. 6.9. Observed axis ratio q and blue absolute magnitude M_B for elliptical galaxies from two different samples, represented by filled and open circles. Bright galaxies (right) on average appear rounder. Contours show probability density; the top contour level is for probability density 4.5 times higher than that at the lowest, with others equally spaced – B. Tremblay and D. Merritt 1996 *AJ* **111**, 2243.

When $A > B > C$, then z is the shortest axis and x the longest; the cross-section in any plane perpendicular to each of these *principal axes* is an ellipse.

Problem 6.5 Use Equation 6.11 to show that, if we view them from random directions, the fraction of oblate elliptical galaxies with true axis ratio B/A that appear more flattened than axis ratio q is

$$F_{\text{obl}}(<q) \equiv \int_{B/A}^{q} f_{\text{obl}}(q')\mathrm{d}q' = \sqrt{\frac{q^2 - (B/A)^2}{1 - (B/A)^2}}. \qquad (6.13)$$

If these galaxies have $B/A = 0.8$, show that the number seen in the range $0.95 < q < 1$ should be about one-third that of those with $0.8 < q < 0.85$. Show that, for smaller values of B/A, an even higher proportion of the images will be nearly circular, with $0.95 < q < 1$. Then, in Figure 6.9, count the fraction of objects with $-21 < M_B < -20$ that appear rounder than $q = 0.95$, and explain why it is unlikely that galaxies in this luminosity range all have oblate shapes.

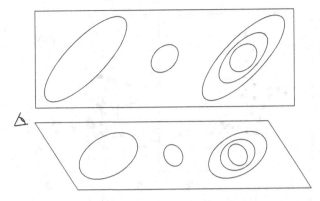

Fig. 6.10. A two-dimensional analogue of isophote twisting in a triaxial galaxy: ellipses on a flat surface (top) are viewed from below left (lower panel). The left ellipse still points to top right; but the long axis of the rounder center ellipse now heads to lower right. The rightmost set of ellipses shows how, if the axis ratios of a triaxial galaxy vary with radius, the isophotes may twist even when the principal axes do not – after J. Kormendy.

6.1.2 Twisty, disky, or boxy?

The isophotes of elliptical galaxies are close to being true ellipses; but the small deviations yield information about how the galaxy departs from a simple axisymmetric form. In Figure 6.1(b), we see the long axis of the isophotes swing from roughly horizontal in the inner galaxy to almost vertical at the outermost isophote. An observed *isophote twist* is generally taken as an indication that the galaxy may be triaxial. The isophotes of an oblate galaxy always form a sequence of nested ellipses; the long axis points along the line where the galaxy's equatorial plane intersects the plane of the sky, normal to the observer's direction. Figure 6.10 shows how this twisting can result if the galaxy is triaxial, with the ratio $A : B : C$ of the axes changing with the radius m, and we view it from a direction other than along one of the principal axes.

The isophotes of some elliptical galaxies differ from exact ellipses in appearing *disky*. They show extra light along the major axis, as though the galaxy contained an equatorial disk embedded within it; see Figure 6.1(c). Measurements of the stellar motions show that this is exactly the case; disks containing up to 30% of the total light are embedded within the elliptical body. By contrast, a galaxy with *boxy* isophotes has more light in the 'corners' of the ellipse; see Figure 6.1(d). To quantify these distortions, we write the equation of the ellipse that best matches a given isophote as

$$x = a \cos t, \quad y = b \sin t, \tag{6.14}$$

where x and y are distances along the major and minor axes, and the parameter t describes the angle around the ellipse. Let $\Delta r(t)$ be the distance between this ellipse and the galaxy's isophote, measured outward from the center. Then, we

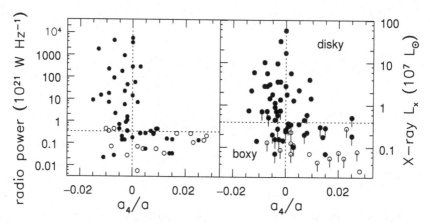

Fig. 6.11. Radio and X-ray power of elliptical galaxies. Boxy galaxies with $a_4 < 0$ tend to be strong sources; disky ellipticals with $a_4 > 0$ are usually weak. Filled circles show bright objects, with $M_B < -19.5$; open circles are dimmer galaxies. Points with downward-extending bars show upper limits on the X-ray emission; luminosities are calculated for $H_0 = 75 \text{ km s}^{-1} \text{ Mpc}^{-1}$ – R. Bender.

can write

$$\Delta r(t) \approx \sum_{k \geq 3} a_k \cos(kt) + b_k \sin(kt); \qquad (6.15)$$

the terms with $k = 0, 1$, and 2 vanish because we have chosen the best-fitting ellipse. Terms a_3 and b_3 would describe slightly egg-shaped isophotes, and they are generally very small. The term b_4 is also usually small, but a_4 is not. If $a_4 > 0$, the isophote is 'disky': it is pushed out beyond the best-fitting ellipse on the major and minor axes, at $t = 0, \pm\pi/2, \pi$ where $\Delta r > 0$. When $a_4 < 0$, the isophote bulges out at $45°$ to the axes, giving a 'boxy' or peanut-shaped appearance.

Boxy galaxies are more likely to show isophote twists. They are often luminous ellipticals, which are also the most likely to be triaxial. Figure 6.11 shows that boxy ellipticals more often have strong radio and X-ray emission; we will see below that they also are the slowest rotating. By contrast, midsized elliptical galaxies are more likely to be disky, oblate, and faster rotating, with less of the X-ray-luminous hot gas. They resemble S0 galaxies where the bulge is so large as to swallow the disk. The larger boxy systems may have formed by merger of smaller galaxies, which would destroy any disks that were present, and can easily leave a triaxial shape. Some astronomers have suggested that we should describe elliptical galaxies by their degree of boxiness, in preference to Hubble's En classification: disky ellipticals would form an intermediate class between luminous boxy systems and the S0 galaxies.

6.2   Motions of the stars

In contrast to disk galaxies, the stars of elliptical galaxies do not follow an orderly pattern of rotation. Instead, most of their kinetic energy is invested in random motion. Just as a spiral galaxy's luminosity is linked to its rotation speed, the more luminous ellipticals have a higher velocity dispersion; we can use this relation to determine distances to galaxies. Many astronomers were surprised to find that elliptical galaxies rotate as slowly as they do; it shows that the stars have not *relaxed* into anything close to a most probable final state. Elliptical galaxies still retain considerable information about their origins.

6.2.1 Measuring stellar velocities

Measuring the orbital velocities of stars within galaxies is relatively difficult. We can quite easily find the speed of a cool or warm gas cloud using the Doppler shift at the bright peak of an emission line, such as Hα from warm ionized gas, or the radio-frequency 21 cm line of neutral atomic hydrogen. But for stars, we must use *absorption* features in the spectra; many of these are not narrow atomic lines, but have an appreciable intrinsic spread in wavelength. We will see below that we must be concerned not just with the position of the line center, but with its width and often its shape as well; measuring these demands a high signal-to-noise ratio and long observations.

The absorption-line spectra of galaxies are usually measured with a long-slit spectrograph. The slit is placed across the galaxy; reflection from a grating then spreads the light in wavelength along the perpendicular direction before it falls onto a detector, most commonly a CCD; see Section 5.1. Sometimes fiber-optic bundles are used instead; the fibers collect light from various positions across the face of the galaxy and deliver it onto the grating. Light entering at the ends of the slit, or through fibers placed far from the galaxy, allows the observer to record the emission of the night sky (often with that of distant street lights). Except near its center, the sky is brighter than the galaxy; so we must subtract its spectrum from what we measure, to recover that of the galaxy.

The light of all the stars in a galaxy is the sum of their individual spectra, each Doppler-shifted in wavelength according to their motion. Their orbital motion causes lines in the summed spectrum to be wider and shallower than those of an individual star. We will see in Section 6.3 that most of the light of elliptical galaxies comes from G and K giant stars. Typically we observe some nearby stars of these types as *templates* for comparison. We take these spectra with the same telescope and spectrograph setup as for the galaxy, so that we can correct for instrumental effects. For example, a spectrograph broadens the lines by an amount $\Delta\lambda$ corresponding to the *spectral resolution*; generally $\lambda/\Delta\lambda \gtrsim 5000$ for these observations, giving a resolution of $60\,\mathrm{km\,s^{-1}}$.

Elliptical galaxies contain little cool dusty gas, so their light is quite close to the sum of what is emitted by all their stars. Let us write the energy received from a typical star, when it is at rest with respect to an observer, as $F_{\lambda,\star}(\lambda)\Delta\lambda$ between wavelengths λ and $\lambda + \Delta\lambda$. We choose Cartesian coordinates x, y, z with z pointing from our position toward the galaxy. Then, if the star moves away from us with velocity $v_z \ll c$, the light that we receive at wavelength λ was emitted at wavelength $\lambda[1 - v_z/c]$. To find the spectrum at position x, y on the galaxy image, we must integrate over all the stars along our line of sight. Approximating the number density of stars at position $\mathbf{x}$ with z velocities between v_z and $v_z + \Delta v_z$ as $f(\mathbf{x}, v_z)\Delta v_z$, the observed spectrum is

$$F_{\lambda,g}(x, y, \lambda) = \int_{-\infty}^{\infty} F_{\lambda,\star}(\lambda[1 - v_z/c]) \left\{ \int_{-\infty}^{\infty} f(\mathbf{x}, v_z)dz \right\} dv_z. \qquad (6.16)$$

If we knew the distribution function $f(\mathbf{x}, \mathbf{v})$ for each type of star, and their spectra $F_{\lambda,\star}(\lambda)$ were exactly the same as those of our template stars, we could use Equation 6.16 to construct the galaxy's spectrum. In practice, we usually guess at a form for the integral of $f(\mathbf{x}, v_z)$ along the line of sight, the term in curly braces, depending on a few parameters. We then choose those parameters to reproduce the galaxy's observed spectrum as well as possible. A common choice is the Gaussian form

$$\int_{-\infty}^{\infty} f(\mathbf{x}, v_z)dz \propto \exp\left[-(v_z - V_r)^2 / \left(2\sigma_r^2\right)\right]; \qquad (6.17)$$

here $\sigma_r(x, y)$, usually written simply as σ, is the velocity dispersion of the stars, while $V_r(x, y)$ is the mean radial velocity at that position. Rotation is evident as a gradient in $V_r(x, y)$ across the face of the system. When the spectrum is measured to high precision, the Gaussian might not provide an adequate approximation. For example, in some disky ellipticals a long 'tail' of rapidly moving stars shows that a fast-rotating disk is buried within the slower-rotating body of the galaxy.

Figure 6.12 shows the mean radial velocity and velocity dispersion measured along the major axis of the cD galaxy NGC 1399 out to a distance of about $5R_e$, where the surface brightness has fallen more than 100-fold below its central level. In a spiral galaxy like our own, the ordered rotation of disk stars is almost ten times faster than their random speeds; but in this elliptical, we see a peak rotation speed $V_{max} \ll \sigma_r$. We would not expect elliptical galaxies to rotate as rapidly as spirals, since they are not disks; but we will see in the next subsection that most of them do not even rotate as fast as they 'ought to' for their rounder shapes.

Problem 6.6 The core radius r_c of NGC 1399 is $\sim 5''$ or 400 pc. Use Equation 3.46 to combine the measured dispersion σ_r from Figure 6.12 with the result of Problem 6.3, to show that the mass-to-light ratio in the central parts is $\mathcal{M}/L_V \sim 7$. This is only slightly larger than that in the Milky Way's globular clusters (Section 3.1), suggesting that the core of this galaxy contains little dark matter.

Further reading: for more on spectroscopic techniques, see Chapter 3 of D. F. Gray, 1992, *The Observation and Analysis of Stellar Photospheres*, 2nd edition (Cambridge University Press, Cambridge, UK). For the analysis, see Chapter 11

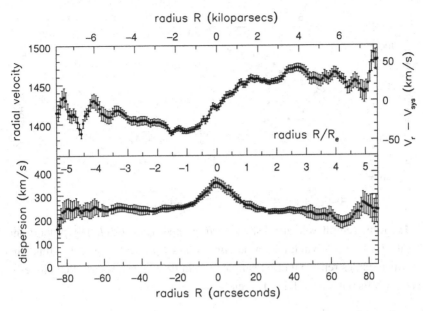

Fig. 6.12. Measured radial velocity V_r and velocity dispersion σ_r on the major axis of cD galaxy NGC 1399; vertical bars show uncertainties. Notice that $(V_r - V_{sys})/\sigma_r \ll 1$; V_r reverses slope in the central few arcseconds – A. Graham.

of J. Binney and M. Merrifield, 1998, *Galactic Astronomy*, 3rd edition (Princeton University Press, Princeton, New Jersey). These are both graduate texts.

6.2.2 The Faber–Jackson relation and the fundamental plane

The range in the velocity dispersion σ for elliptical galaxies is close to what we saw in Section 5.3 for the peak rotation speed of disk galaxies. Just as for spirals, the stars move faster in more luminous galaxies. At the centers of bright ellipticals, the dispersion can reach $500 \, \mathrm{km \, s^{-1}}$, while $\sigma \sim 50 \, \mathrm{km \, s^{-1}}$ in the least luminous objects. The left panel of Figure 6.13 shows that roughly $L \propto \sigma^4$; this is often called the *Faber–Jackson* relation. In the V band, roughly

$$\frac{L_V}{2 \times 10^{10} L_\odot} \approx \left(\frac{\sigma}{200 \, \mathrm{km \, s^{-1}}} \right)^4. \tag{6.18}$$

Like the Tully–Fisher relation for spirals, the Faber–Jackson relation can be used to estimate a galaxy's distance from its measured velocity dispersion. But it is hard to determine the total amount of light we receive from a galaxy, because much of it comes from the faint outer parts; distances derived from the Faber–Jackson relation are not very precise. A better method is to measure the diameter D of the isophote within which the surface brightness averages to a given level. We saw in Figure 6.6 that a galaxy's central brightness $I(0)$ depends on its luminosity L;

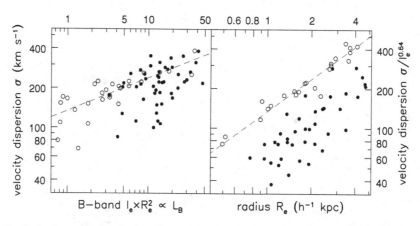

Fig. 6.13. Left, central velocity dispersion σ plotted against $I(R_e)R_e^2$ proportional to luminosity L_B in the B band; the dashed line shows $L_B \propto \sigma^4$. Right, the 'fundamental plane'. Open circles represent elliptical galaxies in the Coma cluster; filled circles show those at redshifts $0.8 < z < 1.2$ – T. Treu.

so there is a corresponding D–σ *relation*. Popular choices for D take an average surface brightness of 20–21 mag arcsec^{-2} in the B band.

    Another possibility is to use the *fundamental plane* relation. The right panel of Figure 6.13 shows that the elliptical galaxies of the Coma cluster all lie close to a plane in the three-dimensional 'space' of central velocity dispersion σ, effective radius R_e, and surface brightness $I_e = I(R_e)$. Approximately, we have

$$R_e \propto \sigma^{1.2} I_e^{-0.8}. \qquad (6.19)$$

Like Figure 6.6, the fundamental plane relation reflects some basic processes, still to be understood, by which elliptical galaxies form. We see the galaxies at $z \sim 1$ as they were when the cosmos was half its present age. They are brighter than those nearby, and do not follow the same fundamental plane; their mass-to-light ratios are roughly five times lower.

Problem 6.7 Assuming that the velocity dispersion σ and the ratio $\mathcal{M}/L$ are roughly constant throughout the galaxy, and that no dark matter is present, show that the kinetic energy $\mathcal{KE} = 3\mathcal{M}\sigma_r^2/2$. Approximating it crudely as a uniform sphere of radius R_e, we have $\mathcal{PE} = -3G\mathcal{M}^2/(5R_e)$ from Problem 3.12. Use Equation 3.44, the virial theorem, to show that the mass $\mathcal{M} \approx 5\sigma^2 R_e/G$. If all elliptical galaxies could be described by Equation 6.1 with the same value of n, explain why we would then have $\mathcal{M} \propto \sigma^2 R_e$ and the luminosity $L \propto I_e R_e^2$, so that $\mathcal{M}/L \propto \sigma^2/(I_e R_e)$.

    (a) Show that, if all ellipticals had the same ratio $\mathcal{M}/L$ and surface brightess $I(R_e)$, they would follow the Faber–Jackson relation.

    (b) Show that Equation 6.19 implies that $I_e \propto \sigma^{1.5} R_e^{-1.25}$, and hence that $\mathcal{M}/L \propto \sigma^{0.5} R_e^{0.25}$ or $\mathcal{M}^{0.25}$: the mass-to-light ratio is larger in big galaxies.

6.2.3 How fast should an elliptical galaxy rotate?

To understand why astronomers expected elliptical galaxies to rotate faster than they do, we use a form of the virial theorem. We start from the equation for the gravitational force on star α at position $\mathbf{x}_\alpha$ due to the other stars in the galaxy, at positions $\mathbf{x}_\beta$, with masses m_β:

$$\frac{d}{dt}(m_\alpha \mathbf{v}_\alpha) = -\sum_{\substack{\beta \\ \alpha \neq \beta}} \frac{Gm_\alpha m_\beta}{|\mathbf{x}_\alpha - \mathbf{x}_\beta|^3}(\mathbf{x}_\alpha - \mathbf{x}_\beta). \tag{3.2}$$

Then, we follow the process that led us to Equation 3.38 of Section 3.1. But instead of taking a scalar product, we multiply the z component of Equation 3.2 by the z coordinate z_α and sum over all the stars, to get

$$\sum_\alpha \frac{d}{dt}(m_\alpha v_{z\alpha})z_\alpha = -\sum_{\substack{\alpha,\beta \\ \alpha \neq \beta}} \frac{Gm_\alpha m_\beta}{|\mathbf{x}_\alpha - \mathbf{x}_\beta|^3}(z_\alpha - z_\beta)z_\alpha, \tag{6.20}$$

where $v_{z\alpha}$ is the z component of $\mathbf{v}_\alpha$, the velocity of star α. But we could also have started with the force on star β, to find

$$\sum_\beta \frac{d}{dt}(m_\beta v_{z\beta})z_\beta = -\sum_{\substack{\alpha,\beta \\ \alpha \neq \beta}} \frac{Gm_\alpha m_\beta}{|\mathbf{x}_\alpha - \mathbf{x}_\beta|^3}(z_\beta - z_\alpha)z_\beta. \tag{6.21}$$

Averaging these two yields an equation similar to Equation 3.42:

$$\frac{1}{2}\frac{d^2 I_{zz}}{dt^2} = 2\mathcal{KE}_{zz} + \mathcal{PE}_{zz}, \tag{6.22}$$

where the z component of the moment of inertia is defined as

$$I_{zz} \equiv \sum_\alpha m_\alpha z_\alpha z_\alpha, \tag{6.23}$$

the kinetic energy associated with motion in the z direction is

$$\mathcal{KE}_{zz} \equiv \frac{1}{2}\sum_\alpha m_\alpha v_{z\alpha} v_{z\alpha}, \tag{6.24}$$

and the zz contribution to the potential energy is

$$\mathcal{PE}_{zz} \equiv -\sum_{\substack{\alpha,\beta \\ \alpha \neq \beta}} \frac{1}{2}\frac{Gm_\alpha m_\beta}{|\mathbf{x}_\alpha - \mathbf{x}_\beta|^3}(z_\alpha - z_\beta)^2. \tag{6.25}$$

In the same way as for Equation 3.44, if all the stars are bound within the galaxy, we can write

$$2\langle \mathcal{KE}_{zz} \rangle + \langle \mathcal{PE}_{zz} \rangle = 0. \tag{6.26}$$

where the angle brackets denote a long-term average. Similar equations hold for the x and y components of the motion; they make up part of the *tensor virial theorem*. This theorem tells us that not only must the average kinetic and potential energies be in balance, but also contributions in the different directions must separately be equal. If a galaxy is highly flattened, like the Milky Way, $\mathcal{PE}_{zz}$ will be much smaller in magnitude than $\mathcal{PE}_{xx}$. Equation 6.26 then tells us that the energy in random speeds in the z direction must be less than that from the random motion and rotation which make up the x component of the kinetic energy.

Suppose that an elliptical galaxy is axisymmetric, with the density of stars described by Equation 6.5, and it rotates about the symmetry axis z. Then we can split the kinetic energy in the x direction into the sum of rotation and random motions. If we make the simplifying assumption that the rotation speed V and the velocity dispersions σ_x and σ_z in these two directions are almost constant throughout the galaxy, then

$$\frac{\langle \mathcal{PE}_{zz} \rangle}{\langle \mathcal{PE}_{xx} \rangle} = \frac{\langle \mathcal{KE}_{zz} \rangle}{\langle \mathcal{KE}_{xx} \rangle} \approx \frac{\sigma_z^2}{\frac{1}{2}V^2 + \sigma_x^2}, \tag{6.27}$$

since the kinetic energy of rotation is split between the x and y directions.

The ratio of the two potential-energy terms turns out to depend only on the axis ratio B/A, or equivalently the ellipticity $\epsilon \equiv 1 - B/A$. It is not affected by how the mass is distributed inside the galaxy; roughly,

$$\frac{\langle \mathcal{PE}_{zz} \rangle}{\langle \mathcal{PE}_{xx} \rangle} \approx (B/A)^{0.9} = (1 - \epsilon)^{0.9}. \tag{6.28}$$

The true rotation speed V is higher than the measured average speed V_{max}, since the stars spend part of their time in motion across the sky, which does not contribute to V_{max}; to allow for this, we write $V_{\text{max}} \approx \pi V/4$. If random motions are *isotropic*, the same in all directions, we have $\sigma_x = \sigma_z = \sigma$, and Equation 6.27 becomes

$$\left(\frac{V_{\text{max}}}{\sigma} \right) = \left(\frac{V}{\sigma} \right)_{\text{iso}} \equiv \frac{\pi}{4} \sqrt{2[(1 - \epsilon)^{-0.9} - 1]} \approx \sqrt{\epsilon/(1 - \epsilon)}; \tag{6.29}$$

the approximation is valid when ϵ is small. According to this relation, even fairly round galaxies should rotate quite fast; for example, $B/A = 0.7$ should imply $V_{\text{max}}/\sigma \approx 0.68$.

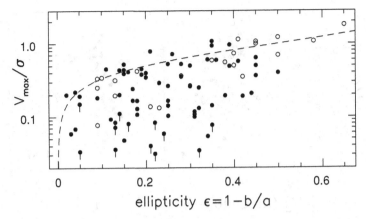

Fig. 6.14. The ratio of measured peak rotation speed V_{max} to central velocity dispersion σ for elliptical galaxies, plotted against apparent ellipticity: filled circles show bright galaxies ($M_B < -19.5$); open circles are dimmer galaxies. Points with downward-extending bars indicate upper limits on V_{max}. The dashed line gives $(V/\sigma)_{iso}$, the fastest rotation expected for a given flattening – R. Bender.

We cannot measure the true flattening B/A for a galaxy, but only the apparent axis ratio b/a. If we view it at an angle i to the direction of the axis z, the measured rotation speed V_{max} is reduced by a factor $\sin i$ from the value it would have if we looked at it from a position in the $x-y$ plane. But, if the galaxy is not too flattened, the following problem shows that $2\epsilon \approx \sin^2 i[1 - (B/A)^2]$, so the right and left sides of Equation 6.29 both decrease approximately as $\sin i$. Thus, if we compute $(V/\sigma)_{iso}$ using the apparent ellipticity $\epsilon_{app} = 1 - b/a$, oblate galaxies with isotropic velocity dispersions should still fall close to the line $V_{max}/\sigma = (V/\sigma)_{iso}$ in a plot like Figure 6.14. But the giant E1 galaxy NGC 1399 is nowhere close to this relation: for $\epsilon_{app} \approx 0.1$ we expect $(V/\sigma)_{iso} = 0.33$, while Figure 6.12 shows that $V_{max}/\sigma \lesssim 0.15$.

Problem 6.8 Show from Equation 6.9 that the apparent flattening $\epsilon_{app} = 1 - b/a$ of an oblate galaxy is given by

$$\sin^2 i(1 - B^2/A^2) = 1 - (b/a)^2 \approx 2\epsilon_{app} \quad \text{when } \epsilon_{app} \ll 1. \quad (6.30)$$

Figures 6.14 and 6.15 show that many ellipticals, especially the brighter ones, rotate much more slowly than they should if the velocity dispersion were isotropic. Dwarfs have been omitted from the figures, so they include only luminous and midsized galaxies. Equation 6.27 tells us that the slow rotation must be compensated by random motion; we deduce that $\sigma_x \gg \sigma_z$. The flattening of these galaxies is caused not by their rotation, but by *velocity anisotropy*. Luminous ellipticals are more likely to have significant anisotropy, whereas midsized ones are fast rotating. The right panel of Figure 6.15 shows what is going on: ellipticals with

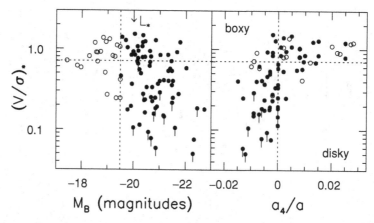

Fig. 6.15. The ratio $(V/\sigma)_\star$ of measured V_{max}/σ to $(V/\sigma)_{iso}$, the rotation expected for an oblate galaxy according to Equation 6.29. Downward-pointing bars show upper limits on V_{max}; filled circles are bright galaxies, with $M_B < -19.5$ for $H_0 = 75\,\text{km s}^{-1}\,\text{Mpc}^{-1}$. Left, luminous galaxies often rotate slowly, falling below the dotted horizontal line at $(V/\sigma)_\star = 0.7$. Right, boxy galaxies, with $a_4 < 0$, are almost all slow rotators; many of these are luminous – R. Bender.

disky isophotes, which are predominantly midsized, rotate rapidly. These are composite systems, with a fast-rotating stellar disk embedded within a slow-rotating outer body. Recall from Section 4.4 that dwarf ellipticals commonly rotate with $V_{max}/\sigma \lesssim 1$, despite a considerably flattened appearance; they are also flattened by velocity anisotropy.

The slow rotation of ellipticals disproved an explanation that had been developed to explain the very smooth appearance of these galaxies. A star like the Sun is a round smooth ball of gas because frequent random encounters or 'collisions' between atoms change their direction of motion. Statistical mechanics tells us that, as a result, the velocities of the atoms will tend to be near their most probable overall state. In this state, the star is slightly flattened along the axis of its rotation, and is symmetric about that axis. The random velocities $\mathbf{v}$ of particles with mass m at any position $\mathbf{x}$ in the star are very close to the Maxwellian distribution:

$$f_M(\mathbf{x}, \mathbf{v}) \propto \exp\left[-m\mathbf{v}^2/(k_B T)\right], \qquad (3.58)$$

where k_B is Boltzmann's constant and T is the gas temperature. Random motions are the same in all directions, since they are set by the single temperature T.

We saw in Section 3.2 that it would take about 10^{13} years for encounters between pairs of stars in a galaxy to change their motions significantly. The smooth roundness of elliptical galaxies must have a different explanation from the smooth roundness of the Sun. So the theory of *violent relaxation* was developed. Aside from periodically rotating bars and spiral patterns, the gravitational potential in normal galaxies today appears to change only slowly over time; so Equation 3.27 tells us that a star's energy $\mathcal{E}$ is almost constant along its orbit. But while lumps of matter were falling together to form the galaxy, the potential $\Phi(\mathbf{x}, t)$ at any point $\mathbf{x}$ would swing up and down by large amounts, changing the star's energy. This process was expected to mix stars among the various orbits, giving them equal random motions in all directions.

The observed slow rotation of elliptical galaxies tells us that violent relaxation was not complete. The stars in these galaxies, unlike the atoms of gas that make up stars, retain information about the way in which they were assembled. In particular, they might not have relaxed enough for the galaxy to develop an axis of symmetry. A further indication of only partial mixing is given by the rotation curve of Figure 6.12. The innermost part of the galaxy NGC 1399 seems to rotate in the opposite direction to the outer regions. These *kinematically decoupled cores* have been found increasingly often, as the precision of measurement improves. Some elliptical galaxies also show a gradient in the mean velocity V_r when measurements are made along the apparent *minor* axis. This state could not last long in an axisymmetric galaxy, but we will see below that it can persist if the potential is triaxial. Minor-axis rotation is generally reckoned as one further piece of evidence that at least some elliptical galaxies have triaxial shapes.

6.2.4 Stellar orbits in a triaxial galaxy

To understand the implications of a triaxial elliptical galaxy, we must study the orbits that stars would follow in such a system. One of the simplest triaxial potentials is the *triaxial harmonic oscillator*

$$\Phi_{HO}(\mathbf{x}) = \frac{1}{2}\left(\omega_x^2 x^2 + \omega_y^2 y^2 + \omega_z^2 z^2\right) \qquad (6.31)$$

with $\omega_x < \omega_y < \omega_z$. This describes the gravitational force inside a homogeneous galaxy described by Equation 6.12 with A, B, and C held fixed, and constant density $\rho(m^2)$ for $m \leq 1$; there is no mass outside that ellipsoidal surface. A star in this potential would follow independent harmonic motion in the x, y, and z directions, with frequencies ω_x, ω_y, and ω_z, respectively; these are complicated functions of the axis ratios $A : B : C$. Unless the frequencies are rational multiples of each other, this orbit completely fills a rectangular block, with sides parallel to the axes: we call it a *box* orbit. We saw in Section 3.3 that a star moving in an axisymmetric potential must stay away from the center unless its angular momentum is exactly zero; by contrast, a star on a box orbit can travel right to the origin. A box orbit has no fixed sense of angular momentum. The sign of L_z reverses as the orbit reaches the 'edge' of the block in the x and y directions, and similarly for L_x and L_y; so each of these is zero on average.

Problem 6.9 For a particle moving in the potential Φ_{HO}, use the x component of Equation 3.3, the force equation, to show that

$$E_x \equiv \frac{1}{2}\left(\dot{x}^2 + \omega_x^2 x^2\right) \tag{6.32}$$

is constant, as are E_y and E_z when defined similarly: this potential has three *integrals of motion*. Write E_x in terms of the maximum extent x_{max} of the orbit along the x axis, and explain why the average x speed is independent of the particle's motion in the y and z directions. (Unlike the distribution function $f(E, L_z)$ that we discussed in Section 3.4, making a model galaxy with $f = f(E_x, E_y, E_z)$ lets us choose the velocity dispersion in the z direction independently of motion in the x–y plane: the triaxial potential makes it easier to set up a galaxy with $\sigma_z \ll \sigma_x$.)

Problem 6.10 Show that the potential of Equation 6.31 corresponds to a uniform density $\rho(\mathbf{x})$ such that $4\pi G\rho = \omega_x^2 + \omega_y^2 + \omega_z^2$. Taking the values $\omega_x = 1, \omega_y = \pi/2$, compute the path of a star that at $t = 0$ has $x = 2, y = 1, z = 0$ with $\dot{x} = 0 = \dot{y} = \dot{z}$; plot its position at intervals of 0.1 for $-10 \le t \le 10$. Show that $L_z = x\dot{y} - y\dot{x}$ reverses sign at $t = 0$.

Orbits in an axisymmetric potential have the rosette shapes of Figure 3.10: these are *loop* orbits. A star on a loop orbit is prevented by its angular momentum from passing through the center, but always circles it in the same sense, keeping a minimum distance. Triaxial potentials in general allow both loop orbits and box orbits, along with some more complicated orbits which are generally less important. Loops can circle any one of the three symmetry axes: a loop around the z axis has a fixed sense of L_z, while L_x and L_y oscillate.

In general, most loop orbits around the longest and the shortest axes of a triaxial potential are stable. A star that is initially close in position and velocity to one of these loops will always remain near it. But a star placed near a loop orbit around the intermediate axis will move away from that orbit at an exponentially increasing rate; these orbits are unstable. In triaxial galaxies, we expect to find some stars following orbits that circle the shortest axis, as in an oblate galaxy; so we will measure a gradient in the mean velocity V_r along the galaxy's major axis. Other stars will loop around the longest axis; if these all circle in the same sense, we would then find a gradient along the *minor* axis, with V_r increasing on one side and decreasing on the other.

Figure 6.16 shows orbits in the x–y plane of the triaxial potential

$$\Phi_L(x, y) = \frac{1}{2}v_0^2 \ln\left(R_e^2 + x^2 + \frac{y^2}{q^2}\right), \tag{6.33}$$

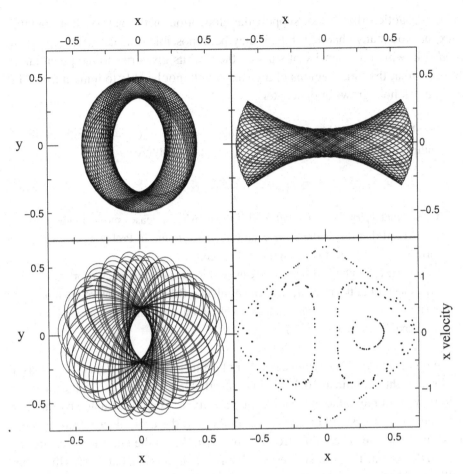

Fig. 6.16. Orbits in the potential of Equation 6.33, with $v_0 = 1$, $q = 0.8$, and $R_e = 0.14$; all have energy $E = -0.337$. The top left panel shows a loop orbit, which avoids the center; at the top right is a box orbit, which passes through it; lower left is a chaotic orbit, produced when a central spherical potential is added. The lower right panel shows a *surface of section*: values of $(x, \dot{x})$ for all three orbits are plotted each time the orbit crosses $y = 0$ in the direction $\dot{y} > 0$.

where $q < 1$; this corresponds to a galaxy elongated in the x direction. The box orbit has roughly a rectangular form, elongated in the same direction as the potential; stars on loop orbits, that avoid the center, are stretched out in the perpendicular direction. To maintain its shape, we might expect a triaxial galaxy to contain many stars on box orbits and only a few on loops; since box orbits have no fixed sense of rotation, we would measure high random motions and only slow average rotation.

The third orbit is a *chaotic* orbit, neither loop nor box. Like a loop it avoids the galaxy center, but it occasionally reverses the sense in which it circles the center. To produce this orbit, we have added to Φ_L a Plummer potential corresponding to a small mass $\mathcal{M} = 0.2$ with core $a_P = 0.1$. This changes the asphericity of the potential close to the center, and disrupts box orbits which venture there. Stars on this chaotic orbit do not support the elongation of the galaxy. If a potential $\Phi(\mathbf{x})$ has too many chaotic orbits, it may be impossible to make a *self-consistent model*, in which the density of stars on their orbits gives rise to the potential Φ. In particular, the inner regions of a galaxy might not be able to remain triaxial if a big black hole grows at the center.

Problem 6.11 Show that, if the density of stars $\rho(r) = \rho_0(r_0/r)^\alpha$, then the mass within radius r follows $\mathcal{M}(<r) \propto r^{3-\alpha}$, and the radial force

$$F_r(r) = -\frac{G\mathcal{M}(<r)}{r^2} \propto r^{1-\alpha}.$$

We see that $F_r(r) \to 0$ at small radii if the density increases more slowly than r^{-1}. Recall that this is the same condition as we found in Problem 6.4 for the surface density $\Sigma(R)$ to remain finite at the center.

(If the forces tend to zero at the center of a triaxial potential, box orbits will exist much as in the triaxial harmonic oscillator. Otherwise, many of the boxes are replaced by chaotic orbits. Galaxies with central luminous cusps are probably not triaxial near their centers.)

A compact way of showing a number of orbits is a *surface of section* plot, as in the lower right of Figure 6.16. Each time the orbit crosses the x axis in the direction $\dot{y} > 0$, we plot the value of x and the x velocity $\dot{x}$. The points made by the loop orbit fall along the small closed curve on the left; the orbit always transits the x axis anticlockwise. The points from the box orbit lie on the largest curve, circling around the origin. This orbit has no fixed sense of circulation, but crosses the x axis with $\dot{y} > 0$ both on $x > 0$ and on $x < 0$; $\dot{x}$ is large whenever x is small, as the particle passes close to the center. The chaotic orbit shows both senses of circulation about the center; but, like the loop, it avoids the origin. The points from this orbit *do not fall on a curve at all*; their more disorderly distribution betrays the difference between a chaotic orbit, and the non-chaotic or *regular* boxes and loops. The discovery and exploration of chaotic properties was one of the major developments of twentieth-century mathematics; we refer interested readers to the books below.

Further reading: at the graduate level, Sections 3.3, 3.4, and 4.3 of J. Binney and S. Tremaine, 1987, *Galactic Dynamics* (Princeton University Press, Princeton, New Jersey). For a clear discussion of chaotic orbits, written for the general reader, see I. Stewart, 1990, *Does God Play Dice? The Mathematics of Chaos* (Blackwell, Cambridge, Massachusetts).

6.3 Stellar populations and gas

Unlike spiral and irregular galaxies, elliptical galaxies conspicuously lack luminous blue stars; the brightest stars are red giants and stars on the asymptotic giant branch (AGB; see Section 1.1). We cannot see individual stars directly in galaxies

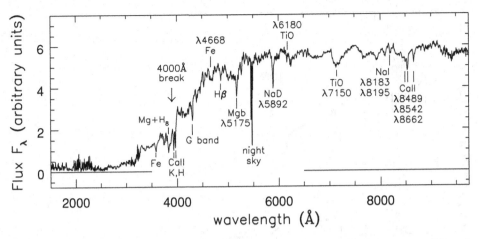

Fig. 6.17. The spectrum of an elliptical galaxy; compare this with the spectra of K and M stars in Figure 1.1, and those of disk galaxies in Figure 5.24 – A. Kinney.

more distant than about 20 Mpc. So, even in the closest ellipticals, we are limited to the AGB stars and those near the tip of the red giant branch. The integrated spectra of ellipticals, such as that in Figure 6.17, show deep absorption lines of heavy elements such as calcium and magnesium, similar to the K-star spectrum in Figure 1.1. There is little light below 3500 Å, showing that they have made very few new stars in the last 1–2 Gyr. Table 1.1 shows that only stars with masses below $2\mathcal{M}_\odot$ will survive as long as 1 Gyr, and these produce most of their light as red stars after they have left the main sequence. So the galaxy's light comes mainly from red giants. Unlike the old stars in the Milky Way's globular clusters, but like those of the Galactic bulge, stars at the centers of elliptical galaxies appear fairly metal-rich, with about the same composition as the Sun. The spectrum shows a break at 4000 Å, since lines of metals absorb much of the light at shorter wavelengths.

Figure 6.18 shows the spectrum of a model galaxy that has made all its stars in a sudden burst lasting 100 Myr. Shortly afterward, the galaxy is bright and very blue; helium absorption lines characteristic of hot O and B stars are prominent. We would also see emission lines from gas ionized by these stars. About 100 Myr after star formation has ceased, it is both dimmer and redder. At 1 Gyr, we see a *post-starburst* spectrum, with the deep Balmer lines characteristic of A stars. Elliptical galaxies with such a spectrum are called 'E + A' systems. They experienced rapid starbirth ~1 Gyr ago followed by a sharp decline to almost nothing. After the first ~2 Gyr cool giants predominate, and the model spectrum starts to resemble that of the elliptical in Figure 6.17. Over time the galaxy fades and slowly reddens, as the 4000 Å break becomes more pronounced.

Problem 6.12 At age 10 Gyr the model galaxy of Figure 6.18 is roughly three times fainter in the B band at 4400 Å than it was at age 1 Gyr, and ten times fainter than it was at age 100 Myr. Figure 6.13 shows that, compared with the

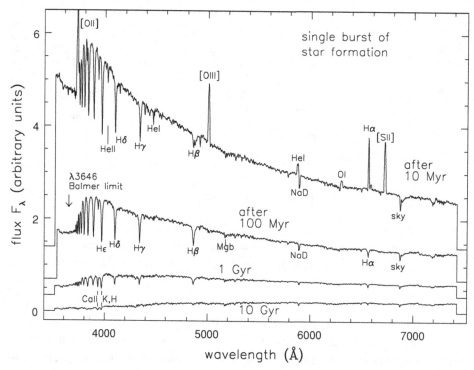

Fig. 6.18. Spectra for a 'galaxy' that makes its stars in a 10^8 yr burst, all plotted to the same vertical scale. Emission lines of ionized gas are strong 10 Myr after the burst ends; after 100 Myr, the galaxy has faded and reddened, and deep hydrogen lines of A stars are prominent. Beyond 1 Gyr, the light dims and becomes slightly redder, but changes are much slower – B. Poggianti.

present-day galaxies of the Coma cluster, the ratio M/L_B is five times less for the systems at $z \sim 1$, roughly 8 Gyr ago. If these distant galaxies made all their stars in a single burst and then simply faded to resemble the galaxies of the Coma cluster, explain why they must have formed <1Gyr before we observe them.

(It is not likely that this random set of ellipticals all formed very close to 9 Gyr ago. More probably, some starbirth continued until close to $z = 1$.)

We saw in the previous section that the central brightness, velocity dispersion, and rotation speed of an elliptical galaxy depend on its luminosity. Figure 6.19 shows that the same is true of the overall color of its light: both at visible wavelengths and in the near-infrared K band at 2.2μm, brighter galaxies are redder and fainter systems bluer. This trend could be explained if small elliptical galaxies were either younger or more metal-poor than large bright ones. Some astronomers argue that almost all the stars of luminous ellipticals date from at least 10 Gyr ago, while those of their smaller counterparts are a few gigayears younger.

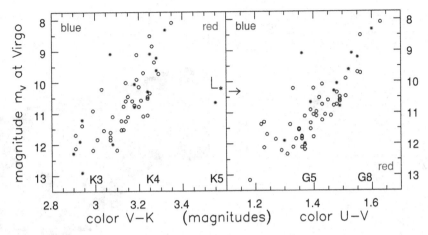

Fig. 6.19. For elliptical galaxies in the Virgo (open symbols) and Coma (closed symbols) clusters, the $U - V$ and $V - K$ colors are plotted against apparent magnitude. Colors of giant stars from Table 1.5 are indicated. Coma galaxies are shown 3.6 magnitudes brighter, as they would appear at the distance of Virgo – data from Bower *et al.* 1992 *MN* **254**, 601.

Luminous galaxies do on average contain more older stars. The right panel of Figure 6.20 shows the stellar birthrates in nearby galaxies from the Sloan Digital Sky Survey. These were estimated from the intensity of the Hα line ionized by young stars, and the strength of the 4000 Å break. The largest galaxies appear to complete their starbirth early: in galaxies more than about twice as luminous as the Milky Way, starbirth is at ∼1% of its cosmic average. We saw this already in Figure 1.16: the most luminous galaxies today are red systems full of old and middle-aged stars. Among midsized galaxies, those of roughly the Milky Way's luminosity or $L \sim L_\star$ of Equation 1.24, in a minority starbirth has slowed as strongly as in the massive galaxies. But most midsized galaxies are making stars at near or above the average rate needed to build their present stellar content over the lifetime of the Universe. Dwarf galaxies less luminous than $M_r \sim -18$ are forming stars even faster than at their average rate. The dwarf elliptical and dwarf spheroidal galaxies, where starbirth has now ceased, are so dim that few are included in this survey.

Other astronomers claim that the relation between color and luminosity that we see in Figure 6.19 can be explained mainly by differences in chemical composition: big ellipticals are richer in heavy elements than the midsized ones. The left panel of Figure 6.20 shows the metal abundance in a large sample of nearby galaxies, calculated from the emission lines of gas at their centers. The most luminous galaxies are clearly richer in metals. Smaller galaxies may lack them because they lost most of the metal-enriched gas shed by their old stars. Larger systems were better able to trap theirs, incorporating the heavy elements into new stars. Figure 1.5 showed us that metal-poor stars of a given mass are bluer, especially while they are burning helium in their cores. So we are not surprised to find that smaller galaxies are bluer.

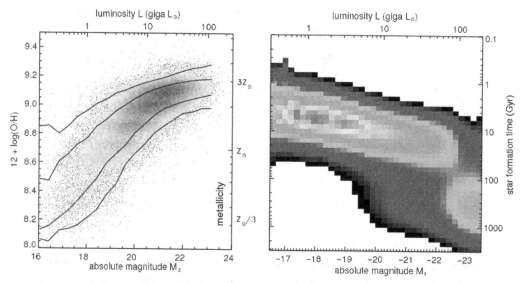

Fig. 6.20. For 83 963 galaxies of the Sloan Digital Sky Survey, the left panel shows that luminous systems are richer in heavy elements. Outer curves show the region where 95% of the galaxies lie; inner curves enclose 60% of them. The Milky Way would correspond to $(3-4) \times 10^{10} L_\odot$ in the z band of Table 1.3; near its center the metal content Z is slightly above $Z_\odot$. Right, time taken to build up a galaxy's present stellar content, at its current rate of starbirth. In the most luminous systems, star formation is now far below its cosmic average. Many small galaxies are making stars considerably faster than that average – C. Tremonti and J. Brinchmann.

When we look at absorption lines such as the Mgb feature at 5175 Å, we find that stars at the centers of elliptical galaxies are more metal-rich than those in the periphery. This is the same pattern as we saw in Figure 4.15 for disk galaxies. The central stars of luminous ellipticals are at least as metal-rich as the Sun; but they do not contain heavy elements in the same proportions. Relatively light atoms such as oxygen, sodium, and magnesium are a few times more abundant relative to iron. We find the same pattern in old metal-poor stars near the Sun (Figure 4.17), and in the interstellar gas of the young galaxy cB58 in Figure 9.16, seen less than 3 Gyr after the Big Bang. The effect is weaker in midsized ellipticals.

As we discussed in Section 4.3, iron in the interstellar gas has been released mainly by supernovae of Type Ia, which occur in binary stars. Many of the stars that explode in this way do so at ages of 1 Gyr or more. Stars near the Sun that have high ratios of oxygen to iron probably formed before many Type Ia supernovae had exploded, or before the gas they ejected could be confined within the Galaxy. But the lighter elements like oxygen are ejected mainly in Type II supernova explosions, of massive stars which run through their lives in 100 Myr or less. These should be added quickly to the gas, and incorporated into new stars.

A similar argument may apply to elliptical galaxies. Perhaps the most luminous ellipticals formed most of their stars so early that Type Ia supernovae had not yet begun to add iron to the galactic gas. Or these galaxies may have lacked Type Ia supernovae, because they made relatively more of the massive stars, or fewer binaries. Another possibility is that, early on, the products of Type Ia supernova explosions were able to leave the galaxy, perhaps mixing with the hot X-ray-emitting gas of a surrounding cluster of galaxies. The gas ejected in Type II supernova explosions moves slower, and so is easier to confine within the galaxy.

Surprisingly, despite their lack of young hot stars, elliptical galaxies (and the bulges of disk systems) are not dark at ultraviolet wavelengths. Metal-rich ellipticals tend to be brightest, suggesting that the sources are not old metal-poor stars on the horizontal branch, like the ones we see in globular clusters. Instead, the emission presumably comes from old metal-rich stars that have left the main sequence and then lost most of their hydrogen envelope, exposing the hot stellar core. No such stars are found near the Sun, since the Milky Way's disk is too young. So we have not been able to study them in detail, and models remain uncertain.

X-ray sources in elliptical galaxies include active galactic nuclei; stellar binaries where material is accreted onto a black hole or neutron star, or burns on the surface of a white dwarf; and hot gas at temperatures $T \gtrsim 10^6$ K, either filling the galaxy or localized in stellar atmospheres, stellar winds, or supernova remnants. Most elliptical galaxies lack an active nucleus, and X-rays come mainly from the hot interstellar medium. Their supernova remnants are not strong X-ray sources. When supernova remnants expand into the cool and warm gas of a galaxy disk where $T \lesssim 10^4$ K, they are highly supersonic and create strong shocks that heat gas to X-ray-emitting temperatures. But in an elliptical galaxy the remnants expand into the hot gas with barely supersonic speeds, developing only weak shocks. When the galaxy is gas-poor, stellar binaries are the most important sources of X-rays.

We should not be surprised that ellipticals today largely lack new stars, since most now contain very little cool gas from which to make any. At first glance, ellipticals seem generally free of dust and gas. According to Hubble, '*small patches of obscuring material are occasionally silhouetted against the luminous background, but otherwise these nebulae present no structural details*'. Closer examination shows that almost all of them have some dust in the nuclear regions, which is presumably mixed with cold gas. But only 5%–10% of normal ellipticals contain enough atomic hydrogen or molecular gas to be detectable – which means that most big elliptical galaxies have less than $(10^8$–$10^9)\mathcal{M}_\odot$ of cool gas, while a large Sc galaxy contains almost $10^{10}\mathcal{M}_\odot$.

A minority of elliptical galaxies, especially those with 'peculiarities' such as outer shells or obvious dust lanes, go against this trend: a few have as much cool gas as a large spiral. Where HI gas is plentiful, it often forms a ring well outside the stellar body of the galaxy, and it frequently orbits about an axis that is nowhere close to the short axis of the galaxy's image or the rotation axis of its stars. The

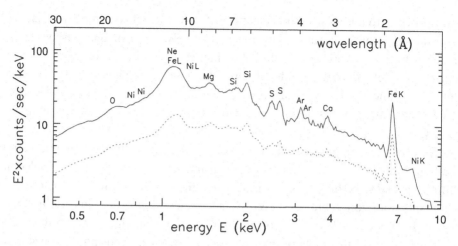

Fig. 6.21. The X-ray spectrum of hot gas at $T \sim 2 \times 10^7$ K around the luminous elliptical M87. The solid line shows emission from gas within $4'$ or 5 kpc of the center; the broken line is for gas between $4'$ and $8'$ radius. All lines except iron L and nickel L are emitted as electrons drop to the lowest-energy orbits, in the K shell – XMM-Newton: K. Matsushita *et al.* 2002 *AAp* **386**, 77.

elliptical NGC 5128 (Centaurus A), which is also a powerful radio galaxy, has a spectacular ring of gas and dust circling its apparent major axis. This pattern contrasts with spirals, but is similar to the gas-rich S0 galaxies that we discussed in Section 5.2. The cool gas is unlikely to be either a remnant of the raw material from which the galaxy's stars were made or gas lost from those stars. Instead, it was probably captured from outside the galaxy.

However, the average elliptical galaxy contains huge quantities of hot ionized gas. Over several gigayears, gas lost from aging stars builds up a substantial reservoir. For a population of stars that is more than a few gigayears old, mass shed during the red giant and AGB phases amounts to roughly $(1-2)\mathcal{M}_\odot$ of gas per year for each $10^{10} L_\odot$ of luminosity. The hot gas is invisible at optical and radio wavelengths because it is too diffuse to emit or absorb much at those energies. At temperatures of $(1-3) \times 10^7$ K, it radiates mainly in X-rays. More luminous galaxies, with higher velocity dispersions, have hotter galactic 'atmospheres'. These extend typically at least 30 kpc from the center; the brightest ellipticals have $(10^9-10^{11})\mathcal{M}_\odot$ of gas, which is 10%–20% of their mass in luminous stars. Small, low-mass ellipticals have much less gas; their weaker gravity is less capable of preventing its escape to intergalactic space.

Figure 6.21 shows the X-ray spectrum of the hot gas around M87. Over this energy range, the galaxy emits $\sim 3 \times 10^{35}$ W, or $10^9 L_\odot$, where $L_\odot$ is the Sun's total (bolometric) luminosity of 3.86×10^{26} W. We see metal lines, meaning that at least part of this gas has been processed by nuclear burning in stars; it has about half the solar proportion of iron. The hot gas around elliptical galaxies is not hugely metal-rich; generally its abundance is $\sim 0.5 Z_\odot$. Material thrown out

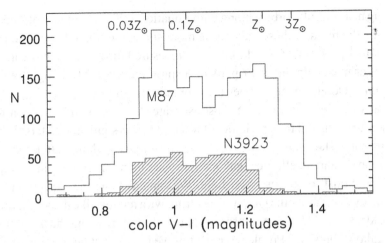

Fig. 6.22. Colors of globular clusters around giant elliptical M87 and shell galaxy NGC 3923, with a rough estimate of metal abundance – A. Kundu and S. Zepf.

by supernova explosions should be far richer; most of the hot gas is probably the outer envelope material shed by aging stars. Those stars have large random motions within the galaxy; as gas from one star runs into that of another at speeds close to the stellar velocity dispersion σ, shocks convert its kinetic energy into heat. Since $\sigma \gtrsim 350\,\mathrm{km\,s^{-1}}$ in bright ellipticals, this provides enough energy to heat the gas to the temperatures that we observe. In giant ellipticals at the centers of rich galaxy clusters, this hot atmosphere merges into the gas of the cluster: see Section 7.2.

Elliptical galaxies are surrounded by swarms of globular clusters: on average, each has about twice as many as a similarly bright disk system. The cD galaxy NGC 1399 has 7000 known globular clusters. The number can vary greatly between galaxies: M87 is about as luminous as NGC 3923, but is four times richer in bright globulars.

Figure 6.22 shows that the clusters around both these galaxies split into a redder and a bluer group. The bluer clusters are older, or poorer in metals. The Milky Way's globular clusters would have a similar color distribution, with the two peaks corresponding to metal-rich clusters of the thick disk and to metal-poor halo clusters; see Section 2.2. The arclike shells around NGC 3923 (Figure 6.5) indicate that another galaxy has merged into it; perhaps the more metal-rich and presumably younger globular clusters formed as part of that process. Some astronomers consider the similar color distribution among M87's clusters to indicate that it too has suffered a merger.

6.4 Dark matter and black holes

We saw in Sections 2.3 and 5.3 that, for spiral galaxies, we could find the mass within radius R of the center by using the measured rotation speed $V(R)$ of gas following near-circular orbits, along with Equation 3.20 for the gravitational force. Generally, the mass of these galaxies is more spread out than the luminous stars; the mass-to-light ratio $\mathcal{M}/L$ is larger when measured further from the center. Most of whatever makes up the outer parts of a spiral emits very little light; we call it *dark matter*. Do elliptical galaxies also contain dark matter?

To answer this question, we must estimate the mass within a given region, from measurements of the gravitational force. If that is significantly larger than the mass present in stars and gas, we ascribe the difference to dark matter. Finding the distribution of stars in elliptical galaxies is easier than for spirals, since ellipticals contain little dust to obscure their light. We can calculate the expected mass-to-light ratio by comparing the galaxy's spectrum with models like that of Figure 6.18. This yields $3 \lesssim \mathcal{M}/L_V \lesssim 5$, close to what we found by using the virial theorem for the galaxy cores; see Problem 6.6. But, if we measure at larger radii, generally we find $\mathcal{M}/L \gtrsim 20$. As in disk galaxies, most of the mass in the outer parts of ellipticals is dark.

6.4.1 Dark halos

For the few elliptical galaxies that have cold gas, we can use the same techniques as for disk systems to find their mass. For example, the luminous E4 galaxy NGC 5266 has a prominent dust lane along its apparent *minor* axis. Mapping in the 21 cm line of neutral hydrogen reveals about $10^{10} \mathcal{M}_\odot$ of HI gas, extending to $4'$ from the center, and apparently following near-circular orbits with $V(R) \approx 250 \, \mathrm{km \, s^{-1}}$. Within a radius of 50 kpc, this galaxy has $\mathcal{M}/L \approx 10$–20 in solar units, which is much larger than we found in Problem 6.6 for the center of NGC 1399, but similar to results for spiral galaxies in Section 5.3. If the systems containing cold gas are otherwise typical, then the outer parts of elliptical galaxies consist mainly of dark matter.

> **Problem 6.13** The redshift of NGC 5266 is $cz \approx 3000 \, \mathrm{km \, s^{-1}}$; if $H_0 = 75 \, \mathrm{km \, s^{-1} \, Mpc^{-1}}$, show that its distance $d \approx 40 \, \mathrm{Mpc}$. Use Equation 3.20 to show that the mass $\mathcal{M}(<4') \approx 7 \times 10^{11} \mathcal{M}_\odot$. The total apparent magnitude $B_T^0 = 12.02$; show that $L_B \approx 4 \times 10^{10} L_\odot$ – this is a big galaxy – so that the mass-to-light ratio $\mathcal{M}/L_B \approx 18$.

The motions of globular clusters can also be used to test for dark matter in the outer reaches of elliptical galaxies. In NGC 1399, velocities were measured for 468 globular clusters out to $9'$ from the center. Like the halo globular clusters of our Milky Way, they appear to follow random orbits with no overall rotation, and the velocity dispersion is almost constant at $\sigma_r \sim 275 \, \mathrm{km \, s^{-1}}$. Within 50 kpc of the center, the mass-to-light ratio $\mathcal{M}/L_V \sim 50$ (see below), several times higher than we found in Problem 6.6. Most of the mass in the galaxy's outer parts is dark.

Problem 6.14 We can use the singular isothermal sphere to represent the globular clusters of NGC 1399. Problem 3.28 shows that, in this model, a circular orbit has speed $V_H = \sqrt{2}\sigma_r$ which is constant with radius. With a distance $d = 20$ Mpc for NGC 1399, show that the measured velocity dispersion corresponds to a rotation speed V_H that implies $\mathcal{M} \sim 2 \times 10^{12} \mathcal{M}_\odot$ within 50 kpc. NGC 1399 has absolute V-magnitude $M_V = -21.7$; show that $\mathcal{M}/L \sim 50$.

6.4.2 Central black holes

Recent spectroscopic observations with high spatial resolution, from the ground and with the Hubble Space Telescope, have helped in the hunt for massive black holes at the centers of nearby galaxies. Stars close to a central black hole should move faster than those further out (as we see in Figure 2.17). If they orbit in random directions, we will see a central rise in the velocity dispersion measured from the absorption lines of the galaxy's spectrum. We should notice this effect when we probe close enough to the center that the circular speed of an orbit around it exceeds the velocity dispersion σ_c of the surrounding stars, meaning that

$$V^2(r) \approx \frac{G\mathcal{M}_{BH}}{r} \gtrsim \sigma_c^2. \tag{6.34}$$

So we must observe within a radius r_{BH} such that

$$r_{BH} \approx 45 \,\text{pc} \times \left(\frac{\mathcal{M}_{BH}}{10^8 \mathcal{M}_\odot}\right) \times \left(\frac{\sigma_c^2}{100 \,\text{km s}^{-1}}\right)^{-2}. \tag{6.35}$$

Figure 6.23 shows the masses inferred for compact objects at the centers of nearby elliptical galaxies and some bulges of disk systems. In the Local Group galaxy M32, we need an extra $2 \times 10^6 \mathcal{M}_\odot$ within the central parsec; in the giant elliptical M87 the excess mass is $\sim 3 \times 10^9 \mathcal{M}_\odot$ within 20 pc. Our own Milky Way's black hole with $\mathcal{M}_{BH} \approx 4 \times 10^6 \mathcal{M}_\odot$ fits neatly onto the trend:

$$\mathcal{M}_{BH} \approx 2 \times 10^8 \mathcal{M}_\odot \times \left(\frac{\sigma_c}{200 \,\text{km s}^{-1}}\right)^{4.86}. \tag{6.36}$$

The largest central masses are found in galaxies where the velocity dispersion σ is highest, and the virial theorem indicates the deepest gravitational potential wells. These are also the most luminous objects, as we should expect from the Faber–Jackson relation and the fundamental plane of Figure 6.13. The right panel shows that these masses are close to the minimum that could have been found in the best current observations. Smaller objects could have remained hidden.

If line-emitting gas is present in orbit about the center so that we can measure the orbital speed $V(r)$, then we can use Equation 3.20 to find how much mass is

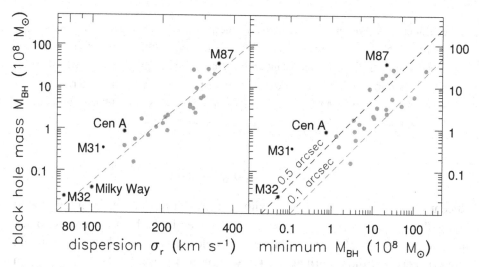

Fig. 6.23. Left, the central compact mass, probably a black hole, grows with velocity dispersion σ_c of the galaxy's central region; the dashed line is from Equation 6.36. Right, the inferred masses are close to the minimum that could be detected: dashed lines show where $r_{BH} = 0.5''$ and $0.1'' - L$. Ferrarese.

present within that orbit. In M87, there is a small disk of gas around the nucleus: its rotation speed, as measured from the gas emission lines, is exactly as expected for material orbiting a central object of $3 \times 10^9 \mathcal{M}_\odot$. We suspect, but cannot prove, that these large masses are black holes. To confirm it would require observations of gas at a few times the Schwarzschild radius, orbiting at speeds close to that of light; that is far from our present capabilities.

Problem 6.15 The Schwarzschild radius R_s of a black hole was defined in Section 2.3. Show that, for $\mathcal{M}_{BH} = 3 \times 10^9 \mathcal{M}_\odot$,

$$R_s \equiv \frac{2G\mathcal{M}_{BH}}{c^2} \approx 10^{10}\,\text{km} \qquad \text{or} \qquad 3 \times 10^{-4}\,\text{pc},$$

corresponding to 4×10^{-6} arcsec at the distance $d \sim 16\,\text{Mpc}$ of M87.

Problem 6.16 Emission lines from gas near the center of M87 have been observed with the Hubble Space Telescope. The observations are best fit with orbital speeds of $1000\,\text{km}\,\text{s}^{-1}$ at $0.1''$ from the center. Assuming that the distance $d = 16\,\text{Mpc}$, show that a mass $\mathcal{M} \sim 2 \times 10^9 \mathcal{M}_\odot$ is present inside this radius. (Why was a space telescope needed to make this measurement?)

Most elliptical galaxies brighter than $M_B \sim -20$ or $L \gtrsim L_\star$ are radio sources, emitting power $P \gtrsim 10^{20}\,\text{W}\,\text{Hz}^{-1}$ at 20 cm wavelength. Although the energy in radio emission is only about a hundred times the Sun's total luminosity, it is still ten times more than we expect to see from the HII regions and supernova remnants that power most of the radio emission of spiral galaxies. Ellipticals usually have a small radio-bright core, no more than a few parsecs across, in the nucleus. Those systems with $P \gtrsim 10^{23}\,\mathrm{W\,Hz^{-1}}$ often show double-lobed structures, miniatures of those of the powerful radio galaxies that we will study in Section 9.1, where we do observe motion at near-light speeds. If the radio sources of normal bright ellipticals really are underpowered versions of what we observe in active nuclei, this may be the best evidence that black holes with mass $\mathcal{M}_{\mathrm{BH}} \gtrsim 10^6 \mathcal{M}_\odot$ lurk at the centers of virtually all elliptical galaxies more luminous than $L \sim L_\star$.

7

Galaxy groups and clusters

About half the galaxies in the Universe are found in groups and clusters, complexes where typically half the member galaxies are packed into a region $\lesssim 1$ Mpc across. Groups and clusters no longer expand with the cosmic flow: mutual gravitational attraction is strong enough that the galaxies are moving inward, or have already passed through the core. Clusters are the denser and richer structures. Within the central megaparsec, they typically contain at least 50 luminous galaxies: those with $L \gtrsim L_\star \sim 2 \times 10^{10} L_\odot$. Poorer associations are called groups; they are generally less massive than $\sim 10^{14} \mathcal{M}_\odot$.

Galaxies and clusters are not simply concentrations of galaxies: the fact that a galaxy is a member significantly affects its development. The most common inhabitants of groups are spiral and irregular galaxies. The Local Group is typical; its three bright members are all spirals. Almost all the large galaxies of the Ursa Major group, featured in Figures 5.6, 5.8, and 5.23, are disk systems; there are no more than two ellipticals among the brightest 79 galaxies. Elliptical and S0 galaxies predominate in dense clusters. Curiously, both the most luminous giant ellipticals *and* the dwarfs with $L \lesssim 3 \times 10^9 L_\odot$ are concentrated more strongly into clusters than the midsized ellipticals.

By contrast with groups, which we discuss in Section 7.1, we will see in Section 7.2 that *most* of the baryonic ('normal') matter in galaxy clusters is not in the galaxies themselves. Galaxy clusters are filled with hot gas at temperatures $T \sim 10^7 - 10^8$ K, and glow brightly in X-rays. As new galaxies join a cluster they fall in through the hot gas, and its pressure strips away their cool atomic and molecular gas. The HI disks of galaxies in dense regions, such as the inner parts of the Virgo cluster, are much smaller than those further out, since they have lost the cool gas in their outer parts. Galaxy groups largely lack this relatively dense hot atmosphere of intergalactic gas. However, far more dilute gas between their galaxies may contain 90% of all the baryons in the Universe. Since this gas is too diffuse to cool by radiation, we can trace it only as it absorbs the light of bright background galaxies.

Problem 7.1 Measured speeds of galaxies in the Coma cluster are $\sigma_r \sim 1000\,\mathrm{km\,s^{-1}}$ near the center, falling to $\sigma_r \sim 800\,\mathrm{km\,s^{-1}}$ at radii $R \gtrsim 1.5\,\mathrm{Mpc}$. How long will it take for an average galaxy at $R \sim 3\,\mathrm{Mpc}$ to cross the cluster? Show that this is almost half of the cosmic time t_H.

Compared with the oldest stars in our Milky Way, galaxy groups and clusters are latecomers to the cosmos. The motions of galaxies within a cluster are only a few times larger than the speeds of stars within the galaxies, but the clusters are a hundred times larger; so galaxies on the outskirts have not had time to travel once through the cluster since the Big Bang. In our Local Group, we saw in Section 4.5 that the Milky Way and M31 are falling together for the first time. So how did our Milky Way, which began to form its stellar disk at least 10 Gyr ago, 'know' that it would spend its life in a sparse group? How did the elliptical galaxies that are already full of old red stars at redshifts $z \gtrsim 2$ develop into typical cluster members, long before the rich clusters in which they live had come together? We will discuss these questions in Section 7.3, but unfortunately we cannot answer them yet.

As in galaxies themselves, most of the mass in groups and clusters appears to be dark. However, our usual methods for estimating masses assume that the system is in equilibrium, so that its overall structure is not changing, and also that it is isolated. Neither of these is true for most galaxy groups and clusters. In Section 7.4 we discuss how to measure masses using the gravitational bending of light, a method that can be applied even when the cluster is still growing and changing.

Further reading: there is no comprehensive undergraduate-level treatment of these topics. A graduate-level text is M. S. Longair, 1998, *Galaxy Formation* (Springer, Berlin). Abell's catalogue of galaxy clusters is given in G. O. Abell, H. G. Corwin, and R. P. Olowin, 1989, *Astrophysical Journal Supplement*, **70**, 1.

7.1   Groups: the homes of disk galaxies

Figure 7.1 shows Stephan's Quintet, a rare *compact group* in which galaxies almost touch. Their mutual gravitational pull has torn long tails of stars from the galaxies' disks. Stars equivalent to an entire bright galaxy are scattered around the barred spiral NGC 7319 and the two overlapping galaxies (a spiral and an elliptical) near it. Around these three galaxies, $10^9 \mathcal{M}_\odot$ of hot gas at $T_X \sim 10^7\,\mathrm{K}$ glows brightly in X-rays. There is about $10^{10} \mathcal{M}_\odot$ of cool gas in this group, but very little is in the galaxies. Instead, a tail of HI over 100 kpc long curls around the hot gas to the south and east. Probably, cool HI gas was stripped from the galaxy disks as they came close. Some of the gas clouds were then heated to X-ray temperatures as they ran into each other.

Fig. 7.1. Stephan's Quintet (compact group Hickson 92) is ~85 Mpc distant and 80 kpc across, or 3.2′ on the sky. North is at the top and east is to the left. NGC 7319, the barred spiral, has an active nucleus: it is a Seyfert 2. The large spiral in the lower center, NGC 7320, is not a group member; it is in the foreground with a much smaller redshift – D. J. Pisano, WIYN telescope.

The observed motion of the galaxies and the stellar tails lets us reconstruct part of the group's history. The spiral NGC 7318b (one of the two overlapping galaxies) has only just arrived in the group, tearing out the bright stellar trail leading from one of the spiral arms in NGC7319. Its cool gas is now being stripped away. To the south, the longer but more diffuse tail of stars curves toward the small galaxy NGC 7320c, 140 kpc to the east; this passed through the group ~500 Myr ago, and will return. Stephan's Quintet has about twice as much gas as the Milky Way, and three times as much stellar light. One could imagine that its galaxies began life much like our Milky Way, losing their gas in repeated collisions and near-misses.

Problem 7.2 Suppose that gas atoms and galaxies in a group move at the same average random speed σ along each direction. At temperature T, the average energy of a gas particle is $3k_B T/2$, where k_B is Boltzmann's constant. If the gas is mainly ionized hydrogen, these particles are protons and electrons; show that, if the atom's kinetic energy $(3m_p/2)\sigma^2$ is shared equally between them, then

$$T \approx \frac{(m_p/2)\sigma^2}{k_B} \approx 5 \times 10^6 \left(\frac{\sigma}{300\,\text{km s}^{-1}}\right)^2 \text{K}. \qquad (7.1)$$

Hot gas in a group or cluster is usually close to this *virial temperature*.

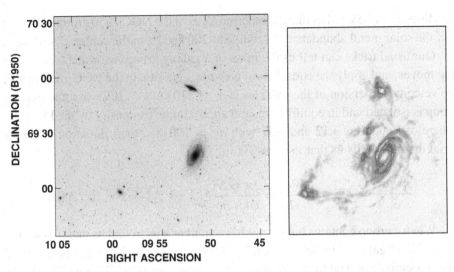

Fig. 7.2. The group around the large Sc spiral galaxy M81, about 3.5 Mpc distant. Left, negative image in visible light; the elongated object north of M81 is starburst galaxy M82; NGC 3077 is to the southeast. Right, map in HI to the same spatial scale – M. Yun.

Problem 7.3 The energies of X-ray photons are measured in kilo-electron-volts (keV), 1000 times the energy that an electron gains in moving through a one-volt potential. In Stephan's Quintet, the gas radiates X-rays with typical energy $k_B T_X = 0.5\,\text{keV}$; show that this implies $T_X \approx 6 \times 10^6$ K. The speeds V_r of the galaxies along our line of sight have a dispersion $\sigma_r \sim 350\,\text{km s}^{-1}$. Show that the kinetic energy of a hydrogen atom moving at speed σ_r is roughly equal to the one-dimensional energy $k_B T/2$ for each of two particles (a proton and an electron) at $T = 7 \times 10^6$ K. Thus, as group galaxies run into each other, the energy of their motion can heat gas to the observed temperature.

Figure 7.2 shows the sparse group around the nearby spiral M81. We see only a few intergalactic stars, but the galaxies are connected by long streamers of atomic hydrogen torn from their disks as they passed close to one another. Because the HI layer of a spiral galaxy extends much further in radius than its main stellar disk, the gas disks are more easily damaged. Here, almost all of the gas that was torn from the galaxies has remained cool. Similarly, in Figure 4.6, we see HI gas from the Magellanic Clouds pulled out into a stream around the Milky Way.

Only half of galaxy groups contain gas hot enough to shine in X-rays, which requires $T \gtrsim 3 \times 10^6$ K. Those groups are generally more populous, and include at least one elliptical galaxy. For example, the group of 15 galaxies within 1 Mpc around the S0 galaxy NGC 1550 (including two ellipticals) radiates $4 \times 10^9 L_\odot$ in X-rays, 100 times more than Stephan's Quintet. X-ray spectral lines of 24-times ionized iron, and of multiply-ionized oxygen, magnesium, and sulfur, show that elements heavier than helium are present in a few tenths of the solar proportions.

Iron lines at ~ 1 keV tell us that the innermost gas of the NGC 1550 group has about half the solar metal abundance $Z_\odot$, while at 200 kpc from the center $Z \approx 0.1 Z_\odot$.

Our usual tricks can tell us the mass of a galaxy group: we watch something that moves, and apply the equations of Newton's gravity. In the NGC 1550 group, the velocity dispersion of the galaxies is $\sigma_r = 310$ km s^{-1}. If we assume that the group is isolated and in equilibrium, and approximate its density by the Plummer sphere of Equations 3.12 and 3.37 with $a_P = 100$ kpc, then Equation 3.44 (the virial theorem) tells us that its mass $\mathcal{M}$ is

$$\frac{3\mathcal{M}\sigma_r^2}{2} = \mathcal{KE} = -\frac{\mathcal{PE}}{2} = \frac{3\pi}{64}\frac{G\mathcal{M}^2}{a_P}, \quad \text{so } \mathcal{M} \approx 2 \times 10^{13} \mathcal{M}_\odot.$$

When a group contains hot gas we have another way to find its mass, as we did for elliptical galaxies in Section 6.4. We first estimate the X-ray temperature T_X from spectral lines. The gas is diffuse, so we can use Equation 2.23 from Section 2.4 to find the density. When $T_X \gtrsim 10^7$ K, gas of roughly the Sun's composition loses energy mainly by *free–free* radiation, also called *thermal bremsstrahlung*. If each cubic centimeter contains n atoms, its luminosity is

$$L_X = n^2 \Lambda(T_X), \quad \text{where } \Lambda \approx 3 \times 10^{-27} T_X^{1/2} \text{erg s}^{-1}. \tag{7.2}$$

(Look back at Figure 2.25 to see how slowly Λ increases with temperature.) So we can work backward from the measured surface brightness I_X to find n.

The equation of *hydrostatic equilibrium* tells us how much gravity is needed to prevent the hot gas from blowing away as a wind. Its pressure p must be counterbalanced by an inward pull: if $\rho(r)$ is the gas density, then, in a spherical galaxy,

$$\frac{dp}{dr} = -\rho(r)\frac{G\mathcal{M}(<r)}{r^2}. \tag{7.3}$$

We write the average mass m of a gas particle as $m = \mu m_p$: for fully ionized hydrogen $\mu = 0.5$, whereas $\mu \approx 0.6$ for gas of solar composition. For a perfect gas at temperature T and density ρ, we have

$$p = \frac{\rho}{\mu m_p}k_B T, \quad \text{so } \mathcal{M}(<r) = \frac{k_B}{\mu m_p}\frac{r^2}{G\rho(r)}\frac{d}{dr}(-\rho T). \tag{7.4}$$

Problem 7.4 Suppose that the hot gas of a group has a uniform temperature T_X, and outside the very center the surface brightness in X-rays drops with radius R as $I_X \propto R^{-\gamma}$. Writing the gas density $\rho(r) \propto r^{-\beta}$, use Equation 7.2 to show that the luminosity per cubic centimeter follows $L(r) \propto r^{-2\beta}$. Now use the result of Problem 6.4 to show that $\gamma = 2\beta - 1$.

In the NGC 1550 group, $T_X = 1.6 \times 10^7$ K and $\rho(r) \propto r^{-1.1}$ out to 200 kpc from the center. Use Equation 7.4 to show that the group's mass is roughly $10^{13} \mathcal{M}_\odot$. The hot gas itself accounts for $8 \times 10^{11} \mathcal{M}_\odot$, while the light of the galaxies adds up to $L_B \approx 8 \times 10^{10} \mathcal{M}_\odot$. The mass-to-light ratio of the stars is unlikely to be more than what we found for the old stars at the center of NGC 1399 in Problem 6.6. Show that the mass of hot gas is at least twice that in all the stars, so the group contains about eight times as much mass as is present in the galaxies and hot gas.

In general, the mass-to-light ratio of groups is in the range $80 \lesssim \mathcal{M}/L \lesssim 300 \mathcal{M}_\odot/L_\odot$. Stars and gas hot enough to shine in X-rays typically make up less than 10% of the total. Groups as a whole appear to contain several times more mass than what we found in Section 5.3 from the rotation curves of individual spirals. Either the dark matter halo of each galaxy extends several times further out than the HI disk gas with which we measure its rotation, or substantial mass is present between the galaxies as a 'group halo'.

Problem 7.5 Suppose that all galaxy groups share a common form for the density $\rho(r)$: for example, the Plummer sphere of Equations 3.12 and 3.37. If all groups have the same radius a_P, and their mass is proportional to the number of members N, show that the virial theorem predicts that $\sigma_r \propto \sqrt{N}$. This is roughly what we see in Figure 7.3. Points for the sparsest groups lie above this relation: show that those groups should have smaller radii.

Figure 7.3 shows that the speeds of group galaxies are $\sigma_r \sim 100$–500 km s^{-1}, not much faster than the motion of stars within the galaxies. Because of their low relative speeds, gravity has more time to pull strongly at the gas and stars in the outer parts of other galaxies as they pass. In the next subsection, we see how this can lead two galaxies to merge and become one.

7.1.1 Close encounters between galaxies: dynamical friction

We use the methods of Section 3.2 to examine what happens as two galaxies pass close to one another: part of their energy of forward *motion* is transferred to *motion* of the stars within them. Thus, as they separate, they travel more slowly than they did on their approach. If the galaxies are then moving too sluggishly to escape from one another, their orbits bring them back together to encounter each other again. Usually, they end by merging.

In Figure 7.4, as in Figure 3.5, an object of mass $\mathcal{M}$ moves at speed V past a star of mass m at a distance b from its path. But now $\mathcal{M}$ is one of our galaxies, while the star m belongs to another: so $\mathcal{M}/m \sim 10^8$–10^{11} (!) As the two galaxies

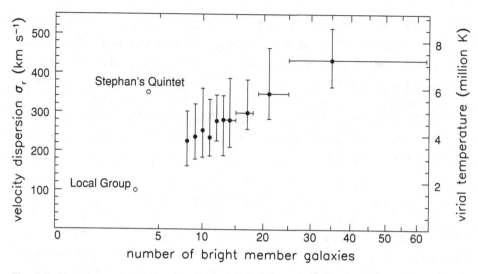

Fig. 7.3. For groups chosen from the 2dF galaxy catalogue, the average velocity dispersion σ_r of the galaxies, and hence the virial temperature of Problem 7.2, increase with the number of members. The vertical bar shows the range in σ_r within which half the galaxies fall – F. van den Bosch and X. Yang.

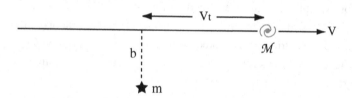

Fig. 7.4. A fast or weak encounter: a galaxy of mass $\mathcal{M}$ moves with speed V past a stationary star of mass m in a second galaxy at distance b from its path.

pass, $\mathcal{M}$ acquires a motion

$$\Delta V_\perp = \frac{2Gm}{bV} \tag{3.51}$$

in the direction perpendicular to its original path. This formula applies when the approach is not too close, so that galaxy $\mathcal{M}$ is contained within a radius $\ll b$. We also require a rapid flyby, in which $\mathcal{M}$ and m do not move significantly toward each other during the encounter. As before, we must have

$$b \gg \frac{2G(\mathcal{M} + m)}{V^2} \equiv 2r_s \tag{3.53}$$

(for $\mathcal{M} = m$, r_s is the *strong-encounter radius* defined by Equation 3.48). The star m in the second galaxy must gain an equal and opposite momentum, so the total kinetic energy in perpendicular motions is

$$\Delta\mathrm{KE}_\perp = \frac{M}{2}\left(\frac{2Gm}{bV}\right)^2 + \frac{m}{2}\left(\frac{2GM}{bV}\right)^2 = \frac{2G^2 m M(M+m)}{b^2 V^2}. \quad (7.5)$$

Note that the object of smaller mass acquires most of the energy. This energy can be provided only from the forward motion of the galaxy M, which consequently changes by an amount $\Delta V_\parallel$. Long before the encounter and long afterward, the potential energy is small; so the kinetic energies are equal, and we have

$$\frac{M}{2}V^2 = \Delta\mathrm{KE}_\perp + \frac{M}{2}(V + \Delta V_\parallel)^2 + \frac{m}{2}\left(\frac{M}{m}\Delta V_\parallel\right)^2. \quad (7.6)$$

If $\Delta V_\parallel \ll V$, we can drop terms in $\Delta V_\parallel^2$ to find that each star m in the 'stationary' galaxy slows M by an amount

$$-\Delta V_\parallel \approx \frac{\Delta\mathrm{KE}_\perp}{MV} = \frac{2G^2 m(M+m)}{b^2 V^3}. \quad (7.7)$$

Because the kinetic energy transferred to m grows as M^2, the forward motion is braked more quickly as the mass M of the passing galaxy increases. The faster M flies by, the less time it has to transfer energy to m, and the less speed it loses.

Suppose that our intruding galaxy M passes through a region of the second galaxy that contains n stars of mass m per cubic parsec. Then we can integrate over all those stars to find the average rate at which it slows:

$$-\frac{dV}{dt} = \int_{b_{\min}}^{b_{\max}} nV \frac{2G^2 m(M+m)}{b^2 V^3} 2\pi b\, db = \frac{4\pi G^2(M+m)}{V^2} nm \ln\Lambda, \quad (7.8)$$

where $\Lambda = b_{\max}/b_{\min}$. As in the discussion following Equation 3.55, we usually take $b_{\min}$ as the radius r_s of Equation 3.48, within which M's deflection is no longer small, and $b_{\max}$ as the distance at which the density of stars becomes much less than it is in the neighborhood of M. This deceleration is described as *dynamical friction*, since it acts to brake any motion relative to the background of stars.

According to Equation 7.8, the slower the galaxy M moves, the larger its deceleration. A high-speed encounter between two galaxies drains less energy from their forward motion than a slow passage. But this is true only as long as we can neglect the random motions of the stars m compared with the forward motion of galaxy M. If not, the drag is reduced. When V is far below the velocity dispersion of the stars in the second galaxy, we have $dV/dt \propto -V$, just as in Stokes' law describing the fall of a parachutist through the air.

Formula 7.8 can be applied, for example, to a small satellite galaxy orbiting within the dark halo of a larger galaxy; its orbit decays, like that of an Earth satellite subject to atmospheric drag, and it spirals inward. The frictional force does not depend on the mass of whatever objects make up the dark halo, but only on their mass density, through the product nm. A massive satellite is slowed more quickly than a small one. So the following problem shows that the Large Magellanic Cloud (LMC) is likely to merge with our Galaxy within a few gigayears, but most of the globular clusters, which are 10^5 times lighter, are in no danger.

Problem 7.6 Far outside the core a_H of the 'dark halo' potential of Equation 2.19, use Equation 3.14 to show that the density $\rho(r) \approx V_H^2/(4\pi G r^2)$. Explain why the force opposing the motion of a satellite of mass $\mathcal{M}$ in a circular orbit at radius r is approximately

$$F_\| = -\frac{G\mathcal{M}^2}{r^2} \ln \Lambda \cdot \mathcal{F}.$$

The force is reduced by a factor $\mathcal{F}$ compared with Equation 7.8 because the particles of the dark halo have their own random motions: $\mathcal{F} \approx 0.4$ if the random speeds of the halo particles are equal to $V_H/\sqrt{2}$ in all directions, as in the singular isothermal sphere of Equation 3.105. As the satellite spirals in, its angular momentum is $L = \mathcal{M} r V_H$. Setting the frictional torque equal to the rate of change in L, show that

$$r\frac{dr}{dt} = -\frac{G\mathcal{M}}{V_H} \ln \Lambda \cdot \mathcal{F}, \quad \text{and so} \quad t_{\text{sink}} = \frac{r^2 V_H}{2G\mathcal{M}} \frac{1}{\ln \Lambda \cdot \mathcal{F}},$$

where t_{sink} is the time that the satellite takes to reach the center of the potential. For the LMC, with $\mathcal{M} \approx 2 \times 10^{10} \mathcal{M}_\odot$, $V_H \approx 200 \, \text{km s}^{-1}$, and $r \approx 50 \, \text{kpc}$, estimate the radius $r_s \approx b_{\text{min}}$, and hence show that $\Lambda \sim 20$. Show that, with $\mathcal{F} \approx 0.4$, this simple formula predicts that the LMC will merge with the Milky Way within 3 Gyr, but a globular cluster with $\mathcal{M} = 10^6 \mathcal{M}_\odot$ will not sink far in that time unless it orbits within a kiloparsec of the Galactic center.

(In fact, we are on shaky ground when we use Equation 7.8 for two galaxies in orbit about each other. To find the rate at which $\mathcal{M}$ is slowed, we added the effects of successive encounters with masses m, assuming that these were independent of each other. But the stars m are bound together in a galaxy and affect each other's motion. Also, as $\mathcal{M}$ orbits that galaxy, it will encounter the same stars m repeatedly, experiencing *resonance* effects that may either weaken or strengthen the 'friction'.)

Dynamical friction removes energy from the forward motion of two passing galaxies, transferring it to the random motions of their stars. Long afterward, though, the combination of rotational motion and random speeds within each galaxy will be *lower* than before the collision. To see why, suppose that, before the encounter, the internal kinetic energy of one of the galaxies was $\mathcal{KE}_0$. By Equation 3.44, the virial theorem, the potential energy $\mathcal{PE}_0 = -2\mathcal{KE}_0$; so its internal energy $\mathcal{E}_0$ must be

$$\mathcal{E}_0 = \mathcal{KE}_0 + \mathcal{PE}_0 = -\mathcal{KE}_0. \tag{7.9}$$

Dynamical friction increases the energy in random stellar motions, and hence the galaxy's internal energy, by $\Delta\mathcal{KE}$. The system is less strongly bound, so it expands. Long after the encounter, when it is once again in virial equilibrium, the kinetic energy is *less* than before:

$$\mathcal{KE}_1 = -(\mathcal{E}_0 + \Delta\mathcal{KE}) = \mathcal{KE}_0 - \Delta\mathcal{KE}. \tag{7.10}$$

The stars that acquire the most energy escape from the galaxy; those receiving less remain loosely attached, as a bloated outer envelope. In a spiral or S0 galaxy, the energy added to vertical motions thickens the stellar disk.

If two galaxies pass by one another at the high speeds typical of rich galaxy clusters, they are unlikely to slow each other enough to become a bound pair. They separate, leaving both somewhat dishevelled. But in groups the galaxies travel more slowly, not much faster than the stellar motions within them; so a close passage is far more disturbing. Almost all galaxies that we now see in the process of merging are group members.

Many group galaxies show evidence of near-miss encounters. These can push a galaxy into making a bar or spiral that would not have done so if it had been alone. Like M81, many galaxies with spectacular two-armed spirals have close companions. Mergers and near-misses also force disk gas in toward the galaxy's center. Because the changing gravity pulls material away from its near-circular orbits, gas streams converge from different directions. As we discussed in Section 5.5, this causes shocks that remove energy from the gas, letting some of it fall inward. If it reaches the galaxy's center, it can grow a massive black hole or fuel activity in the nucleus. Both NGC 7319 in Stephan's Quintet and M81 are Seyfert galaxies, a type of active nucleus that we discuss in Section 9.1.

7.1.2 Galaxy mergers and starbursts

If an encounter has drained enough energy from the galaxies' orbital motion to leave them bound to each other, they will probably merge. Figure 7.5 shows three views of what appears to be such a merger in progress, and Figure 7.6 shows a computer simulation of the collision that may have led to it. This is a *gravitational N-body simulation*, in which roughly 10^6 particles attract each other by the gravitational force of Equation 3.2. In fact, we cannot solve these equations exactly; it is too much work to calculate the forces between 10^{12} different

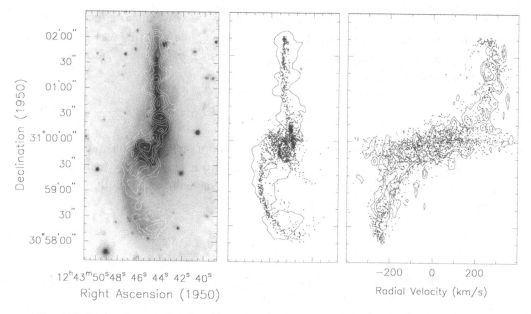

Fig. 7.5. The galaxy pair NGC 4676, known as 'the Mice'. Left, an *R*-band optical image, with white contours showing neutral hydrogen gas in the tails; center, results of a gravitational *N*-body simulation following two disks of 'stars'. Positions of the stars are shown on top of the outer H I contour. Right, velocities of the stars compared with the gas at each position along the tail – J. Hibbard and J. Barnes.

pairs of particles, particularly when they come very close. This simulation uses a 'tree code', which divides space into cells by a grid, and approximates the pull of distant material by regarding all the mass in any grid cell as making up a single particle. Each particle has its own grid, which is coarser for more distant material, while the forces of near-neighbor particles are calculated one by one. Even this is too time-consuming for the largest simulations. Then, we simply lay down a single grid for the whole calculation, and average the forces from all particles within each cell. Using a grid also has the same beneficial effect as the 'softening' that we discussed in Section 3.2, in suppressing unwanted two-body relaxation.

In this collision, the disk of one galaxy happens to lie close to the plane in which the two systems orbit one another, and rotates in the same sense; so we see particularly strong effects. On the side of the disk nearer to the intruding system, the rotation cancels out much of the relative motion of the two bodies. Stars and gas here spend a long time close to the perturber. They receive both angular momentum and energy, pushing them onto elongated orbits which can take them out to very large radii. The material that has been torn off the disks will remain visible as 'tidal tails' for several gigayears before it falls back onto the merged remnant of the galaxies.

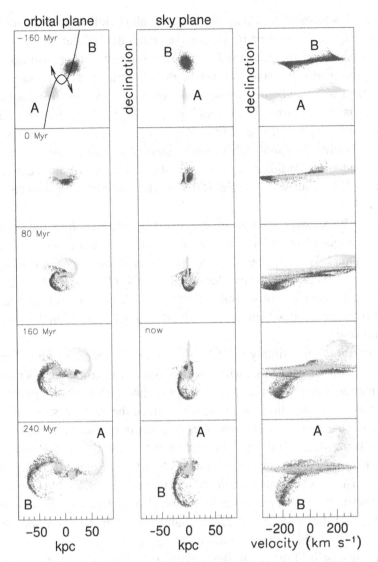

Fig. 7.6. Galaxy pair NGC 4676: the computer simulation of Figure 7.5. Left, motion in the galaxies' initial orbital plane; center, view along our line of sight; right, radial velocity V_r of particles at each declination. Time is in megayears from closest first approach, distance in kiloparsecs, assuming that $H_0 = 75 \, \mathrm{km \, s^{-1} \, Mpc^{-1}}$. Material at the tips of the tails is still travelling outward at the end of the run; it will not fall back onto the galaxies for many gigayears – J. Hibbard and J. Barnes.

Because the close passage or merger of two gas-rich galaxies compresses their gas, it often causes a *starburst*: stars form so rapidly as to exhaust the gas supply within a few hundred million years. Galaxies with tidal tails and other indications of a recent merger often have $(10^9 – 10^{10}) \mathcal{M}_\odot$ of dense molecular gas at the center. Much of it was probably brought inward from the main disk during the encounter.

As this gas is squeezed by gravity, the likely result is violent star formation. After this dies away, it may leave a bright compact inner disk of stars.

The uppermost spectrum in Figure 5.24 is from a blue starburst galaxy. Its newly-born massive stars are bright in the ultraviolet, and ionize the gas around them to produce the strong emission lines. The small galaxy M82 in Figure 7.2 is also a starburst. It is making $(2-4)\mathcal{M}_\odot\,\mathrm{yr}^{-1}$ of new stars, about the same as a large spiral like the Milky Way, but in a region only 600 pc across. The radio-bright knots show where young stars have ionized the gas, producing free–free radiation. At this rate, its supply of $2 \times 10^8 \mathcal{M}_\odot$ of atomic and molecular gas will be consumed within 100 Myr, the *free-fall* or dynamical time of Equation 3.23.

On the back cover are two images of M82, showing how the starburst is blowing gas out of the galaxy as a strong wind. The top image was taken in infrared light at 3–9 μm. The new stars have heated dust in the escaping wind, which glows 'red' at 8 μm while the stellar disk is 'blue'. The lower one shows dust lanes crossing the whitish stellar disk; the purple color shows emission in the Hα line from the wind, which removes about as much mass from the galaxy as it turns into new stars.

Starbursts are not simply scaled-up versions of the star-forming regions in galaxies like our own. In M82, where we can directly measure the gas pressure, it is 100 times greater than that in the solar neighborhood. More of their gas is actively star-forming: emission lines of HCN, which trace the dense gas of star-forming cores (see Table 1.8) are about five times stronger in starbursts relative to the CO emission from more diffuse gas. In Section 9.4, we will find that many distant galaxies have the chaotic appearance and high surface brightness characteristic of starbursts. The very fact that we can observe them at all means that they are much more powerful than most local starbursts, and also relatively unobscured by dust. These gigantic starbursts at $z \gtrsim 2$, less than 5 Gyr after the Big Bang, probably built up the dense old regions of present-day galaxies.

If a starburst is shrouded in the dusty gas from which the stars were born, that dust intercepts most of the stellar light, re-radiating the energy at infrared wavelengths. Figure 7.7 shows that about 90% of M82's starlight suffers this fate. A galaxy like the Milky Way radiates about $10^{10}L_\odot$ in the far-infrared, but a powerful starburst surrounded by dust will be seen as a *luminous infrared galaxy* with $L_{\mathrm{FIR}} \gtrsim 10^{11}L_\odot$, while only a few percent of its starlight escapes directly. Even brighter sources with $L_{\mathrm{FIR}} > 10^{12}L_\odot$ are known as *ultraluminous* infrared galaxies or ULIRGs. Almost all ULIRGs are in the late stages of a merger.

If we assume that starbursts form the same relative numbers of stars of different masses as in the Milky Way (the initial mass function of Section 2.1 is the same), then the energy emitted at infrared wavelengths tells us how fast new stars are made. Measuring the luminosity L_{FIR} between 10 μm and 1 mm in units of the Sun's bolometric luminosity of 3.86×10^{26} W, new stars are formed roughly at

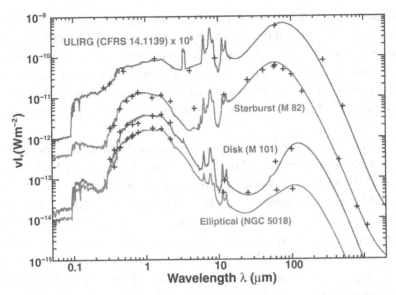

Fig. 7.7. Normal elliptical and disk galaxies are brightest in the visible and near-infrared, at $\lambda < 2\,\mu$m. Most dust grains are cooler than 30 K, and their emission peaks beyond 100 μm. In the starburst M82 and the ultraluminous infrared galaxy, dust intercepts far more of the light, and it is hotter, radiating mainly at $\lambda < 100\,\mu$m. See Figure 2.24 for details of emission lines – ISO: P. Chanial and G. Lagache.

the rate

$$\dot{\mathcal{M}}_\star \sim \frac{L_{\mathrm{FIR}}}{6 \times 10^9 L_\odot} \mathcal{M}_\odot \, \mathrm{yr}^{-1} \,. \tag{7.11}$$

So the most powerful ULIRGs, radiating $L_{\mathrm{FIR}} \gtrsim 10^{13} L_\odot$, give birth to $\sim$1000$\mathcal{M}_\odot$ of new stars each year. About half of these have an active nucleus (see Section 9.1) in addition to a ferocious starburst.

Arp 220, a merging pair of galaxies with characteristic tidal tails, is only 75 Mpc away, giving us a close-up view of a ULIRG. It is making $\sim$200$\mathcal{M}_\odot\,\mathrm{yr}^{-1}$ of new stars, but only a few percent of their light escapes directly. The rest is absorbed by dust, so we see $L_{\mathrm{FIR}} \sim 1.5 \times 10^{12} L_\odot$. Radio and infrared observations that can see through the dust reveal two nuclei, each surrounded by a gas disk $\sim$20 pc in radius with $6 \times 10^8 \mathcal{M}_\odot$ of gas at densities $n(\mathrm{H}_2) \sim 10^3\,\mathrm{cm}^{-3}$. These in turn are embedded in a kiloparsec-sized disk with $>10^9 \mathcal{M}_\odot$ of dense gas. Typically a ULIRG contains $(5\text{--}10) \times 10^9 \mathcal{M}_\odot$ of dense molecular gas, mainly in a rotating disk or ring in the central kiloparsec. The Milky Way has half as much gas (see Table 2.4), but it is spread over ten times the area and forms stars a hundred times more slowly.

Within a gigayear after two disk galaxies have merged, the excitement is mostly over. The hot massive stars made from gas compressed during the merger have burned out, and the most conspicuous remaining sign of the original disks is the tidal tails extending outward. Often the inner part of the combined galaxy has become rounder, and the former disk stars now make up a system with large random motions and little rotation. Its structure looks very much like an elliptical galaxy. Many astronomers would argue that *all* bright ellipticals have their origin in such a violent galactic collision: we will discuss this further in Section 7.3.

What does the future hold for a galaxy group? The Local Group is quite isolated, with no substantial galaxies poised to fall into it. In Section 4.5 we saw that M31 and the Milky Way are likely to collide, and will probably merge. The Local Group will probably become a 'fossil group', with a single isolated galaxy that has eaten all the luminous systems, so that only small satellites remain. The cannibal might end up as an elliptical, or a giant S0 or Sa galaxy like the Sombrero in Figure 5.5. While new galaxies fall into a group, as they are still doing in Stephan's Quintet, their gravitational pull increases the random speeds of the other galaxies. That replaces some of the energy sapped by dynamical friction, and delays merging. We will see in the next section that rich clusters grow by absorbing clumps of galaxies. Many groups will end by dissolving in larger systems before their component galaxies merge.

7.2   Rich clusters: the domain of S0 and elliptical galaxies

About 5%–10% of luminous galaxies live in clusters. The nearest populous clusters are 15–20 Mpc from us, in the constellations of Virgo in the northern sky and Fornax in the south. We list some of their properties in Table 7.1, comparing them with richer clusters and with Stephan's Quintet. George Abell's 1958 catalogue and its 1989 supplement list 4073 rich clusters having at least 30 giant member galaxies within a radius of $\sim 1.5h^{-1}$ Mpc.

Of the 1300 catalogued members of the Virgo cluster, only 150 are brighter than apparent total B-magnitude $B_T = 14$, equivalent to $L \gtrsim 10^9 L_\odot$ if the distance $d = 16$ Mpc. Figure 7.8 shows that most of the galaxies are dwarfs. In the densest part of the Virgo cluster, near the giant elliptical M87, the averaged B-band surface brightness is $\sim 5 \times 10^{11} L_\odot$ Mpc^{-2}, or about $0.5 L_\odot$ pc^{-2}. The *core radius*, where the surface density of galaxies falls to half its central value, is approximately $r_c = 1.7°$ or 0.5 Mpc. Making the very rough approximation that the cluster is spherical, in the core we have $3 \times 10^{11} L_\odot$ Mpc^{-3}. Within the central 6°, the cluster's total luminosity is $L \approx 1.3 \times 10^{12} L_\odot$, giving an average density of $4 \times 10^{10} L_\odot$ Mpc^{-3}. At the core of the Fornax cluster is the cD galaxy NGC 1399. Fornax has only a fifth as many bright members as Virgo but is more compact; at its center we find one of the highest galaxy densities anywhere in the local Universe. Continuing the trend that we found among galaxy groups, Fornax has less hot gas in proportion to its starlight than does the more luminous Virgo cluster.

Table 7.1 Nearby galaxy clusters, compared with a distant cluster and a nearby group

	Virgo	Fornax	Coma A1656	Perseus A426	RDCS 1252.9–2927	Stephan's Quintet
Distance (Mpc)	16	20	100	80	[a]$z = 1.24$	85
Number of galaxies $>10^9 L_\odot$	150	30	450	350	120	4
B-band starlight L_B ($10^{10} L_\odot$)	130	20	500	300	[b]300	7
Velocity dispersion σ_r (km s^{-1})	700–800	350	~1000	1300→600	800	350
Core r_c (kpc)	400	200	200	250	100	25
X-ray L_X ($10^{10} L_\odot$)	1.3	0.03	25	50	20	0.005
Temperature T_X (10^7 K)	2	1–2	9	7	7	0.6
[c]Hot gas M_X ($10^{10} M_\odot$)	2 000	≳60	7 000	7 000	2 000	≳0.07
[c]Mass M ($10^{10} M_\odot$)	20 000	5 000	40 000	50 000	20 000	100
M/L_B	150	250	80	180	[b]~200	14
M/M_X	10	80	6	7	10	>1000

[a] At $z = 1.24$ the benchmark cosmology gives $d_L = 8.5$ Gpc, $d_A = 1.7$ Gpc.
[b] Luminosity in the z band at 9000 Å. [c] Mass within 1 Mpc of the cluster's center.
X-ray telescopes are less sensitive to gas with $T_X \lesssim 10^7$ K, so masses M_X may be underestimated.

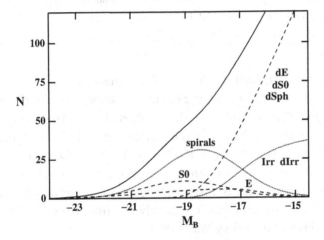

Fig. 7.8. The Virgo cluster: numbers of galaxies of various types between absolute magnitude M_B and $M_B + 1$. The luminosity function $\Phi(L)$ depends on galaxy type; the Schechter function of Equation 1.24 is only an average. Here, most bright galaxies with $M_B \lesssim -20$ are spirals; there are many faint ellipticals and even fainter dwarf galaxies. The heavy solid curve shows the total – H. Jerjen.

Problem 7.7 Use the result of Problem 6.3 for NGC 1399's central surface brightness, and assume that all this light comes from the core of radius $r_c = 2''$ or 650 pc, to find the core luminosity density. Show that the luminosity density at the center of the Virgo cluster is 2500 times greater than average for the Universe as a whole (see Equation 1.25), but the core of NGC 1399 is 10^7 times denser yet. Stars are packed far more tightly into galaxies than galaxies are packed into clusters.

Fig. 7.9. The core of the Perseus cluster. Left, an *R*-band negative image showing the huge cD galaxy NGC 1275 on the lower left, with numerous dwarf galaxies, other bright ellipticals, and S0s. Right, an image in a narrow band including the Hα line; bright filaments of glowing gas surround NGC 1275 – C. Conselice, WIYN telescope.

The Coma cluster, Abell 1656, and Perseus cluster, Abell 426, are even bigger than Virgo. Coma is three times more luminous; although it is $70h^{-1}$ Mpc from us, it stretches more than 4° across the sky, or about 7 Mpc. The cluster looks round and quite symmetric; at the center is a pair of very luminous elliptical galaxies. Perseus is the brightest cluster in the X-ray sky. Near the center, 10–15 bright galaxies appear strung out into a line. At one end is the huge elliptical NGC 1275 shown in Figure 7.9, which is also a *radio galaxy* (see Section 9.1). But Perseus has been less well studied because it is only 13° from the Galactic plane, so we must view it through the Milky Way's dust.

Most clusters have irregular or lumpy shapes. Like big cities, they grow as they absorb surrounding groups and clusters. But unlike suburbs, the orbits of newly-acquired groups take them 'downtown' through the core of the cluster. Clusters become more massive, but their radii increase only slowly. In the Virgo cluster, roughly a third of the galaxies form a clump around the elliptical M49, while the rest surround M87 about 1 Mpc to the north (see Figure 8.2). In Fornax, 15% of the galaxies clump around NGC 1316, about 3° (1 Mpc) from the center. These include two of the cluster's most vigorously star-forming galaxies, together with $5 \times 10^8 \mathcal{M}_\odot$ of HI gas, which will probably be stripped out when it joins the main cluster. The Coma cluster looks smooth on the sky, but measuring velocities shows that systems bunch around the three brightest galaxies. The clump around the cD galaxy NGC 4839 is clearly visible in Figure 7.10, about 1° southwest of the center where the X-ray contours bulge out. At this rate, infalling clumps add at least 10% to the Coma cluster's mass every 2–3 Gyr. But once the scale $a(t)$ of the

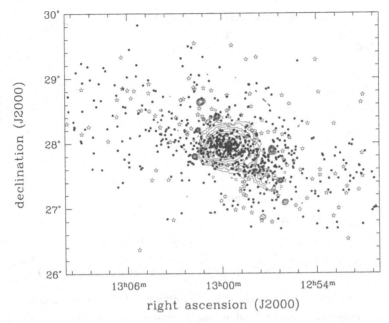

declination (J2000)

right ascension (J2000)

Fig. 7.10. The Coma cluster: solid dots show elliptical galaxies; open stars are spirals. Contours show the intensity of X-rays: they bulge to the south-west where a clump of gas and galaxies surrounds the cD NGC 4839. The diffuse emission is from hot cluster gas, the point sources are distant active galaxies – M. van Haarlem.

Universe has reached 2–3 times its present size, we expect cosmic acceleration caused by the dark energy (see Section 8.2) to overpower the cluster's gravity, so that infall almost ceases.

The Virgo cluster is a fairly loose cluster, and contains many spiral galaxies which are still forming new stars. By contrast, spirals are almost excluded from the densest parts of rich clusters. Figure 7.10 illustrates this *morphology–density relation*: Coma's core contains only elliptical galaxies, while the spirals are relegated to the outer suburbs. Perseus has an even stronger deficiency of star-forming galaxies. Figure 7.11 shows quite generally that, in dense places such as rich clusters, the most luminous galaxies are red, with spectra that show no sign of recently born hot stars. There are hardly any luminous blue galaxies, and we see star formation only in unusual central galaxies, such as NGC 1275 in Figure 7.9. In the sparse 'void' regions (illustrated in Section 8.1 below), a fairly luminous galaxy is as likely to be a blue star-former as to be 'red and dead', while dwarf galaxies are overwhelmingly blue star-forming irregulars. In clusters, most of the dwarfs are of types dE and dSph, which contain only old and middle-aged stars.

Most cluster galaxies can no longer make stars because they have lost all their cool gas. As galaxies fall in to join a cluster, they rush through the hot cluster gas much faster than the sound speed in the galaxy's gas ($\sim 10\,\mathrm{km\,s^{-1}}$ in the disk of a galaxy like ours: recall Problem 2.21 of Section 2.4). A shock develops

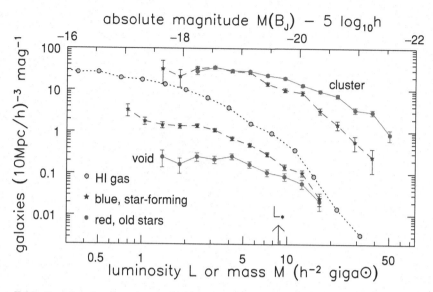

Fig. 7.11. Luminosity functions of blue star-forming galaxies (stars) and red galaxies (filled dots). Bright red galaxies predominate in clusters, but dim blue galaxies are more frequent in the low-density field or void regions. Clouds of neutral hydrogen, found in low-density regions, resemble blue galaxies in that small systems are far more common than big ones – D. Croton, 2dF survey; M. Zwaan, HIPASS.

at the boundary between the galaxy's gas and the hot intergalactic gas, which prevents them from moving freely through each other. The stars and dark matter are unaffected, so the gas can be left behind while the rest of the galaxy falls in. In spiral galaxies nearest to the cores of the Virgo and Coma clusters, the H I disks have smaller radii; these inner spirals have already lost the outer parts of their neutral hydrogen layers.

Just as in galaxy groups, there are stars between the galaxies of rich clusters, but they are hard to find because they are so thinly spread. The Hubble Space Telescope barely picks out red giants at the distance of Virgo; in most clusters, only special objects such as an exploding nova or supernova, or planetary nebulae, can be seen individually. Some intergalactic novae have been spotted in the Fornax cluster. In a planetary nebula, a low-mass star has exhausted its nuclear fuel and blown off its outer layers, baring the hot stellar core. The core's ultraviolet radiation ionizes the ejected gas, which glows strongly in emission lines, particularly the oxygen line at 5007Å. By searching for objects shining brightly in this line, planetary nebulae have been found between the galaxies in several nearby clusters. In the Virgo cluster, their numbers imply that 10%–20% of all the stars lie between the galaxies. The 'vagabond' stars could have been torn from galaxies during a merger, or they may be remnants of loosely bound small galaxies torn apart by tides (recall Section 4.1).

Just as for the stars in individual elliptical galaxies, we can use the virial theorem to find the mass of a galaxy cluster from the motions of the galaxies within it. The rate at which the spiral galaxies in the outskirts of the Virgo cluster fall in toward the center depends on the mass at smaller radii: we infer $M \approx 2 \times 10^{14} M_\odot$, so the mass-to-light ratio $M/L \sim 150 M_\odot / L_\odot$. The mass is only approximate, because the clusters are neither spherical nor in dynamical equilibrium.

Problem 7.8 In the Coma cluster, the core radius measured from the galaxy density is $r_c = 200$ kpc from Table 7.1. Make a crude model by assuming the cluster to be a Plummer sphere; from Problem 3.2, show that $a_P \approx 300$ kpc. Taking the average velocity dispersion σ_r from Table 7.1, use the result of Problem 3.11 and follow the method of Problem 3.13 to show that $M \sim 7 \times 10^{14} M_\odot$, implying $M/L \sim 150 M_\odot / L_\odot$.

As for groups, we can use Equation 7.4 to find the cluster masses from observations of hot gas. In the Virgo cluster, this yields $M \sim 3 \times 10^{14} M_\odot$ within 6° or 1.7 Mpc of the cluster's center. It is more than we found from the motions of the galaxies, because the gas can be traced to larger radii. A recent study of 32 clusters using this method found that the mass-to-light ratio mostly lies in the range $180 \lesssim M/L \lesssim 300$. These ratios are much greater than the range $5h \lesssim M/L \lesssim 25h$ that we found for individual spiral galaxies in Section 5.3, or for ellipticals in Problems 6.6 and 6.14. Like the hot gas (see below), most of a cluster's dark matter must lie between the galaxies rather than within them.

7.2.1 Hot gas in clusters of galaxies

Galaxy clusters do not really deserve their name: they are huge accumulations of X-ray-luminous hot gas, cohabiting with a few galaxies. The mass of hot gas is roughly equal to that in stars in a poorer cluster like Fornax; it rises to a ratio of $10:1$ in the richest systems like Coma. Table 7.1 shows that the largest clusters have been least efficient at turning their gas into stars. By contrast with groups like Stephan's Quintet, there is so much hot gas in rich clusters that it cannot all have been stripped from the galaxies. The hot gas fills the whole cluster, and can often be traced to larger radii than can the stellar light.

Figure 7.12 shows that the most luminous clusters contain the hottest gas, with temperatures of up to 10^8 K. As in groups, the cluster gas is roughly at the virial temperature predicted from the velocity dispersion σ_r of the galaxies. Because new clumps of galaxies are continually added to a cluster, both its X-ray luminosity and the virial temperature should grow over time. Models predict that the temperature increases faster, so that a cluster with a particular luminosity should be cooler at earlier times. But we see no sign of this effect in Figure 7.12, even for clusters at $z \sim 1$ when the Universe was half its present age. Perhaps winds from early starburst galaxies and active nuclei also added energy to the gas.

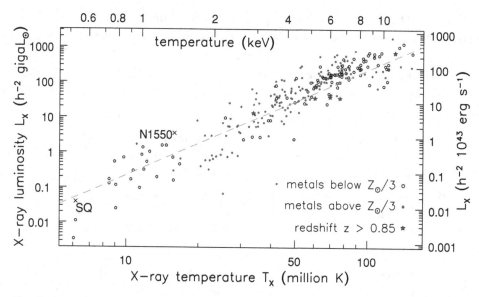

Fig. 7.12. The X-ray luminosity L_X of a galaxy cluster or group, in units of the Sun's bolometric luminosity, increases with the gas temperature T_X: the dashed line shows $L_X \propto T_X^3$. Stephan's Quintet and the NGC 1550 group follow the same trend. This relation has changed little since redshift $z \approx 1$. In most clusters with $T_X > 3 \times 10^7$ K, the gas has roughly one-third of the solar content of iron – D. Horner.

Problem 7.9 Suppose that all galaxy groups and clusters have the same average density, according to Problem 8.2 below. If the gas in a cluster is heated to the virial temperature of Problem 7.2, show that the cluster's mass $\mathcal{M} \propto T_X^{3/2}$. If hot gas makes up a fixed fraction of the cluster's mass, the average density n is the same for all of them. Use Equation 7.2 to show that we expect $L_X \propto \mathcal{M}\sqrt{T_X}$ so that $L_X \propto T_X^2$. (Figure 7.12 shows that L_X increases even more steeply with temperature; the gas in galaxy groups and poor clusters must be less dense than that in rich clusters.)

The *cooling time* t_{cool} measures how fast the hot gas radiates away its thermal energy. A cubic centimeter of completely ionized hydrogen containing n atoms has thermal energy $2n \times (3k_B T/2)$, so from Equation 7.2

$$t_{\mathrm{cool}} = \frac{3nk_B T}{3 \times 10^{-27}n^2\sqrt{T}} \, \mathrm{s} \approx 14\left(\frac{10^{-3}\,\mathrm{cm}^{-3}}{n}\right)\left(\frac{10^7\,\mathrm{K}}{T}\right)^{1/2} \mathrm{Gyr}. \qquad (7.12)$$

The gas density in galaxy clusters varies from a few times $10^{-2}\,\mathrm{cm}^{-3}$ in the center to $10^{-4}\,\mathrm{cm}^{-3}$ in the outer parts. In the cores of rich clusters, around the most luminous galaxies, the gas should cool within the cosmic time t_H. Unless its energy is replenished, that gas should cool and become dense, giving birth to new stars. But we do not see such vigorous star formation in the central galaxy. So we believe that the gas is reheated, perhaps by hot winds from supernovae, or the outflows from radio sources that we will discuss in Section 9.1. Gas in the outskirts will remain hot for >10 Gyr, so it cannot form stars.

Problem 7.10 Around NGC 1399 at the center of the Fornax cluster, the X-ray emission from hot gas extends out to radii of 100 kpc. The gas temperature $T_X \approx 1.5 \times 10^7$ K; within $45''$ or 5 kpc of the center, the density $n_H \approx 0.02$ cm^{-3}, falling approximately as $n_H \propto r^{-1}$. Show that $t_{cool} \approx 0.5$ Gyr in the core, increasing to 10 Gyr at $r = 100$ kpc.

Galaxy clusters often appear lumpier in an X-ray image than at optical wavelengths. On the lower right of Figure 7.10, we see X-rays from the clump around the galaxy NGC 4839 in the Coma cluster. Here, the gas is only half as hot as the main cluster. At the 'neck' in the X-ray contours, temperatures roughly double to 10^8 K where the clump gas meets the main cluster. In the Perseus cluster, we see low-density 'bubbles' in the central X-ray gas, which probably were blown by energetic outflows from the radio galaxy NGC 1275. Most galaxy clusters today are still growing; even their centers are not always near equilibrium, since they are disturbed as newly-added galaxies fall through them.

The hot cluster gas contains iron and other heavy elements in roughly one-third of the solar proportion: see Figure 7.12. Gas near the center is more metal-rich; neither infalling clumps of galaxies nor the outflows from active galaxies have been able to mix it with the rest of the cluster gas. Although these heavy elements must have been 'cooked' by nuclear burning in stars, *most* of a cluster's supply of metals is now in the hot gas. Perhaps they were made in galactic stars, and the force of supernova explosions carried metal-bearing material out into the cluster. Or perhaps many early stars formed in smaller clumps which later fell together to make galaxies, releasing metal-enriched gas as they merged. However, the stars of the galaxies themselves are as metal-rich as the Milky Way. So, if the metals were made inside the galaxies, these systems must have been very efficient at producing heavy elements.

The hot gas itself accounts for roughly a tenth of the mass of a luminous cluster, but much less in systems with $T_X < (2–3) \times 10^7$ K. If we take $\mathcal{M}/L_B = 4$–8 for the stars of the galaxies, then stars and hot gas together make up roughly a sixth of the cluster's mass. This is close to the cosmic average for the benchmark cosmology of Section 1.5, so galaxy clusters appear to be a 'fair sample' of the mix of dark and luminous material. By contrast, baryons are more concentrated into galaxies: recall from Problem 2.18 that at most half the mass inside the Sun's orbit around the Milky Way can be dark. In galaxy groups, we find too little luminous gas and stars for the amount of dark matter. Some baryons may have been lost, blown out during episodes of violent star formation; others may be hidden as very dilute gas between the galaxies.

Table 7.2 The cosmic baryon budget

Where it is	Density ($10^{-3}\rho_{\rm crit}$)
Total (benchmark cosmology)	45
Intergalactic gas	
Diffuse and ionized	≈ 40
Damped Lyman-α clouds	1
Hot gas in clusters and Es	1.8
Stars and stellar remnants	
Stars in Es and bulges	1.5
Stars in disks	0.55
Dead stars	0.48
Brown dwarfs	0.14
Cool gas in galaxies	0.78

We can find only 10% of the baryons, which make up $\lesssim$ 15% of
the matter, which is $\lesssim$30% of the critical density – M. Fukugita and
P. J. E. Peebles 2004 *ApJ* **616**, 643.

7.2.2   Where have all the baryons gone?

Throughout this book, we have seen that the stars and gas within galaxies form
only a fraction of their mass. Table 7.2 shows that luminous and dead stars,
together with the cool gas in galaxy disks, make up only $(3\text{--}4) \times 10^{-3}\rho_{\rm crit}$. But
Equation 1.40 tells us that *baryons* (neutrons and protons) in the Universe account
for a fraction Ω_B of the critical density, where $0.03 \lesssim \Omega_B \lesssim 0.07$. The stars and
the cool gas together contribute only 10% of this total. Where is the rest?

    The hot X-ray-emitting gas in clusters of galaxies holds nearly as much mass as
all the stars. In groups, there is cool gas between the galaxies: recall the HI streamers of the M81 group, in Figure 7.2. The amount of ionized gas that they might
accomodate is unknown; but it could be large, since galaxy groups are more numerous than rich clusters. If gas is heated close to the virial temperature of Figure 7.3
as the group forms, then Equation 7.12 shows that, if $n < 10^{-3}$ cm^{-2}, its low density will prevent it from cooling. It would remain ionized up to the present; since
it does not absorb in the Lyman-α line, it would be almost impossible to detect.

    In Section 9.3 we will discuss the intergalactic gas, which we can investigate
through the absorption lines it produces in the spectra of distant objects. The
densest gas is in the damped Lyman-α clouds, which are mainly neutral and atomic,
and contain more cool gas than there is in today's galaxies. But they leave us far
from balancing the account. Most of the baryons are likely to be in the Lyman-α
forest, the most diffuse gas, in which hydrogen is almost completely ionized.

7.3   Galaxy formation: nature, nurture, or merger?

Clusters are not simply places where galaxies are more densely packed: cluster galaxies themselves are different. Figure 7.11 shows that the largest red

galaxies, the ellipticals and cD galaxies, live in dense regions of rich clusters, while starforming spirals and irregulars inhabit less crowded regions. This segregation is puzzling, because the stars of elliptical galaxies may be almost as old as the Universe, while we will see in Section 8.1 that galaxy clusters are still coming together today. As a galaxy forms, how can it know whether it will end up 'downtown' in the cluster core, or as a spiral in the suburbs? We do not understand this process yet, and have only some hints.

In Section 4.3 we took a brief look at the origin of galaxies. Early in cosmic history, dark matter was evenly spread, mixed with baryons in the form of hot gas. In slightly denser regions, clumps of matter were pulled together under their own gravity. The larger clumps absorbed smaller ones that fell into them, much as we see in galaxy clusters today. As clumps collided, their gas was compressed and heated. If it was dense enough to cool, the gas lost energy and fell inward. The particles of dark matter could not lose energy, so the central parts of a galaxy consist mainly of cool gas and luminous stars. Where the heated gas was too diffuse it did not cool, and its pressure prevented it from flowing inward.

Perhaps elliptical galaxies are not as old as their stars. We know that some nearby ellipticals have recently eaten other galaxies. They show faint shells and other asymmetries in the outer parts (Section 6.1); stars in the core rotate differently from those in the main galaxy (Section 6.2); cold gas orbits at seemingly random inclination (Section 6.3). We will see in Section 9.4 that 5%–10% of fairly bright galaxies at $z \sim 1$ appear to be in the throes of a major merger. At this rate, between one and two thirds of today's fairly luminous galaxies (with $L \gtrsim 0.4L_\star$) underwent a major merger within the past 5 Gyr. Did all elliptical galaxies form by merger of smaller galaxies, so that their stars were born before the galaxy had completely assembled? If so, we could understand why they live in rich clusters, which grow by adding smaller clumps of galaxies. The clumps are excellent sites for mergers because their galaxies have fairly low random speeds, as we discussed in Section 7.1.

What should a merger remnant look like, long after the initial disturbance and subsequent vigorous starbirth? That depends on whether the colliding systems contained gas, or only stars and dark matter. When two galaxies of about the same size merge, their stellar disks are destroyed. When cool gas is present, a disk where new stars will be born can be rebuilt. Without cool gas, no new stars can form; the remnant will have only old and middle-aged stars. For these, the virial theorem makes a strong prediction: the merged system should become less dense.

Problem 7.11 Suppose that a galaxy is made from N identical fragments, each of mass $\mathcal{M}$ and size R. In each one, the average separation between any two stars (or particles of dark matter) is $R/2$; so the potential energy $\mathcal{PE} \approx -G\mathcal{M}^2/R$, as in Equation 3.56. Use Equation 3.44 to show that each fragment has energy $\mathcal{PE}/2$, so that, while they are well separated and moving toward each other only slowly, the total energy $\mathcal{E} \approx -G\mathcal{M}^2 N/(2R)$. Long after the merged galaxy has come to virial equilibrium, its energy is $-G(\mathcal{M}N)^2/(2R_g)$; show that its new size $R_g = NR$, and its density is only $1/N^2$ as large as that of the original fragments. Explain why Figure 6.6 supports the idea that giant elliptical galaxies (filled circles) arise by repeated merger.

But we cannot just combine the stellar populations of two smaller gas-free galaxies to form a massive elliptical. Recall from Figure 6.20 that stars at the center of a luminous elliptical galaxy are more abundant in heavy elements than those in any part of a smaller galaxy. If the merging systems contain some cool gas, then new stars can be made. As we discussed in Section 7.1, rapidly changing gravitational forces will drive gas inward to form a central disk and trigger star formation. If more than one generation of stars is born, later stars incorporate heavy elements released by the earliest, so the galaxy can develop a dense metal-rich center. In the disorder of a collision such as that in Figure 7.6, the spin of the inward-driven gas need not be aligned with the stars around it. The galaxy's core then differs in its rotation from the rest of the system, as it does in NGC 1399: see Figure 6.12.

Thus merging two systems that contain some cool gas could produce a 'disky' elliptical galaxy of moderate luminosity with a bright center. Because of the small dense inner disk, containing the youngest stars, these galaxies would rotate rapidly, as Figure 6.15 shows that disky ellipticals do. Some of these in turn might collide to form the most luminous ellipticals. Gravitational N-body simulations show that the result is likely to be triaxial, with a 'boxy' shape. Indeed, Figure 6.11 shows that boxy systems are the most luminous.

But we still have some potentially awkward facts to explain. Astronomers who do not agree that most elliptical galaxies result from merger will point to the fundamental plane of Figure 6.13, relating a galaxy's luminosity to its core size and central brightness. Why should galaxies merge in such a way as to produce this relation? What is more, a galaxy's luminosity should depend mainly on the total of stars and gaseous raw material assembled throughout its history, whereas color is sensitive to the metal abundance. So why are the two linked, as Figure 6.19 shows them to be?

If the most luminous galaxies are formed by multiple generations of merger, then early in cosmic history they should be very rare indeed. But we will see in Section 9.4 that very luminous red galaxies were fairly common as far back as redshift $z \sim 2$, when the Universe was less than 5 Gyr old. These systems contain up to twice the mass of stars that the Milky Way has today. They lack young blue stars or the deep Balmer absorption lines of A stars, so those stars formed at least a gigayear earlier. A few galaxies at $z \gtrsim 6$, when the cosmic age was not even 1 Gyr, already show a strong 4000 Å break. Models such as that in Figure 6.18 tell us that their stars are at least 100 Myr old, while the galaxies appear to have grown to a fifth or more of the Milky Way's stellar

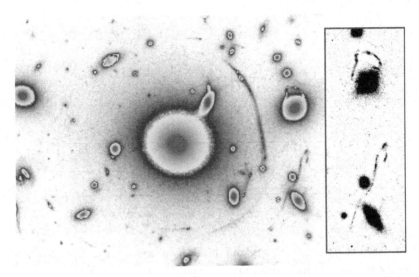

Fig. 7.13. Cluster Abell 383 at $z = 0.188$: north is to the left. The bright arc $16''$ in radius curving around the central cD galaxy is the image of a background galaxy at $z = 1.01$. The inset shows details of complex arcs where light from another distant galaxy passes close to individual cluster galaxies – G. P. Smith *et al.* 2001 *ApJ* **552**, 493.

mass. This is quite unexpected. If these early galaxies formed by merger, both the merging process and early starbirth must have been far more efficient than either is today.

7.4 Intergalactic dark matter: gravitational lensing

Gravity acts on light itself, bending the paths of photons. If we look at two galaxies that are neighbors on the sky, then the light of the more distant system swerves toward the nearer galaxy as it passes by on its way to us. Figure 7.13 shows the cluster Abell 383 at redshift $z = 0.188$. The bright arc curving around the huge cD galaxy is the image of a background galaxy at $z = 1.01$. The gravity of the entire cluster has bent its light, forming multiple images that run together to make up the arc. Another distant galaxy gives rise to the broken arc further out that twists past two fainter cluster members. The cluster's gravity has deflected more light toward us than we would otherwise receive, so the arcs are brighter than the original sources. They are also distorted and magnified in size. This effect is called *gravitational lensing*, even though the light is not brought to a focus.

Gravitational lensing provides a powerful method to study the dark matter between galaxies, because the effect does not depend on whether a cluster has reached equilibrium, or is still growing and changing. We will begin our discussion with the simple case in which the lensing mass is compact, and can be treated like a single point. Then we consider how to calculate the bending from an extended

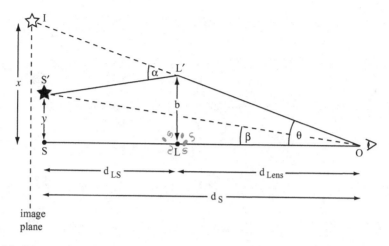

Fig. 7.14. The gravity of the mass $\mathcal{M}$ at L bends light from a distant source at S′ toward the observer at O; the source appears instead at position I.

'lens' such as a galaxy or cluster. Lensing allows us to study distant objects that would otherwise be far too faint. Some of the best-studied galaxies at high redshift, such as cB58 in Figure 9.16, are strongly brightened in this way. But we must beware of lensing when taking a census of any type of distant object.

7.4.1 Microlensing: light bent by a compact object

The Sun's bending of light provided one of the first tests of Einstein's General Relativity. During the total solar eclipse of 1919, an expedition led by Arthur Eddington took photographs of the sky around the obscured Sun. They found that light was bent as shown in Figure 7.14, so that nearby stars appeared to have moved very slightly away from the calculated position of the Sun's center. Einstein predicted that light passing at distance b from a mass $\mathcal{M}$ is bent by an angle α given by

$$\alpha \approx \frac{4G\mathcal{M}}{bc^2} = \frac{2R_s}{b}. \tag{7.13}$$

Here R_s is the *Schwarzschild radius* $2G\mathcal{M}/c^2$, about 3 km for an object of the Sun's mass. The approximation holds as long as the bending is small, with $\alpha \ll 1$. This formula prescribes exactly *twice* the bending that we get by applying Equation 3.51 to particles travelling at speed c.

    Using this formula, we can calculate where the image of a distant source will appear, if a point object of mass $\mathcal{M}$ is placed in front of it to act as a *gravitational lens*. If the lens L in Figure 7.14 had been absent, we would have seen the star S′ at an angle β on the sky from the direction to L; $\beta \approx y/d_S$ if the distance $d_S \gg y$. Because the light is bent by an amount α, the star appears instead at an angle θ to that direction; $\theta \approx x/d_S$ if $d_S \gg x$. If x is the distance between the line OL and I, the star's apparent position in the *image plane*, then Figure 7.14 shows that, when bending is small, $x - y = \alpha d_{LS}$. Finally, the impact parameter $b = \theta d_{Lens}$ as long as $d_S \gg b$. Using Equation 7.13 for α, we divide by d_S to find

$$\theta - \beta = \frac{\alpha d_{LS}}{d_S} = \frac{1}{\theta}\frac{4GM}{c^2}\frac{d_{LS}}{d_{Lens}d_S} \equiv \frac{1}{\theta}\theta_E^2; \qquad (7.14)$$

the angle θ_E is called the *Einstein radius*. We have a quadratic equation for the angular distance θ between L and the star's image:

$$\theta^2 - \beta\theta - \theta_E^2 = 0, \quad \text{so} \quad \theta_\pm = \frac{\beta \pm \sqrt{\beta^2 + 4\theta_E^2}}{2}. \qquad (7.15)$$

A star exactly behind the lens, with $\beta = 0$, will be seen as a circle of light on the sky, with radius θ_E. When $\beta > 0$, the image at θ_+ is further from the lens, with $\theta_+ > \beta$, and it lies outside the Einstein radius: $\theta_+ > \theta_E$. These exterior images were the ones seen around the eclipsed Sun. The image at θ_- is inverted; it lies within the Einstein radius, on the opposite side of the lens.

Problem 7.12 Show that, when $d_{Lens} = 1\,\mathrm{AU} \ll d_S$, then $\theta - \beta \approx \alpha$. When light from a distant star S is bent by the Sun's gravity, show that the Einstein radius $\theta_E \approx 40''$. The Sun's disk has a diameter of $30'$ or $0.5°$; show that starlight that just grazes the Sun's surface would be bent by about $2''$, making the star appear further from the Sun's center.

Quite regularly, one star in the Milky Way's bulge is gravitationally lensed by another in the disk. The images θ_+ and θ_- are then too close to distinguish individually. But we can tell that a star is lensed because it appears to become brighter (see below). Because of the small size of the Einstein ring, gravitational lensing by compact objects in the halo of a galaxy is often called *microlensing*.

Problem 7.13 If the lens L is an object of mass $\mathcal{M}_\odot$ at distance d_{Lens} from us, show that the Einstein radius for a star at distance $d_S = 2d_{Lens}$ is

$$\theta_E = \sqrt{\frac{R_s}{d_{Lens}}} \approx 2 \times 10^{-3}\sqrt{\frac{1\,\mathrm{kpc}}{d_{Lens}}}\ \mathrm{arcsec}. \qquad (7.16)$$

If our point-mass lens is in front of a small extended patch of brightness, its image will be *two* patches around $\theta_\pm$. Gravitational lensing leaves the surface brightness $I(\mathbf{x})$ unchanged, so the apparent brightness of each image of a source is just proportional to its area. Consider a region S' that is a segment of an annulus

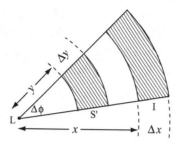

Fig. 7.15. Magnification of an image by gravitational lensing.

centred on L, between radius y and $y + \Delta y$, as in Figure 7.15. An image I of S' occupies the same angle $\Delta\phi$, but distances from the center are expanded or contracted: $x/y = \theta/\beta$ while $\Delta x/\Delta y = d\theta/d\beta$. The ratio of areas is

$$\frac{\mathcal{A}_{\pm}(\text{image})}{\mathcal{A}(\text{source})} = \left|\frac{\theta}{\beta}\frac{d\theta}{d\beta}\right| = \frac{1}{4}\left(\frac{\beta}{\sqrt{\beta^2 + 4\theta_{\mathrm{E}}^2}} + \frac{\sqrt{\beta^2 + 4\theta_{\mathrm{E}}^2}}{\beta} \pm 2\right). \quad (7.17)$$

Thus the image θ_+ that is further from L is always brighter than the source; it is also stretched in the tangential direction. The closer image, θ_-, is dimmer unless $\beta^2 < (3 - 2\sqrt{2})\,\theta_{\mathrm{E}}^2\,/\,\sqrt{2}$ or $\beta \lesssim 0.348\,\theta_{\mathrm{E}}$. This also holds when the lens is an extended system such as a galaxy or cluster: at least one of the images is brighter than the source would be without lensing.

> **Problem 7.14** From Equation 7.17, show that $\mathcal{A}_+ + \mathcal{A}_- > \mathcal{A}(\text{source})$: more light reaches us in the two images, taken together, than we would have received from the source S if the lens had been absent. When $\beta = \theta_{\mathrm{E}}$, show that the increase is roughly 40%, while the total brightness doubles when $\beta \approx 0.7\theta_{\mathrm{E}}$.

> **Problem 7.15** Show that, if the distance d_S to the source is fixed, then the area $\pi\theta_{\mathrm{E}}^2 d_{\mathrm{Lens}}^2$ inside the Einstein radius is largest when the lens is midway between the source and the observer: $d_S = 2d_{\mathrm{Lens}}$. If the objects that might act as lenses are uniformly spread in space, a source is most likely to be strongly brightened by a lens that is about halfway between it and the observer.

7.4.2   Lensing by galaxies and clusters

When the lens is an entire galaxy or cluster, we can think of it as a collection of point masses. We first rewrite Equation 7.13 to define a *lensing potential* ψ_{L}:

$$\alpha(b) \equiv \frac{d\psi_{\mathrm{L}}}{db}, \quad \text{where} \quad \psi_{\mathrm{L}} = \frac{4G\mathcal{M}}{c^2}\ln b. \quad (7.18)$$

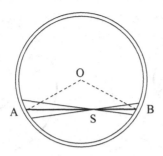

Fig. 7.16. Gravitational bending of light by a circular ring of uniform density.

We can then calculate the bending by summing up the effect of all the mass within the lens. If the lens is compact compared with both the distance d_{Lens} from us and the distance d_{LS} from it to the light source, then the deflection depends only on the *surface density* $\Sigma(\mathbf{x})$ of the lens. In general we must specify the light ray's closest approach to the cluster's center by a vector $\mathbf{b}$; integrating over the cluster gives the vector deflection α:

$$\alpha(\mathbf{b}) \equiv \nabla \psi_{\text{L}}(\mathbf{b}), \quad \text{where} \quad \psi_{\text{L}}(\mathbf{b}) = \frac{4G}{c^2} \int \Sigma(\mathbf{b}') \ln|\mathbf{b} - \mathbf{b}'| \mathrm{d}S'. \quad (7.19)$$

This has a very similar form to Equation 3.4 for the gravitational potential $\Phi(\mathbf{x})$, but the integral is now in two dimensions, and a logarithmic term replaces $1/|\mathbf{x} - \mathbf{x}'|$.

In general we must calculate $\psi_{\text{L}}(\mathbf{b})$ with a computer, from the distribution of matter given by $\Sigma(\mathbf{b})$. But suppose that the lens galaxy or cluster is axisymmetric, so that the surface density depends only on the projected distance R from the center. We can show that the bending of a ray passing at radius b then depends only on the mass $\mathcal{M}(<b)$ *projected* within that circle:

$$\alpha(b) = \frac{4G}{bc^2} \int_0^b \Sigma(R) 2\pi R \, \mathrm{d}R = \frac{4G}{c^2} \frac{\mathcal{M}(<b)}{b}. \quad (7.20)$$

To prove this, we adapt the arguments that we used in Section 3.1 to show that the gravitational force at distance r from a spherical object is the same as if all the matter at radii less than r had been concentrated at the center of the sphere. Readers who are willing to trust our assertion should skip beyond Equation 7.24 for applications of the result.

First, we show that a light ray passing through a uniform circular ring is not bent at all. In Figure 7.16, a ray through point S is pulled in opposite directions by the sections of the ring falling within the cone at A and at B. The sides AO and BO of the triangle AOB are equal; so the angle OAB between the line AB and the normal OA at A is the same as the angle OBA at B. Thus the arcs lying within the narrow cone have lengths (and masses) in the ratio SA/SB. Since the bending is inversely proportional to the distance from S, the deflections by the matter at

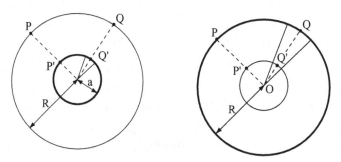

Fig. 7.17. Gravitational bending of light passing outside a uniform circular ring.

A and B are equal and opposite, cancelling each other out. So the ray is not bent; the lensing potential ψ_L must be constant inside the ring. It is easiest to calculate ψ_L at the center; for a ring of mass $\mathcal{M}$ and radius a,

$$\psi_L(R < a) = \frac{4G\mathcal{M}}{c^2} \ln a. \tag{7.21}$$

Next, we must prove that a ray passing outside the ring is bent in the same way as by a point mass at its center. We use Figure 7.17 to show that the lensing potential $\psi_L(P)$ at a point P lying at distance $R > a$ from the center of a uniform ring of mass $\mathcal{M}$ is the same as the potential $\psi'_L(P')$ *inside* a ring of the same mass, but with radius R. In the picture on the left, the mass in an arc $\Delta\theta$ around Q' contributes an amount

$$\Delta\psi_L(P) = \frac{4G}{c^2}\frac{\mathcal{M}\,\Delta\theta}{2\pi} \ln|\mathbf{x}(P) - \mathbf{x}(Q')| \tag{7.22}$$

to the potential at P. From the picture on the right, the mass within the same wedge at Q contributes to the lensing potential at P' an amount

$$\Delta\psi'_L(P') = \frac{4G}{c^2}\frac{\mathcal{M}\Delta\theta}{2\pi} \cdot \ln|\mathbf{x}(P') - \mathbf{x}(Q)|. \tag{7.23}$$

But $PQ' = P'Q$, so these are equal; integrating over the whole ring,

$$\psi_L(P) = \psi'_L(P') = \frac{4G\mathcal{M}}{c^2} \ln R : \tag{7.24}$$

the bending angle for light passing outside the ring at radius R is the same as if all its mass had been concentrated at the center. So we arrive at Equation 7.20: if the lens is axisymmetric, light is bent just as if all the material projected within that radius had been exactly at the center.

We can use Equation 7.20 to find how an axisymmetric galaxy cluster bends the light of a distant galaxy behind it, as in Figure 7.14. If the cluster had been absent, we would have seen the galaxy at S', at an angle β to the cluster's direction

OL. Instead, we see its image I at an angle θ to OL. Equation 7.14 is now modified to read

$$\theta - \beta = \alpha(\theta)\frac{d_{LS}}{d_S} = \frac{1}{\theta} \cdot \frac{4GM(<b)}{c^2} \frac{d_{LS}}{d_{Lens}d_S}. \tag{7.25}$$

Remembering that $b = \theta d_{Lens}$, we can rewrite this in terms of Σ_{crit}, the *critical density* for lensing:

$$\beta = \theta\left[1 - \frac{1}{\Sigma_{crit}}\frac{M(<b)}{\pi b^2}\right], \quad \text{where } \Sigma_{crit} \equiv \frac{c^2}{4\pi G}\frac{d_S}{d_{Lens}d_{LS}}. \tag{7.26}$$

The quantity $M(<b)/(\pi b^2)$ is just the average surface density within radius b; usually the surface density $\Sigma(R)$ declines from a peak at the center, so this average will fall as well. If the central density exceeds Σ_{crit}, then the image of a source at $\beta = 0$, exactly in line with the cluster's center, will be a thin circular *Einstein ring*, of angular size $\theta_E = b_E/d_{Lens}$, where b_E is the radius where the average density falls to the critical value:

$$\frac{M(<b_E)}{\pi b_E^2} = \Sigma_{crit}. \tag{7.27}$$

If the central surface density is less than Σ_{crit}, then the cluster cannot produce multiple images of any source behind it; no ring is seen.

Problem 7.16 Show that, when the lensing object is a point mass, Equations 7.27 and 7.14 give the same result for the Einstein radius θ_E.

Problem 7.17 If a lens at distance d_{Lens} bends the light of a much more distant galaxy, so that d_S and $d_{LS} \gg d_{Lens}$, show that the critical density is

$$\Sigma_{crit} \approx 2 \times 10^4 \left(\frac{100\,\text{Mpc}}{d_{Lens}}\right)M_\odot\,\text{pc}^{-2}, \tag{7.28}$$

and that the mass projected within angle θ_E of the center is

$$M(<\theta_E) \approx \left(\frac{d_{Lens}}{100\,\text{Mpc}}\right)\left(\frac{\theta_E}{1''}\right)^2 10^{10}M_\odot. \tag{7.29}$$

Use the results of Problems 6.3 and 6.6 to show that in the cD galaxy NGC 1399 we have roughly $6 \times 10^4 M_\odot\,\text{pc}^{-2}$. At the center of the Virgo cluster we see $0.5L_\odot\,\text{pc}^{-2}$, which for $M/L \sim 150$ corresponds to $< 100 M_\odot\,\text{pc}^{-2}$. We expect to see lensed arcs around individual luminous galaxies, but clusters with giant arcs like Abell 383 are far more massive than Virgo, and often at $d_{Lens} > 1\,\text{Gpc}$.

Problem 7.18 The 'Einstein Cross' is a set of five images of a quasar at redshift $z = 1.695$, shining through a spiral galaxy, 2237+0305, at $z = 0.039$. Show that $d_{\text{Lens}} \approx 120h^{-1}$ Mpc, and that, since $d_S \ll d_{\text{Lens}}$, $\Sigma_{\text{crit}} \approx 1.4 \times 10^4 h \mathcal{M}_\odot \, \text{pc}^{-2}$. Four of the images lie close to a circle, with radius $0.9''$: taking this value for θ_E, show that $b_E \approx 0.5h^{-1}$ kpc, and $\mathcal{M}(<b_E) \approx 10^{10}h^{-1}\mathcal{M}_\odot$, which is appropriate for the center of a bright galaxy.

Problem 7.19 For the 'dark halo' potential of Equation 2.19, show that

$$\Sigma(R) = \frac{V_H^2}{4G} \frac{1}{\sqrt{R^2 + a_H^2}}, \text{ so } \mathcal{M}(<b) = \frac{a_H V_H^2}{G} \frac{\pi}{2} \left[\sqrt{1 + (b/a_H)^2} - 1\right]. \quad (7.30)$$

Use Equation 7.20 to show that, when $d_S, d_{LS} \gg d_{\text{Lens}}$, a light ray passing far enough from the center that $b \gg a_H$ is bent by an angle

$$\alpha \approx \frac{2\pi V_H^2}{c^2} \text{ radians}, \text{ or } 5'' \times \left(\frac{V_H}{600 \, \text{km s}^{-1}}\right)^2. \quad (7.31)$$

A distant source with $\beta = 0$ directly behind the cluster will appear as a ring of radius $\theta_E \approx \alpha$. Figure 7.13 shows Abell 383 with a large tangential arc $16''$ from the center, corresponding to $V_H \approx 1800 \, \text{km s}^{-1}$. What is the kinetic energy of a particle in circular orbit in this potential? To what virial temperature (Equation 7.1) does this correspond? Show that this is similar to the observed X-ray temperature $T_X \approx 6 \times 10^7$ K.

Problem 7.20 Show that, for the Plummer sphere of Problem 3.2,

$$\mathcal{M}(<b) = \frac{\mathcal{M}b^2}{a_P^2 + b^2}. \quad (7.32)$$

In clusters where both lensed arcs and the X-ray emission of hot gas are observed, we can compare masses estimated with the two methods. Equation 7.29 shows that, to explain arcs with radii of tens of arcseconds in very distant clusters, we need masses rising beyond $10^{14}\mathcal{M}_\odot$. The mass required to account for the positions of the arcs agrees well with what is indicated by the X-ray observations. Disagreement is most common for lumpy clusters, where clumps of galaxies are still falling in and may heat the cluster gas, or where two clumps lie behind each other.

The light of the various images reaches us by different paths, and rays that left the source together do not arrive simultaneously. A strongly bent ray must travel further, and time passes more slowly deep in the gravitational potential. If a distant quasar is lensed by a centrally concentrated galaxy, light arrives fastest at

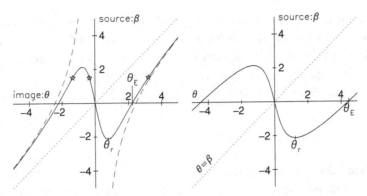

Fig. 7.18. Positions of a lensed galaxy and its image. Left, for a Plummer sphere with $\Sigma(0) = 6\Sigma_{crit}$ (solid curve); angles are in units of a_P/d_{Lens}. Rays passing far from the center are bent in almost the same way as by an equal point mass (dashed curve): stars show images of a source at $\beta = 1.5$. Right, for a 'dark halo' potential; angles are in units a_H/d_{Lens}. With more mass at large radii the curve approaches the dotted line $\theta = \beta$ (no bending) much more slowly.

the images furthest from the center: those rays have suffered the least bending, and have spent less time in the depths of the potential. Time delays have been measured by observing quasars that vary their light output, so the different images brighten in sequence.

The positions and shapes of the images tell us how mass is distributed within the lensing galaxy or cluster. Figure 7.18 shows how we can apply Equation 7.26 to find image positions in the simple cluster models of Equations 7.30 and 7.32. The images of a source directly behind the cluster are a point at $\theta = 0$, and an Einstein ring at the radius θ_E where the curve $\beta(\theta)$ crosses the axis $\beta = 0$. An object close to the center forms three images: one on the same side of the cluster and two on the opposite side. The image nearest to the center is inverted, since $d\theta/d\beta < 0$. As the source is taken further from the axis OS, the two images closest to the center move together: as $d\beta/d\theta \to 0$, at $\beta = \beta_r$, they fuse and then disappear. A source further out produces only a single image. The *odd-image theorem* guarantees that, if the lens has a finite surface density everywhere, then a single source produces an odd number of images.

If the galaxy is almost centred behind the cluster, β is small, and Equation 7.17 tells us that its images can be highly magnified and hence very bright. Two of them form arcs near the circle $\theta = \theta_E$, called the *tangential critical curve*. If the cluster is not exactly round, these can each split into a number of separate images. Images near the center where $\theta \approx 0$ will always be faint.

When the source is just inside the radius β_r, then $d\beta/d\theta \to 0$, and Equation 7.17 tells us that the image is again extremely luminous. Since now the image is extended radially, the curve $\beta = \beta_r$ is called the *radial caustic*. Figure 7.18 shows that these bright streaks must lie on a circle inside the

Einstein radius, at $\theta = \theta_r$; this is called the *radial critical curve*. In the left panel, the cluster has the density of a Plummer sphere, and $\theta_E = \sqrt{5}a_P/d_{Lens}$. The right panel shows critical points for a cluster with the potential of the dark halo of Equation 2.19. The Einstein radius θ_E is larger in relation to θ_r than for the Plummer model, because its mass is less concentrated to the center.

Problem 7.21 For the Plummer model, show that the surface density (Equation 3.13) drops to half its central value at the core radius $r_c = 0.644a_P$. What is r_c for the 'dark halo' potential of Problem 7.19? Taking the central density to be $2\Sigma_{crit}$, $4\Sigma_{crit}$, and finally $10\Sigma_{crit}$, find the Einstein radius θ_E for the Plummer model in units of r_c/d_{Lens}. Now do the same for the dark halo. (For a wide range of central density, and for many cluster-like potentials, $d_{Lens}\theta_E$ is a few times the core radius.)

Problem 7.22 In the galaxy cluster Abell 383 of Figure 7.13, a radial arc lies $2'' \approx \theta_r$ from the center. The large tangential arc of Figure 7.13 is at $16'' \approx \theta_E$. Find d_{Lens}, assuming $H_0 = 70\,\text{km}\,\text{s}^{-1}\,\text{Mpc}^{-1}$ (if you have read Section 8.3: use the angular size distance d_A and the benchmark cosmology). Show that $d_{Lens}\theta_E \sim 50\,\text{kpc}$; the cluster's core radius r_c will be a few times smaller. From Equation 7.29, show that the mass projected within the tangential arc is $\mathcal{M}(<\theta_E) \sim 2 \times 10^{13}\,\mathcal{M}_\odot$. The luminosity within θ_E corresponds to $L_V \sim 4 \times 10^{11}\,L_\odot$, so $\mathcal{M}/L \sim 50$. (A model placing both radial and tangential arcs at the observed positions has $r_c \approx 40\,\text{kpc}$; in the X-ray-emitting gas, the core is twice as large.)

7.4.3 Weak gravitational lensing

When galaxies lie behind a cluster but well outside its Einstein radius, their images are weakly magnified and slightly stretched in the tangential direction. Figure 7.15 shows why the image of a round galaxy will be an ellipse, with tangential and radial axes in the ratio $x/y : \Delta x/\Delta y$ or $|d\beta/d\theta| : |\beta/\theta|$. The *shear* γ measures the difference in the amount that the image has been compressed in the two directions. For an image at distance $b \gg \theta_E d_{Lens}$ from the cluster's center we have

$$\gamma \equiv \frac{1}{2}\left(\frac{d\beta}{d\theta} - \frac{\beta}{\theta}\right) = \frac{\bar{\Sigma}(<b) - \Sigma(b)}{\Sigma_{crit}}. \tag{7.33}$$

Here $\bar{\Sigma}(<b) = \mathcal{M}(<b)/(\pi b^2)$ is the average surface density of matter projected within radius b, and $\Sigma(b)$ is the surface density at that radius.

Problem 7.23 Derive Equation 7.33 from Equation 7.26. Why must we specify $\theta \gg \theta_E$? Show that, if the surface density $\Sigma(R)$ is constant, then we have $\theta \propto \beta$. Use Figure 7.15 to explain why images of a round galaxy will be circular: the shear is zero. This *mass-sheet degeneracy* means that we can always add a uniform sheet of matter to our model cluster without changing the shapes of the images.

Measuring the average shape of many galaxies in the background that have been weakly distorted allows us to estimate the shear, and hence the distribution of mass in the outer parts of galaxy clusters. Generally the results agree with what we found from the hot gas. But, for some clusters, weak lensing indicates a higher mass. These may be composites, with one cluster behind the other. Some distant clusters have even been discovered by the signature of their weak lensing alone.

Weak lensing has also been used to study the halos of individual bright galaxies. To do this, we measure the average shape of more distant galaxies that lie close to the nearby lensing system on the sky, and then average these results over many lens-galaxies. Recent studies show that the dark halo of a typical bright elliptical or S0 galaxy, with $L \sim L_*$, extends to at least 200 kpc and has mass $\gtrsim 10^{12}h^{-1}\mathcal{M}_\odot$; so the mass-to-light ratio $\mathcal{M}/L > 50$ in solar units. So dark halos are larger and more massive than we were able to show by measuring rotation in the outer parts of disk galaxies. Blue galaxies that are still making stars are brighter for a given dark-halo mass. The halo appears to be slightly aspherical, distorted in the same sense as the galaxy's stellar body.

Unlike galaxies themselves, groups and clusters do not separate cleanly from one another: we will see in the next chapter how they are connected by walls and filaments of luminous galaxies. Weak lensing lets us measure how the dark matter is distributed in this structure, but the calculation is more complex. The light of a distant source is bent many times as it passes by each of these large structures on its way to us, so we cannot make the approximation of Equation 7.19, that all the bending takes place at a single lens-distance d_{Lens}. Instead, we must use computers to calculate it.

Further reading: a detailed presentation of gravitational lensing is given by P. Schneider, J. Ehlers, and E. E. Falco, 1992, *Gravitational Lenses* (Springer, New York); this is an advanced text. A popular book is N. Cohen, 1989, *Gravity's Lens: Views of the New Cosmology* (Wiley, New York).

8

The large-scale distribution
of galaxies

Since the early 1980s, multi-object spectrographs, CCD detectors, and some dedicated telescopes have allowed the mass production of galaxy redshifts. These large surveys have revealed a very surprising picture of the luminous matter in the Universe. Many astronomers had imagined roughly spherical galaxy clusters floating amongst randomly scattered field galaxies, like meatballs in sauce. Instead, they saw galaxies concentrated into enormous walls and long filaments, surrounding huge voids that appear largely empty. The galaxy distribution has been compared to walls of soapy water, surrounding bubbles of air in a basinful of suds; linear filaments appear where the walls of different soap bubbles join, and rich clusters where three or more walls run into each other. A more accurate metaphor is that of a sponge; the voids are interlinked by low-density 'holes' in the walls. Sometimes we think of the filaments as forming a *cosmic web*.

For a star like the Sun in the disk of our Milky Way, the task of finding where it formed is essentially hopeless, because it has already made many orbits about the galaxy, and the memory of its birthplace is largely lost. But the large structures that we discuss in this chapter are still under construction, and the regions where mass is presently concentrated reveal where denser material was laid down in the early Universe. The peculiar motions of groups and clusters of galaxies, their speeds relative to the uniformly expanding cosmos, are motions of infall toward larger concentrations of mass. So the problem of understanding the large-scale structure that we see today becomes one of explaining small variations in the density of the early Universe.

We begin in Section 8.1 by surveying the galaxies around us, mapping out both the local distribution and the larger structures stretching over hundreds of megaparsecs. The following sections discuss the history of our expanding Universe, within which the observed spongy structures grew, and how the expansion and large-scale curvature affect our observations of galaxies. In Section 8.4 we discuss fluctuations in the cosmic microwave background and what they tell us about the initial irregularities that might have given rise to the galaxies that we

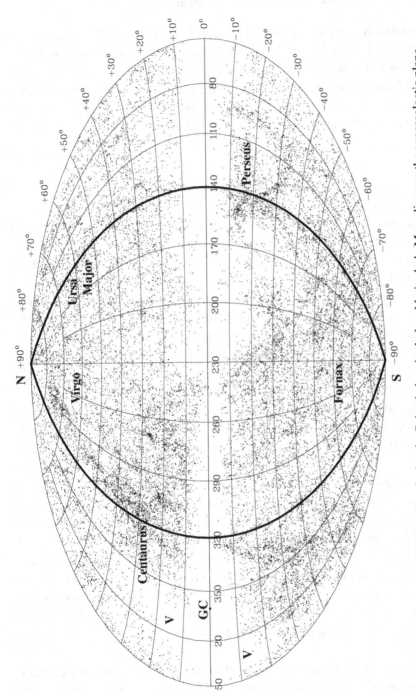

Fig. 8.1. Positions of 14 650 bright galaxies, in Galactic longitude l and latitude b. Many lie near the supergalactic plane, approximately along the Great Circle $l = 140°$ and $l = 320°$ (heavy line); V marks the Local Void. Galaxies near the plane $b = 0$ of the Milky Way's disk are hidden by dust – T. Kolatt and and O. Lahav.

see today. We will see how we can use the observed peculiar motions of galaxies to estimate how much matter is present. The final section asks how dense systems such as galaxies developed from these small beginnings.

Further reading: B. Ryden, 2003, *Introduction to Cosmology* (Addison Wesley, San Francisco, USA) and A. Liddle, 2003, *An Introduction to Modern Cosmology*, 2nd edition (John Wiley & Sons, Chichester, UK) are undergraduate texts on roughly the level of this book. On the graduate level, see M. S. Longair, 1998, *Galaxy Formation* (Springer, Berlin) and T. Padmanabhan, 1993, *Structure Formation in the Universe* (Cambridge University Press, Cambridge, UK).

8.1   Large-scale structure today

As we look out into the sky, it is quite clear that galaxies are not spread uniformly through space. Figure 8.1 shows the positions on the sky of almost 15 000 bright galaxies, taken from three different catalogues compiled from optical photographs. Very few of them are seen close to the plane of the Milky Way's disk at $b = 0$, and this region is sometimes called the *Zone of Avoidance*. The term is unfair: surveys in the 21 cm line of neutral hydrogen, and in far-infrared light, show that galaxies are indeed present, but their visible light is obscured by dust in the Milky Way's disk. Dense areas on the map mark rich clusters: the Virgo cluster is close to the north Galactic pole, at $b = 90°$. Few galaxies are seen in the Local Void, stretching from $l = 40°$, $b = -20°$ across to $l = 0$, $b = 30°$.

The galaxy clusters themselves are not spread evenly on the sky: those within about 100 Mpc form a rough ellipsoid lying almost perpendicular to the Milky Way's disk. Its midplane, the *supergalactic plane*, is well defined in the northern Galactic hemisphere ($b > 0$), but becomes rather scruffy in the south. The pole or Z axis of the supergalactic plane points to $l = 47.4°$, $b = 6.3°$. We take the supergalactic X direction in the Galactic plane, pointing to $l = 137.3°$, $b = 0$, while the Y axis points close to the north Galactic pole at $b = 90°$, so that $Y \approx 0$ along the Zone of Avoidance. The supergalactic plane is close to the Great Circle through Galactic longitude $l = 140°$ and $l = 320°$, shown as a heavy line in Figure 8.1. It passes through the Ursa Major group of Figures 5.6 and 5.8, and the four nearby galaxy clusters described in Section 7.2; the Virgo cluster at right ascension $\alpha = 12^h$, declination $\delta = 12°$; Perseus at $\alpha = 3^h$, $\delta = 40°$; Fornax at $\alpha = 4^h$, $\delta = -35°$; and the Coma cluster at $\alpha = 13^h$, $\delta = 28°$, almost at the north Galactic pole.

Figure 8.2 shows the positions of the elliptical galaxies within about 20 Mpc, tracing out the Virgo and Fornax clusters. The distances to these galaxies have been found by analyzing *surface brightness fluctuations*. Even though they are too far away for us to distinguish individual bright stars, the number N of stars falling within any arcsecond square on the image has some random variation.

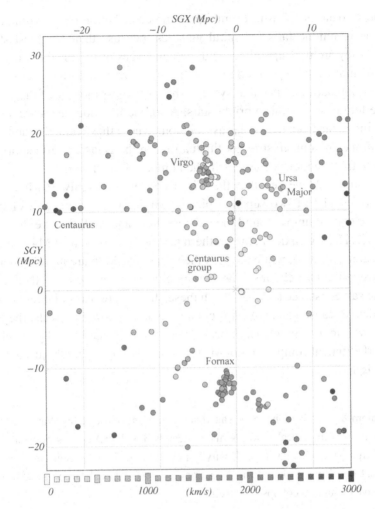

Fig. 8.2. Positions of nearby elliptical galaxies on the supergalactic $X–Y$ plane; the origin is the Milky Way. Shading indicates recession velocity V_r – J. Tonry.

So the surface brightness of any square fluctuates about some average value. The closer the galaxy, the fewer stars lie within each square, and the stronger the fluctuations between neighboring squares: when N is large, the fractional variation is proportional to $1/\sqrt{N}$. So if we measure the surface brightness fluctuations in two galaxies where we know the relative luminosity of the bright stars that emit most of the light, we can find their relative distances.

This method works only for relatively nearby galaxies in which the stars are at least 3–5 Gyr old. In these, nearly all the light comes from stars close to the tip of the red giant branch. As we noted in Section 1.1, these have almost the same luminosity for all stars below about $2\mathcal{M}_\odot$. These stars are old enough that they have made many orbits around the center, so they are dispersed smoothly through the galaxy. The observations are usually made in the I band near 8000 Å, or in the K band at 2.2 μm, to minimize the contribution of the younger bluer stars. The technique fails for spiral galaxies, because their brightest stars are younger: they are red supergiants and the late stages of intermediate-mass stars, and their luminosity depends on the stellar mass, and so on the average age of the stellar population. Since that average changes across the face of the galaxy, so does the luminosity of those bright stars. Also, the luminous stars are too short-lived to move far from the stellar associations where they formed. Their clumpy distribution causes much stronger fluctuations in the galaxy's surface brightness than those from random variations in the number of older stars.

In Figure 8.2, we see that the Virgo cluster is roughly 16 Mpc away. It appears to consist of two separate pieces, which do not coincide exactly with the two velocity clumps around the galaxies M87 and M49 that we discussed in Section 7.2. Here, galaxies in the northern part of the cluster, near M49, lie mainly in the nearer grouping, while those in the south near M86 are more distant; M87 lies between the two clumps. The Fornax cluster, in the south with $Y < 0$, is at about the same distance as Virgo. Both these clusters are part of larger complexes of galaxies. Because galaxy groups contain relatively few ellipticals, they do not show up well in this figure. The Local Group is represented only by the elliptical and dwarf elliptical companions of M31, as the overlapping circles just to the right of the origin.

Problem 8.1 In Section 4.5 we saw that the motions of the Milky Way and M31 indicate that the Local Group's mass exceeds $3 \times 10^{12} \mathcal{M}_\odot$: taking its radius as 1 Mpc, what is its average density? Show that this is only about $3h^{-2}$ of the critical density ρ_{crit} defined in Equation 1.30 – the Local Group is only just massive enough to collapse on itself.

Problem 8.2 The free-fall time $t_{ff} = 1/\sqrt{G\rho}$ of Equation 3.23 provides a rough estimate of the time taken for a galaxy or cluster to grow to density ρ. In Problem 4.7 we saw that, for the Milky Way, with average density of $10^5 \rho_{crit}$ within the Sun's orbit, this minimum time is ~3×10^8 years or $0.03 \times t_H$, the expansion age given by Equation 1.28. Show that a cluster of galaxies with density $200\rho_{crit}$ can barely collapse within the age of the Universe. This density divides structures like the Local Group that are still collapsing from those that might have settled into an equilibrium.

To probe further afield, we use a 'wedge diagram' like Figure 8.3 from the 2dF survey, which measured redshifts of galaxies in two large slices across the sky. If we ignore peculiar motions, Equation 1.27 tells us that the recession speed $V_r \approx cz \approx H_0 d$, where H_0 is the Hubble constant; the redshift is roughly proportional to the galaxy's distance d from our position at the apex of each wedge. So this figure gives us a somewhat distorted map of the region out to $600h^{-1}$ Mpc.

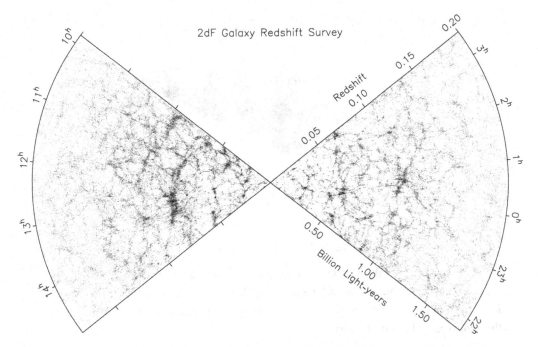

Fig. 8.3. A 'wedge diagram' of 93 170 galaxies from the 2dF survey with the Anglo-Australian 4-meter telescope, in slices $-4° < \delta < 2°$ in the north (left wedge) and $-32° < \delta < -28°$ in the south (right) – M. Colless *et al.* 2001 *MNRAS* **328**, 1039.

The three-dimensional distribution of galaxies in Figure 8.3 has even more pronounced structure than Figure 8.1. We see dense linear features, the walls and stringlike filaments of the cosmic web; at their intersections there are complexes of rich clusters. Between the filaments we find large regions that are almost empty of bright galaxies: these *voids* are typically $\gtrsim 50h^{-1}$ Mpc across. The galaxies appear to thin out beyond $z = 0.15$ because redshifts were measured only for objects that exceeded a fixed apparent brightness. Figure 8.4 shows that, at large distances, just the rarest and most luminous systems have been included. When a solid angle Ω on the sky is surveyed, the volume between distance d and $d + \Delta d$ is $\Delta \mathcal{V} = \Omega d^2 \Delta d$, which increases rapidly with d. Accordingly, we see few galaxies nearby; most of the measured objects lie beyond 25 000 km s^{-1}.

Problem 8.3 The Local Group moves at 600 km s^{-1} relative to the cosmic background radiation. At this speed, show that an average galaxy would take $\sim 40h^{-1}$ Gyr to travel from the center to the edge of a typical void. Whatever process removed material from the voids must have taken place very early, when the Universe was far more compact.

In Figures 8.3 and 8.4, the walls appear to be several times denser than the void regions. But ignoring the peculiar motions has exaggerated their narrowness and

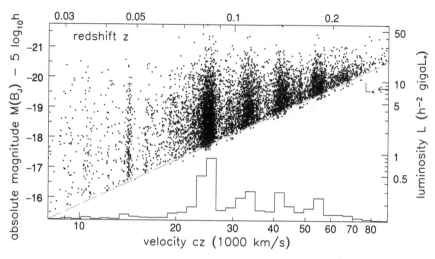

Fig. 8.4. Luminosity (absolute magnitude in B_J) of 8438 galaxies near 13^h20^m in Figure 8.3, showing the number at each redshift. The luminosity L_* of a typical bright galaxy is taken from Figure 1.16; the dashed line at apparent magnitude $m(B_J) = 19.25$ shows the approximate limit of the survey.

sharpness; they would appear less pronounced if we could plot the true distances of the galaxies. The extra mass in a wall or filament attracts nearby galaxies in front of the structure, pulling them toward it and away from us. So the radial velocities of those objects are increased above that of the cosmic expansion, and we overestimate their distances, placing them further from us and closer to the wall. Conversely, galaxies behind the wall are pulled in our direction, reducing their redshifts; these systems appear nearer to us and closer to the wall than they really are. In fact, the walls are only a few times denser than the local average.

By contrast, dense clusters of galaxies appear elongated in the direction toward the observer. The cores of these clusters have completed their collapse, and galaxies are packed tightly together in space. They orbit each other with speeds as large as $1500 \, \mathrm{km \, s^{-1}}$, so their distances inferred from Equation 1.27 can have random errors of $15h^{-1}$ Mpc. In a wedge diagram, rich clusters appear as dense 'fingers' that point toward the observer.

Problem 8.4 How long is the narrow 'finger' in the left panel of Figure 8.5 near $z = 0.12$ and 12^h30^m? Show that this represents a large galaxy cluster with $\sigma_r \approx 1500 \, \mathrm{km \, s^{-1}}$.

Figure 8.5 shows wedge diagrams for red galaxies, with spectra that show little sign of recent star formation, and blue galaxies, with spectral lines characteristic

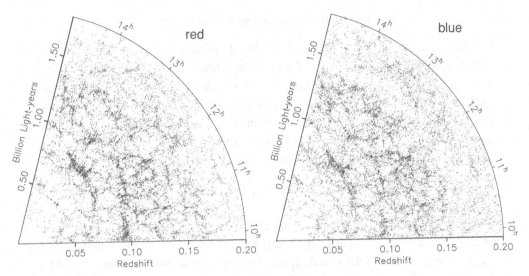

Fig. 8.5. About 27 000 red galaxies (left) with spectra like those of elliptical galaxies, and the same number of star-forming blue galaxies (right), in a slice $-32° < \delta < -28°$ from the 2dF survey. These are luminous galaxies, with $-21 < M(B_J) < -19$. The elliptical and S0 galaxies cluster more strongly than the spiral-like systems.

of young massive stars and the ionized gas around them. In Figure 8.4 we see clumps of galaxies around 13^h20^m at $z = 0.05$, 0.08, 0.11, and 0.15. These are quite clear in the left wedge of Figure 8.5, but much weaker in the right wedge. Similarly, the void at 12^h and $z = 0.08$ is emptier in the left wedge – why? The left-wing galaxies are red elliptical and S0 systems, while on the right are spirals and irregulars. As we saw in Section 7.2, elliptical galaxies live communally in the cores of rich clusters, where spiral galaxies are rare. Accordingly, the clustering of the red galaxies is stronger than that of the bluer systems of the right panel. Whether galaxies are spiral or elliptical is clearly related to how closely packed they are: we see a *morphology–density relation*.

In fact, we should not talk simply of 'the distribution of galaxies', but must be careful to specify which galaxies we are looking at. We never see all the galaxies in a given volume; our surveys are always *biassed* by the way we choose systems for observation. For example, if we select objects that are large enough on the sky that they appear fuzzy and hence distinctly nonstellar, we will omit the most compact galaxies. A survey that finds galaxies by detecting the 21 cm radio emission of their neutral hydrogen gas will readily locate optically dim but gas-rich dwarf irregular galaxies, but will miss the luminous ellipticals which usually lack HI gas. The Malmquist bias of Problem 2.11 is present in any sample selected by apparent magnitude. Even more insidious are the ways in which the bias changes with redshift and apparent brightness. Mapping even the luminous matter of the Universe is no easy task.

8.1.1 Measures of galaxy clustering

One way to describe the tendency of galaxies to cluster together is the *two-point correlation function* $\xi(r)$. If we make a random choice of two small volumes ΔV_1 and ΔV_2, and the average spatial density of galaxies is n per cubic megaparsec, then the chance of finding a galaxy in ΔV_1 is just $n\Delta V_1$. If galaxies tend to clump together, then the probability that we then also have a galaxy in ΔV_2 will be greater when the separation r_{12} between the two regions is small. We write the joint probability of finding a particular galaxy in both volumes as

$$\Delta P = n^2[1 + \xi(r_{12})]\Delta V_1\,\Delta V_2; \tag{8.1}$$

if $\xi(r) > 0$ at small r, then galaxies are clustered, whereas if $\xi(r) < 0$, they tend to avoid each other. We generally compute $\xi(r)$ by estimating the distances of galaxies from their redshifts, making a correction for the distortion introduced by peculiar velocities. On scales $r \lesssim 10h^{-1}$ Mpc, it takes roughly the form

$$\xi(r) \approx (r/r_0)^{-\gamma}, \tag{8.2}$$

with $\gamma > 0$. When $r < r_0$, the *correlation length*, the probability of finding one galaxy within radius r of another is significantly larger than for a strictly random distribution. Since $\xi(r)$ represents the deviation from an average density, it must at some point become negative as r increases.

Figure 8.6 shows the two-point correlation function $\xi(r)$ for galaxies in the 2dF survey. The correlation length $r_0 \approx 5h^{-1}$ Mpc; it is $6h^{-1}$ Mpc for the ellipticals, which are more strongly clustered, and smaller for the star-forming galaxies. The slope $\gamma \approx 1.7$ around r_0. For $r \gtrsim 50h^{-1}$ Mpc, which is roughly the size of the largest wall or void features, $\xi(r)$ oscillates around zero: the galaxy distribution is fairly uniform on larger scales.

Unfortunately, the correlation function is not very useful for describing the one-dimensional filaments or two-dimensional walls of Figure 8.3. If our volume ΔV_1 lies in one of these, the probability of finding a galaxy in ΔV_2 is high only when it also lies within the structure. Since $\xi(r)$ is an average over all possible placements of ΔV_2, it will not rise far above zero once the separation r exceeds the thickness of the wall or filament. We can try to overcome this by defining the three-point and four-point correlation functions, which give the joint probability of finding that number of galaxies with particular separations; but this is not very satisfactory. We do not yet have a good statistical method to describe the strength and prevalence of walls and filaments.

The Fourier transform of $\xi(r)$ is the *power spectrum* $P(k)$:

$$P(\mathbf{k}) \equiv \int \xi(\mathbf{r})\exp(i\mathbf{k}\cdot\mathbf{r})d^3\mathbf{r} = 4\pi \int_0^\infty \xi(r)\frac{\sin(kr)}{kr}r^2\,dr, \tag{8.3}$$

so that small k corresponds to a large spatial scale.

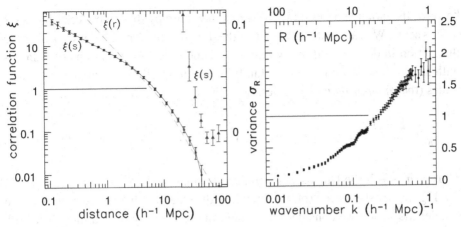

Fig. 8.6. Left, the correlation function $\xi(s)$ for the 2dF galaxies, at small (circles, left logarithmic scale) and large (triangles, right linear scale) separations; vertical bars show uncertainties. $\xi(s)$ is calculated assuming that Hubble's law holds exactly: the 'fingers' of Figure 8.5 reduce $\xi(s)$ on scales $r \lesssim 1$ Mpc, but infall to the walls makes clustering look stronger on scales near r_0. The dashed line shows $\xi(r)$, corrected for these effects. Right, the variance σ_R of Equation 8.4, describing how much the average density varies between regions of size R – S. Maddox and S. Cole.

Since $\xi(r)$ is dimensionless, $P(k)$ has the dimensions of a volume. The function $\sin(kr)/kr$ is positive for $|kr| < \pi$, and it oscillates with decreasing amplitude as kr becomes large; so, very roughly, $P(k)$ will have its maximum when k^{-1} is close to the radius where $\xi(r)$ drops to zero. In Figure 8.17 we show $P(k)$ calculated by combining the 2dF galaxy survey with observations of the cosmic microwave background.

Problem 8.5 Prove the last equality of Equation 8.3. One method is to write the volume integral for $P(k)$ in spherical polar coordinates r, θ, ϕ and set $\mathbf{k} \cdot \mathbf{r} = kr\cos\theta$. Show that, because $\xi(r)$ describes departures from the mean density, Equation 8.1 gives $\int_0^\infty \xi(r)r^2 \, dr = 0$, and hence $P(k) \to 0$ as $k \to 0$.

Problem 8.6 Show that the power spectrum $P(k) \propto k^n$ corresponds to a correlation function $\xi(r) \propto r^{-(3+n)}$. Hence $\gamma \approx 1.5$ implies $n \approx -1.5$. Figure 8.17 shows that when k is large $P(k)$ declines roughly as $k^{-1.8}$, about as expected.

Another way to describe the non-uniformity of the galaxy distribution is to ask how likely we are to find a given deviation from the average density. We can write the local density at position $\mathbf{x}$ as a multiple of the mean level $\bar{\rho}$, as $\rho(\mathbf{x}) = \bar{\rho}[1 + \delta(\mathbf{x})]$, and let δ_R be the fractional deviation $\delta(\mathbf{x})$ averaged within a sphere of radius R. When we take the average $\langle \delta_R \rangle$ over all such spheres, this must be zero. Its variance $\sigma_R^2 = \langle \delta_R^2 \rangle$ measures how clumpy the galaxy distribution is on this scale. We can relate it to $k^3 P(k)$, the dimensionless number prescribing the fluctuation in density within a volume k^{-1} Mpc in radius. If clumps of galaxies with size $k^{-1} \approx R$ are placed randomly in relation to those on larger or smaller scales (the *random-phase* hypothesis), we have

$$\sigma_R^2 \approx \frac{k^3 P(k)}{2\pi^2} \equiv \Delta_k^2; \quad \text{so, if} \quad P(k) \propto k^n, \quad \text{then} \quad \sigma_R \propto R^{-(n+3)/2}. \quad (8.4)$$

Thus, if $n > -3$, the Universe is lumpiest on small scales.

Figure 8.6 shows σ_R (or Δ_k) for the 2dF galaxies. It increases with k: the smaller the region we consider, the greater the probability of finding a very high density of galaxies. We often parametrize the clustering by σ_8, the average fluctuation on a scale $R = 8h^{-1}$ Mpc: Figure 8.6 shows that $\sigma_8 \approx 0.9$. The wiggles at $k \sim 0.1$ correspond to the 'baryon oscillations' that influence the cosmic background radiation: see Section 8.5. If $P(k) \propto k^n$ and $\sigma(\mathcal{M})$ is the variance in density over a region containing a mass $\mathcal{M} \approx 4\pi R^3 \bar{\rho}/3$, then, since $\mathcal{M} \propto R^3$, we have $\sigma(\mathcal{M}) \propto \mathcal{M}^{-(n+3)/6}$. Cosmological models for the development of structure can predict $P(k)$; we return to this topic in Section 8.5.

Problem 8.7 The quantity $\langle \Delta_k^2 \rangle^{1/2}$ gives the expected fractional deviation $|\delta(\mathbf{x})|$ from the mean density in an overdense or diffuse region of size $1/k$. Write $\delta(\mathbf{x})$ and $\Phi(\mathbf{x})$ as Fourier transforms and use Equation 3.9, Poisson's equation, to show that these lumps and voids cause fluctuations $\Delta\Phi_k$ in the gravitational potential, where $k^2|\Delta\Phi_k| \sim 4\pi G\bar{\rho}\langle \Delta_k^2 \rangle^{1/2}$. Show that, when $P(k) \propto k$, the *Harrison–Zel'dovich* spectrum, $|\Delta\Phi_k|$ does not depend on k: the potential is equally 'rippled' on all spatial scales.

In this section we have seen that the present-day distribution of galaxies is very lumpy and inhomogeneous on scales up to $50h^{-1}$ Mpc. But measurements of the cosmic background radiation show that its temperature is the same in all parts of the sky to within a few parts in 100 000. As we saw in Section 1.5, before the time of recombination when $z = z_{\text{rec}} \approx 1100$, the cosmos was largely filled with ionized gas. Since light could not stream freely through the charged particles, the gas was opaque and glowing like a giant neon tube. The Universe became largely neutral and transparent only after recombination. Because the cosmic background radiation is smooth today, we know that the matter and radiation were quite smoothly distributed at that time. How could our present highly structured Universe of galaxies have arisen from such uniform beginnings? To understand what might have happened, we must look at how the Universe expanded following the Big Bang, and how concentrations of galaxies could form within it.

8.2 Expansion of a homogeneous Universe

Because the cosmic background radiation is highly uniform, we infer that the Universe is isotropic – it is the same in all directions. We believe that on a large scale the cosmos is also homogeneous – it would look much the same if we lived in any other galaxy. Then, mathematics tells us that the length s of a path linking any two points at time t must be given by integrating the expression

$$\Delta s^2 = \mathcal{R}^2 \left(\frac{\Delta \sigma^2}{1 - k\sigma^2} + \sigma^2 \, \Delta \theta^2 + \sigma^2 \sin^2\theta \, \Delta \phi^2 \right), \tag{8.5}$$

where σ, ϕ, θ are spherical polar coordinates in a *curved space*. The origin $\sigma = 0$ looks like a special point, but in fact it is not. Just as at the Earth's poles where lines of longitude converge, the curvature here is the same as everywhere else, and we can equally well take any point to be $\sigma = 0$. The constant k specifies the *curvature* of space. For $k = 1$, the Universe is *closed*, with positive curvature and finite volume, analogous to the surface of a sphere; $\mathcal{R}$ is the radius of curvature. If $k = -1$, we have an *open* Universe, a negatively curved space of infinite volume, while $k = 0$ describes familiar unbounded flat space.

Near the origin, where $\sigma \ll 1$, the formula for Δs is almost the same for all values of k; on a small enough scale, curvature does not matter. If we look at a tiny region, the relationships among angles, lengths, and volumes will be the same as they are in flat space. We can call the comoving coordinate σ of Equation 8.5 an 'area radius', because at time t the area of a sphere around the origin at radius σ is

$$\mathcal{A}(\sigma, t) = 4\pi \mathcal{R}^2 \sigma^2. \tag{8.6}$$

Problem 8.8 In ordinary three-dimensional space, using cylindrical polar coordinates we can write the distance between two nearby points (R, θ, z) and $(R + \Delta R, \theta + \Delta \theta, z + \Delta z)$ as $\Delta s^2 = \Delta R^2 + R^2 \, \Delta \theta^2 + \Delta z^2$. The equation $R^2 + z^2 = \mathcal{R}^2$ describes a sphere of radius $\mathcal{R}$: show that, if our points lie on this sphere, then the distance between them is

$$\Delta s^2 = \Delta R^2 (1 + R^2/z^2) + R^2 \, \Delta \theta^2 = \mathcal{R}^2 \left(\frac{\Delta \sigma^2}{1 - \sigma^2} + \sigma^2 \, \Delta \theta^2 \right), \tag{8.7}$$

where $\sigma = R/\mathcal{R}$. Integrate from a point P, at radius R and with $z > 0$, to the 'north pole' at $z = \mathcal{R}$ to show that the distance $s = \mathcal{R} \arcsin \sigma$. Show that the circumference $2\pi R$ of the 'circle of latitude' through P is always larger than $2\pi s$, but approaches it when $s \ll \mathcal{R}$. When $k = 1$ in Equation 8.5, any surface of constant ϕ is the surface of a sphere of radius $\mathcal{R}$.

The cosmic expansion is described by setting $\mathcal{R} = \mathcal{R}(t)$, allowing the radius of curvature to grow with time. Apart from their small peculiar speeds, galaxies remain at points with fixed values of σ, ϕ, θ; so these are called *comoving coordinates*. The separation d of two galaxies is just the length s of the shortest path between them. So, if they stay at fixed comoving coordinates, d expands proportionally to $\mathcal{R}(t)$: Hubble's law is just one symptom of the expansion of curved space. Equation 8.5 tells us that the two systems are carried away from each other at a speed

$$V_r = \dot{d} = \frac{\dot{\mathcal{R}}(t)}{\mathcal{R}(t)} d \equiv H(t)d. \tag{8.8}$$

Here $H(t)$ is the *Hubble parameter*, which at present has the value H_0. Equation 8.8 describes the average motion of the galaxies; we defer discussion of the peculiar motions that develop from the gravitational pull of near neighbors until Section 8.4 below.

General Relativity tells us that the distance between two events happening at different times and in different places depends on the motion of the observer. But all observers will measure the same *proper time* τ along a path through space and time connecting the events, which is found by integrating

$$\Delta \tau^2 = \Delta t^2 - \Delta s^2/c^2. \tag{8.9}$$

Light rays always travel along paths of zero proper time, $\Delta \tau = 0$. If we place ourselves at the origin of coordinates, then the light we receive from a galaxy at comoving distance σ_e has followed the radial path

$$\frac{c\,\Delta t}{\mathcal{R}(t)} = -\frac{\Delta \sigma}{\sqrt{1 - k\sigma^2}}. \tag{8.10}$$

The light covers less comoving distance per unit of time as the scale length $\mathcal{R}(t)$ of the Universe grows. We can integrate this equation for a wavecrest that sets off at time t_e, arriving at our position now at time t_0:

$$c \int_{t_e}^{t_0} \frac{dt}{\mathcal{R}(t)} = \int_0^{\sigma_e} \frac{d\sigma}{\sqrt{1 - k\sigma^2}}. \tag{8.11}$$

Suppose that another wavecrest sets off later, by a time Δt_e. We receive it at time $t_0 + \Delta t$, given by the same equation with the new departure and arrival times. The galaxy's comoving position σ_e, and the integral on the right-hand side of Equation 8.11, have not changed. So the left-hand side also stays constant:

$$\int_{t_e + \Delta t_e}^{t_0 + \Delta t} \frac{dt}{\mathcal{R}(t)} = \int_{t_e}^{t_0} \frac{dt}{\mathcal{R}(t)}, \quad \text{so} \quad \frac{\Delta t_e}{\mathcal{R}(t_e)} = \frac{\Delta t}{\mathcal{R}(t_0)}, \tag{8.12}$$

as long as $\Delta t \ll \mathcal{R}(t)/\dot{\mathcal{R}}(t)$. Thus all processes in the distant galaxy appear to be slowed down by the factor $\mathcal{R}(t_0)/\mathcal{R}(t_e)$. If Δt_e is the time between two consecutive crests emitted with wavelength $\lambda_e = c\,\Delta t_e$, that light is received with $\lambda_{obs} = c\,\Delta t$. So the wavelength grows along with the scale length $\mathcal{R}(t)$, while the frequency, momentum, and energy of each photon decay proportionally to $1/\mathcal{R}(t)$. The measured redshift z of a distant galaxy tells us how much expansion has taken place since the time t_e when its light set off on its journey to us. This is the *cosmological redshift* of Equation 1.34:

$$1 + z = \frac{\lambda_{obs}}{\lambda_e} = \frac{\mathcal{R}(t_0)}{\mathcal{R}(t_e)}. \tag{8.13}$$

We often use redshift $z(t)$ as a substitute for time t_e or comoving distance σ_e. The time t_e corresponding to a given redshift depends on how fast the Universe has expanded, while Equation 8.11 tells us the comoving distance σ_e from which the light would have started.

The rate at which the Universe expands is set by the gravitational pull of matter and energy within it. We first use Newtonian physics to calculate the expansion, and then discuss how General Relativity modifies the result. Consider a small sphere of radius r, at a time t when our homogeneous Universe has density $\rho(t)$; we take $r \ll \mathcal{R}(t)$, so that we can neglect the curvature of space. Everything is symmetric about the origin $r = 0$, so we appeal to Newton's first theorem in Section 3.1: the gravitational force at radius r is determined only by the mass $\mathcal{M}(<r)$ within the sphere. If our sphere is large enough that gas pressure forces are much smaller than the pull of gravity (see Section 8.5 below), then Equation 3.20 gives the force on a gas cloud of mass m at that radius:

$$m\frac{d^2 r}{dt^2} = -\frac{Gm\mathcal{M}(<r)}{r^2} = -\frac{4\pi Gm}{3}\rho(t)r. \tag{8.14}$$

Our sphere of matter is expanding along with the rest of the Universe, so its radius $r(t) \propto \mathcal{R}(t)$. The mass m of the cloud cancels out, giving

$$\ddot{\mathcal{R}}(t) = -\frac{4\pi G}{3}\rho(t)\mathcal{R}(t); \tag{8.15}$$

the higher the density, the more strongly gravity slows the expansion.

Nothing enters or leaves our sphere, so the mass within it does not change: $\rho(t)\mathcal{R}^3(t)$ is constant. Multiplying by $\dot{\mathcal{R}}(t)$ tells us how the kinetic energy decreases as the sphere expands:

$$\frac{1}{2}\frac{d}{dt}[\dot{\mathcal{R}}^2(t)] = -\frac{4\pi G}{3}\frac{\rho(t_0)\mathcal{R}^3(t_0)}{\mathcal{R}^2(t)}\dot{\mathcal{R}}(t), \tag{8.16}$$

where the time t_0 refers to the present day. Integrating, we have

$$\dot{\mathcal{R}}^2(t) = \frac{8\pi G}{3}\rho(t)\mathcal{R}^2(t) - kc^2, \tag{8.17}$$

where k is a constant of integration. Although we derived it using Newtonian theory, Equation 8.17 is also correct in General Relativity, which tells us that the constant k is the same one as in Equation 8.5. According to thermodynamics, as heat ΔQ flows into a volume $\mathcal{V}$ its internal energy $\mathcal{E}$ must increase, or it expands and does work against pressure:

$$\Delta Q = \Delta \mathcal{E} + p\,\Delta \mathcal{V} = \mathcal{V}\,\Delta(\rho c^2) + (\rho c^2 + p)\Delta \mathcal{V}, \tag{8.18}$$

where the density ρ includes all forms of matter and energy. The cosmos is uniform, so no volume $\mathcal{V}$ gains heat at the expense of another:

$$\Delta Q = 0 = \Delta \rho + \left(\rho + \frac{p}{c^2}\right)\frac{\Delta \mathcal{V}}{\mathcal{V}}, \quad \text{or} \quad \frac{d\rho}{dt} = -3\frac{\dot{\mathcal{R}}(t)}{\mathcal{R}(t)}\left(\rho + \frac{p}{c^2}\right). \tag{8.19}$$

Differentiating Equation 8.17 and substituting for $d\rho/dt$ yields

$$\ddot{\mathcal{R}}(t) = -\frac{4\pi G}{3}\mathcal{R}(t)\left[\rho(t) + \frac{3p(t)}{c^2}\right]. \tag{8.20}$$

This change to Equation 8.15 shows that, in General Relativity, the pressure p *adds* to the gravitational attraction. Equations 8.17 and 8.20 describe the *Friedmann models*, telling us how the contents of the Universe determine its expansion.

For cool matter, the pressure $p \sim \rho c_s^2$, where the sound speed $c_s \ll c$. So we can safely neglect the pressure term in Equation 8.20, and Equation 8.19 tells us that the density follows $\rho(t) \propto \mathcal{R}^{-3}(t)$. For radiation, and particles moving almost at the speed of light, pressure is important: $p \approx \rho c^2/3$, where ρ is the energy density divided by c^2. We have $\rho(t) \propto \mathcal{R}^{-4}(t)$ from Equation 8.19. For any mixture of matter and radiation, the term $\rho + 3p/c^2$ must be positive, so *the expansion always slows down*. While matter and radiation account for most of the energy, $\rho(t)\mathcal{R}^2(t)$ decreases as $\mathcal{R}(t)$ grows. Thus, in a closed Universe with $k = 1$, the right-hand side of Equation 8.17 becomes negative at large $\mathcal{R}$. But $\dot{\mathcal{R}}^2$ cannot be negative, so the distance between galaxies cannot grow forever; $\mathcal{R}(t)$ attains some maximum before shrinking again. If the Universe is open with $k \geq 0$, the expansion will never halt.

General Relativity also allows a *vacuum energy*, with constant density $\rho_{\text{VAC}} = \Lambda/(8\pi G)$. Since ρ_{VAC} does not change, the rightmost term of Equation 8.19 must also be zero, so the pressure $p_{\text{VAC}} = -\Lambda c^2/(8\pi G)$. Instead of a pressure pushing inwards on the contents of our sphere, this term represents a tension pulling outwards. The vacuum energy contributes a positive term to the right-hand side of Equation 8.20, speeding up the expansion. If the Universe expands far enough, the vacuum energy must become the largest term, and $\mathcal{R}(t)$ then grows exponentially.

Problem 8.9 By substituting into Equation 8.20, show that, when vacuum energy dominates the expansion, we have $\mathcal{R}(t) \propto \exp(t\sqrt{\Lambda/3})$.

There are reasons to believe that very early, at $t \lesssim 10^{-32}$ s, ρ_{VAC} might have been much larger than the density of matter or radiation. During this period, $\mathcal{R}(t)$ *inflated*, growing exponentially by a factor $\sim e^{70} \approx 10^{30}$. The almost uniform cosmos that we now observe would have resulted from the expansion of a tiny near-homogeneous region. Because this patch was so small, the curvature of space within it would have been negligible; hence devotees of inflation expect our present Universe to be nearly flat, with $k = 0$. We will see in Section 8.4 that the measured temperature fluctuations in the cosmic background radiation imply that this is in fact so.

Further reading: for a discussion of the physics behind inflation, see Chapter 11 of Ryden's *Introduction to Cosmology*.

In the borderline case when space is flat and $k = 0$, Equation 8.17 requires that the density ρ is equal to the critical value

$$\rho(t) = \rho_{\text{crit}}(t) \equiv \frac{3H^2(t)}{8\pi G}. \tag{8.21}$$

At the present day, the critical density $\rho_{\text{crit}}(t_0) = 3.3 \times 10^{11} h^2 \mathcal{M}_\odot \, \text{Mpc}^{-3}$. We can measure the mass content of the Universe as a fraction of this critical density, defining the *density parameter* $\Omega(t)$ as

$$\Omega(t) \equiv \rho(t)/\rho_{\text{crit}}(t), \tag{8.22}$$

and writing Ω_0 for its present-day value. Equation 8.17 then becomes

$$H^2(t)[1 - \Omega(t)] = -kc^2/\mathcal{R}^2(t). \tag{8.23}$$

If the Universe is closed, with $k = 1$, then $\Omega(t) > 1$ and the density always exceeds the critical value, whereas if $k = -1$, we always have $\Omega(t) < 1$. If the density is now equal to the critical value, then Equation 8.23 tells us that $\Omega(t)$ must be unity at all times. The present value $\Omega(t_0)$ is often written as Ω_{tot} or (especially when $\Lambda = 0$) as Ω_0.

We already saw in Section 1.5 that normal (baryonic) matter, largely neutrons and protons, makes up only 4%–5% of the critical density. Including the dark matter, in Section 8.4 below we arrive at only $(0.2-0.3)\rho_{\text{crit}}$. Radiation contributes hardly anything. So space can be flat only if there is a nonzero vacuum energy. This is often called the *dark energy*. It probably has a different physical origin from

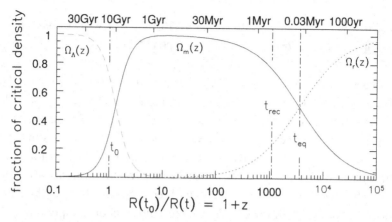

Fig. 8.7. For the benchmark cosmology with $\Omega_r = 8.42 \times 10^{-5}$, $\Omega_m = 0.3$, and $\Omega_\Lambda = 0.7$, the fraction of the critical density contributed at each time by radiation (dotted line), matter (solid), and the dark energy (dashed). For this model, matter–radiation equality occurs at $z_{eq} = 3570$, $t_{eq} = 0.05$ Myr; recombination is complete at $z_{rec} = 1100$, $t_{rec} = 0.35$ Myr. The present age $t_0 = 13.5$ Gyr.

the vacuum energy causing inflation at early times. We describe the contributions of matter and dark energy at the present time by writing

$$\Omega_m \equiv \rho_m(t_0)/\rho_{crit}(t_0) \text{ and } \Omega_\Lambda \equiv \rho_{VAC}/\rho_{crit}(t_0). \tag{8.24}$$

In the *benchmark* model illustrated in Figure 8.7, $\Omega_m = 0.3$, $\Omega_\Lambda = 0.7$, and $H_0 = 70 \text{ km s}^{-1} \text{ Mpc}^{-1}$; thus space is flat. The baryon density $\Omega_B = 0.045$, with *cold dark matter* (see Section 8.5) making up the remainder of Ω_m. A warning: although Ω_m, Ω_Λ, and similar quantities do not carry the subscript 0, they always refer to the present day.

After inflation ended, the Universe was *radiation-dominated*. It was so extremely hot that its energy was almost entirely due to radiation and *relativistic particles*, moving so close to light speed that their energy, momentum, and pressure are related in the same way as for photons. From Equation 8.19, we know that the energy density $\rho_r c^2$ is proportional to $\mathcal{R}^{-4}(t)$. (If few photons are created or destroyed, the number per unit volume is proportional to $1/\mathcal{R}^3(t)$, while by Equation 8.12 the energy of each falls as $1/\mathcal{R}(t)$.) For a blackbody spectrum, the temperature $T \propto 1/\mathcal{R}(t)$: recall Problem 1.17.

As expansion proceeded the density of matter followed $\rho_m(t) \propto \mathcal{R}^{-3}(t)$, falling more slowly than the radiation density. At the time t_{eq} of *matter–radiation equality*, about a million years after the Big Bang, its energy density $\rho_m(t)c^2$ exceeded that in radiation, and the Universe became *matter-dominated*. Figure 8.7 shows that, in the benchmark model, the matter density has only recently dropped below that of the dark energy.

Problem 8.10 The cosmic background radiation is now a blackbody of temperature $T = 2.73$ K: show that its energy density $\rho_r c^2 = 4.2 \times 10^{-13}$ erg cm^{-3}, so $\Omega_r = 4.1 \times 10^{-5} h^{-2}$. From Equation 1.30, the matter density $\rho_m = 1.9 \times 10^{-29} \Omega_m h^2$ g cm^{-3}. Show that the energy density $\rho_m c^2$ was equal to that in radiation at redshift $z_{eq} \approx 40\,000 \Omega_m h^2$. This is well before the redshift of *recombination*, when the gas becomes largely neutral. If the neutrinos ν_e, ν_μ, and ν_τ have masses $m_\nu \ll k_B T_{eq}/c^2$, where T_{eq} is the temperature at the time t_{eq}, then at earlier times they are relativistic. The energy density of 'radiation' is increased by a factor of 1.68, and equalization is delayed until $z_{eq} \approx 24\,000 \Omega_m h^2$, or $z \approx 3600$ in the benchmark model.

To measure the cosmic expansion relative to the present day, we define the dimensionless *scale factor* $a(t) \equiv \mathcal{R}(t)/\mathcal{R}(t_0)$. We use Equation 8.23 to rewrite $k/\mathcal{R}^2(t_0)$ in terms of the present-day quantities H_0 and $\Omega_{tot} = \Omega(t_0)$. Then Equation 8.17 becomes

$$\frac{kc^2}{\mathcal{R}^2(t_0)} = H_0^2(1 - \Omega_{tot}) = a^2(t)\left[H^2(t) - \frac{8\pi G}{3}\rho(t)\right]. \qquad (8.25)$$

Adding up the contributions to $\rho(t)$ and recalling that $1 + z = 1/a(t)$, we have

$$H^2(t) = H_0^2[\Omega_r(1 + z)^4 + \Omega_m(1 + z)^3 + (1 - \Omega_{tot})(1 + z)^2 + \Omega_\Lambda]. \qquad (8.26)$$

Problem 8.11 Blackbody radiation and relativistic particles provide most of the energy density at $t \ll t_{eq}$. Show that Equation 8.26 then implies that $H(t) = \dot{a}/a \propto a^{-2}$, so $\dot{a} \propto 1/a(t)$; hence $\mathcal{R}(t) \propto t^{1/2}$, and $H(t) = 1/(2t)$. Early on, the leftmost term of Equation 8.25 is tiny, so $H^2(t) \approx 8\pi G\rho(t)/3$: show that the temperature $T(t)$ is given by Equation 1.38.

Problem 8.12 Use Equation 8.20 to show that, in the benchmark model, cosmic expansion has speeded up since $\rho_m(t) = 2\rho_{VAC}$, at redshift $z = 0.67$. According to this model, we live during that small fraction of cosmic history in which normal gravity and the cosmological constant have roughly equal influence on the expansion. At $z \geq 1$, the Λ term hardly affects the expansion, whereas when $\mathcal{R}(t) \geq 2\mathcal{R}(t_0)$, the gravitational pull of matter will become irrelevant.

Problem 8.13 Show that, when cool matter accounts for most of the energy density, and the Universe is flat with $k = 0$, we have

$$\dot{a} \propto a^{-1/2}, \quad \text{and} \quad a(t) \propto t^{2/3}. \qquad (8.27)$$

Show that, if $k = 0$, then Equation 8.27 holds when $z_{eq} \gg 1 + z \gg (\Omega_\Lambda/\Omega_m)^{1/3}$, whereas at very late times $a(t) \propto \exp(t\sqrt{\Lambda/3})$. If the curvature is negative, with $\Omega_{tot} = \Omega_r + \Omega_m + \Omega_\Lambda < 1$, the third term of Equation 8.26 exceeds the second when $1 + z < (1 - \Omega_{tot})/\Omega_m$. Expansion then proceeds almost at a constant speed, with $a(t) \propto t$; it is barely slowed by the matter, and not accelerated by dark energy until $(1 + z)^2 < \Omega_\Lambda/(1 - \Omega_{tot})$.

Equation 8.27 is a good description of the expansion over much of cosmic history. In particular, most of the structure of galaxy clusters and voids that we see today developed after the Universe became matter-dominated, but before the dark energy became important.

Problem 8.14 Even if the cosmos has infinite volume, we can observe only a finite part of it because light travels at a finite speed. From Equation 8.11, light reaching us at time t has travelled no further than the distance σ_H given by

$$c \int_0^t \frac{dt'}{\mathcal{R}(t')} \equiv \int_0^{\sigma_H} \frac{d\sigma}{\sqrt{1 - k\sigma^2}}, \tag{8.28}$$

so we cannot see beyond our *horizon*, at comoving radius σ_H. Explain why only points within $\sigma_H(t)$ of each other can exchange signals or particles before time t. Use Equation 8.27 to show that, while the Universe was matter-dominated, $\mathcal{R}(t)\sigma_H \approx 3ct$; at the time t_{rec} of recombination $\mathcal{R}(t_{rec})\sigma_H \approx 0.43$ Mpc in the benchmark model of Figure 8.7. We will see below that a sphere of matter with this diameter would cover only about $2°$ on the sky. So it is very surprising that the cosmic microwave background has almost the same spectrum across the whole sky.

An inflationary cosmology can explain this *horizon problem*. When $\mathcal{R}(t) \propto \exp(t\sqrt{\Lambda/3})$ after some initial time t_i, show that when $t \gg t_i$

$$\int_{t_i}^t \frac{dt'}{\mathcal{R}(t')} \approx \sqrt{\frac{3}{\Lambda}} \frac{1}{\mathcal{R}(t_i)} \quad \text{so} \quad \mathcal{R}(t)\sigma_H \approx c\sqrt{\frac{3}{\Lambda}} \frac{\mathcal{R}(t_f)}{\mathcal{R}(t_i)}. \tag{8.29}$$

For $t_i \sim 10^{-35}$ s and $\sqrt{3/\Lambda} \sim t_i$ as in most inflationary models, show that, when inflation ends at $t_f \approx 70t_i$, the horizon distance $\mathcal{R}(t_f)\sigma_H \approx ct_f \times e^{70}/70 \approx 3.6 \times 10^{28} ct_f$ or ~ 7 km. After inflation, the Universe is radiation-dominated, so $a(t) \propto t^{1/2}$. Show that, by the time of matter–radiation equality at ~ 0.35 Myr, this distance has grown by $z(t_{eq})/z(t_f)$ to about 30 Gpc. Inflation expands a region within which light signals could be exchanged, until it is much larger than the entire Universe that we can see today.

Further reading: on cosmic horizons, see Chapter 10 of the book by Padmanabhan.

8.2.1 How old is that galaxy? Lookback times and ages

Because light travels at a finite speed, we see a younger cosmos as we look toward more distant galaxies at higher redshifts. As we observe a galaxy at redshift z, at what time t_e did those photons leave on their journey toward us? Equation 8.13 tells us that they arrive today at time t_0 with redshift $1 + z = \mathcal{R}(t_0)/\mathcal{R}(t_e)$. Light from another galaxy at a smaller redshift $z - \Delta z$ must have been emitted slightly later by a small interval Δt_e, when $\mathcal{R}(t_e)$ was larger by $\Delta \mathcal{R} = \dot{\mathcal{R}}(t_e)\Delta t_e$. Since

$$\Delta z = -\Delta \mathcal{R} \frac{\mathcal{R}(t_0)}{\mathcal{R}^2(t_e)}, \quad \text{we have} \quad \Delta t_e = -\frac{1}{H(t_e)} \frac{\Delta z}{1 + z}, \tag{8.30}$$

where $H(t) \equiv \dot{\mathcal{R}}(t)/\mathcal{R}(t)$ is the Hubble parameter. Integrating the second relation gives us the *lookback time* $t_0 - t_e$:

$$t_0 - t_e = \int_0^z \frac{1}{H(t)} \frac{dz'}{1 + z'}. \tag{8.31}$$

At redshifts $z \lesssim z_{eq}$, when the density of radiation is no longer important, we can calculate this time simply for two special cases: a flat Universe with $k = 0$, and one with matter only, so that $\Lambda = 0$. When the densities of dark energy and of matter must add to give $\Omega_m + \Omega_\Lambda = 1$,

$$t_e = \int_z^\infty \frac{1}{H(z')} \frac{dz'}{1 + z'} = \int_z^\infty \frac{dz'}{H_0(1 + z')[\Omega_m(1 + z')^3 + \Omega_\Lambda]^{1/2}}, \tag{8.32}$$

which integrates to

$$t_e = \frac{2}{3H_0\sqrt{\Omega_\Lambda}} \ln\left(\frac{1 + \cos\theta}{\sin\theta}\right), \quad \text{where} \quad \tan^2\theta \equiv \frac{(1 - \Omega_\Lambda)}{\Omega_\Lambda}(1 + z)^3. \tag{8.33a}$$

Even if $k \neq 0$, this formula is accurate to within a few percent if we replace Ω_Λ by $0.3\Omega_\Lambda + 0.7(1 - \Omega_m)$, so long as this quantity is positive. For $\Lambda = 0$, we have

$$t_e = \int_z^\infty \frac{1}{H_0\sqrt{1 + \Omega_0 z}} \frac{dz'}{(1 + z')^2},$$

where $\Omega_0 = \Omega_m$ is the present-day value of the density parameter $\Omega(t)$. We have exact expressions when $\Omega_0 = 1$ and the density is equal to the critical value, or in an empty Universe with $\Omega_0 = 0$:

$$t_0 - t_e = \begin{cases} \dfrac{1}{H_0} \displaystyle\int_0^z \dfrac{dz'}{(1 + z')^{5/2}} = \dfrac{2}{3H_0}\left[1 - \dfrac{1}{(1 + z)^{3/2}}\right] & \text{if } \Omega_0 = 1, \\[4mm] \dfrac{1}{H_0} \displaystyle\int_0^z \dfrac{dz'}{(1 + z')^2} = \dfrac{1}{H_0} \dfrac{z}{(1 + z)} & \text{for } \Omega_0 = 0. \end{cases}$$

$$\tag{8.33b}$$

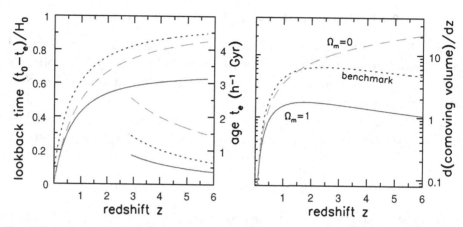

Fig. 8.8. Left, lookback time $t_0 - t_e$ to an object seen with redshift z, in units of the Hubble time t_H; the right-hand scale and right curve-segments show t_e itself. Right, comoving volume per unit redshift, $\mathcal{R}^{-3}(t)d\mathcal{V}/dz$, in units $(c/H_0)^3$. The solid line is for $\Lambda = 0$, $\Omega_m = 1$; the dashed line shows $\Lambda = 0 = \Omega_m$; the dotted line is for the benchmark model.

Early in the matter-dominated era, the curvature term that we left out of Equation 8.32 will be tiny and the Ω_Λ term is also small, so

$$t_e \approx \frac{2}{3H_0\sqrt{\Omega_m}} \frac{1}{(1+z)^{3/2}} \quad \text{when} \quad \frac{\Omega_m}{\Omega_r} \gg 1+z \gg \left(\frac{\Omega_\Lambda}{\Omega_m}\right)^{1/3} \text{ and } \frac{1}{\Omega_m}. \quad (8.34)$$

Problem 8.15 Nearby, Hubble's law tells us that an object at distance d recedes with speed $V_r = cz = H_0d$. The lookback time is just the time that light takes to cover this distance, so $t_0 - t_e = d/c = z/H_0$. Show that both parts of Equation 8.33 agree with this in the limit $z \ll 1$.

Problem 8.16 Show that, if $\Omega = 1$ in a matter-dominated Universe, then $H(t) = 2/(3t)$, so that the time t_0 since the Big Bang is two-thirds of our simple estimate $t_H = 1/H_0$ in Equation 1.28. From Equation 8.21, show that the density falls as $1/t^2$.

Equation 8.33 also tells us t_0, the present age of the Universe. When $\Lambda = 0$, this is always less than our simple estimate $t_H = 1/H_0$ in Equation 1.28. For an empty Universe with $\Lambda = 0$, $t_0 = H_0^{-1} = t_H$, whereas for $\Omega_0 = 1$ the age is only two-thirds as large. For a fixed value of H_0, Figure 8.8 shows that the lookback time in gigayears to a given redshift z is longer in the case $\Omega_0 = 0$ than for $\Omega_0 = 1$, but the galaxies are older: the time t_e is longer. When $\Lambda > 0$ and $k = 0$ so the Universe is flat, t_0 exceeds the Hubble time $t_H = 1/H_0$ if the matter density is low, with $\Omega_m \lesssim 0.2$. For example, $\Omega_\Lambda = 0.9$, $\Omega_m = 0.1$ yields $t_0 = 1.3t_H$. In the benchmark model, the age is $0.964t_H$ or 13.5 Gyr.

Problem 8.17 Galaxies have now been observed at redshifts $z \gtrsim 5$: how old are they? Show that, at $z = 5$, $t_e = 0.17t_0 = 1.6h^{-1}$ Gyr if the Universe is nearly empty so $\Omega_0 \approx 0$, but only $0.07t_0 = 0.44h^{-1}$ Gyr if $\Omega_0 = 1$, while for the benchmark model $t_e = 0.8h^{-1}$ Gyr. The time available to assemble the earliest galaxies is very short! Show that, at redshift $z = 3$, the time t_e is $2.4h^{-1}$ Gyr if $\Omega_0 = 0$, $0.82h^{-1}$ Gyr for $\Omega_0 = 1$, and $0.07t_H = 1.5h^{-1}$ Gyr for the benchmark model.

8.3 Observing the earliest galaxies

Our view of very distant objects is complicated by the cosmic expansion, and by the curvature of the space through which their light must travel. Because of the expansion, distant galaxies look bigger than we would expect: at redshifts z larger than $1/\Omega_m$ a given object covers more of the sky when it is *further* from us. But the starlight of these big galaxies is rapidly dimmed as it spreads out in the expanding cosmos. Their ultraviolet and visible light is also shifted into the infrared, where strong emission lines from the Earth's atmosphere (see Figure 1.15) make it more difficult to observe from the ground. To estimate how densely galaxies are scattered through space, we need to take account of both cosmic expansion and the curvature of space.

8.3.1 Luminosity, size, and surface brightness

For nearby objects, the apparent brightness F of a star or galaxy at distance d is related to its luminosity L by

$$F = L/(4\pi d^2). \tag{1.1}$$

In an expanding Universe, this no longer holds. Objects appear dimmer because the cosmic expansion saps photons of their energy, and the area of the sphere over which the light rays must spread is expanding. This is just as well, because in a homogeneous static cosmos the light of distant stars would accumulate without limit, making the sky bright everywhere: this is *Olbers' paradox*.

Problem 8.18 Suppose that galaxies of luminosity L are spread uniformly through space. Show that the number $N(>F)$ that you observe to have apparent brightness larger than F varies as $F^{-3/2}$, while the number between F and $F + \Delta F$ varies as $N(F) \propto F^{-5/2}$. Explain why $N(F) \propto F^{-5/2}$ even when the galaxies have a range of luminosities, as long as they are spread uniformly and the luminosity function is the same everywhere.

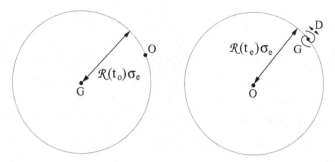

Fig. 8.9. Left: as observer O receives the light of galaxy G, its energy is spread over a sphere of area $4\pi \mathcal{R}^2(t_0)\sigma_e^2$. Right: as seen by observer O, galaxy G with diameter D covers $D/\mathcal{R}(t_e)\sigma_e$ radians on the sky.

Show that the total light from all galaxies that appear brighter than $F_\star$ is proportional to $\int_{F_\star}^{\infty} F N(F)\mathrm{d}F$. Whenever $N(F)$ rises faster than $1/F$ as $F \to 0$, this grows without limit as $F_\star$ decreases.

To see how we should modify Equation 1.1, let σ_e be the comoving distance or 'area radius' of a galaxy G that we observe today with redshift z. The apparent brightness of a distant galaxy is diminished by one power of $1 + z$ because each photon carries less energy, and by another because those photons arrive at a slower rate. The left panel of Figure 8.9 shows that G's light is now spread over a sphere with area $4\pi \mathcal{R}^2(t_0)\sigma_e^2$. The flux of energy F that we receive, in W m^{-2}, is related to the total or *bolometric* luminosity L by

$$F = \frac{L}{4\pi \mathcal{R}^2(t_0)\sigma_e^2(1+z)^2} \equiv \frac{L}{4\pi d_L^2}, \quad \text{where} \quad d_L = (1+z)\mathcal{R}(t_0)\sigma_e. \quad (8.35)$$

We call d_L the *luminosity distance* of the galaxy.

Similarly, we must modify Equation 1.2, telling us how large an object will appear on the sky. Suppose that galaxy G is D kpc across. The right panel of Figure 8.9 shows that, at time t_e, it covered a fraction $D/[2\pi \mathcal{R}(t_e)\sigma_e]$ of the circumference of the sky. So it extends over an angle

$$\alpha \text{ (in radians)} = \frac{D}{\mathcal{R}(t_e)\sigma_e} \equiv \frac{D}{d_A}, \quad \text{so } d_A = \mathcal{R}(t_e)\sigma_e = \frac{\mathcal{R}(t_0)\sigma_e}{1+z}. \quad (8.36)$$

Thus the *angular-size distance* d_A is less than the luminosity distance d_L by a factor $(1 + z)^2$. We must use the distance d_A when calculating the gravitational bending of light using Equations 7.14 and 7.25. A warning: some authors refer to $\mathcal{R}(t_0)\sigma_e$ as the angular-size distance; this is $1 + z$ times larger than our d_A.

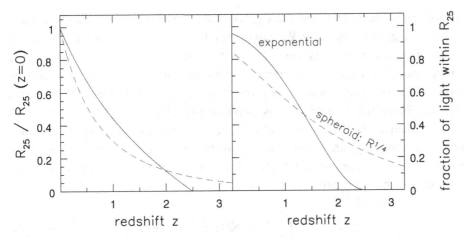

Fig. 8.10. Left, the measured radii R_{25} of two model galaxies: an exponential disk (solid curve) and an $R^{1/4}$ spheroid (Equation 6.1: dashed curve). At $z = 0$ we take $R_{25} = 5h_R$ for the disk and $R_{25} = 4R_e$ for the spheroid, then for each redshift we plot the radius where the measured surface brightness reaches this level. Right, the fraction of the total light within this radius. At small redshifts, the exponential disk shrinks less than the spheroid, but when $z \gtrsim 1$ it is more strongly affected.

Problem 8.19 Show that a galaxy of known luminosity L and diameter D at redshift z will appear larger by a factor of $(1 + z)^2$ than one would expect by using Equation 1.1 to calculate its distance from its measured apparent brightness, and then finding the apparent diameter from Equation 1.2.

The surface brightness $I(\mathbf{x})$ of a galaxy is the flux we receive from each square arcsecond as it appears on the sky. If we integrate over all wavelengths to measure the bolometric surface brightness, then, instead of Equation 1.23, we have

$$I(\mathbf{x}) = \frac{F}{\alpha^2} = \frac{L/(4\pi d_L^2)}{D^2/d_A^2} = \frac{L}{4\pi D^2}\left(\frac{d_A}{d_L}\right)^2 = \frac{L}{4\pi D^2}\frac{1}{(1+z)^4}. \qquad (8.37)$$

The bolometric surface brightness of a nearby galaxy, with $z \ll 1$, does not diminish with distance. If we neglect changes caused by the birth of new stars or by stellar aging, then a given isophote always corresponds to a fixed radius in the galaxy, and encloses the same fraction of the galaxy's light, regardless of the distance. But at redshifts beyond a few tenths $I(\mathbf{x})$ drops rapidly, making photometry increasingly difficult and expensive.

Figure 8.10 shows how an isophote defined by a fixed surface brightness shrinks, and the amount of light within it decreases. We must correct for this missing light when comparing the luminosities of distant galaxies with those of nearby systems. The correction would depend only on the redshift z if the luminosity from each square parsec of the galaxy had remained unchanged. But we will see in Section 9.4 that many galaxies at $z \gtrsim 0.5$ were significantly brighter than their counterparts today. We can interpret the brightening with models to tell us how the light of new stars should fade through time; Figure 6.18 illustrates a simple model in which all the galaxy's stars are born at once. To turn this into a prediction about luminosity at a given redshift, we need Equation 8.31, which involves the expansion rate $H(z)$ as well as the redshift.

To calculate the distances d_L and d_A to a galaxy G seen at redshift z, we must know its comoving 'area' distance σ_e. We start by asking how far its light must go to reach us. According to Equation 8.5, at time t the distance from the origin to a point with area radius σ is

$$R(t)\chi \equiv \int_0^\sigma ds = R(t) \int_0^\sigma \frac{d\sigma'}{\sqrt{1 - k\sigma'^2}}. \tag{8.38}$$

This defines the comoving 'distance radius' χ, with

$$\chi(\sigma) = \begin{cases} \arcsin(\sigma) & \text{for } k = 1, \\ \sigma & \text{for } k = 0, \\ \operatorname{arcsinh}(\sigma) = \ln(\sigma + \sqrt{1 + \sigma^2}) & \text{for } k = -1. \end{cases} \tag{8.39}$$

The two distances are the same if the Universe is flat, with $k = 0$. In a closed Universe, with $k = 1$, the area radius σ is smaller than the distance radius χ, just as the length of a circle of latitude on the Earth is less than 2π times the distance to the pole. Conversely, if $k = -1$, and the Universe is open, we have $\sigma > \chi$: the perimeter of a circle, or the area of a sphere, is larger than we would expect, compared with the distance from the center to its boundary.

Moving at light speed, in time Δt galaxy G's light travels a distance $c\,\Delta t = R(t)\Delta\chi$ toward us. Using Equation 8.30 to relate Δt to the redshift, the total distance is

$$\chi_e = \int_{t_e}^{t_0} \frac{c\,dt}{R(t)} = \int_0^z \frac{c}{R(t_0)} \frac{dz'}{H(t)}. \tag{8.40}$$

This integral can be done exactly when $\Lambda = 0$:

$$\frac{R(t_0)H_0}{c}\chi_e = \begin{cases} \displaystyle\int_0^z \frac{dz'}{(1 + z')^{3/2}} = 2\left(1 - \frac{1}{\sqrt{1 + z}}\right) & \text{for } \Omega_0 = 1, \\ \displaystyle\int_0^z \frac{dz'}{(1 + z')} = \ln(1 + z) & \text{for } \Omega_0 = 0. \end{cases} \tag{8.41}$$

Now, substituting for χ_e in Equation 8.39 gives the comoving radius σ_e:

$$R(t_0)\sigma_e = \begin{cases} \dfrac{2c}{H_0}\left[1 - \dfrac{1}{\sqrt{1 + z}}\right] & \text{for } \Omega_0 = 1, \\ \dfrac{c}{H_0}\dfrac{z(1 + z/2)}{1 + z} & \text{for } \Omega_0 = 0. \end{cases} \tag{8.42a}$$

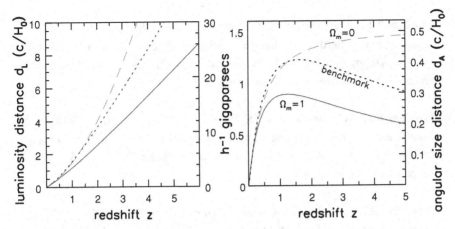

Fig. 8.11. Left, the luminosity distance d_L; and right, the angular-size distance d_A, for a source at redshift z; both are in units of $c/H_0 \approx 3h^{-1}$ Gpc. The solid curve ($\Omega_0 = 1$) and the dashed line ($\Omega_0 = 0$) are for $\Lambda = 0$; the dotted line is for the benchmark model.

An excessive quantity of algebra yields the more general *Mattig formula*

$$\mathcal{R}(t_0)\sigma_e = \frac{c}{H_0} \frac{2}{\Omega_0^2(1+z)} [\Omega_0 z + (\Omega_0 - 2)(\sqrt{1 + \Omega_0 z} - 1)]. \qquad (8.42b)$$

Notice that at large redshift we have $\mathcal{R}(t_0)\sigma_e \to 2c/(H_0\Omega_0)$. When the density is small, with $\Omega_0 \ll 1$, a more convenient form is

$$\mathcal{R}(t_0)\sigma_e = \frac{c}{H_0} \frac{z}{1+z} \frac{1 + \sqrt{1 + \Omega_0 z} + z}{1 + \sqrt{1 + \Omega_0 z} + \Omega_0 z/2}. \qquad (8.42c)$$

For the flat Universe with $\Omega_{tot} = 1$ there is no general solution, but for $\Omega_m \gtrsim 0.1$

$$\mathcal{R}(t_0)\sigma_e \approx 2c / \left(H_0 z \, \Omega_m^{0.4}\right) \text{ as } z \to \infty. \qquad (8.43)$$

Figure 8.11 shows d_L and d_A for some special cases. Nearby, both grow linearly with the redshift, according to Problem 8.21. But the luminosity distance d_L always increases faster than this linear relation, more strongly so when the density Ω_{tot} is low: this rescues us from Olbers' paradox. The angular-size distance d_A grows more slowly; it reaches a maximum at $z \sim 1/\Omega_0$ (or $z \sim 1/\Omega_m$ in the flat model with $\Lambda = 0$), and then declines. In our benchmark model, at redshifts $z \gtrsim 1.5$ a source looks *larger* as it is moved further away.

Problem 8.20 Show from Equation 8.39 that $\sigma \approx \chi$ when both are much less than unity, and that both d_A and d_L are then much less than $c/H_0 \approx 2990h^{-1}$ Mpc. (We do not have to worry about the curvature of space when measuring nearby galaxies, just as we are not concerned with the Earth's curvature when crossing a town.)

Problem 8.21 Use Equation 8.40 to show that, when the redshift $z \ll 1$, both d_L and d_A tend to the distance that we would normally calculate from the redshift $z: d_A = \mathcal{R}(t_e)\sigma_e \to c/H_0 \approx d_L$. From Equation 8.42b, show that, when $\Lambda = 0$ and the redshift is large, $d_A \to 2c/(H_0 z \Omega_m)$.

Problem 8.22 With $M_{bol,\odot} = 4.75$, $L_\star = 2 \times 10^{10} L_\odot$, and $H_0 = 70$ km s^{-1} Mpc^{-1}, show that, if $\Lambda = 0$, an $L_\star$ galaxy at redshift $z = 3$ would have $m_{bol} = 25.2$ if $\Omega_0 = 1$, or $m_{bol} = 26.6$ for $\Omega_0 = 0$, whereas $m_{bol} = 26.1$ in the benchmark cosmology.

Problem 8.23 Show that, in a low-density Universe with $\Omega_0 = 0$, $d_A \to c/(2H_0)$ at large redshifts. Use Figure 8.11 to show that the angle subtended by an object 10 kpc across is always at least $1.4h$ arcsec for $\Omega_0 = 0$, and $\gtrsim 3h$ arcsec if $\Omega_0 = 1$. (The star-forming regions of Figure 9.14 have radii of 0.1–0.2 arcseconds; so these bright patches are $\lesssim 1$ kpc across.)

8.3.2 Galaxy spectra and photometric redshifts

In practice we do not measure bolometric luminosities; we measure the apparent brightness in a specific band of wavelength or frequency. Cosmic expansion changes a galaxy's color: when we observe in a particular waveband, the light we now see was radiated in a bluer part of the spectrum. Suppose that, at time t_e, a galaxy that we see at redshift z has luminosity $L_\lambda(\lambda, t_e)\Delta\lambda$ in the wavelength range from λ to $\lambda + \Delta\lambda$. The apparent brightness F_{BP} in a bandpass BP that transmits all the light between wavelengths λ_1 and λ_2 is given by Equation 8.35, if we take the luminosity L to be the energy emitted at wavelengths that will pass through our bandpass when we receive it. So

$$
\begin{aligned}
F_{BP} &= \frac{1}{4\pi d_L^2} \int_{\lambda_1/(1+z)}^{\lambda_2/(1+z)} L_\lambda(\lambda, t_e) d\lambda \\
&= \frac{1}{4\pi d_L^2} \frac{1}{1+z} \int_{\lambda_1}^{\lambda_2} L_\lambda[\lambda/(1+z), t_e] d\lambda.
\end{aligned}
\tag{8.44}
$$

For a nearby galaxy, Equation 1.15 tells us how to calculate its apparent magnitude at any distance. But at large redshift we must add two more terms: the apparent magnitude m_{BP} in this bandpass is

$$
m_{BP} = M_{BP} + 5 \log_{10}\left(\frac{d_L}{10\,\text{pc}}\right) + k_{BP}(z) + e_{BP}(z).
\tag{8.45}
$$

The absolute magnitude M_{BP} is the apparent magnitude that the galaxy would have if seen from a distance of 10 pc, emitting as it does at the present time t_0.

The term $k(z)$, historically known as the *k correction*, represents the effect of shifting the galaxy's light in wavelength. The *evolutionary* term $e(z)$, which we will discuss in Section 9.3, allows for changes in the galaxy's luminosity between the time that its light was emitted and the present day.

From Equation 8.44, we have

$$k_{\mathrm{BP}}(z) \equiv 2.5 \log_{10}(1 + z) - 2.5 \log_{10} \left\{ \frac{\int_{\lambda_1}^{\lambda_2} L_\lambda[\lambda/(1 + z), t_0]\mathrm{d}\lambda}{\int_{\lambda_1}^{\lambda_2} L_\lambda(\lambda, t_0)\mathrm{d}\lambda} \right\}, \qquad (8.46)$$

where $L_\lambda(\lambda, t_0)$ is the present-day spectrum. Of course, we cannot measure the spectrum of a distant galaxy as it is today; that light is still on its way to us. But we can calculate how a present-day galaxy would appear if we observed it at a redshift z. Figure 8.12 shows that, if we observe in the B band at 4400 Å, an elliptical galaxy will dim rapidly as its redshift increases, since our passband moves into the ultraviolet region where its stars emit little light. A starburst galaxy will fade much less, because its hot young stars are bright at short wavelengths. At $z \lesssim 1$ both galaxies fade much less in the red light of the I band at 8000 Å than in the B band. So the $B - I$ color becomes progressively redder at higher redshift. At redshifts $z \gtrsim 2$, the light of both galaxies has largely moved into the infrared region.

Problem 8.24 If a galaxy emits a spectrum $L_\nu \propto \nu^{-\alpha}$, show that $L_\lambda \propto \lambda^{\alpha-2}$, and that $k(z) = (\alpha - 1) \times 2.5 \log_{10}(1+z)$. The k correction is zero if νL_ν is nearly constant so that $\alpha \approx 1$, as it is for many quasars (see Section 9.1). When the spectrum declines more rapidly than $L_\nu \propto \nu^{-1}$ toward high frequencies, $k(z) > 0$, and the object appears dimmer.

A *photometric redshift* is an estimate of a galaxy's redshift made by comparing its apparent brightness in several bandpasses with that predicted by a diagram like Figure 8.12. For example, an elliptical galaxy at $z \sim 0.5$ has already become very red in the $B - I$ color, but it is less so in $V - I$. At $z \sim 1$, it is fading rapidly in the I band, so the $I - H$ color starts to redden. With 17 filters at wavelengths from 3640 Å to 9140 Å, the COMBO-17 team could estimate redshifts to $\Delta z \approx 0.05$ over the range $0.2 < z < 1.2$. The most spectacular use of photometric redshifts has been to find galaxies at $z > 3$. These *Lyman break galaxies* almost disappear at wavelengths less than $912(1 + z)$ Å, where intergalactic atoms of neutral hydrogen absorb nearly all their light (see Section 9.4).

Problem 8.25 Explain why, if we base a galaxy survey on images in the B band, then at $z \gtrsim 0.5$ we will fail to include many of the systems with red spectra similar to present-day ellipticals.

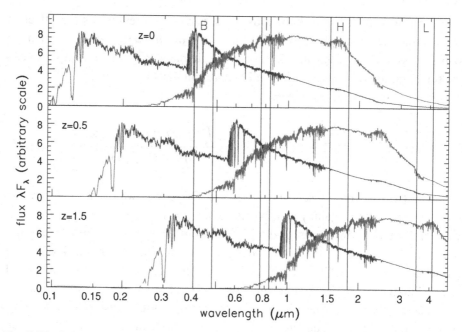

Fig. 8.12. Spectra of two model galaxies: the stars of the bluer system formed in a single burst 100 Myr ago, while those of the redder galaxy are all 4 Gyr old. Vertical lines show B, I, H, and L wavelength regions from Figure 1.7. The top panel shows the emitted light of each galaxy, while the lower panels illustrate how the redshift affects the relative brightness in each bandpass. The energy in each wavelength region is proportional to the area under the curve – S. Charlot.

8.3.3 How many galaxies? Space densities

To trace the formation of galaxies through cosmic history, we must take account of the expansion in counting the number within any given volume. The number of objects that we will see between redshifts z and $z + \Delta z$ is proportional to the corresponding volume of space $\Delta \mathcal{V}$. This is just the product of the area $\mathcal{A}(\sigma_e, t_e) = 4\pi \mathcal{R}^2(t_e)\sigma_e^2$ of the sphere containing the galaxy at the time its light was emitted and the distance $c|\Delta t_e|$ that the light travels toward us in the time corresponding to this interval in redshift. From Equations 8.30 and 8.31, we have

$$\frac{\mathcal{A}c|\Delta t_e|}{\Delta z} \approx \frac{d\mathcal{V}}{dz} = \frac{4\pi c \mathcal{R}^2(t_0)\sigma_e^2}{H(z)(1+z)^3}, \tag{8.47}$$

where we replaced $\mathcal{R}(t_e)$ by $\mathcal{R}(t_0)/(1+z)$ in the last step.

The volume $\Delta \mathcal{V}$ at redshift z will expand to fill a volume $\Delta \mathcal{V}(1+z)^3$ by the present day: we refer to $\Delta \mathcal{V}(1+z)^3$ as the *comoving* volume. If the number of galaxies in the Universe had always remained constant, then the *comoving density*, the number in each unit of comoving volume, would not change. If there are presently n_0 of a particular galaxy type in each cubic gigaparsec, then between redshift z and $z + \Delta z$ we would expect $(\mathrm{d}\mathcal{N}_-/\mathrm{d}z)\Delta z$ of them, where

$$\frac{\mathrm{d}\mathcal{N}_-}{\mathrm{d}z} = n_0(1+z)^3 \frac{\mathrm{d}\mathcal{V}}{\mathrm{d}z} = n_0 c \frac{4\pi \mathcal{R}^2(t_0)\sigma_e^2}{H(z)}. \tag{8.48}$$

Comparing the measured number of galaxies $\mathrm{d}\mathcal{N}/\mathrm{d}z$ at each redshift with $\mathrm{d}\mathcal{N}_-/\mathrm{d}z$ from Equation 8.48 tells us how the comoving density has changed.

The left panel of Figure 8.8 shows the comoving volume $(1+z)^3\, \mathrm{d}\mathcal{V}/\mathrm{d}z$ between redshifts z and $z + \Delta z$. It is much larger in the open Universe with $\Lambda = \Omega_0 = 0$ than it is in the flat model with $\Omega_0 = 1$. So we expect to see relatively more galaxies at high redshift in the open model. The benchmark model has slightly more volume at low redshift than the $\Lambda = \Omega_0 = 0$ model, but less at $z \gtrsim 2$.

> **Problem 8.26** Use Equations 8.23 and 8.42 to show that, if $\Omega_0 = 0$, then redshift $z = 5$ corresponds to $\mathcal{R}(t_0)\sigma_e = 2.92c/H_0$, whereas for $\Omega_0 = 1$, $\mathcal{R}(t_0)\sigma_e = 1.18c/H_0$. For any given density $n(z)$, use Equation 8.48 to show that, if $\Omega_0 = 0$, then at $z = 5$ we would expect to find roughly 15 times as many objects within a small redshift range Δz as we would see if $\Omega_0 = 1$. What is this ratio at $z = 3$?

Quasars, the extremely luminous 'active' nuclei of galaxies which we discuss in the following chapter, are so bright that we can see them across most of the observable Universe. They also have strong emission lines that make it easy to measure their redshifts. Figure 1.16 told us that each cubic gigaparsec now contains $\sim 10^6$ galaxies with $L \approx L_\star$, where $L_\star \approx 2 \times 10^{10} L_\odot$ is the luminosity of a bright galaxy defined by Equation 1.24. At present, each cubic gigaparsec contains about one very luminous quasar with $L \gtrsim 100L_\star$; bright quasars are much rarer than luminous galaxies. But Figure 8.13 shows that, at redshifts $z \approx 2$, the brightest quasars were about 100 times more common than they are today. There was roughly one quasar for every 10 000 present-day giant galaxies. What happened to them?

If quasars represent the youth of a galactic nucleus, then at least one in 10 000 luminous galaxies must have been bright quasars in the past. The fraction could be as high as 100% if the nuclear activity lasts much less than a gigayear. A period with a quasar nucleus might be a normal part of a galaxy's early development.

We will see in Section 9.1 that quasars shine by the energy released as gas falls into a hugely massive black hole of $\gtrsim 10^9 \mathcal{M}_\odot$. In Section 6.4 we found that today's luminous galaxies harbor massive black holes at their centers; perhaps these remain from an early quasar phase. It takes time to build up the black hole as it consumes gas, so quasars are rare during the first quarter of cosmic history, before $z \sim 2$. It is perhaps more surprising that we begin to see them already at $z > 6$, less than a gigayear after the Big Bang.

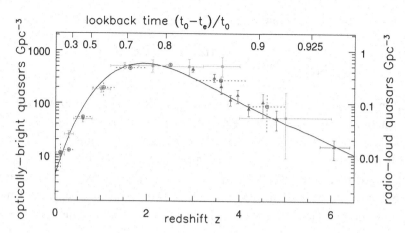

Fig. 8.13. The curve shows the density of very radio-loud ($\nu L_\nu > 3 \times 10^{10} L_{bol,\odot}$ at 2.7 GHz) quasars, triangles show optically-bright quasars ($L \gtrsim 100 L_\star$); both are most common at redshifts $z \sim 2$. Numbers of quasars bright in soft X-rays (filled dots) and hard X-rays (open dots) follow the same pattern. Density per comoving Gpc^3 is calculated using the benchmark model – J. Wall.

8.4   Growth of structure: from small beginnings

The cosmic background radiation is almost, but not quite, uniform: across the sky, its temperature differs by a few parts in 10^5. These tiny differences tell us how lumpy the cosmos was at the time t_{rec} of recombination, when the radiation cooled enough for neutral atoms to form. Quantum fluctuations in the field responsible for inflation left their imprint as irregularities in the density of matter and radiation. Most versions of inflation predict that fluctuations should obey the random-phase hypothesis, and that the power spectrum $P(k) \propto k$ (see Problem 8.7): we will call these the *benchmark initial fluctuations*. Equation 8.4 then tells us that the density varies most strongly on small spatial scales or large k.

   The largest features, extending a degree or more across the sky, tell us about that early physics. Smaller-scale irregularities are modified by the excess gravitational pull toward regions of high density and by the pressure of that denser gas. Observing them tells us about the geometry of the Universe and its matter content. After recombination, dense regions rapidly became yet denser as surrounding matter fell into them. By observing the *peculiar motions* of infalling galaxies, we can probe the large-scale distribution of mass today and compare it with what is revealed by the light of the galaxies.

8.4.1   Fluctuations in the cosmic microwave background radiation

How did the distribution of matter affect the cosmic background radiation as we observe it today? To reach us from an overdense region, radiation has to climb out of a deeper gravitational potential. In doing this, it suffers a *gravitational redshift* proportional to $\Delta\Phi_g$, the excess depth of the potential: its temperature T changes by ΔT, where $\Delta T/T \sim \Delta\Phi_g/c^2$. The temperature is reduced where the potential is unusually deep, since $\Delta\Phi_g$ is negative there. But time also runs more slowly within the denser region by a fraction $\Delta t/t = \Delta\Phi_g/c^2$, so we see the gas at an earlier time when it was hotter. The radiation temperature decreases as $T \propto 1/a(t)$, so

$$\frac{\Delta T}{T} = -\frac{\Delta a}{a} = -\frac{2}{3}\frac{\Delta t}{t} = -\frac{2}{3}\frac{\Delta\Phi_g}{c^2}, \tag{8.49}$$

where we have used $a \propto t^{2/3}$ from Equation 8.27. This partly cancels out the gravitational redshift to give $\Delta T/T \sim \Delta\Phi_g/(3c^2)$. At these early times, the average density $\bar{\rho}$ is very nearly equal to the critical density of Equation 8.21. If our region has density $\bar{\rho}(1+\delta)$ and radius R, its excess mass is $\Delta\mathcal{M} = 4\pi\bar{\rho}R^3\delta/3$. We can write

$$3c^2\frac{\Delta T}{T} = \Delta\Phi_g \sim -\frac{2G\Delta\mathcal{M}}{R} = -\frac{8\pi}{3}G\bar{\rho}R^2\delta \approx -\delta(t)[\bar{H}(t)R]^2. \tag{8.50}$$

Radiation reaching us from denser regions is *cooler*.

The best current measurements of the cosmic microwave background on scales larger than $0.3°$ are from the WMAP satellite, which was launched in June 2001. WMAP confirmed that the background radiation has the form of a blackbody everywhere on the sky: only its temperature differs slightly from point to point. This is exactly what we would expect if it was affected by non-uniformities in the matter density. We can describe the temperature variations by choosing some polar coordinates θ, ϕ on the sky. As we look in a given direction, we can write the difference ΔT from the mean temperature T using the spherical harmonic functions Y_l^m:

$$\Delta T(\theta, \phi) = \sum_{l>1} \sum_{-l \le m \le l} a_l^m Y_l^m(\theta, \phi). \tag{8.51}$$

Since Y_l^m has l zeros as the angle θ varies from 0 to π, the a_l^m measure an average temperature difference between points separated by an angle $(180/l)°$ on the sky. Apart from the $l = 1$ terms which reflect our motion relative to the background radiation, all the a_l^m must average to zero; their squared average measures how strongly T fluctuates across the sky. Theorists aim to predict $C_l = \langle|a_l^m|^2\rangle$ averaged over all the m-values, since this does not depend on which direction we chose for $\theta = 0$, our 'north pole'. Figure 8.14 shows $\Delta_T(l)$, defined by $\Delta_T^2 = T^2 l(l+1)C_l/(2\pi)$.

Gravity, which makes denser regions even denser, and pressure forces that tend to even out the density, have modified the fluctuations left behind after inflation. These forces cannot propagate faster than light, so they act only within the *horizon*

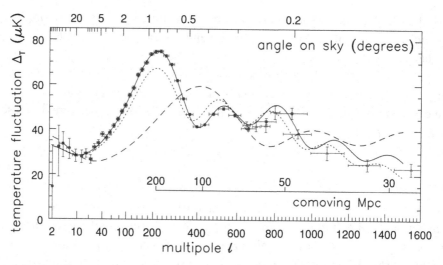

Fig. 8.14. Temperature fluctuations Δ_T in the cosmic microwave background: triangular points combine data from many experiments, circles are from WMAP. Horizontal bars show the range of angular scales. Curves show predictions for the benchmark model (solid), for a flat model with half as many baryons (dotted), and for $\Omega_0 = 0.3$, $\Lambda = 0$ (dashed). The second, third, and subsequent peaks correspond to regions that sound waves could cross twice, three times, etc., before recombination. When l is small, we have few a_l^m to average over (only five for $l = 2$), and vertical bars indicate larger uncertainties – M. Tegmark, CMBFAST.

scale of Problem 8.14. When the gas became transparent, the comoving distance σ_H to the horizon was

$$\mathcal{R}(t_{rec})\sigma_H = 3ct_{rec} = \frac{2c}{H(t_{rec})} \approx \frac{2c}{H_0\sqrt{\Omega_m}(1 + z_{rec})^{3/2}}, \qquad (8.52)$$

where we used Equation 8.26 in the last step. (Why can we ignore Ω_Λ?) A region of this size will expand to $184/(h^2\Omega_m)^{1/2}$ Mpc by today. Because inflation left us with $P(k) \propto k$, on larger scales (at small l), we should expect $\Delta_T(l)$ to rise smoothly as l increases.

The angle θ_H that the horizon covers on the sky depends on Ω_Λ and Ω_m through the angular-size distance d_A. When $\Lambda = 0$ and $\Omega_0 z \gg 1$, Equation 8.42 tells us that $d_A \to 2c/(H_0 z \Omega_0)$. So only points separated by less than the angle

$$\theta_H \approx \frac{\mathcal{R}(t_{rec})\sigma_H}{d_A(t_{rec})} \approx \sqrt{\frac{\Omega_0}{z_{rec}}} \approx 2° \times \sqrt{\Omega_0} \qquad (8.53)$$

can communicate before time t_{rec}. The lower the matter density, the smaller this angle should be. Detailed calculation shows that, if $\Omega_0 = 1$, Δ_T is largest on scales just less than a degree, where we see the main *acoustic peak* in Figure 8.14.

A model with $\Omega_0 = 0.3$ places the peak at roughly half this angle. Setting $\Omega_m + \Omega_\Lambda = 1$ changes the way that the distance radius d_A depends on the matter density. From Equation 8.43 we have $d_A \propto 1/\Omega_m^{0.4}$ at large redshift, so the angular size of the ripples is almost independent of Ω_m. The observed position of the acoustic peak is the most powerful current evidence in favor of dark energy.

We will see in the following section that, before recombination, irregularities built up most strongly in the dark matter. The mixed fluid of baryons and radiation then simply fell into the denser regions under gravity. The maximum distance through which that mixed fluid can fall by t_{rec} sets the position of the first peak in Δ_T, at $l = 220$. In the benchmark cosmology, this distance corresponds to 105 Mpc today, or a sphere containing $2.5 \times 10^{16} M_\odot$. The second peak, at $l = 540$, corresponds to a smaller lump of dark matter, where the fluid has time to fall in and be pushed out again by its own increased pressure. The third peak corresponds to 'in–out–in', the fourth to 'in–out–in–out', and so on: hence the label of 'acoustic peaks'. The more dark matter is present, the stronger its gravity causes its irregularities to become, and so the greater the height of the main peak. The mass of the baryonic matter 'helps' the baryon–radiation fluid to fall into dense regions of dark matter, but hinders its 'bouncing' out again; this strengthens the odd-numbered peaks relative to the even peaks. The benchmark model, with $\Omega_B \approx 0.045$, $\Omega_m = 0.3$, $\Omega_\Lambda = 0.7$, and $H_0 = 70 \, \text{km s}^{-1} \, \text{Mpc}^{-1}$, gives correct predictions for the abundance of deuterium and lithium (see Section 1.5), the motions of the galaxies, and fluctuations in the cosmic background radiation.

Further reading: see Chapter 6 of the book by Padmanabhan.

8.4.2 Peculiar motions of galaxies

One way that we can explore the largest structures is to map out the galaxies, as in Figure 8.3; but this samples only the luminous matter. Another is to look at the peculiar motions of galaxies, their deviation from the uniform flow described by Equation 8.8. Peculiar motions grow because of the extra tug of gravity from denser regions. In the Local Group, the Milky Way and the Andromeda galaxy M31 approach each other under their mutual gravitational attraction (Section 4.5), while groups of galaxies fall into nearby clusters (Section 7.2). We saw how to use these motions to weigh the groups and clusters. Similarly, we can use the observed peculiar motions on larger scales to reconstruct the distribution of mass, most of which is dark.

We can see the peculiar motions of the nearby elliptical galaxies in Figure 8.2. Although the Fornax cluster is roughly as far away as the Virgo cluster, the galaxies of Fornax on average are moving more rapidly away from us. It appears that the Local Group, and the galaxies nearby, are falling toward the complex of galaxies around Virgo. To examine the *Virgocentric infall*, in Figure 8.15 we look at the average radial velocity with which each *group* of galaxies in Figure 8.2 recedes

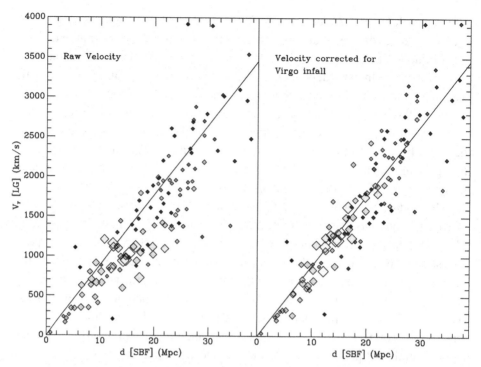

Fig. 8.15. Diamonds show average recession speed V_r, measured relative to the Local Group, for groups of galaxies in Figure 8.2. The two largest white symbols are two clumps within the Virgo cluster; others decrease in size to show distance from Virgo. Left, velocity V_r falls further below the linear trend, the closer the group is to Virgo; right, after correction for Virgocentric infall – J. Tonry.

from us. The peculiar velocities of individual galaxies are affected by their orbits within the group, but averaging over the whole group should reveal the larger-scale motions. These velocities are plotted in the left panel. The two largest white symbols represent the two clumps of Virgo cluster galaxies, around M86 and M49. The other big symbols, indicating groups close to Virgo, fall below the general linear trend.

The right panel of Figure 8.15 shows the result of subtracting out Virgocentric inflow according to a simple model, which predicts an infall speed of 270 km s^{-1} at our position. We now see roughly $V_r \propto d$. Within about 25 Mpc of Virgo, most of the plotted values deviate from the linear trend by less than 100 km s^{-1}. Peculiar motions complicate our attempts to measure H_0. If we tried to do this by finding distances and velocities of galaxies in the direction of Virgo, we would underestimate the Hubble constant, because Virgocentric inflow partially cancels out the cosmic expansion. But if we had observed galaxies in the opposite direction, our value for H_0 would be too high.

The best-measured peculiar motion is that of the Local Group, determined from the Sun's velocity relative to the cosmic microwave radiation: recall Section 1.5. The Local Group now moves with $V_{pec} \approx 630$ km s^{-1} in the direction $(l, b) \approx (276°, 30°)$. Most of that peculiar motion seems to be caused by the gravitational pull of very distant matter, tugging at both us and the Virgo cluster. The velocities of galaxies furthest from the Virgo cluster, which are mostly on the opposite side of the sky, lie mainly above the sloping line in the right panel of Figure 8.15. This is what we would expect if more distant matter were pulling Virgo and Fornax apart.

Both the local velocity dispersion and the Local Group's motion toward the Virgo cluster are significantly less than our motion relative to the cosmic microwave background. The flow of galaxies through space is 'cold' on small scales: galaxies within tens of megaparsecs of each other share a large fraction of their peculiar velocity.

Problem 8.27 Here you use Monte Carlo simulation to show that the peculiar velocities of nearby galaxies must be very close to that of the Milky Way, or Hubble could never have discovered the cosmic expansion from his sample of 22 local galaxies.

Your model sky consists of galaxies in regions A ($1\,\mathrm{Mpc} < d < 3\,\mathrm{Mpc}$), B ($3\,\mathrm{Mpc} < d < 5\,\mathrm{Mpc}$), C ($5\,\mathrm{Mpc} < d < 7\,\mathrm{Mpc}$), and D ($7\,\mathrm{Mpc} < d < 9\,\mathrm{Mpc}$). If the density is uniform, and you have four galaxies in region B, how many are in regions A, C, and D (round to the nearest integer)? For simplicity, put all the objects in region A at $d = 2\,\mathrm{Mpc}$, those in B at $4\,\mathrm{Mpc}$, those in C at $6\,\mathrm{Mpc}$, and those in D at $8\,\mathrm{Mpc}$. Now assign peculiar velocities at random to the galaxies. For each one, roll a die, note the number N on the upturned face, and give your galaxy a radial velocity $V_r = H_0 d + (N - 3.5) \times 350\,\mathrm{km\,s^{-1}}$, taking $H_0 = 70\,\mathrm{km\,s^{-1}\,Mpc^{-1}}$. (If you like to program you can use more galaxies; place them randomly in space, and choose the peculiar velocities from a Gaussian random distribution with zero mean and standard deviation of $600\,\mathrm{km\,s^{-1}}$.)

Plot both V_r and also the average velocity in each region against the distance d; is there a clear trend? How many of your model galaxies have negative radial velocities? How does your plot compare with the right panel of Figure 8.15? Hubble found *no* galaxies beyond the Local Group that are approaching us.

8.4.3 How do peculiar velocities build up?

Peculiar velocities tend to die away as the Universe expands, because a moving galaxy keeps overtaking others, until it reaches the region where its motion matches that of the cosmic expansion. We can imagine two nearby comoving observers, P and Q, at rest relative to the background radiation; they recede from each other only because of cosmic expansion. A galaxy passes observer P heading toward Q with a peculiar motion V_{pec}, and arrives there after a time $\approx d/V_{pec}$. If P and Q are close enough that $V_{pec} \gg H(t)d$, their separation remains almost constant as the galaxy travels between them. But relative to observer Q, the galaxy moves only at speed $V_{pec} - H(t)d$. The galaxy's speed, relative to a comoving observer at its current position, has decreased at the rate

$$\frac{dV_{pec}}{dt} = -\frac{H(t)d}{d/V_{pec}} = -V_{pec}\frac{\dot{R}(t)}{R(t)}. \tag{8.54}$$

Integrating this shows that $V_{pec} \propto 1/R(t)$; a galaxy's peculiar velocity V_{pec} falls in exactly the same way as the momentum of a photon is reduced according to Equation 8.12.

If peculiar velocities simply decreased according to Equation 8.54, then shortly after recombination at $z \approx 1100$, the material of the Local Group would have been moving at nearly the speed of light. But this would have caused shocks in the gas and huge distortions in the cosmic microwave background. In fact, the peculiar motions of the galaxies were generated quite recently, by their mutual gravitational attraction. When some part of the Universe contains more matter than average, its increased gravity brakes the expansion more strongly. Where there is less matter than average, the expansion is faster; the region becomes even more diffuse relative to its surroundings. So the galaxies move relative to the cosmic background: they acquire peculiar motions.

To calculate how this happens, suppose that the average density of matter is $\bar{\rho}_m(t)$, and the average expansion is described by the scale factor $\bar{a}(t)$ and the Hubble parameter $\bar{H}(t)$. Locally, within the volume we are studying we can write

$$\rho_m(t) = \bar{\rho}_m(t)[1 + \delta(t)], \quad \text{and} \quad a(t) = \bar{a}(t)[1 - \epsilon(t)]. \tag{8.55}$$

If our region is approximately spherical, the matter outside will not exert any gravitational force within it; it will behave just like part of a denser, more slowly expanding, cosmos. Where $\Omega_m[1 + \delta(t)] > 1$ so the local density exceeds the critical value, expansion can be halted to form bound groups and clusters of galaxies.

Life is much simpler if we stay in the *linear* regime, where δ and ϵ of Equation 8.55 are much less than unity. We saw in the discussion following Equation 8.4 that this applies to structures with sizes larger than about $8h^{-1}$ Mpc: their density differs by only a small fraction from the cosmic average. When we substitute the expressions for $\rho_m(t)$ and $a(t)$ into Equation 8.25, we can then ignore terms in $\delta^2, \delta\epsilon, \epsilon^2$, and higher powers of these variables. Remembering that $a(t)H(t) = \dot{a}(t)$ and that terms involving only barred average quantities will cancel out, Equation 8.25 becomes

$$\Delta\left[H_0^2(1 - \Omega_{tot})\right] + 2\frac{d\bar{a}}{dt}\frac{d}{dt}[\bar{a}\epsilon(t)] + \frac{8\pi G}{3}\bar{\rho}(t)\bar{a}^2(t)[\delta(t) - 2\epsilon(t)] = 0. \tag{8.56}$$

Here the first term represents the change in the present density and expansion rate within our denser region.

We saw from Figure 8.7 that, for most of the period during which galaxy clusters and groups were forming, dark energy was not important, and we can simply use Equation 8.27 to describe the average expansion $\bar{a}(t)$. While the Universe is matter-dominated, ρa^3 is constant, so $\delta = 3\epsilon$. Then

$$\delta \propto t^{2/3} \propto \bar{a}(t) \text{ as long as } 1 + z \gg (1 - \Omega_{tot})/\Omega_m, (\Omega_\Lambda/\Omega_m)^{1/3} \qquad (8.57)$$

is a 'growing' solution to Equation 8.56 (substitute back to check!). Early on, the contrast δ grows proportionally to $\mathcal{R}(t)$. If $\Omega_m + \Omega_\Lambda < 1$, then at some point the first condition on z is violated, and the average motion becomes $\bar{a} \propto t$: matter coasts outward with constant speed. Once the matter exerts too little gravity to have any effect on the expansion, δ remains fixed: the structure *freezes out*. In a flat model with $\Omega_{tot} = 1$, growth continues until $1 + z \sim (\Omega_\Lambda/\Omega_m)^{1/3}$. In the benchmark model, large structures continued to grow until very recently, at $z \sim 0.3$. In a low-density Universe with $\Omega_m = 0.3$ and $\Omega_\Lambda = 0$, they would have ceased to become denser around redshift $z \sim 2$.

Further reading: On peculiar motions, see Chapter 4 of Padmanabhan's book.

8.4.4 Weighing galaxy clusters with peculiar motions

Any denser-than-average region pulled the surrounding galaxies more strongly toward it. While the fractional deviations $\delta(\mathbf{x}, t)$ from uniform density remain small, Equation 8.57 tells us that, over a given time, $\delta(\mathbf{x})$ increased by an equal factor everywhere. Because the pull on a galaxy from each overdense region increases in the same proportion, its acceleration, and hence its peculiar velocity, is always parallel to the local gravitational force. So, by measuring peculiar motions, we can reconstruct the force vector, and hence the distribution of mass.

To see how this works, we can write the velocity $\mathbf{u}(\mathbf{x}, t)$ of matter at point $\mathbf{x}$ as the sum of the average cosmic expansion directly away from the origin and a peculiar velocity $\mathbf{v}$:

$$\mathbf{u}(\mathbf{x}, t) = \bar{H}(t)\mathbf{x} + \mathbf{v}(\mathbf{x}, t). \qquad (8.58)$$

The equation of mass conservation relates the velocity field $\mathbf{u}(\mathbf{x}, t)$ to the density, which we write as $\rho(\mathbf{x}, t) = \bar{\rho}(t)[1 + \delta(\mathbf{x}, t)]$:

$$\left(\frac{\partial \rho}{\partial t}\right)_{\mathbf{x}} + \nabla_{\mathbf{x}} \cdot \rho\,\mathbf{u} = 0. \qquad (8.59)$$

Remembering that terms involving only the barred average quantities will cancel out, and dropping terms in δ^2, $\delta\mathbf{v}$, and $\mathbf{v}^2$, we have

$$\left(\frac{\partial \delta}{\partial t}\right)_{\mathbf{x}} + \bar{H}(t)\mathbf{x} \cdot \nabla_{\mathbf{x}}\delta + \nabla_{\mathbf{x}}\mathbf{v} = 0. \qquad (8.60)$$

Setting $\mathbf{x} = \bar{a}(t)\mathbf{r}$, we switch to the coordinate $\mathbf{r}$ comoving with the average expansion. The time derivative following a point at fixed $\mathbf{r}$ is

$$\left(\frac{\partial}{\partial t}\right)_{\mathbf{r}} = \left(\frac{\partial}{\partial t}\right)_{\mathbf{x}} + \bar{H}(t)\,\mathbf{x}\cdot\nabla_{\mathbf{x}}, \tag{8.61}$$

and, since $\bar{a}(t)\nabla_{\mathbf{r}} = \nabla_{\mathbf{x}}$, Equation 8.60 simplifies to

$$\left(\frac{\partial\delta}{\partial t}\right)_{\mathbf{r}} + \nabla_{\mathbf{x}}\mathbf{v} = 0. \tag{8.62}$$

For a small enough volume, if we assume that the Universe beyond is homogeneous and isotropic, we can use Newton's laws to calculate the gravitational potential Φ_g corresponding to local deviations from the average density $\bar{\rho}$. The gravitational force $\mathbf{F}_g = -\nabla\Phi_g = d\mathbf{v}(\mathbf{x}, t)/dt$, so we have $0 = d(\nabla\times\mathbf{v})/dt$. Peculiar motions that have grown in this way from small initial fluctuations thus have $\nabla\times\mathbf{v}\approx 0$, and we can define a *velocity potential* Φ_v such that $\mathbf{v} = \nabla_{\mathbf{x}}\Phi_v$. Rewriting Equation 8.62 in terms of Φ_v gives

$$\nabla_{\mathbf{x}}^2\Phi_v = -\left(\frac{\partial\delta}{\partial t}\right)_{\mathbf{r}}. \tag{8.63}$$

Equation 3.9, Poisson's equation, tells us that

$$\nabla_{\mathbf{x}}^2\Phi_g = -\nabla_{\mathbf{x}}\cdot\mathbf{F}_g = 4\pi G\bar{\rho}\delta(\mathbf{x}, t) \tag{8.64}$$

– which looks suspiciously like the equation for Φ_v. Equation 8.57 assures us that all perturbations grow at the same rate: if δ is twice as large, then so is $\dot{\delta}$. Thus $\delta(\mathbf{x}, t) \propto \partial\delta(\mathbf{x}, t)/\partial t$, and the right-hand sides of Equations 8.63 and 8.64 are proportional to each other. Then, as long as both $\mathbf{v}(\mathbf{x}, t)$ and $\mathbf{F}_g$ diminish to zero as $|\mathbf{x}|$ increases, they must also be proportional: *the peculiar velocity is in the same direction as the force resulting from local concentrations of matter.* On dividing the right-hand side of Equation 8.63 by that of 8.64, we find

$$\frac{|\mathbf{v}(\mathbf{x}, t)|}{|\mathbf{F}_g|} = \frac{\bar{H}(t)f}{4\pi G\bar{\rho}(t)}, \quad \text{where } f \equiv \frac{\bar{a}(t)}{\delta}\left(\frac{\partial\delta}{\partial t}\right)_{\mathbf{r}}\bigg/\frac{d\bar{a}}{dt}. \tag{8.65}$$

From Equation 8.57, in a matter-dominated Universe we have $f = 1$ for $\Omega_m \approx 1$, and $f \to 0$ as $\Omega_m \to 0$. In general, $f(\Omega) \approx \Omega^{0.6}$ is a good approximation. Using Equation 3.5 for the force, we can write the peculiar velocity as

$$\mathbf{v}(\mathbf{x}, t) = \frac{\bar{H}(t)f(\Omega)}{4\pi}\int\frac{\delta(\mathbf{x}')(\mathbf{x} - \mathbf{x}')}{|\mathbf{x} - \mathbf{x}'|^3}\,d^3\mathbf{x}'. \tag{8.66}$$

Problem 8.28 Show that, if the density is uniform apart from a single overdense lump at $\mathbf{x} = 0$, then distant galaxies move toward the origin with $\mathbf{v}(x, t) \propto 1/x^2$.

Problem 8.29 In the expanding (comoving) coordinate $\mathbf{r}$, show that

$$\mathbf{v}(\mathbf{r}, t) = \frac{\bar{H}(t) f(\Omega) \bar{a}(t)}{4\pi} \int \frac{\delta(\mathbf{r}')(\mathbf{r} - \mathbf{r}')}{|\mathbf{r} - \mathbf{r}'|^3} d^3\mathbf{r}'. \qquad (8.67)$$

Show that, while it is early enough that we can use Equation 8.57 for $\delta(\mathbf{r})$, the peculiar velocity $\mathbf{v} \propto t^{1/3}$. (Why did we have to transform to comoving coordinates to apply Equation 8.57?)

So, if we can measure the overdensity $\delta(\mathbf{x})$ of the nearby rich galaxy clusters, and the peculiar velocities of the galaxies around them, we should be able to test Equation 8.66, and solve for the matter density Ω_m. First, we determine the average peculiar motion $\mathbf{v}(\mathbf{x})$ of our galaxies. We must assume that the Universe is homogeneous and isotropic on even larger scales, so that forces from galaxies outside our survey volume will average to zero. Inverting Equation 8.66 should then yield the product $f(\Omega_0) \cdot \delta(\mathbf{x})$, from which we can find Ω_m.

But the mass distributions predicted from measured peculiar velocities do not match the observed clustering of galaxies very well. Alternatively, we could say that the forces calculated from the galaxies at their observed positions do not yield the measured peculiar motions. The pull of matter outside the volume of our present surveys appears to be significant. In particular, we still do not know that concentration of matter is responsible for most of the Local Group's peculiar motion of $\sim 600 \, \text{km s}^{-1}$. Work is under way on this problem, and galaxy surveys are being extended as techniques for finding distances improve.

Locally, we can use the crude model of Figure 8.15 for the Virgocentric infall to estimate the mass density Ω_m. Let $d_V \approx 16 \, \text{Mpc}$ be the distance of the Local Group from the center of the Virgo cluster. Within a sphere of radius d_V about the cluster's center, the density of luminous galaxies is roughly 2.4 times the mean; if the mass density is increased by the same factor, then the overdensity $\delta \approx 1.4$. Although Equation 8.65 was derived for $\delta \ll 1$, we can use it to make a rough calculation of $f(\Omega)$.

Assuming that the Virgo cluster is roughly spherical, the additional gravitational pull on the Local Group is $F_g \approx 4\pi G d_V \bar{\rho} \delta / 3$, just as if all the cluster's mass had been concentrated at its center. So our peculiar motion toward Virgo is

$$|\mathbf{v}_{LG}| \approx \frac{(H_0 d_V) \Omega_m^{0.6} \delta}{3} \approx 270 \, \text{km s}^{-1}. \qquad (8.68)$$

Cosmic expansion is pulling the cluster away from us at a speed $H_0 d_V \approx 1200 \, \text{km s}^{-1}$, so this yields $\Omega_m \approx 0.3$, in reasonable agreement with the benchmark model.

8.4.5 Tidal torques: how did galaxies get their spin?

The Sun rotates for the same reason that water swirls around the plug-hole as it runs out of a sink. The material originally had a small amount of angular momentum $\rho \mathbf{x} \times \mathbf{v}$ about its center in a random sense. This is approximately conserved as the fluid is drawn radially inward, so as $|\mathbf{x}|$ decreases the rotation described by $\mathbf{v}$ must speed up. But galaxies and clusters do not owe their rotation to early random motions; this peculiar motion arises from irregular lumps of matter pulling on each another by gravity, as illustrated in Figure 4.13.

In Problem 8.29 we saw that, while the Universe is matter-dominated, peculiar velocities grow as $t^{1/3}$, while the distance d between galaxies follows $a(t) \propto t^{2/3}$. So angular momentum builds up as $d \times v \propto t$ as long as we remain in the linear regime with $\delta(t) \ll 1$. It stops increasing when the dense region starts to collapse on itself, as we will discuss in Section 8.5. The denser the initial lump, the sooner it collapses and the less time it has to spin up. But tidal torques are stronger in denser regions, so, in a cosmos filled with cold dark matter, objects acquire the same average angular momentum in relation to their mass and energy.

To measure how important a galaxy's angular momentum is, we note that a galaxy of radius R, mass $\mathcal{M}$, and angular momentum $\mathcal{L}$ will rotate with angular speed $\omega \sim \mathcal{L}/(\mathcal{M}R^2)$. The angular speed ω_c of a circular orbit at radius R is given by $\omega_c^2 R \sim G\mathcal{M}/R^2$. The energy $\mathcal{E} \sim -G\mathcal{M}^2/R$ (see Problem 3.36 and recall the virial theorem). So the ratio

$$\lambda = \frac{\omega}{\omega_c} = \frac{\mathcal{L}}{\mathcal{M}R^2} \times \frac{R^{3/2}}{\sqrt{G\mathcal{M}}} = \frac{\mathcal{L}|\mathcal{E}|^{1/2}}{G\mathcal{M}^{5/2}} \tag{8.69}$$

tells us how far the galaxy is supported against collapse by rotation, rather than pressure or random motion of its stars. Gravitational N-body simulations show that the distribution of galaxies we observe would not spin up collapsing lumps very strongly: we expect $0.01 < \lambda < 0.1$. This is similar to what we see in elliptical galaxies, but disk galaxies like our Milky Way have $\lambda \approx 0.5$. The parameter λ can increase if material loses energy to move inward, as a gas disk can do by radiation.

This argument already tells us that the Milky Way has a dark halo – otherwise its disk would not have time to form. Without a halo, $\mathcal{L}$ and $\mathcal{M}$ do not change as the proto-disk moves inward, so its radius must shrink 100-fold to increase $\mathcal{E}$ by the same factor. Disk material near the Sun must originate 800 kpc from the center, but the mass $\mathcal{M}(<R)$ within that radius would be just what now lies between the Sun's orbit and the Galactic center. Equation 3.20 shows that the orbital period would then be 1000 times longer than that in the Sun's current orbit, or about 240 Gyr. The Galaxy would shrink at roughly the same rate; it would take several times longer than the age of the Universe to make the disk.

But in Problem 3.5 we saw that the Milky Way has $\mathcal{M}/L \gtrsim 50$; at least 90% of its mass is dark. Because the dark halo cannot lose energy and shrink, the gas that is to become the disk originates closer to the center by a factor $\mathcal{M}(\text{disk})/\mathcal{M}(\text{total})$. So our disk had to collapse only to a tenth of its original size to reach $\lambda \approx 0.5$. Since the infall and orbital speeds are set by the dark halo, they would have been near today's values. Shrinking at $200\,\text{km s}^{-1}$ from a radius of $80\,\text{kpc}$, the disk could have formed in $\lesssim 2\,\text{Gyr}$.

Further reading: see Chapter 8 of Padmanabhan's book.

8.5   Growth of structure: clusters, walls, and voids

The galaxy clusters and huge walls that we see in Figure 8.3 are visible because the density of luminous matter in them is a few times greater than that in the surrounding regions. If galaxies trace out the mass density, then the fractional variations in density are now large: in the language of Equation 8.55, $\delta(t_0) \gtrsim 1$. How did the small fluctuations that we examined in Section 8.4 develop into the structure that we now see?

8.5.1   Pressure battles gravity: the Jeans mass

Objects like stars are supported by gas pressure, which counteracts the inward pull of gravity. The larger a body is, the more likely it is that gravity will win the fight against the outward forces holding it up. In life, the giant insects of horror movies would be crushed by their own weight. For a spherical cloud of gas, we can estimate the potential energy $\mathcal{PE}$ using the result of Problem 3.11 for a uniform sphere of radius r and density ρ. We then compare it with the thermal energy $\mathcal{KE}$. The sound speed c_s in a gas is close to the average speed of motion of the particles along one direction, so we can write

$$\mathcal{PE} \equiv -\frac{1}{2} \int \rho(\mathbf{x})\Phi(\mathbf{x}) \mathrm{d}^3 \mathbf{x} \approx -\frac{16\pi^2}{15} G\rho^2 r^5, \quad \text{and} \quad \mathcal{KE} \approx \frac{3c_\mathrm{s}^2}{2} \frac{4\pi r^3 \rho}{3}. \quad (8.70)$$

In equilibrium the virial theorem, Equation 3.44, requires $|\mathcal{PE}| = 2\mathcal{KE}$; we might expect the cloud to collapse if the kinetic energy is less than this. That always happens if the cloud is big enough: $\mathcal{KE} < |\mathcal{PE}|/2$ when

$$2r \gtrsim \sqrt{\frac{15}{\pi}} \sqrt{\frac{c_\mathrm{s}^2}{G\rho}} \approx \lambda_\mathrm{J}, \quad \text{where} \quad \lambda_\mathrm{J} \equiv c_\mathrm{s}\sqrt{\frac{\pi}{G\rho}}. \quad (8.71)$$

The length λ_J is called the *Jeans length*. When a gas cloud is compressed, its internal pressure rises and tends to cause expansion, but the inward pull of gravity also strengthens. If its diameter is less than λ_J, the additional pressure more than offsets the increased gravity: the cloud re-expands. In a larger cloud gravity wins, and collapse ensues.

Early on, while the Universe is radiation-dominated, the density $\rho_r = a_B T^4/c^2$ is low and the pressure is high, with $c_s = c/\sqrt{3}$. So Equation 8.71 gives

$$\lambda_J = c^2 \left(\frac{\pi}{3Ga_B T^4} \right)^{1/2} \propto T^{-2}. \tag{8.72}$$

The *Jeans mass* $\mathcal{M}_J$ is the amount of matter in a sphere of diameter λ_J:

$$\mathcal{M}_J \equiv \frac{\pi}{6} \lambda_J^3 \rho_m, \tag{8.73}$$

where ρ_m refers *only* to the matter density. In the radiation-dominated period we have $\mathcal{M}_J \propto \rho_m T^{-6}$, with $T \propto 1/\mathcal{R}(t)$ and ρ_m decreasing as $\mathcal{R}^{-3}$. So the Jeans mass grows as $\mathcal{M}_J \propto \mathcal{R}^3(t)$; the mass enclosed in a sphere of diameter λ_J increases as the Universe becomes more diffuse. At the time t_{eq} when the density of matter is equal to that of radiation, the temperature is T_{eq} and $\rho_m = \rho_r = a_B T_{eq}^4/c^2$. Radiation still provides most of the pressure, so $p \approx c^2 \rho_r/3$ and

$$\mathcal{M}_J(t_{eq}) = \frac{\pi}{6} \rho_m(t_{eq}) \left(\frac{\pi c^4/3}{Ga_B T_{eq}^4} \right)^{3/2} = \frac{\pi^{5/2}}{18\sqrt{3}} \frac{c^4}{G^{3/2} a_B^{1/2}} \frac{1}{T_{eq}^2}. \tag{8.74}$$

If equality occurs at the redshift $1 + z_{eq} = 24\,000\Omega_m h^2$ of Problem 8.10, then

$$\mathcal{M}_J(T_{eq}) = 3.6 \times 10^{16} (\Omega_m h^2)^{-2} \mathcal{M}_\odot. \tag{8.75}$$

This is 100 times more than the Virgo cluster, or roughly the mass that we would find today in a huge cube $50/(\Omega_m h^2)$ Mpc on a side. This is approximately the spatial scale of some of the largest voids and complexes of galaxy clusters in Figure 8.3. Overdense regions with masses below $\mathcal{M}_J$ could not collapse because the outward pressure of radiation was too strong. Instead, radiation gradually diffused out of them, taking the ionized gas with it, and damping out small irregularities.

After this time, matter provides most of the mass and energy, but the pressure comes mainly from the radiation: so $\rho \approx \rho_m$, but $p \approx c^2 \rho_r/3$. If a small box of the combined matter–radiation fluid is squeezed adiabatically, then, just as in the cosmic expansion, the change $\Delta\rho_m$ in the matter density is related to $\Delta\rho_r$ by $4\,\Delta\rho_m/\rho_m = 3\,\Delta\rho_r/\rho_r$. So the sound speed

$$c_s^2 = \frac{\partial p}{\partial \rho} = \frac{c^2\,\Delta\rho_r/3}{\Delta\rho_m} = \frac{c^2}{3} \frac{4\rho_r}{3\rho_m} \propto \frac{1}{\mathcal{R}(t)}, \quad \text{so } \lambda_J = c_s \sqrt{\frac{\pi}{G\rho_m}} \propto \mathcal{R}(t) \quad (8.76)$$

and the Jeans mass of Equation 8.73 stays constant.

By a redshift $z_{rec} \sim 1100$ when the temperature $T_{rec} \approx 3000\,\mathrm{K}$, hydrogen atoms had recombined. Radiation streamed freely through the neutral matter, and no longer contributed to the pressure. The sound speed dropped to that of the matter:

$$c_s(t_{rec}) \approx \sqrt{\frac{k_B T}{m_p}} \approx 5\,\mathrm{km\,s^{-1}}. \tag{8.77}$$

Just afterward, the Jeans mass is

$$M_J = \frac{\pi}{6}\rho_m \left(\frac{\pi k_B T_{rec}}{G\rho_m m_p}\right)^{3/2} \approx 5 \times 10^4 (\Omega_m h^2)^{-1/2} M_\odot; \tag{8.78}$$

it has fallen abruptly by a factor of $\sim 10^{12}$.

Radiation continues to transfer some heat to the matter, keeping their temperatures roughly equal until $z \sim 100$. Now the Jeans mass $M_J \propto T^{3/2}\rho_m^{-1/2}$, and, because the radiation cools as $T_r \propto \mathcal{R}^{-1}$, that decrease offsets the drop in density ρ_m to keep M_J nearly constant. If the first dense objects formed with roughly this mass, similar to that of a globular cluster, they could subsequently have merged to build up larger bodies. Once it is no longer receiving heat, the matter cools according to $T_m \propto \mathcal{R}^{-2}$. To see why, think of the perfect gas law relating temperature to volume, or recall that expansion reduces the random speeds of atoms according to Equation 8.54. So the Jeans mass falls further; after recombination, gas pressure is far too feeble to affect the collapse of anything as big as a galaxy.

But how can we make objects the size of galaxies or galaxy clusters, that are too small to grow before recombination? Equation 8.57 tells us that the fraction δ by which their density exceeds the average grows with time as $t^{2/3}$ or $\mathcal{R}(t)$. To reach $\delta \gtrsim 1$ before the present, we would need $\delta(t_{rec}) \gtrsim 10^{-3}$ at $z_{rec} = 1100$. But, aside from the highest peak in Figure 8.14, $\Delta_T < 50\,\mathrm{mK}$ or 2×10^{-5} times the average temperature. This is far too small; so why do we see any galaxies and galaxy clusters today?

8.5.2 WIMPs to the rescue!

The dark matter far outweighs the neutrons and protons. Although we have yet to detect the particles themselves, dark matter is most probably composed of weakly interacting massive particles (WIMPs). Like neutrinos, WIMPS lack strong and electromagnetic interactions – or they would not be 'dark' – and they have some small but nonzero mass. WIMPs can collapse into galaxy-sized lumps early on, because, unlike the baryons, they are unaffected by radiation pressure.

To describe this collapse, we can follow the same calculation as for the Jeans length and Jeans mass, but for WIMPs with density ρ_w and typical random speeds c_w. Instead of Equation 8.73, we find that a dense region has too little kinetic energy and falls in on itself if it contains a mass larger than

$$\mathcal{M}_{J,\text{wimp}} = \frac{\pi}{6}\rho_w\left(\frac{\pi c_w^2}{G\rho_w}\right)^{3/2}. \tag{8.79}$$

While the WIMPs are relativistic, their Jeans mass is high and grows with time just as in the radiation-dominated case. A slightly overdense region that is not actively collapsing simply disperses, as WIMPs stream out of it at light speed. All structure smaller than the horizon scale of Problem 8.14 is erased in this way.

But as soon as the speed c_w of random motions drops appreciably below $c/\sqrt{3}$, the Jeans mass starts to fall. Very roughly, all dense clumps of WIMPs that are larger than the horizon scale at the time when they cease moving relativistically now begin to collapse. Since inflation left behind fluctuations with a power spectrum $P(k)$ rising with k, lumps just larger than this will have the highest densities. The more massive the WIMPs, the smaller the horizon scale when they cease to move relativistically, and the smaller and denser the structures that form.

Neutrinos, with masses of a few electron-volts, remain relativistic until almost t_{eq}, when the comoving size of the horizon is ~16$(h^2\Omega_m)^{-1}$ Mpc. Such light particles are called *hot dark matter*. If the dark matter is hot, we still have difficulty in understanding how something as small as a galaxy or even a galaxy cluster can form. WIMPs massive enough that their sound speed c_w fell below the speed of light long before the time t_{eq} of matter–radiation equality are called *cold dark matter*. The most popular WIMP candidates have masses $\gtrsim 1$ GeV; their random motions drop well below light speed when $T < 10^{13}$ K, only 10^{-6} s after the Big Bang, when the mass $\mathcal{M}_H$(WIMP) within the horizon was less than $\mathcal{M}_\odot$.

As it escaped from the contracting clouds of WIMPs, the radiation took the normal matter with it. So both of these should be quite evenly spread at recombination, and the temperature of the cosmic background radiation should be nearly the same across the whole sky. At recombination, as the matter became neutral and was freed of the radiation pressure, it fell into the already-dense clumps of WIMPs. Fluctuations in the density of normal matter could then grow far more rapidly than Equation 8.57 allows, building up the galaxies and clusters.

If the dark matter is cold, galaxies themselves would be built from successive merger of these smaller fragments. We call this the *bottom-up* picture because galaxies form early, and then fall together to form clusters and larger structures. Figure 8.16 shows results from a gravitational *N*-body simulation following the way that gravity amplifies small initial ripples in an expanding Universe of cold

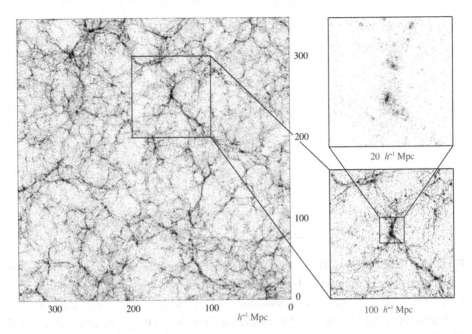

Fig. 8.16. A slice $20h^{-1}$ Mpc thick, through a gravitational N-body simulation with cold dark matter, viewed at the present day. Side frames show magnified views of dense clumps; galaxy groups would form in these 'dark halos' – D. Weinberg.

dark matter. The figure shows a stage of the calculation representing the present day. Notice the profusion of small dense clumps linked by the filamentary cosmic web, and that smaller structures look like denser, scaled-down copies of larger ones. The densest regions, shown in the side boxes, have ceased to expand and have fallen back on themselves. Gas would accumulate there, cooling to form clusters of luminous galaxies.

Figure 8.17 combines the information from WMAP in Figure 8.14 with that from the 2dF galaxy survey in Figures 8.3–8.5 to estimate the power spectrum $P(k)$ for matter today. Using a model close to the benchmark cosmology, Dr. Sánchez deduced from the irregularities in the cosmic microwave background what the distribution of WIMPS and baryons must have been at the time t_{rec}. He then calculated how the concentrations of WIMPS became denser according to Equation 8.57, while baryons fell into them. The results agree with $P(k)$ measured from the galaxies of 2dF in the region where they overlap. On these scales, luminous galaxies are distributed in the same way as the dark matter, and both are well described by the model curve. Does this mean that we have now solved all the problems of cosmology? One might hope for a physical understanding of the dark energy, which is now simply inserted as a term in the Friedmann equations. But for the large structures that we have discussed in this chapter, the benchmark cosmology and benchmark initial fluctuations give an excellent account of what we can observe.

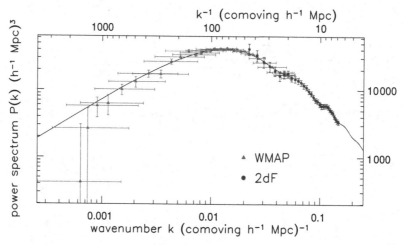

Fig. 8.17. Data from WMAP (triangles) and the 2dF galaxy survey (dots) are combined to trace the power spectrum $P(k)$. The smooth curve shows the prediction from a flat ($k = 0$) model similar to the benchmark cosmology. The wiggle at $k \approx 0.1$ is an acoustic peak on a scale of $\sim 10\,\text{Mpc}$, too small to be measured by WMAP – A. Sánchez: from 2006 *MNRAS* **366**, 189.

8.5.3 How early can galaxies and clusters form?

To find out how long a galaxy or cluster takes to reach its present density, we can use the 'top-hat' model, thinking of the overdense protocluster as a uniform sphere. In a homogeneous Universe, the matter beyond that sphere does not exert any forces within it. So we are free to make our sphere more or less dense than its surroundings, and the Friedmann equations still hold. In the following problem, we use this fact to examine the collapse of a denser-than-usual region that is destined to become a galaxy or cluster.

Problem 8.30 Suppose that the time t_0 refers to a moment when the Universe is matter-dominated, and $\Omega_m > 1$ in our spherical protocluster. By substituting into Equation 8.26, show that the parametric equations

$$\frac{\mathcal{R}(t)}{\mathcal{R}(t_0)} = \frac{\Omega_m}{2(\Omega_m - 1)}(1 - \cos\eta),$$

$$H_0 t = \frac{\Omega_m}{2(\Omega_m - 1)^{3/2}}(\eta - \sin\eta) \tag{8.80}$$

describe a solution. (This is the same as Equation 4.24 of Section 4.5, for $e = 0$ – why?) Show that $\mathcal{R}(t)$ is largest when $\eta = \pi$, at the *turn-around* time $t_{ta} = \pi\Omega_m/[2H_0(\Omega_m - 1)^{3/2}]$, and that the sphere collapses to high density at time $2t_{ta}$.

At time t_0, suppose that this denser region is expanding at the same rate as its surroundings, and that t_0 is early enough that we can apply Equation 8.27: $\mathcal{R}(t) \propto t^{2/3}$ so that $\rho(t) \propto 1/t^2$, and $t_0 H_0 = 2/3$. Using the result of Problem 8.16, show that, between t_0 and t_{ta}, the density ρ_{out} outside the sphere drops such that

$$\frac{\rho_{\text{out}}(t_{\text{ta}})}{\rho_{\text{out}}(t_0)} = \left[\frac{9\pi^2}{16}\frac{\Omega_{\text{m}}^2}{(1-\Omega_{\text{m}})^3}\right] \text{ while inside } \frac{\rho_{\text{in}}(t_{\text{ta}})}{\rho_{\text{in}}(t_0)} = \left(\frac{\Omega_{\text{m}}-1}{\Omega_{\text{m}}}\right)^3. \quad (8.81)$$

So $\rho_{\text{in}}(t_{\text{ta}})/\rho_{\text{out}}(t_{\text{ta}}) = (3\pi/4)^2$. As it turns around and begins to collapse, this sphere is roughly 5.6 times denser than its surroundings.

Just as the free-fall time of Equation 3.23 is the same for all particles in a sphere of uniform density, the collapse time $2t_{\text{ta}}$ is the same throughout this sphere. So, in our simple model, all the particles reach the center at the same moment. In the real cosmos, they would have small random motions which prevent this. The dark matter and any stars present will undergo *violent relaxation* (see Section 6.2) as they settle into virial equilibrium. Gas can lose energy by radiating heat away as it is compressed. Once our protocluster settles into equilibrium, the virial theorem tells us that its energy $\mathcal{E}_1 = \mathcal{PE}_1 + \mathcal{KE}_1 = -\mathcal{PE}_1/2$. The final energy $\mathcal{E}_1$ can be no greater than the total energy $\mathcal{E}_0 = \mathcal{PE}_0$ when it was at rest at time t_{ta}, poised between expansion and contraction: so we must have $\mathcal{PE}_1 < 2\mathcal{PE}_0$.

Problem 8.31 Use Equation 3.33 for the potential energy $\mathcal{PE}$ of a galaxy of stars to show that, if the distances between stars all shrink by a factor f, so the density increases as $1/f^3$, then $\mathcal{PE}$ increases as $1/f$.

If we make the too-simple approximation that the collapse is *homologous*, so that all distances between particles shrink by an equal factor, Problem 8.31 tells us that our protocluster's final radius is no more than half as large as it was at turn-around, and the density is at least eight times greater. Meanwhile, the cosmos continues to expand, and its average density has dropped at least four times since t_{ta} (why?). So at the time that it reaches virial equilibrium, our cluster is $4 \times 8 \times 5.6 \approx 180$ times denser than the critical density for the Universe around it: recall Problem 8.2. In a galaxy cluster, we define the radius r_{200} such that, within it, the average density is 200 times the critical density; r_{200} is sometimes called the *virial radius*. At larger radii, the cluster cannot yet be relaxed and in virial equilibrium. Even the relaxed core will be disturbed when new galaxies fall through it as they join the cluster.

We can use the 'top-hat' model to estimate when the galaxies and clusters could have formed. Within the Sun's orbit, the Milky Way's density averages to $10^5 \rho_{\text{crit}}$. So when it collapsed at time $2t_{\text{ta}}$, the *average* cosmic density was no more than 500 times the present critical density; $\Omega_m = 0.3$, so this is 1700 times the present average density. The average density varies as $(1 + z)^3$, so the collapse was at $1 + z \leq (1700)^{1/3} \approx 12$. It could have been later, since the gas can radiate away energy, and become denser than the virial theorem predicts. But it could not have taken place any earlier.

Problem 8.32 In Problem 7.7 we found that, in the core of the Virgo cluster, luminous galaxies are packed 2500 times more densely than the cosmic average. Assuming that dark and luminous matter are well mixed in the cluster, show that its core could not have assembled before redshift $z = 1.3$. How early could the central region of NGC 1399, from Problem 6.4, come together?

8.5.4   Using galaxies to test model cosmologies

How well does the benchmark cosmology with cold dark matter account for real galaxies? Its huge success is to explain why the cosmic microwave background is so smooth, while the distribution of galaxies is so lumpy. We can even explain the shape of the power spectrum $P(k)$ in Figure 8.17, which describes the non-uniformity. That power spectrum requires that the smallest lumps of matter are now densest, as we saw in Figure 8.6. Problem 8.30 shows that they are the first to stop expanding and collapse on themselves. So we might expect that all galaxies will contain some very dense regions, which should have made stars early on. Even the smallest galaxies should have some very old stars – as we saw in Section 4.4 for the dwarf galaxies of the Local Group.

Structures that collapsed most recently should be larger and less dense than those that formed earlier. Using Equation 8.21 for the critical density, we can find the mass $\mathcal{M}_{200}$ measured within the virial radius r_{200}:

$$\mathcal{M}_{200} = \frac{4}{3}\pi r_{200}^3 \times 200\rho_{\text{crit}} = \frac{100 r_{200}^3 H^2(t)}{G}, \tag{8.82}$$

while the speed of a circular orbit at that radius is

$$V_c^2(r_{200}) = \frac{G\mathcal{M}_{200}}{r_{200}}, \quad \text{so } \mathcal{M}_{200}(t) = \frac{V_c^3(r_{200})}{10GH(t)}. \tag{8.83}$$

So, if we measure rotational or random speeds near the radius r_{200} (or if they do not change very much with radius), the mass or luminosity should increase steeply with those measured speeds.

We see this pattern in the Tully–Fisher relation for disk galaxies (Figure 5.23), the fundamental plane for elliptical galaxies (Figure 6.13), and the relation between temperature and X-ray luminosity for gas in galaxy clusters (Figure 7.12).

In the past the Hubble parameter $H(t)$ was larger, so we expect that temperatures and speeds were higher for a given mass or luminosity. Figure 6.13 shows this effect, but in Figure 7.12 galaxy clusters at $z \sim 1$ follow the same relation as local objects. Adding gas to simulations like that of Figure 8.16 does result in model galaxies that follow Tully and Fisher's dependence of mass on rotation speed. But the 'galaxies' fail to gather enough gas from large distances, so the disk has too little angular momentum and its radius is too small.

The slope of $P(k)$ in Figure 8.17 means that small objects will be far more numerous than large ones. The smallest have roughly the solar mass, since this is the mass $\mathcal{M}_H(\text{WIMP})$ within the horizon when the random motions of the WIMPs drop below near-light speeds. The halo of a galaxy like the Milky Way is made by merging many thousands of smaller objects, most of which are torn apart. Those that fall in relatively late survive today as distinct objects: satellite dark halos. In models such as that in Figure 8.16 a Milky-Way-sized dark halo will have ~ 300 dark satellites massive enough to have $V_c > 10 \,\mathrm{km\,s^{-1}}$. But the real Milky Way only has ten or so luminous satellites. Choosing 'warm' dark matter, for which the random motions remained relativistic until the mass within the horizon $\mathcal{M}_H(\text{WIMP}) \gtrsim 10^9 \mathcal{M}_\odot$, would erase all but the largest satellites. Some theories of particle physics include a 'sterile neutrino' with mass $\sim 1 \,\mathrm{keV}$, which would have this property. Other possibilities are that their first few stars blew all the remaining gas out of most of these dark halos, or that fierce ultraviolet and X-ray radiation from the first galaxies heated it so far that it could not cool to make new stars. The Milky Way would then have 10 luminous satellites and 290 dark ones.

Another facet of the same difficulty is that we see large objects 'too early' in cosmic history. In Section 9.4 we will find that some massive galaxies have formed more than $10^{11} \mathcal{M}_\odot$ of stars, corresponding to more than $10^{12} \mathcal{M}_\odot$ of dark matter, less than 2 Gyr after the Big Bang, at $z \gtrsim 3$. The benchmark model with cold dark matter predicts that such early 'monster' galaxies should be extremely rare; it is not clear whether the model already conflicts with the observations.

If the dark matter is cold, then all galaxies should have very dense cores. Figure 8.6 shows that σ_R, the variation in density on lengthscale R, rises at small R or large k. So the first regions to collapse will be the smallest and also the densest. Equation 8.83 shows that the velocities of particles within them will also be low, because $H(t)$ is large. So the positions and velocities of these first objects are tightly grouped: the density in phase space is high. Simulations like Figure 8.16 show that, as the galaxy is built, such objects fall together into a very dense center: the density of WIMPs follows $\rho_w \propto r^{-\alpha}$, where $1 \lesssim \alpha \lesssim 1.5$. The Navarro–Frenk–White model of Equation 3.24 was developed to describe this central *cusp*. The prevalence of dwarfs in galaxy clusters (see Figure 7.8) shows that they must indeed be robust, and hence dense, to avoid being torn apart by tidal forces. But the stars we observe are never as concentrated as this model requires the WIMPS to be, and the rotation curves of spiral galaxies in Figure 5.21 also seem to rise more gently than this form of ρ_w would allow.

However, we know that galaxy centers contain mostly normal baryonic matter, and that gas physics is complex – we do not even know how to predict the masses of stars formed locally in our own Milky Way. So it is no surprise that we cannot yet use basic physics to calculate exactly how the galaxies should form. In the next chapter we turn to observations of the distant Universe, and to what we can learn by viewing galaxies and protogalactic gas as they were 8–10 Gyr ago as the Milky Way began to form its disk, and even earlier when our oldest stars were born.

Active galactic nuclei and the early history of galaxies

We begin this chapter by discussing galaxies with an *active nucleus*, a compact central region from which we observe substantial radiation that is *not* the light of stars or emission from the gas heated by them. Active nuclei emit strongly over the whole electromagnetic spectrum, including the radio, X-ray, and γ-ray regions where most galaxies hardly radiate at all. The most powerful of them, the quasars, easily outshine their host galaxies. With luminosities exceeding $10^{12} L_\odot$, many are bright enough to be seen most of the way across the observable Universe. But the emitting region may be no bigger than the solar system; its power source is probably the energy released by gas falling into a central black hole. Very luminous active nuclei, such as the quasars, were far more common when the Universe was 20%–40% of its present age than they are today; nuclear activity seems to be characteristic of a galaxy's early life.

In many bright quasars, narrow twin jets are seen to emerge from the nucleus; they are probably launched and kept narrow by strong magnetic fields that build up in the surrounding disk of inflowing matter. In some cases, the jets appear to move outward faster than the speed of light. This is an illusion: the motion is slower than, but close to, light speed. In Section 9.2 we discuss these and similar 'superluminal' jets from stellar-mass objects: microquasars, which are neutron stars and black holes accreting mass from a binary companion, and γ-ray bursts, the final explosion of a very massive star.

In Section 9.3 we consider gas lying between us and a distant galaxy or quasar, which produces absorption lines in its spectrum. Most of the absorbing material is very distant from the quasar, and simply lies along our line of sight to it. The denser gas is probably in the outer parts of galaxies, while the most tenuous material, only a few times denser than the cosmic average, follows the filamentary 'cosmic web' of the dark matter. Surprisingly, this gas is not pristine hydrogen and helium; even when it lies far from any galaxy, it is polluted with the heavy elements which result from nuclear burning in stars.

In the last section of this final chapter, we turn to the question of how today's galaxies grew out of the primeval mixture of hydrogen and helium. Roughly halfway in time back to the Big Bang, galaxies appear fairly normal although starbirth was more vigorous than it is at present. Beyond a redshift $z \sim 2$, they are furiously star-forming, often very dusty, and can no longer be classified according to the scheme of Figure 1.11. The most distant observed systems are seen at $z \sim 6$, less than a gigayear after the Big Bang. New and more sensitive telescopes in the infrared and millimeter-wave regions promise us a much clearer view of the birth of the galaxies.

9.1   Active galactic nuclei

Twinkle, twinkle, little star,
We know exactly what you are:
Nuclear furnace in the sky,
You'll burn to ashes, by and by.

But twinkle, twinkle, quasi-star,
Biggest puzzle from afar;
How unlike the other ones,
Brighter than a trillion suns.
Twinkle, twinkle, quasi-star,
How we wonder what *you* are . . .
       after G. Gamow and N. Calder

Many galactic nuclei are very luminous at optical, ultraviolet, and X-ray wavelengths. Others are far dimmer than their host galaxies in these spectral regions, but are strong radio sources. What they have in common is a large energy output from a very small volume, and internal motions that are *relativistic*, with speeds $> 0.1c$ and often much larger.

The optical and ultraviolet spectrum of a quasar typically shows strong broad emission lines characteristic of moderately dense gas (Figure 9.1). The widths of the lines correspond to the Doppler shifts expected from emitting gas travelling at speeds $\sim 10\,000\,\mathrm{km\,s^{-1}}$. These emitting clouds are moving much faster than the galaxy's stars, which typically orbit at a few hundred kilometers per second. Many active nuclei are variable, changing their luminosity substantially within a few months, days, or even hours. The emission lines also strengthen and decline, within a few days or weeks. To allow such fast variability, both broad lines and continuum radiation must come from a region no more than a few light-weeks across.

This tiny volume contains a huge mass. We can use Equation 3.20 to calculate the gravitational force required to prevent the clouds that produce the broad emission lines from escaping out of the nucleus. For velocities $V \sim 10^4\,\mathrm{km\,s^{-1}}$, and radii $r = 0.01$ pc or about two light-weeks, the inferred mass is $\sim 10^8 \mathcal{M}_\odot$. In the nearby radio galaxy M87, we have $\sim 3 \times 10^9 \mathcal{M}_\odot$ within 10 pc of the center

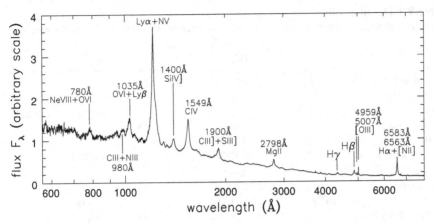

Fig. 9.1. The ultraviolet and optical spectrum of an 'average' radio-quiet quasar – R. Telfer *et al.* 2002 *Ap J* **565**, 773.

(Problem 6.16). The only way to pack the mass of a hundred million suns into a region little bigger than the solar system is as a *black hole*. We then expect the active nucleus to generate its power within a few times the *Schwarzschild radius* R_s. For a mass $\mathcal{M}_{BH}$, this is

$$R_s = \frac{2G\mathcal{M}_{BH}}{c^2} \approx 3 \times \frac{\mathcal{M}_{BH}}{\mathcal{M}_\odot} \text{ km}. \tag{9.1}$$

Problem 9.1 Show that, for a black hole with the Earth's mass, $R_s \approx 1$ cm, whereas if $\mathcal{M}_{BH} = 10^8 \mathcal{M}_\odot$, $R_s \approx 2$AU or 15 light-minutes. What is R_s for the black hole in the Seyfert galaxy NGC 4258, of Problem 5.15?

Broad emission lines from a galactic nucleus were first reported in 1907, in the early days of galaxy spectroscopy, but no systematic study was made until 1943. Then, Carl Seyfert published a list of 12 galaxies in which the nuclear spectrum showed strong broad emission lines of ions that could be excited only by photons more energetic than those of the young stars that ionize HII regions. These were later divided into the *Seyfert 1* class, with very broad emission lines like those of Figure 9.1, and *Seyfert 2* spectra with lines $\lesssim$1000 km s^{-1} wide. Most of Seyfert's galaxies were spirals, but his list included the huge cD galaxy NGC 1275 at the center of the Perseus cluster of galaxies, which is an elliptical; see Figure 7.9. Table 9.1 shows that 1%–2% of luminous galaxies have Seyfert nuclei.

In the 1950s, as radio astronomy blossomed, many of the strongest radio sources were found to be associated with luminous elliptical galaxies; these are now called *radio galaxies*. In many of these, twin radio-bright lobes, each up to 1 Mpc across, straddle the galaxy. The radio emission is *nonthermal*, produced by energetic particles moving through magnetic fields. For some years, radio

Table 9.1 Densities of normal and active galaxies

Type	Locally (Gpc^{-3})	At $z \sim 1$ (*Gpc^{-3})	$z \sim 2$–3 (*Gpc^{-3})	$z \sim 4$–5 (*Gpc^{-3})
Luminous galaxies: $L > 0.3L_\star$ (Fig. 1.16)	7 000 000	20 000 000		
Lyman break galaxies: $L > 0.3L_\star$			1 000 000	(300 000)
LIRGs: $L_{FIR} > 10^{11} L_\odot$	30 000	3 000 000		
ULIRGs: $L_{FIR} > 10^{12} L_\odot$	<10 000	2 000 000		
Massive galaxies: $L > 2 - 3L_\star$	400 000^a		200 000^b	(10 000^c)
Seyfert galaxies	100 000			
Radio galaxies: $L_r > 2 \times 10^8 L_\odot$	1 000			
X-ray AGN: $L_X > 8 \times 10^{10} L_\odot$	100		5 000	
$\quad L_X > 2.5 \times 10^9 L_\odot$	20 000	100 000	30 000	
Quasars: $L > 25L_\star$	90			
$\quad L > 100L_\star$ (Fig. 8.13)	20		600	50
Radio-loud quasars: $L_r > 5 \times 10^8 L_\odot$	4			
$\quad L_r > 3 \times 10^{10} L_\odot$ (Fig. 8.13)	0.004		0.6	0.05

*Densities per *comoving* Gpc^3 with benchmark cosmology; $L_\star \approx 2 \times 10^{10} L_\odot$ from Figure 1.16. Values in ( ) are known to no better than a factor of 3–5.
^{a}Local galaxies from 2dF; ^{b} 'red and dead' galaxies at $z \sim 1.5$; ^{c} submillimeter-detected galaxies.

astronomers were puzzled by finding some galaxies with radio-bright compact nuclei and others with huge lobes. Better radio maps revealed tiny central cores at the nuclei of radio galaxies, linked to the outer lobes by bright linear jets that carried energy outward.

The first *quasars* (for 'quasi-stellar radio source') were discovered in the following decade, as 'radio galaxies with no galaxy'. They appeared pointlike in optical photographs; only their enormous redshifts betrayed that they were not Galactic stars. Rather, they were gigaparsecs distant, and hence extremely luminous. Subsequently 'radio-quiet' quasars, called *quasi-stellar objects*, or *QSOs*, were found by searching for objects that appeared stellar, but emitted too strongly at infrared or ultraviolet wavelengths relative to their brightness in visible light. Radio-quiet QSOs outnumber radio-loud quasars by at least a factor of 30; both are now believed to be variants of the same type of object, so we use the term 'quasar' to include the QSOs. In the 1980s, deep images of nearby quasars showed us that they were in fact the bright nuclei of galaxies, so luminous as to outshine the surrounding stars. Most astronomers now regard quasars as more powerful versions of a Seyfert nucleus. Quasars cover a very wide range in luminosity: Table 9.1 shows that the most powerful are also the rarest.

BL Lac objects are quasars with very weak emission lines; they may be the most extreme form of active nucleus. They are named after their prototype, which was originally thought to be a variable star, and designated BL Lacertae. The light output of these objects can fluctuate enormously within a few days; one was seen to double its brightness within three hours. Both radio and optical emission are strongly polarized. Quasars with the same pattern of variability, but having stronger emission lines, are called 'optically violently variable' (OVV) quasars; these and the BL Lac objects are collectively known as *blazars*. All known blazars are radio-loud. Blazars appear as the most luminous objects in the Universe: if their light were emitted equally in all directions – but see below for reasons why we do not think this is the case – their total output would exceed $10^{14} L_\odot$.

Active nuclei all derive their energy in the same way: gas gives up potential energy as it falls into a black hole. Here, we briefly sketch some of the physical processes involved in turning that energy into the radiation we observe, and explain how a single basic model might explain the diversity observed among active galaxies.

Further reading: B. M. Peterson, 1997, *Active Galactic Nuclei* (Cambridge University Press, Cambridge, UK) reviews the observations. For radio galaxies, see B. F. Burke and F. Graham-Smith, 1997, *An Introduction to Radio Astronomy* (Cambridge University Press, Cambridge, UK). For relevant physics, see M. S. Longair, *High Energy Astrophysics*, 2nd edition: Volume 1, *Particles, Photons and Their Detection* 1992; Volume 2, *Stars, the Galaxy and the Interstellar Medium* 1994 (Cambridge University Press, Cambridge, UK); a graduate-level text is F. H. Shu, 1991, *The Physics of Astrophysics*, Volume 1: *Radiation* (University Science Books, Mill Valley, California).

9.1.1 Seyfert galaxies

Figure 9.2 shows the Seyfert 2 galaxy NGC 4258. In visible light, we see the spiral arms and bright nucleus of a galaxy of type Sbc. The radio map shows emission from bright knots in spiral arms, but also two narrow jets emerging from the nucleus, bending into an 'S' shape, and terminating in twin bright lobes on either side of the galaxy. The nucleus is a bright pointlike source in both radio and X-ray bands. The radio emission is strongly polarized, which tells us that it is *synchrotron* radiation, given off as electrons spiral around lines of magnetic field at speeds close to that of light. In this galaxy the radio jets are unusually strong; they overlap with thin helical jets of ionized gas, from which we see emission lines in the optical and ultraviolet. NGC 4258 shows two features common to Seyfert galaxies: radiation that does not appear to originate from stars, and the directed outflow of matter and energy. As we saw in Section 5.5, the nucleus is surrounded by a small disk of fast-rotating gas which we see edge-on; it probably harbors a black hole with a mass exceeding $10^7 M_\odot$. Seyfert galaxies and quasars shine brightly at infrared, ultraviolet, and X-ray wavelengths, as well as in visible light; but most are not strong radio sources. The quantity νL_ν is roughly constant from the infrared to the X-rays; equal energy is emitted in each interval over which the frequency increases by a factor of ten. The luminosity drops at γ-ray energies, above $\sim 100\,\mathrm{keV}$. Seyfert 2 nuclei tend to be less luminous than the Seyfert 1 nuclei

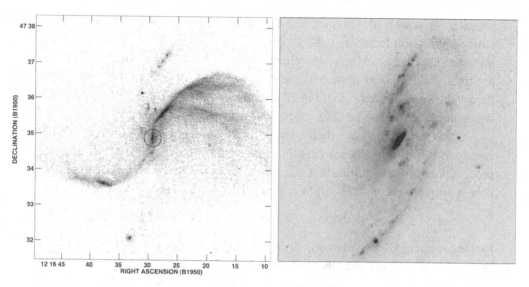

Fig. 9.2. Seyfert 2 galaxy NGC 4258 (Sbc). Left, a radio map at 20 cm shows oppositely directed twin jets (within the circle), channelling radio-bright plasma from the nucleus to lobes at east and west, and HII regions in the spiral arms. Right, an image in the U band at 3700 Å shows the bright center, and brilliant knots of young stars in the spiral arms. At distance $d \approx 7$ Mpc, $1' = 2$ kpc – G. Cecil.

in the spectral regions from infrared to soft X-rays, but have similar power in γ-rays. Seyfert nuclei have $M_V \gtrsim -22.5$ or $L \lesssim 10^{11} L_\odot$; more luminous objects would be classified as quasars. The X-ray power ranges from $\sim 2 \times 10^8 L_\odot$ to $10^{11} L_\odot$.

The active nucleus is probably powered by gas that falls into a central black hole. Because it inevitably has some angular momentum, infalling gas forms an *accretion disk*. Viscosity causes the disk gas to spiral slowly inward, heating up and radiating away its gravitational potential energy, until it reaches the last stable orbit around the black hole (see Problem 3.20) and falls in. Theoretically, up to 42% of $\mathcal{M}c^2$, the rest energy of the material, can be extracted from a mass $\mathcal{M}$ falling into a black hole. In practice, astronomers do not expect more than $\sim 0.1 \mathcal{M}c^2$ to emerge as radiation. This is still much more efficient than nuclear burning, which releases less than 1% of $\mathcal{M}c^2$. Magnetic fields are pulled inward with the flow of the hot ionized gas. Close to the black hole, the field can become strong enough to channel twin jets of relativistic plasma, moving out along the spin axis at speeds close to that of light.

Some of the infrared flux and all the radio emission comes from particles accelerated to relativistic energies in the jet; paradoxically, we can use long-wavelength radio waves to trace extremely energetic processes. Electrons in the jet scatter some radio or visible-light photons, boosting them to γ-ray energies. The X-ray and ultraviolet emission might come from the hot innermost part of the

disk, or from the jet; the visible light probably originates further out in the disk
or jet. Additional infrared light may be emitted by surrounding dust grains heated
by the nuclear radiation.

The light of a Seyfert nucleus is intense enough to exert considerable pressure
on gas around it. If that outward push is too strong, no gas can fall into the
center, and the nucleus runs out of fuel. So we have a limit on the luminosity
that it could sustain. For a spherically symmetric object, we can calculate at what
point radiation pressure just balances the inward force of gravity. We assume
that the gas near the nucleus is fully ionized hydrogen, and we calculate the
outward force due to Thomson scattering by the electrons; scattering from protons
is much less efficient because of their larger mass. The cross-section σ_T of each
electron is

$$\sigma_T = \frac{e^4}{6\pi\,\epsilon_0^2 c^4 m_e^2} \text{ (SI)} \quad \text{or} \quad \frac{8\pi\,e^4}{3c^4 m_e^2} \text{ (cgs)} = 6.653 \times 10^{-25} \text{ cm}^2, \quad (9.2)$$

where e is the charge on the electron and m_e is its mass. If the central source
emits photons carrying luminosity L, these have momentum L/c, so an electron
at radius r receives momentum $\sigma_T L/(4\pi r^2 c)$ each second.

The electrons cannot move outward unless they take the protons with them;
electrostatic forces are strong enough to prevent the positive and negative charges
from separating. So we must compare the combined outward force on the proton
and the electron with the inward force of gravity on both of them. If the central
object has mass $\mathcal{M}$, radiation pressure and gravity balance when

$$\frac{G\mathcal{M}(m_e + m_p)}{r^2} \approx \frac{G\mathcal{M}m_p}{r^2} = \frac{\sigma_T L}{4\pi r^2 c}, \quad (9.3)$$

where m_p is the proton mass. The *Eddington luminosity* L_E is the largest value of
L that still allows material to fall inward:

$$L_E = \frac{4\pi G\mathcal{M}m_p c}{\sigma_T} \approx 1.3 \times 10^{31} \frac{\mathcal{M}}{\mathcal{M}_\odot} \text{ W} \approx 30\,000 \times \frac{\mathcal{M}}{\mathcal{M}_\odot} L_\odot, \quad (9.4)$$

where $L_\odot$ is the Sun's bolometric luminosity of 3.86×10^{26} W. Stars like the
Sun come nowhere near the Eddington luminosity, though the brightest supergiants approach it. Although part of the radiation of a Seyfert nucleus comes
out in a directed jet, its total luminosity is unlikely to be more than a few times
greater than L_E. If $L \sim 10^9 L_\odot$ then Equation 9.4 shows that the central mass
must exceed $10^7 \mathcal{M}_\odot$, to avoid blowing away all the gas that could fuel the active
nucleus.

> **Problem 9.2** As a mass m of gas falls into a black hole, at most $0.1mc^2$ is likely to emerge as radiation; the rest is swallowed by the black hole. Show that the Eddington luminosity for a black hole of mass $\mathcal{M}$ is equivalent to $2 \times 10^{-9}\mathcal{M}c^2\,\mathrm{yr}^{-1}$. Explain why we expect the black hole's mass to grow by at least a factor of e every 5×10^7 years.

The spectrum of a Seyfert 1 nucleus is similar to the quasar spectrum shown in Figure 9.1; broad emission lines from a wide range of ions are present. Some of these, such as the Balmer lines of hydrogen and lines of singly ionized species such as MgII, can be excited by ultraviolet photons; they are also seen in the HII regions around hot stars. Others, such as the multiply ionized species NV and OVI, require higher energies. The relative strengths of the various lines can be understood if they are photoionized by radiation from the nucleus; its soft X-rays excite the high-ionization lines.

Figure 9.3 illustrates a basic model for an active nucleus. In the *broad-line region*, gas forms dense clouds with $n_H \gtrsim 10^{10}$ atoms cm^{-3}. From most Seyfert nuclei, we see continuum radiation with wavelengths $\lambda < 912\,\text{Å}$, shortward of the Lyman limit. These photons would be absorbed if they had to travel through the broad-line emitting gas; so the clouds must cover up only a small fraction of the central source. The emission lines we observe are the sum of Doppler-shifted components from many individual clouds close to the nucleus, each moving at thousands of kilometers per second. As the continuum radiation waxes and wanes, so do the broad emission lines. High-ionization lines follow the continuum with a delay of a few days. Those of low ionization respond later, within a few weeks, showing that they originate further from the nucleus.

The narrow emission lines, such as [OII] at 3727 Å and [OIII] at 5007 Å, come from *forbidden* transitions; see Section 1.2. Forbidden lines are seen only when the density $n_H \lesssim 10^8$ atoms cm^{-3}; at normal laboratory densities, collisions would knock the ion out of its excited state before a photon could be emitted. The forbidden lines of Seyfert galaxies and quasars have widths corresponding to velocities below $1000\,\mathrm{km\,s^{-1}}$. Forbidden lines have not been observed to vary as the nucleus brightens, indicating that they originate further from the nucleus than the broad lines. The *narrow-line region* is generally a few kiloparsecs across, although in some objects ionized gas has been seen hundreds of kiloparsecs from the center. It is probably a combination of gas glowing in response to the active nucleus and material ionized by massive stars nearby.

Further reading: on the emission-line spectra of active nuclei, see D. E. Osterbrock and G. J. Ferland, 2005, *Astrophysics of Gaseous Nebulae and Active Galactic Nuclei*, 2nd edition (University Science Books, Mill Valley, California).

In Seyfert 2 nuclei, most of the emission lines have roughly the same width, $\lesssim 1000\,\mathrm{km\,s^{-1}}$. Some strong lines, such as Hα, may show very faint broad wings.

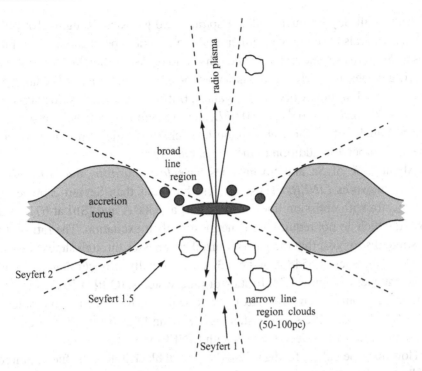

Fig. 9.3. A simple model for an active nucleus. Energetic twin jets emerge at near-light speeds along the spin axis of the central accretion disk. Radiation from the disk and jet photoionizes the dense fast-moving clouds of the broad-line region, which is often $\lesssim 1$ pc across. The more diffuse and slower-moving gas of the narrow-line region is at larger radii. Observers looking directly down the jet would see a brilliant Seyfert 1 nucleus; but when it is viewed sideways, through the opaque accretion torus (gray), we have a Seyfert 2 galaxy.

Intermediate classes are used to indicate their strength; a galaxy with fairly weak broad wings might be labelled a Seyfert 1.8 or 1.9. Some Seyfert 2 galaxies, including NGC 4258, have been observed in polarized light: the spectrum then resembles that of a Seyfert 1, with broad emission lines. Reflected light is generally polarized; that is why polaroid sunglasses reduce the glare of light reflected from snow or water. Seyfert 2 galaxies probably have a hidden broad-line region, which we can see only by the reflection of its light in a layer of dust or gas. Figure 9.3 illustrates how a galaxy could appear as either a Seyfert 1 or a Seyfert 2, depending on the viewing angle. This object would be a Seyfert 1 nucleus for observers looking down on the central disk. For those viewing the galaxy close to the plane of the inner disk (as we do for NGC 4258), the continuum source and the broad-line region are hidden by the doughnut-shaped *accretion torus*; they would see a Seyfert 2 nucleus. Because lower-energy X-rays from the nucleus are more easily absorbed by the gas torus, the spectra of Seyfert 2 galaxies show a larger proportion of energetic 'hard' X-rays, those with energies above a few keV, than is found in spectra of Seyfert 1 galaxies.

Almost all Seyfert nuclei inhabit spiral or S0 galaxies. Roughly 10% of all Sa and Sb spirals have them, so either all these galaxies spend about 10% of their lives as Seyferts, or one in ten of them has a long-lasting Seyfert nucleus. Most Seyfert galaxies are fairly luminous with $L > 0.3L_\star$, where $L_\star$ of Equation 1.24 represents the luminosity of a sizable galaxy. But NGC 4395, a tiny Sd galaxy with $M_B = -17.1$ or $L_B \sim 10^9 L_\odot \sim 0.05L_\star$, has a Seyfert 1 nucleus. The spectra of Seyfert 2 nuclei often show absorption lines characteristic of hot massive stars; there is a starburst in addition to the nuclear activity.

About 25% of Sa and Sb galaxies have *low-ionization nuclear emission regions*, known as *LINERs*. These are less luminous than Seyfert 2 nuclei, and have spectra with emission lines such as [OI] at 6300 Å and [SII] at 6716 Å and 6731 Å, which do not require high energies for their excitation. The ratios of the line strengths suggest that the gas is ionized as it passes through shock waves. In LINERs [NII] lines at 6548 Å and 6583 Å are normally stronger than Hα, unlike for the galaxies of Figure 5.24. In star-forming systems, [OIII] at 5007 Å is strong relative to Hβ only when [NII]/Hα is weak, while in active nuclei both ratios are normally $>1/3$. In large surveys such as the Sloan Digital Sky Survey and 2dF, we use these ratios to select galaxies with LINER or Seyfert nuclei.

How does the galaxy feed gas into the central black hole? The fuel required is usually less than the mass lost by aging stars in a sizable galaxy. Large quantities of molecular gas, above $10^8 M_\odot$, have been found in the central regions of some nearby Seyfert galaxies. But several nearby disk galaxies, including our Milky Way, have gas at their centers, and nuclear black holes exceeding $10^6 M_\odot$ – with little or no nuclear activity. The presence of dilute gas or stars near the black hole is insufficient to fuel activity. Large concentrations of massive stars could move the interstellar gas around, aiding the accretion. Intense star formation is often found in Seyfert nuclei, supporting this idea. But many radio galaxies conspicuously lack any sign of starbirth.

Problem 9.3 Show that $10^{12} L_\odot$ corresponds to an energy output of $0.1 M_\odot c^2$ per year. As they age, stars like those in the solar neighborhood eject about $M_\odot$ per year of gas for each $10^{10} L_\odot$ of stars. If all the gas lost by stars in our Galaxy could be funnelled into the center, and 10% of its mass released as energy, how bright would the Milky Way's nucleus be?

9.1.2 Radio galaxies

If our eyes could see in radio wavelengths, many of the brightest objects in the sky would not be within our Milky Way; they would be the luminous active nuclei of galaxies halfway across the Universe. Normal stars, and normal galaxies, are not powerful radio sources. The Milky Way's optical luminosity exceeds $10^{10} L_\odot$; but

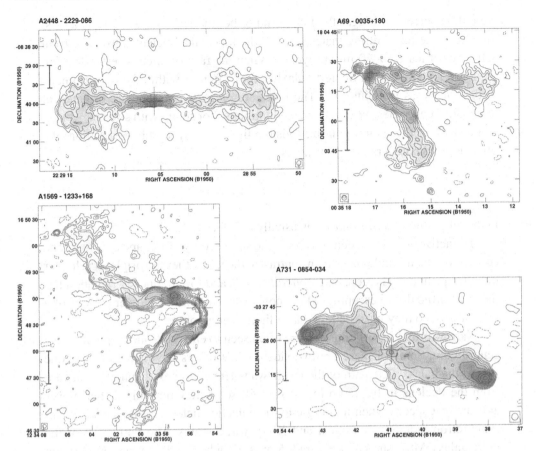

Fig. 9.4. Four radio galaxies, observed at 20 cm: galaxy luminosity L is measured in the R band, radio power P in units of 10^{25} W Hz^{-1} at 20 cm. Clockwise from top left: a twin jet with $L \approx 6L_\star$, $P \approx 1$; a narrow-angle tail source ($L \approx 3L_\star$, $P \approx 1$); an edge-brightened classical double ($L \approx 1.4L_\star$, $P \approx 7$); and a wide-angle tail ($L \approx 2L_\star$, $P \approx 1.7$). The scale bar shows 50 kpc, assuming $H_0 = 75$ km s^{-1} Mpc^{-1} and $\Omega_0 = 1$ – M. Ledlow.

its radio output is only about 10^{30} W, or about $2500L_\odot$ when measured in terms of the Sun's bolometric luminosity $L_{\mathrm{bol},\odot} = 3.86 \times 10^{26}$ W. Seyfert galaxies are 100–1000 times more luminous in the radio waveband, while galaxies with radio power in excess of about 10^{34} W or $\sim 10^8 L_\odot$ are labelled *radio galaxies*. The most powerful radio galaxies and quasars radiate up to 10^{38} W or $\sim 10^{12} L_\odot$. The emission is highly polarized synchrotron radiation. Radio galaxies are much rarer than Seyfert nuclei: Table 9.1 shows that there is only one for every 10^4 normal galaxies.

Radio galaxies have a distinctive structure, with twin radio-bright *lobes* on either side of the galaxy. The galaxy in the lower right corner of Figure 9.4 is a classical radio galaxy, brightest at the outer edges of the twin lobes. The stronger the radio source, the bigger the lobes tend to be; the largest are $\sim$3 Mpc across. To allow time for emitting material to fill the lobes, the nucleus must have been active for at least 10–50 million years. When the radio source is less powerful, the lobes are smaller; in Seyfert galaxies, they often fit within the optical image of the galaxy, as in Figure 9.2. The lobes are *optically thin* and are brightest at low radio frequencies. Within them are luminous 'hot spots' with sizes of $\sim$1 kpc. About 10% of these emit polarized visible light, also synchrotron radiation. We often approximate a radio spectrum as

$$L_\nu \propto \nu^{-\alpha};\qquad\qquad (9.5)$$

in the lobes, the *spectral index* α is usually $0.7 \lesssim \alpha \lesssim 1.2$.

The active nucleus is seen as a *core* radio source, only a few parsecs across. The cores have spectral index $\alpha \sim 0$; in contrast to the lobes, they are brightest at higher radio frequencies. The cores are *optically thick*, and low-frequency radiation has the most difficulty in escaping. Many cores vary in luminosity over periods of a year or less; so they must be less than a light-year across.

Narrow bright *jets* of emission are often seen to emerge from deep within the central core. Some of these are two-sided, while others are visible on only one side of the galaxy. The path of the jet shows where energy is channelled outward from the nucleus to the radio lobes; we will see below that this matter moves at near-light speeds when it is close to the galactic nucleus. Some jets also emit synchrotron radiation at optical and X-ray wavelengths. The optical jet of the radio galaxy M87, shown in Figure 9.5, was already noted as a 'curious straight ray' in a 1918 report by H. D. Curtis. Much later, M87 was discovered to be a radio galaxy; the radio jet coincides with the optical jet and is also bright in X-rays.

Galaxies with large radio lobes turn out to be giant ellipticals and cD galaxies. Often, they are the brightest galaxies in a cluster. Most radio galaxies appear fairly normal in visible light, although some, especially the more powerful, are very peculiar objects indeed. Many have blue colors, and signs of recent star formation at the center. Their nuclei can show an emission-line spectrum similar to that of a Seyfert: an example is the bizarre elliptical NGC 1275 in Perseus (Figure 7.9). When a radio source is present in a less luminous elliptical that is orbiting within the cluster, its motion through the hot cluster gas can sweep the jets sideways into a 'C' shape. The top right and lower left panels of Figure 9.4 show a narrow-tail and a wide-tail source, respectively.

A relative paucity of cool gas appears to favor strong radio emission: radio galaxies are always ellipticals, while Seyfert galaxies are generally spirals. Seyfert galaxies are weaker radio sources than the radio galaxies. The core produces a larger fraction of their emission, and, if twin lobes are present, they are only a few kiloparsecs across, as in Figure 9.2. It is as though the radio lobes are 'smothered' by the dense gas around a Seyfert nucleus.

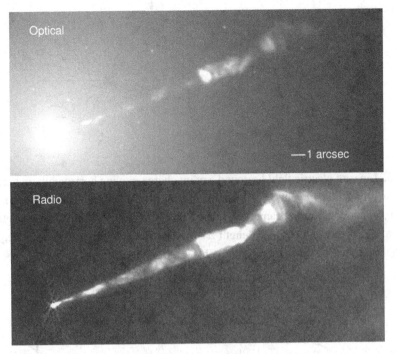

Fig. 9.5. A one-sided jet in the elliptical galaxy M87. Top, in visible light near 8000 Å, from the Hubble Space Telescope, the jet emerges from the glare of the galaxy's center; round white spots are globular clusters. Below, the image at 2 cm shows the radio-bright plasma; 1 arcsec ≈ 80 pc – J. Biretta.

9.1.3 Synchrotron emission from radio galaxies

The energy stored in the lobes of a giant radio galaxy is enormous. To estimate it, we use results from the books by Longair and by Shu, which readers should consult for more detail. Longair uses SI units, and Shu the cgs system which is still common in astronomical publications. An accelerated charge q radiates away its energy $\mathcal{E}$ at the rate

$$-\frac{d\mathcal{E}}{dt} = \frac{q^2|\mathbf{a}|^2}{6\pi\epsilon_0 c^3} \text{ (SI)} \quad \text{or} \quad \frac{2q^2|\mathbf{a}|^2}{3c^3} \text{ (cgs)}, \tag{9.6}$$

where $\mathbf{a}$ is the acceleration in the frame where the charge is instantaneously at rest (see formula 3.9 of Longair's book or Chapter 16 of Shu's). The radiation is polarized, with its electric vector perpendicular to the direction of the acceleration: we can think of the charge dragging its electric field lines along as it moves, as in Figure 9.6. In a uniform magnetic field $\mathbf{B}$, an electron spirals around the field lines with frequency

$$\nu_L = \frac{eB}{2\pi m_e} \text{ (SI)} \quad \text{or} \quad \frac{eB}{2\pi m_e c} \text{ (cgs)}. \tag{9.7}$$

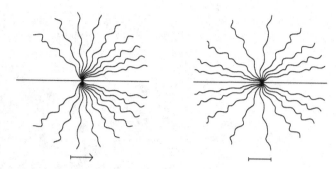

Fig. 9.6. Lines of electric field **E** around a point charge moving horizontally in harmonic motion with angular frequency ω. Left, the charge is centred and moving to the right at speed $c/2$; its radiation is beamed forward. Right, the charge is at rest in its rightmost position. The arrow and bar have length $2\pi c/\omega$, the wavelength of light with frequency ω; wiggles in the field lines have roughly this scale.

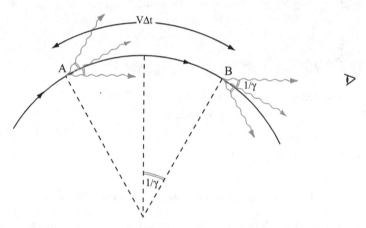

Fig. 9.7. An electron spirals with speed $V \approx c$ around a magnetic field pointing into the page; its radiation is beamed forward in the direction of travel.

The acceleration of an electron moving with speed V at an angle θ to the field lines is then $a = 2\pi \nu_L V \sin\theta$, so we can use Equation 9.6 to calculate the energy that it loses through *cyclotron* radiation at frequency ν_L. The radiation is emitted in a dipole pattern, so its intensity is highest in the direction of that component of the electron's motion that is perpendicular to the field. Because of their larger mass, protons are less strongly accelerated, and their radiation is weaker by the factor $(m_p/m_e)^2 \approx 3 \times 10^6$.

As its speed $V \to c$, the electron's inertia increases. Its orbital frequency drops to ν_L/γ, where $\gamma \equiv 1/\sqrt{1 - V^2/c^2}$, but the frequency of its radiation increases by a factor γ^2. When the electron moves relativistically, with $\gamma \gg 1$, it emits *synchrotron radiation*; almost all of the radiation propagating ahead of it is squeezed into a narrow cone, within an angle $1/\gamma$ of the forward direction. In Figure 9.7, the emission is beamed toward us only during the small interval

$\Delta t \sim 2/v_L$ when the electron is between points A and B. The arrival of that energy is squeezed into an even shorter time. Photons from point B are emitted later than those from A by a time Δt, but their arrival is delayed only by

$$\Delta t(1 - V/c) \approx \Delta t/(2\gamma^2) \sim 1/(\gamma^2 v_L). \tag{9.8}$$

Thus the frequency of the light received is not v_L/γ, but roughly $\gamma^2 v_L$. A more accurate calculation (see Chapter 18 of both Longair's book and Shu's book) shows that most power is emitted close to the frequency

$$v_c = \frac{3}{2}\gamma^2 v_L = 4.2\gamma^2 \left(\frac{B}{1\,\text{G or }10^{-6}\,\text{T}}\right)\,\text{MHz}. \tag{9.9}$$

Problem 9.4 Show that, in a radio lobe where $B \approx 10\,\mu\text{G}$, an electron radiating at 5 GHz must have $\gamma \sim 10^4$.

Equation 9.6 tells us how fast the electron loses its energy. We first compute its *four-velocity* $\tilde{u}$ and four-vector acceleration $\tilde{a} = d\tilde{u}/d\tau$, where τ is the *proper time* of Equation 8.9. In the instantaneous restframe of the spiralling electron $\tilde{a} = (0, \mathbf{a})$, so $|\mathbf{a}|^2$ of Equation 9.6 is equal to $|\tilde{a} \cdot \tilde{a}|$. But $\tilde{a} \cdot \tilde{a}$ is a Lorentz invariant, the same for all uniformly moving observers; we can compute it in our observer's frame. There, $dt/d\tau = \gamma$; apart from radiative losses, γ remains constant as the electron circles the field lines, so we have

$$\tilde{u} \equiv \frac{d\tilde{x}}{d\tau} = \frac{d\tilde{x}}{dt}\frac{dt}{d\tau} = \gamma\begin{pmatrix} 1 \\ \mathbf{v} \end{pmatrix} \quad \text{and} \quad \frac{d\tilde{u}}{d\tau} = \gamma^2\begin{pmatrix} 0 \\ d\mathbf{v}/dt \end{pmatrix} = \tilde{a}. \tag{9.10}$$

Since $|d\mathbf{v}/dt| = 2\pi(v_L/\gamma)V\sin\theta$, the product $|\tilde{a} \cdot \tilde{a}| = (2\pi\gamma v_L V \sin\theta)^2$; the radiated energy $-d\mathcal{E}/dt \propto \gamma^2$. (Compare this with Longair's derivation of formula 18.5.)

To calculate how fast the energy of an average electron decays, we assume that they move in random directions, so that $\sin^2\theta$ averages to $2/3$. When $\gamma \gg 1$, the energy loss is

$$-\frac{d\mathcal{E}}{dt} = \frac{4}{3}\sigma_T c U_{\text{mag}}\gamma^2, \quad \text{where } U_{\text{mag}} = \frac{B^2}{2\mu_0}\,\text{(SI), or } \frac{B^2}{8\pi}\,\text{(cgs)}. \tag{9.11}$$

Roughly half the electron's energy $\mathcal{E} = \gamma m_e c^2$ is gone after a time

$$t_{1/2} = \frac{\mathcal{E}}{2}\left|\frac{dt}{d\mathcal{E}}\right| \approx 170\left(\frac{10^{-5}\,\text{G or }10^{-11}\,\text{T}}{B}\right)^2\left(\frac{1000}{\gamma}\right)\,\text{Myr}. \tag{9.12}$$

For electrons radiating at a fixed frequency v, the energy $\mathcal{E} = \gamma m_e c^2$ is proportional to $B^{-1/2}$, so we can write

$$t_{1/2} \approx 34 \left(\frac{10^{-5}\,\text{G or } 10^{-11}\,\text{T}}{B} \right)^{3/2} \left(\frac{10^9\,\text{Hz}}{v} \right)^{1/2} \text{Myr.} \qquad (9.13)$$

For the fields of 10^{-5} G believed to be typical in radio lobes, the life of electrons emitting at 5 GHz is about 10 million years. But those responsible for the higher-frequency optical and X-ray synchrotron radiation of jets and hotspots lose their energy much more rapidly. In ambient magnetic fields of $\sim 10^{-4}$ G, electrons radiating visible light at 10^{15} Hz lose half their energy in 10^3–10^4 years, and those producing X-rays last no more than 100 years. Since the emitting regions lie many kiloparsecs from the galaxy's center, electrons must be boosted to high energies when they are well outside the nucleus. They are probably accelerated as they pass through shock waves in the jet, and are then scattered by tangled magnetic fields. Further X-ray and γ-ray emission can be produced by the *synchrotron-self-Compton* process. As radio-frequency photons scatter off relativistic electrons, the photon energy is increased by a factor $\sim \gamma^2$.

Equation 9.11 relates the power output L_v at frequency v from a volume $\mathcal{V}$, filled with a number density n of electrons radiating at that frequency, to the total energy U_e in the emitting electrons. We have

$$L_v \propto n\mathcal{V}\mathcal{E}^2 B^2, \quad \text{while } U_e = n\mathcal{V}\mathcal{E} \propto L_v B^{-3/2}. \qquad (9.14)$$

The energy required to produce the observed emission is the sum of the electron energy U_e, which is lower in a stronger magnetic field, and the magnetic energy $U_{\text{mag}} \propto \mathcal{V}B^2$, which becomes larger. The source's total energy is minimized close to *equipartition*, when $U_{\text{mag}} \approx U_e$. For the lobes of giant radio galaxies, this amounts to 10^{59}–10^{61} erg; 10^{60} erg represents the luminosity of 10^{10} suns over a gigayear, or the emission of a powerful radio source for 10^7 years. The energy stored in the compact core and jets is much less, only 10^{52}–10^{58} erg.

We have hardly any other information on the field strength and electron energy in radio sources, and we tend to assume that they are close to these equipartition values. Typically $B = 1$–$10\,\mu$G or 10^{-11}–10^{-12} T for the giant radio lobes, about the same strength as near the Sun's position in the disk of the Milky Way. Fields in the radio jets are about ten times higher. In the very compact cores, the equipartition magnetic fields are about 0.1 G. So electrons emitting at 5 GHz have $\gamma \sim 100$, and can radiate for only about 100 years.

9.1.4 Quasars

Quasars are active nuclei so bright as to outshine their host galaxies: all but the closest appear quasi-stellar in optical images. Their optical luminosities are

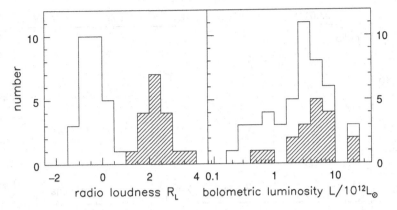

Fig. 9.8. Left, radio loudness $R_{\rm L} = \log_{10}[L_\nu(5\,{\rm GHz})/L_\nu(B\,{\rm band})]$ for a sample of 137 quasars. Right, radio-loud objects (those with $R_{\rm L} > 1$, shaded) are rarely found among the less luminous quasars with $L < 10^{12}L_\odot$ – J. McDowell.

$L_V \gtrsim 5L_\star$ or $10^{11}L_\odot$; fainter objects would be labelled Seyfert 1 nuclei. Quasars are the most luminous objects known and have been observed with redshifts $z > 6$, when the Universe was no more than one-seventh of its present size. Curiously, the spectra of quasars look very similar at all redshifts. It is difficult to estimate the composition of the broad-line clouds, but the relative strength of the lines shows that they have at least the solar abundance of heavy elements. Just less than a gigayear after the Big Bang, the central parts of some galaxies have already formed, and a first generation of stars has polluted the gas with metals.

The spectral energy distribution of a quasar is very different from that of a normal galaxy, with substantial power all the way from the radio to γ-rays. Only a few percent of quasars are strong radio sources (see Table 9.1). In the radio-quiet remainder, Figure 9.8 shows that the radio power is $\lesssim 1\%$ of its level in the loud variety. The central core and the jets of a radio-loud quasar are typically 10–100 times stronger than in radio galaxies; the core accounts for a larger fraction of the emission compared with the extended lobes. The quasars showing the highest optical polarization, and some of the blazars (see below), emit most of their energy as γ-rays: Figure 9.9 shows that νF_ν in γ-rays can be ten times as large as in the radio, millimeter, optical, or X-ray parts of the spectrum.

In the same way as for Seyfert 2 nuclei, a quasar can be hidden from us by the dense gas of the accretion torus shown in Figure 9.3, which conceals the clouds producing the broad emission lines. In these *Type 2 quasars*, we do not see the inner torus that gives out most of the optical and ultraviolet light, and the emission lines are less than 2000 km s^{-1} broad. We have found only the most luminous of them, from their powerful X-ray emission and very strong lines of [OIII] at 5007 Å. In both Seyfert 2 galaxies and Type 2 quasars, the intense radiation shining through the dense gas of the torus can power water masers beamed toward us and radiating at 22.2 GHz, like that in NGC 4258.

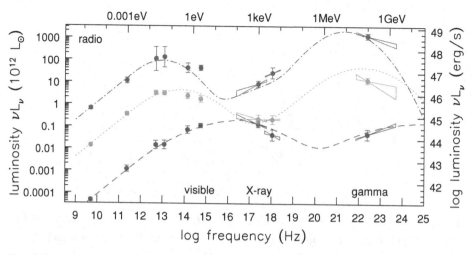

Fig. 9.9. Average spectral energy distributions for blazars, grouped by radio power: the most radio-bright are also the most luminous in γ-rays. The lower-energy peaks in the ultraviolet and X-ray regions represent synchrotron radiation from electrons in the jet; these photons scatter from the same spiralling electrons to produce the γ-ray peak. When the electrons are more energetic, both peaks move to higher frequency – G. Fossati.

Figure 8.13 shows that the very brightest active nuclei of all kinds were most common at $z \sim 2$, roughly 3 Gyr after the Big Bang. Then we see $30-100$ times more quasars than in the local Universe, while at $z \sim 5$ the density was only a few times larger than at present. Less luminous active nuclei show a pattern more like star-forming galaxies (look ahead to Figure 9.17), being most common closer to $z \sim 1$.

Nuclear activity is characteristic of a galaxy's youth, and the masses of black holes in galaxies today tell us that it must be a passing phase. Equation 9.4 tells us that in a quasar shining at $10^{12} L_\odot$ the central black hole has $M_{BH} \gtrsim 3 \times 10^7 M_\odot$, and must put on weight at $\gtrsim M_\odot$ per year while it maintains that power. So, if activity lasts for at least 100 Myr, a mass of at least $10^9 M_\odot$ should remain. No nearby galaxy has a black hole as massive as $10^{10} M_\odot$ (Figure 6.23): so quasars probably do not remain as bright as $10^{12} L_\odot$ for longer than $\sim$1 Gyr. But, just as for radio galaxies, activity is likely to continue for at least 100 Myr.

How does the active nucleus 'know' about the galaxy in which it lives, so that the black hole can grow large enough to produce the relation that we see in Figure 6.23 between its mass and the random stellar speeds? Probably they are connected through star formation. Despite their intense radiation, quasars contain dense molecular gas. Radio emission from CO molecules has been found in a few dozen objects at redshifts $z > 2$; half of them are quasars, along with a few radio galaxies. In the high-redshift quasar J1148 at $z = 6.4$, roughly $10^{10} M_\odot$ of molecular gas orbits 2.5 kpc from the center at almost 300 km s^{-1}. On average, quasars detected in CO contain $\sim (10^{10}-10^{11}) M_\odot$ of molecular gas. J1148 emits $10^{13} L_\odot$ in the far-infrared. Equation 7.11 shows that, if this all came from dust heated by young stars, the system would form $600\mathcal{M}_\odot\,\mathrm{yr}^{-1}$ of stars, using up the available gas within $\sim 20\,\mathrm{Myr}$; but the active nucleus probably provides some of that energy.

> **Problem 9.5** Show that the mass within the orbiting molecular gas of J1148 is $\mathcal{M}_{\mathrm{orbit}} \sim 5 \times 10^9 \mathcal{M}_\odot$. The quasar's bolometric luminosity is roughly $L_{\mathrm{bol}} = 4 \times 10^{40}$ W; use Equation 9.4 to show that the central mass $\mathcal{M}_{\mathrm{BH}} \gtrsim 3 \times 10^9 \mathcal{M}_\odot$. The black hole and the molecular gas account for almost all of $\mathcal{M}_{\mathrm{orbit}}$; the galaxy probably lacks the massive bulge that we would expect on the basis of Figure 6.23.

Some quasars show very broad absorption lines with widths up to $10\,000\,\mathrm{km\,s}^{-1}$, at redshifts that imply that the absorbing material moves toward us at speeds $\sim 0.1c$. The most prominent lines are those of ions such as SiIV, CIV, NV, and OVI, which require high energies for their excitation. The absorbing gas is dense, with 10^{19}–10^{21} atoms cm^{-2}. Few radio-loud quasars show broad absorption lines: they may lack gas in the required form. We do not know whether the 15% of quasars with broad absorption lines are different from those without, or whether every quasar has clouds of dense gas along the line of sight to 15% of all possible observers.

We do not know exactly how the quasar throws this absorbing gas outward at such high speeds. For example, the broad absorption might be produced if our line of sight passed by chance through one of the broad-line *emitting* clouds. Another possibility is that absorption takes place in supernova remnants, in a dense star cluster around the quasar; this would explain why the gas is metal-rich. Or the absorbing material may be propelled outward by the pressure of the quasar's radiation.

9.2 Fast jets in active nuclei, microquasars, and γ-ray bursts

The central compact radio cores of quasars and radio galaxies are only a few parsecs across, but they can be mapped by very long baseline interferometry (see Section 5.2) to reveal features less than a milli-arcsecond across. The majority show a bright inner core with an elongated feature or a series of blobs stretching for 10–50 pc away from it. Where the outer, kiloparsec-scale jet is one-sided, the central elongated feature always lies on the same side as the jet. Figure 9.10 shows the jet of BL Lac. Close to the inner core, the jet of blobs is often curved through tens of degrees, but its outer parts are aligned with the larger-scale jet. So the material of the outer jets, which can be hundreds of kiloparsecs long, must have been focussed into a narrow beam within a parsec of the galaxy center.

Nearly all compact cores are variable, changing their luminosity over days, weeks, or months. Times of peak brightness coincide with the appearance of new blobs, which travel out along corkscrew paths as they fade, as in Figure 9.10. In about half the well-studied cores, motion is *superluminal*: the blobs appear to

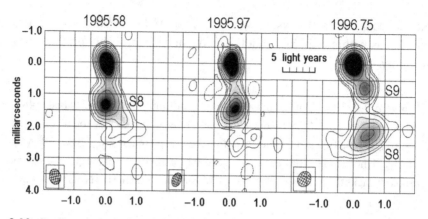

Fig. 9.10. Radio maps at 22 GHz of the blazar BL Lac; the scale bar is 5 light-years long, assuming that $H_0 = 67 \, \mathrm{km \, s^{-1} \, Mpc^{-1}}$. Blob S8 moves in a corkscrew path away from the core at apparent speed $\sim 3c$. The hatched ellipse shows the telescope beam; a pointlike source would appear with roughly this size and shape – G. Denn.

move away from the core with transverse speeds of $(3-50)c$. These high apparent speeds arise because the emitting gas is moving toward us at speeds close to that of light. The one-sidedness of the parsec-scale jet is only apparent; the approaching side is enormously brightened by relativistic beaming.

In the 1990s astronomers learned that the compact remnants of stars can also shoot out narrow jets of material at near-light speeds. In *microquasars*, jets emerge as mass captured from a binary companion forms an accretion disk around a black hole or neutron star. Supernovae marking the violent death of a very massive star can produce twin relativistic jets that are seen as *gamma-ray bursts*.

9.2.1 Superluminal motion and relativistic beaming

To understand these apparently superluminal speeds, consider an observer who sees a blob of jet material approaching at speed V, on a course making an angle θ with the line of sight (Figure 9.11). The blob passes point S at time $t = 0$, and point T at a time Δt later. Radiation emitted at T reaches our observer later than radiation from S; but, because T is closer, the interval between the two arrivals is only

$$\Delta t_{\mathrm{obs}} = \Delta t(1 - V \cos \theta / c). \tag{9.15}$$

In this time, the blob has travelled a distance $V \Delta t \sin \theta$ across the sky, so its apparent transverse speed is

$$V_{\mathrm{obs}} = \frac{V \sin \theta}{1 - V \cos \theta / c}. \tag{9.16}$$

As $V \to c$, the blob's motion can appear faster than light.

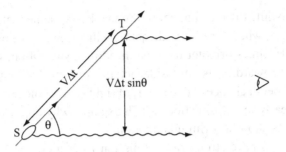

Fig. 9.11. Luminous blobs ejected at angle θ to the line of sight can appear to move superluminally across the sky if their speed $V \approx c$.

Problem 9.6 Defining $\gamma = 1/\sqrt{1 - V^2/c^2}$, show that $V_{\text{obs}} \le \gamma V$, with equality when $\cos \theta = V/c$, and that V_{obs} can exceed c only if $V > c/\sqrt{2}$.

Expansion speeds in blazars are most often around $(5-10)c$; thus the blobs must move outward with $\gamma \gtrsim 5-10$, which is well below the average energy that Equation 9.9 gave us for the radiating electrons. We will observe superluminal motion only in jets that point within an angle $1/\gamma$ of our direction, which is less than $10°$ in most cases. But, because the radiation is beamed in the direction of the jet's motion, those cores where the jet points toward us will appear much brighter.

To calculate this brightening, we recall that, for the stationary observer, atomic clocks aboard an emitting blob appear to run slow by a factor of γ. But, by Equation 9.15, its forward motion multiplies observed time intervals by the factor $(1 - V \cos \theta/c)$. So radiation emitted over a time Δt_{blob} with frequency ν_e in the restframe of the blob arrives during an interval Δt_{obs} at frequency ν_{obs}, where

$$\Delta t_{\text{obs}} = \Delta t_{\text{blob}}[\gamma(1 - V \cos \theta/c)], \text{ and } \nu_{\text{obs}} = \nu_e[\gamma(1 - V \cos \theta/c)]^{-1}. \quad (9.17)$$

Thus, when $\gamma \gg 1$ and the jet moves toward us, so that $\theta \approx 0$, we have $\Delta t_{\text{obs}} \sim \Delta t_{\text{blob}}/(2\gamma)$. Just as in our discussion of synchrotron radiation, the observer sees all the emission squeezed into a narrow cone, within an angle $1/\gamma$ of the direction of motion. If the radiation is isotropic in the blob's restframe, this brightens an approaching blob by a factor $\sim(2\gamma)^2$. The photons are blueshifted according to Equation 9.17, which also expands their frequency range.

Gathering all the factors, we find that the flux $F_\nu(\nu)$ received at frequency ν from a single blob, which in its restframe emits a power $L_\nu \propto \nu^{-\alpha}$, is amplified by $\sim(2\gamma)^{3+\alpha}$ when it moves directly toward the observer. The blob appears dimmed by the same factor when receding. If oppositely directed twin jets are made up of a series of identical blobs, and each radiates for a fixed time as measured in its restframe, then Equation 9.17 shows that the observed lifetime of approaching blobs is shortened. Thus the jet pointed directly at us is brightened only by the factor $(2\gamma)^{2+\alpha}$, while the receding jet is dimmed by the same factor. For compact cores, $\alpha \sim 0$; thus, when $\gamma \sim 5$–10, the jet travelling toward the observer appears $(2\gamma)^4 \sim 10^4$–10^6 times brighter than that heading away. Often, we see only the material heading toward us, as a one-sided jet; the receding half is almost invisible. We have a much better chance of detecting as bright sources those objects for which the jet points nearly in our direction, which explains why about half of observed bright radio cores show superluminal motion.

Blazars always have strong radio emission from a compact core. They are generally found in ellipticals; probably they are radio galaxies where we look directly down the jet. Its relativistic motion makes the approaching half of the jet appear so bright as to outshine the accretion disk and the line-emitting regions. Blazars are strongly variable; some have doubled their optical and radio brightness over a few days. This fast variation is natural, since by Equation 9.17 the time for any change in the jet's luminosity is shortened by the factor $\gamma(1 - V \cos\theta/c)$. The visible light of most quasars is polarized at levels of a few percent, indicating that part of it may be synchrotron radiation from the jet. But in blazars, polarization can be as high as 20%; much more of the radiation comes from the jet. Only one in every thousand quasars is a blazar; but, because they are brightened by forward beaming, these objects account for a few percent of observed bright quasars.

9.2.2 Microquasars: relativistic jets in stellar binaries

In the 1990s, astronomers were startled to discover radio jets emerging at close to light speed from stellar-mass black holes. In a *microquasar*, a massive star transfers mass onto an accretion disk around a black-hole or neutron-star companion. These systems were long known as X-ray sources, and about 10% of them are also radio-loud; so the fast jets should not have come as a surprise. The jets are quite narrow, with opening angles less than $10°$, and the radio emission is synchrotron, just as in active nuclei. At least two of the dozen or so known microquasars show superluminal motion.

Because the central mass is only a few $\mathcal{M}_\odot$, the accretion disk around a microquasar is hotter than that in an active nucleus, and emits most of its power as X-rays rather than ultraviolet light. To calculate its temperature, we look at the energy lost by a gas cloud of mass m in circular orbit around a mass $\mathcal{M}_{BH}$, moving from radius r to $r - \Delta r$. Its initial energy is

$$\mathcal{E} = -\frac{Gm\mathcal{M}_{BH}}{2r}, \quad \text{so} \quad \Delta\mathcal{E} = -\frac{Gm\mathcal{M}_{BH}}{2r^2}\Delta r. \tag{9.18}$$

This energy is transferred to the portion of the disk between r and $r+\Delta r$, which must get rid of it. If it does that by radiating as a blackbody at temperature T, its luminosity is $\sigma_{SB}T^4$ per unit area. Remembering that a disk has two sides, when a mass $\dot{\mathcal{M}}$ flows inward per unit time we have

$$\frac{G\dot{\mathcal{M}}\mathcal{M}_{BH}}{2r^2}\Delta r = \sigma_{SB}T^4 \times 4\pi r \,\Delta r, \quad \text{so } T^4(r) = \frac{G\dot{\mathcal{M}}\mathcal{M}_{BH}}{8\pi r^3 \sigma_{SB}}. \qquad (9.19)$$

If the inflowing mass is converted to energy with some efficiency $\epsilon \approx 10\%$, we can measure $\dot{\mathcal{M}}$ in units of the mass $L_E/(\epsilon c^2)$ required to radiate at the Eddington luminosity of Equation 9.4. Writing r in units of the Schwarzschild radius $R_s = 2G\mathcal{M}_{BH}/c^2$, we have

$$T^4(r) = \frac{1}{G\mathcal{M}_{BH}}\left(\frac{r}{R_s}\right)^{-3}\left(\frac{\dot{\mathcal{M}}}{L_E/(\epsilon c^2)}\right)\frac{c^5 m_p}{16\epsilon\sigma_T\sigma_{SB}}. \qquad (9.20)$$

So the temperature near the inner edge of a disk accreting near its Eddington limit decreases with the mass of the black hole, as $\mathcal{M}_{BH}^{-1/4}$.

These arguments are not quite right, since we have ignored the fact that matter can fall in only when angular momentum is carried outward in the disk. Since energy is transported along with it, the disk at $r \gg R_s$ radiates three times as much energy as is given by Equation 9.19, while the inner disk radiates less.

> **Problem 9.7** Recall from Problem 3.20 that the last stable orbit around a non-rotating black hole is at $3R_s$. Show that, when $\mathcal{M}_{BH} = \mathcal{M}_\odot$ and $\dot{\mathcal{M}}$ is near the Eddington limit, the inner edge of the disk is at $T \approx 3 \times 10^7$ K, corresponding to a photon energy of 2.6 keV. Assuming that the orbital speed of material there is given by Equation 3.20, show that the orbital period $P \approx 3 \times 10^{-4}$ s.

The material is already collimated into a narrow jet within 10 AU of the central source, and the jet travels for a few parsecs rather than megaparsecs. We can watch as bright clumps travel out from the nucleus and fade within a month or two rather than over decades or centuries, as in an active nucleus. These clumps seem to be ejected when the source enters an X-ray-bright state – perhaps as material in the accretion disk swirls rapidly inward and onto the central object. The jet may speed up, so we see bright knots of radio emission as the faster material runs into the slower plasma ahead of it. Microquasars are a remarkable demonstration that gravitational and electromagnetic forces can operate on all spatial scales.

Further reading: Chapter 16 of M. S. Longair, *High Energy Astrophysics*, 2nd edition (Cambridge University Press, Cambridge, UK).

9.2.3 Fast jets from exploding stars: gamma-ray bursts

Gamma-ray bursts (GRBs) were discovered in the 1960s by satellites watching for nuclear bomb tests. They are short, intense spurts of γ-rays, with peak energies around 1 MeV and a low-energy tail of X-rays. They last typically 500 s or less, and appear to arrive from random directions in the sky. Since γ-ray telescopes cannot normally pinpoint a source more closely than to within $\sim 1°$ on the sky, none was identified with an optically-visible object until the Beppo-SAX satellite began observing in 1997. Once it detected a burst, Beppo-SAX could point immediately to take an X-ray image with $3'$ resolution, good enough to tell optical astronomers where to look. 'Long' bursts lasting 2 s or more are typically found in star-forming galaxies. We have now identified the optical counterparts of about a hundred GRBs, at redshifts up to 6.3.

A long burst lasts 50 s on average, but most are made up of many short sub-pulses that peak and fade within 1–100 ms. So the emitting region must be smaller than a light-millisecond or 300 km, the Schwarzschild radius for a $100\mathcal{M}_\odot$ black hole. The bursts must be caused by objects of stellar mass, rather than the giant black holes found at galaxy centers. We never find repeated GRBs from the same spot, which suggests that they originate from some catastrophic event: a massive star exploding, or perhaps one black hole merging with another, or with a neutron star.

We see roughly one GRB for every 100 000 supernovae, but this is likely to be a small fraction of the total. Like blazars, GRBs beam their luminous material in narrow jets. If they radiated equally in all directions, the energy of a typical burst would be an astounding 10^{52}–10^{54} erg, or nearly $\mathcal{M}_\odot c^2$. The radiation from a normal supernova, and the energy of motion in its ejected outer layers, amount to only $\sim 10^{51}$ erg. The jet's opening angle θ is typically $1°$–$20°$, so the energy required from each burst is reduced by a factor θ^2 to only 10^{50}–10^{51} erg, which is within the energy budget of a supernova. But we fail to detect most bursts because the jet is not pointing in our direction.

The radiating material is expanding at close to light speed. Otherwise, the γ-rays would be so tightly packed that they could not escape, but would lose their energy in producing particle–antiparticle pairs. To avoid that fate, the 'fireball' must expand with Lorenz factor $\gamma \equiv \sqrt{1 - V^2/c^2} > 100$. Emission from material heading directly toward us is boosted by a factor $(2\gamma)^4$, so nearly all the energy we receive is from a small cone where the outward motion is directed within an angle $1/\gamma$ of our line of sight. As outgoing material slows and γ drops, this cone widens to include a larger fraction of the radiating volume. Equation 9.17 shows that we see the burst compressed in time by $1/(2\gamma)$, compared with an observer moving with the outflowing gas. The gamma-rays are probably synchrotron radiation, produced by fast electrons accelerated by shocks in the outflowing relativistic material.

Along with the X-rays, we often see an afterglow in optical and radio bands. Absorption lines at visible wavelengths let us measure the redshift of the host galaxy. The light is usually polarized, suggesting that it is synchrotron radiation from electrons accelerated in shocks as the jet runs into the surrounding gas. The visible afterglow starts to dim more rapidly after 1–100 days, which may mark the time when outflowing material has slowed so that $1/\gamma \sim \theta$, so we receive radiation from the entire approaching jet. The jet also begins to expand sideways (since the opening angle is now comparable to the Mach number $\mathcal{M} \sim 1/\gamma$), causing its density and emissivity to drop. The radio afterglow becomes strong only a day or so after the burst, and can remain fairly bright for a year or more. During this time the outward motion slows to well below c and the expansion becomes spherical, so we can reliably estimate the total energy in the outward-moving material. Many bursts seem to have roughly the same amount of energy in the relativistic outflow, roughly 10^{51} erg divided between prompt emission and the afterglow.

After some long bursts, the declining curve of visible light shows a bump that rises to a maximum after $\sim 20(1 + z)$ days and then fades. In some well-observed cases, broad emission lines identify the bump as the light of a Type Ic supernova that exploded at the same time as the burst. Such an explosion is thought to mark the end of life for a rotating star that had a mass of $(20-40)\mathcal{M}_\odot$ when it was on the main sequence. As the star's iron core collapses, enough other material may join it that the core cannot become a stable neutron star, but forms a black hole instead. The rest of the star's material has too much angular momentum to fall directly into the black hole. Most of it is expelled in the explosion, but $\sim 0.1\mathcal{M}_\odot$ is thought to form a short-lived accretion disk around the black hole, which channels twin outflowing jets as shown in Figure 9.3.

Short GRBs, lasting less than 2 s, have a much fainter afterglow. The first was identified only in 2005; GRB 050724 was relatively close, in an elliptical galaxy at $z = 0.26$. The burst itself was at least ten times fainter than a typical long burst, and the afterglow at least a hundred times dimmer. This may have been a merger between two neutron stars (expected to be 10 000 times rarer than a supernova), or even between a neutron star and a black hole. These bursts would be shorter than one from a collapsing massive star because both partners are already very compact.

Gamma-ray bursts may have been relatively more frequent in the early Universe, since it is easier to make very massive stars when metals are absent from the gas. (If these first-generation stars were all massive, and have now ended their lives, this would explain why we see none in the Milky Way today.) Gamma-rays are neither scattered by dust nor readily absorbed by the interstellar gas, so they can help us to probe starbirth in the most distant galaxies. Some of the bursts detected by our present γ-ray telescopes are likely to be at redshifts $z \sim 10$ or higher, too distant for us to see the afterglow at longer wavelengths. If a better understanding allows us to estimate redshifts from the X-ray and γ-ray emission, we could use GRBs to trace the earliest massive stars.

Further reading: J. I. Katz, 2002, *The Biggest Bangs: The Mystery of Gamma Ray Bursts* (Oxford University Press, Oxford, UK) is written for the general reader.

9.3 Intergalactic gas

In the spectra of most quasars, we see multiple systems of absorption lines, at significantly *lower* redshifts than the emission. Most of the lines are narrow; their widths correspond to internal motions slower than $100 \, \mathrm{km \, s}^{-1}$, although their redshifts imply that the gas is travelling away from the quasar and toward us at almost light speed. In fact, this gas is nowhere near the quasar, but simply intercepts its light. Counting the baryons in galaxies and clusters today shows that their formation was not very efficient: recall Table 7.2. Most of the baryons remain in diffuse clouds, that we see only when they intercept the light of a distant source. This is the reservoir from which gas flowed into the nascent galaxies, and which continues to feed them to the present day.

Absorption lines are usually identified in groups that show a simple pattern, such as the Lyman series of hydrogen, or doublets of magnesium and carbon. A small wavelength difference makes it easier, since a single measured spectrum is likely to show both lines of a pair. The MgII lines at 2796.3 Å and 2803.5 Å fall into the visible window when the absorbing cloud's redshift is $0.2 \lesssim z_{\mathrm{abs}} \lesssim 1.5$, and the CIV lines at 1548.2 Å and 1550.8 Å do so for $1.1 \lesssim z_{\mathrm{abs}} \lesssim 3.5$. The OVI lines at 1031.9 Å and 1037.6 Å are harder to observe, since they can be lost among the lines of the Lyman-α forest (see below). Once the absorption redshift is known, isolated lines of other elements can often be found.

The intergalactic gas ranges from clouds of largely neutral material, which are as dense as present-day galactic disks, to diffuse gas where the fraction of neutral atoms is 10^{-3} or less. The number $\mathcal{N}$ of absorption lines with $N(\mathrm{HI})$ hydrogen atoms along the line of sight approximately follows $\mathcal{N} \propto N(\mathrm{HI})^{-1.5}$: the very diffuse clouds are the most common, but the densest of them contain almost all of the neutral gas. The absorbing material is not pristine, but already contains heavy elements produced by nuclear burning in stars.

9.3.1 Neutral gas: damped Lyman-α clouds

If the column density of neutral gas exceeds $N(\mathrm{HI}) \approx 2 \times 10^{20} \, \mathrm{cm}^{-2}$, the Lyman-$\alpha$ (Lyα) line is optically thick with prominent damping wings: these are called damped Lyα clouds. This gas density corresponds to $1.5 \mathcal{M}_\odot \, \mathrm{pc}^{-2}$, which is typical for the outer HI disks of galaxies today: see Section 5.2. A *Lyman-limit* cloud has $N(\mathrm{HI}) \gtrsim 2 \times 10^{17} \, \mathrm{cm}^{-2}$; it absorbs almost all photons that have enough energy to ionize a hydrogen atom, so the quasar's measured flux drops nearly to zero at wavelengths just short of $912(1 + z_{\mathrm{abs}})$ Å. The damped Lyα clouds are largely neutral, but most of the hydrogen in the Lyman-limit clouds is ionized. The spectrum of Figure 9.12 shows a damped Lyα line with $z_{\mathrm{abs}} = 2.827$; the column density is close to $2 \times 10^{20} \, \mathrm{cm}^{-2}$. Damped Ly$\alpha$ clouds in front of radio-loud quasars can also be detected by their absorption in the 21 cm line of HI. Counting

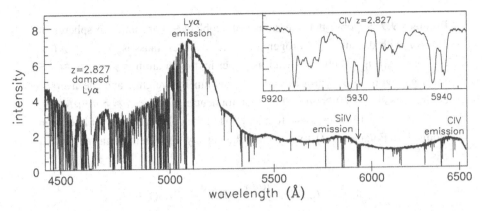

Fig. 9.12. The spectrum of quasar $1425 + 6039$ with $z_{em} = 3.173$: broad Lyα emission at 1216 Å is redshifted to the visible region. At shorter wavelengths, narrow absorption lines of the Lyα forest are dense. The squarish profile at 4650 Å is a damped line of Lyα, at $z_{abs} = 2.827$. The arrow shows absorption at the same redshift in the CIV doublet with rest wavelength near 1550 Å: the inset reveals distinct absorption components from multiple gas clouds – L. Lu and M. Rauch.

the number of clouds in the spectrum of a typical quasar tells us that they contain enough dense gas to make the disks of present-day spiral and irregular galaxies.

Problem 9.8 Suppose that there are $n(z) = n_0(1 + z)^3$ damped Lyα clouds per Mpc^3 at redshift z, each with cross-sectional area σ. Explain why we expect to see through $n(z)\sigma l$ of them along a length l of the path toward the quasar. Use Equation 8.47 to show that between z and $z + \Delta z$ the path $\Delta l = c\Delta z/[H(z)(1 + z)]$, so the number per unit redshift is

$$\frac{d\mathcal{N}}{dz}\Delta z = \frac{n(z)\sigma c \,\Delta z}{H(z)(1 + z)} \equiv \frac{n_0\sigma c}{H_0}\Delta z \frac{dX(z)}{dz}, \qquad (9.21)$$

where (using Equation 8.26 for the second equality)

$$\frac{dX(z)}{dz} = \frac{H_0(1 + z)^2}{H(z)} = \frac{(1 + z)^2}{\sqrt{\Omega_m(1 + z)^3 + (1 - \Omega_{tot})(1 + z)^2 + \Omega_\Lambda}}.$$

Show that, if the Universe is flat and $\Omega_{tot} = 1$, then at early times

$$X(z) = \frac{2}{3\sqrt{\Omega_m}}[(1 + z)^{3/2} - 1] \quad \text{while } (1 + z)^3 \gg \Omega_\Lambda/\Omega_m. \qquad (9.22)$$

Locally we find $d\mathcal{N}/dz \approx 0.045$; if the cross-section σ does not change, show that we expect $d\mathcal{N}/dz \approx 0.16$ at $z = 3$. At $z = 5$ we observe $d\mathcal{N}/dz \approx 0.4$. Show that this is roughly twice what we expect if σ is constant: this result indicates that there were more absorbing clouds, or each was larger.

Problem 9.9 Suppose that the clouds of Problem 9.8 are uniform spheres of radius r with density n_H hydrogen atoms cm^{-3}. Their mass is given by $\mathcal{M} = (4/3)\pi r^3 n_H \mu m_H$, where the mean mass per hydrogen atom is μm_H ($\mu \approx 1.3$ for a gas with 75% hydrogen and 25% helium by weight), and the average column density $N(\text{HI}) \approx r n_H$. Show that, for neutral clouds, $\mathcal{M} \approx \sigma \mu m_H N(\text{HI})$, where $\sigma = \pi r^2$ is the cross-section and $N(\text{HI})$ the column density of hydrogen atoms. Use Equation 9.21 to show that the density ρ_g of neutral gas at redshift z is

$$\rho_g(z) \equiv n(z)\mathcal{M} = \frac{\mu m_H}{c} N(\text{HI}) \frac{d\mathcal{N}}{dz} H(z)(1 + z).$$

If this gas survived unchanged to the present day, explain why it would now represent a fraction Ω_g of the critical density of Equation 8.21, where

$$\Omega_g \equiv \frac{n_0 \mathcal{M}}{\rho_{\text{crit}}(t_0)} = \left[\frac{\mu m_H H_0}{\rho_{\text{crit}}(t_0)c} \right] N(\text{HI}) \frac{d\mathcal{N}}{dz} \bigg/ \frac{dX}{dz}. \tag{9.23}$$

This fraction depends on the present-day Hubble constant as $\Omega_g(z) \propto h^{-1}$ – why? Show that the term in square brackets is $1.2 \times 10^{-23} h^{-1}$ cm^2. Taking $d\mathcal{N}/dz = 0.2$, and an average $N(\text{HI}) = 10^{21}$ cm^{-2} at $z \sim 2$, show that for the benchmark cosmology $dX/dz = 3.0$ and $\Omega_g(z = 2) \approx 10^{-3}$.

We saw in Table 7.2 that the neutral atomic and molecular gas now in galaxy disks corresponds to $\Omega_g \sim 8 \times 10^{-4}$, while stars in the disks of galaxies now make up $\Omega_\star \sim 6 \times 10^{-4}$. So there is about enough neutral gas in the damped Lyα clouds at $z \sim 2$ to make the galaxy disks that we see today. But, if the benchmark cosmology is correct, Ω_g has remained at this level over 6 Gyr between redshifts $0.7 \lesssim z \lesssim 5$, when half of the stars in disks like the Milky Way's were born. So we think that the damped Lyα clouds have been replenished, as more tenuous ionized gas flowed into them and became dense enough to recombine.

Are the damped Lyα clouds simply gas in the disks of galaxies? We do not really know. The HI disks of nearby galaxies are large and numerous enough to account for the absorption features at $z \lesssim 1$. At least half of them should be caused by less luminous galaxies with $L \leq 0.2L_\star$, since these have relatively large HI disks. When $z_{\text{abs}} < 1.5$, we can often see a bright patch of stars within an arcsecond or so of the quasar on the sky, at the same redshift as the damped Lyα cloud. These systems are a mix of star-forming disk galaxies, irregulars, and compact knots of star formation, with luminosities $0.02L_\star < L < 3L_\star$. At higher redshifts, Problem 9.8 shows that the galaxies must be larger or more numerous than they are today, to provide enough absorption lines. *Lyman break galaxies* at $2 \lesssim z \lesssim 3$, selected because they are bright in the ultraviolet, usually give rise to Lyα absorption if they lie within 300 kpc of the path to a quasar. Typically, the closer the quasar's light passes to the galaxy, the denser is the absorbing gas. Curiously, a third of Lyman break galaxies seem to produce no Lyα absorption at all.

Problem 9.10 At present, the density of $L_\star$ galaxies is roughly $n_\star$ of Equation 1.24, or $0.02h^3$ Mpc^{-3}. Figure 1.16 shows that about half of them are disk galaxies, so, taking $h = 0.7$, we now have $n_0 \approx 0.003$ Mpc^{-3} bright spiral galaxies. At $z \sim 3$, we see $d\mathcal{N}/dz \approx 0.25$. If galaxies had already assembled their disks by then, we would expect the density $n(z) \approx n_0(1 + z)^3$. For the benchmark cosmology, show that $dX/dz = 3.6$, and use Equation 9.21 to show that we must have $\sigma = 2500$ kpc^2, so that the absorbing material extends to radii ~ 30 kpc. Figure 5.15 shows that the HI disk of a galaxy like the Milky Way with $\mathcal{M}(\text{HI}) = 10^{10}\mathcal{M}_\odot$ extends almost to this radius.

9.3.2 Metals in the intergalactic gas

Damped Lyα clouds contain metals and dust, as well as hydrogen and helium: we see lines of low ions such as MgII, ZnII, and CrII along with high-ionization lines such as CIV and SiIV. In Figure 9.12, there is a cluster of CIV lines at $z_{\text{abs}} = 2.83$ near 5920 Å; they come from the same gas as is producing the damped Lyα absorption. In less dense clouds with $N(\text{HI}) \gtrsim 10^{16}$ cm^{-2}, the strongest metal lines are of low-ionization species such as MgII, SiII, and OI, which in present-day galaxies are found along with neutral hydrogen in their disks. CIV and SiIV lines, which are characteristic of the diffuse hot gas of today's galactic halos, become more common when $N(\text{HI})$ is yet lower. Such complexes of metal lines are generally 300–500 km s^{-1} wide, but the widths of individual components range down to 10 km s^{-1}, corresponding to $T \lesssim 10^5$ K. A plasma in thermal equilibrium at this temperature would contain very few CIV ions; so the absorbing gas is probably ionized by intergalactic radiation from quasars, or by hot stars born within it.

All damped Lyα clouds cause absorption in MgII, but the reverse is not true. At a given redshift, lines of MgII are about ten times as common as damped Lyα systems, while CIV lines are yet more frequent. So the gas responsible for the MgII absorption must cover a larger area than that producing the damped Lyα features. Strong MgII absorption generally occurs within $40h^{-1}$ Mpc, and weaker absorption up to $80h^{-1}$ Mpc, of a galaxy. Gas causing CIV absorption usually extends 50–100 kpc, but sometimes is seen 200 kpc away. Where gravitational lensing has produced multiple images of a quasar separated by ~ 1 arcsec on the sky, the two light paths are ~ 10 kpc apart as they pass through the absorber. Strong lines of CIV in the two quasar spectra differ when the paths to the images are separated by more than 20–50 kpc, giving the rough size of these complexes of absorbing clouds.

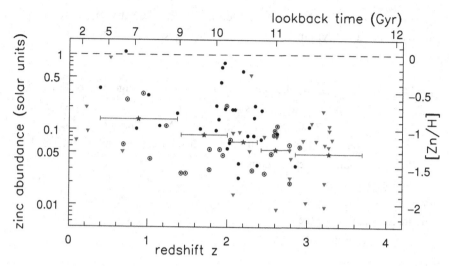

Fig. 9.13. Abundance of zinc relative to hydrogen in gas clouds causing damped Lyα absorption: circled dots show the densest clouds with $N(\text{HI}) \geq 10^{21} \, \text{cm}^{-2}$, filled dots show less dense clouds, and downward triangles show upper limits. Horizontal bars show average abundances weighted by $N(\text{HI})$; lookback time refers to the benchmark model. Zinc is not readily incorporated into dust grains, so its abundance in the gas indicates the total fraction of metals – after Kulkarni *et al.* 2005 *ApJ* **618**, 68.

Just as in galactic halos, OVI absorption mainly comes from a thin layer around partially ionized warm clouds that absorb in CIV and MgII. Sometimes the OVI lines are broad, as expected when $T \gtrsim 3 \times 10^5$ K and OVI is collisionally ionized at densities $n \geq 10^{-3} \, \text{cm}^{-3}$, over 100 times the cosmic mean. This gas may have been heated by shocks in outflowing winds from star-forming galaxies.

Where we can also measure the column density N_H of hydrogen in an absorbing cloud, we can use the strength of the metal lines to estimate the fraction of heavy elements in the gas. Figure 9.13 shows that the average metal abundance in the damped Lyα clouds has slowly risen over time. But it was still less than one-tenth of the solar value 8–10 Gyr ago, when the Milky Way's disk made many of its stars. The gas of Figure 9.13 could not have been the gas from which the Milky Way's disk formed: Table 2.1 and Figure 4.14 show that almost all the disk stars have $Z > 0.1Z_\odot$. But we saw in Figure 4.15 that galaxy disks are richer in heavy elements toward their centers, while the points in Figure 9.13 are more likely to represent a quasar shining through the metal-poor periphery of one.

In the Milky Way, we noted in Section 2.4 that any cloud denser than $N(\text{HI}) = 10^{20} \, \text{cm}^{-2}$ becomes largely molecular. Hydrogen molecules in damped Lyα clouds would easily be seen since they absorb the quasar's ultraviolet radiation; but in fact H_2 is rare. Probably the gas contains too little dust to allow molecules to form easily. In some damped Lyα clouds elements such as iron, that are readily incorporated into dust, are rarer with respect to those like zinc and silicon that are not. These clouds must have begun to produce a little dust, but the ratio of dust to gas is ~ 30 times smaller than Equation 1.22 gives for the Milky Way.

Absorption lines of CII*, excited C$^+$, at 1335.7 Å indicate that far-ultraviolet photons are heating the dust grains, just as they do in the Milky Way. The heat is transmitted to the surrounding gas, and carbon excited to CII* re-radiates most of it away at 158 μm (see Table 2.5). From the strength of the CII* absorption we can estimate how much heat is re-radiated; it is far more than can be supplied by the intergalactic radiation field. These damped Lyα clouds must have their own source of ultraviolet photons – they are forming stars. In a few cases, we have even seen the Lyα emission of those stars. The inferred pace of starbirth is rapid enough at redshifts $z \sim 2$–4 that it would use up all the gas of the damped Lyα clouds within $\sim 2\,$Gyr – further evidence that those clouds are replenished from the reservoir of more diffuse gas.

If these metal lines indeed trace gas that now lies in galaxies or galaxy groups, their redshifts should be clumped in much the same way as for the galaxies. As expected, if a CIV absorption line is detected, it is more probable to find a second line nearby in velocity, but the effect disappears for separations beyond 500 km s^{-1}. The same is true for lines of OVI that lie closer together than 750 km s^{-1}. These speeds are somewhat higher than the rotation speeds of galaxies, or the velocity dispersion within groups of galaxies. But this is roughly the thickness of the velocity peaks in Figure 8.4, which correspond to looking through a wall or filament of galaxies.

9.3.3 The Lyman-α forest

At column densities below $N(\text{HI}) \sim 2 \times 10^{17}$ cm^{-2}, ultraviolet photons penetrate through a gas cloud, and most of the hydrogen is ionized. When $N(\text{HI}) \lesssim 3 \times 10^{14}$ cm^{-2}, we usually detect only the Lyα lines of hydrogen, although deep spectroscopy can reveal weak CIV and OVI lines. The dense profusion of hydrogen absorption lines at wavelengths just short of the quasar's Lyα emission is the *Lyman-α forest*. While we detect the forest through its neutral hydrogen, almost all the gas is ionized; HI forms only a tiny fraction, often less than 0.1% of the total. But these clouds may be the Universe's main repository of neutrons and protons.

In Figure 9.12 we see that forest clouds can remove a substantial fraction of the quasar's light just shortward of its Lyα emission line, leaving the average intensity lower than that on the long-wavelength side of the emission line: the *Gunn–Peterson effect*. Between the lines of the Lyα forest in Figure 9.12, the light level is the same on both sides of the emission line. The Lyα line saturates at $N(\text{HI}) \sim 10^{14}$ cm^{-2}, so there must be hardly any HI gas between the absorbing clouds. This is typical for quasars at redshift $z \lesssim 5.8$. But in many objects with $z \gtrsim 6$, we see a *Gunn–Peterson trough*: all the light just shortward of the Lyα emission line is missing, having been absorbed by diffuse neutral gas at $z \gtrsim 5.8$.

The earliest quasars and star-forming galaxies shone on gas that had become neutral at the time of recombination, at $z_{\text{rec}} \approx 1100$. They formed HII regions, 'islands' of ionized gas like those around hot massive stars in the Milky Way. As the Universe expanded the densest gas clumped together into clouds, while more stars and quasars were born to ionize the smaller amount of diffuse gas that remained. The disappearance of the Gunn–Peterson trough tells us that at $z \sim 6$ those completely ionized regions linked up to surround tiny islands of denser neutral gas. Quite abruptly, over ~ 100 Myr, the Universe became transparent to ultraviolet radiation; it was *reionized*. After that, the ultraviolet light of a quasar was absorbed only where it encountered a cloud of neutral gas.

What is the source of the ultraviolet photons that reionized the gas? Not the luminous quasars: Figure 8.13 tells us that before $z \sim 6$ these were very rare, and contribute little. Can star-forming galaxies do the job? That depends on how much of their ultraviolet light escapes to intergalactic space, and how much denser the gas around them is than the cosmic average. Problem 2.24 shows that, as the surrounding density drops, a given galaxy or quasar can ionize a larger mass of gas. Large galaxies such as the ultraviolet-luminous objects of Figure 9.17 form in dense regions (see Figure 7.11). Thus handicapped, their stars are inadequate to the task. But dim galaxies are far more common than luminous ones (recall Figure 1.16), and predominate where the galaxy density is low; so the gas around them will be less dense. If all galaxies formed stars in proportion to their dark matter, most would individually be too faint for us to see, but together they could reionize the gas. However, we argued in Section 8.5 that most small clumps of dark matter should not make stars, or else the Milky Way would have far more satellites than we observe. Another suggestion is that the X-ray and ultraviolet light of numerous faint quasars provides the energy for reionization.

We can estimate the level of ionizing radiation between the galaxies by looking at how the Lyα forest thins out near the quasar's emission redshift. Clouds with $z_{\text{abs}} \approx z_{\text{em}}$ lie close enough to be affected by the quasar's radiation, which boosts ionization so that very little HI remains. The redshift interval where the forest lines are sparse shows where that radiation makes a significant addition to the general intergalactic background. Measuring this *proximity effect* shows that over the range $1.6 < z < 4$ the photons close to 912 Å that can ionize a hydrogen atom amount to $\sim 3 \times 10^{-22}$ erg cm^{-2} s^{-1} Hz^{-1} sr^{-1}. This corresponds to $\nu I_\nu \sim 2$nWm^{-2}sr^{-1}, somewhat higher than today (see Figure 1.19).

With so many ionizing photons, only one hydrogen atom in 10^3 or 10^4 of the diffuse gas of the Lyα forest would remain neutral. The most diffuse clouds that we observe are only a few times denser than the cosmic mean. Their gas is too rarified to radiate energy so that it could cool and become denser; it is still fairly evenly mixed with dark matter in the filaments of the cosmic web of Figure 8.16. Observations of quasars lying close to each other on the sky with absorption lines at the same redshift show that at $z \sim 2$ a filament of the Lyα forest gas can stretch for $\gtrsim 0.5h^{-1}$ Mpc.

Most of the diffuse ionized gas appears to contain elements heavier than hydrogen and helium. One recent study found carbon and oxygen at about a

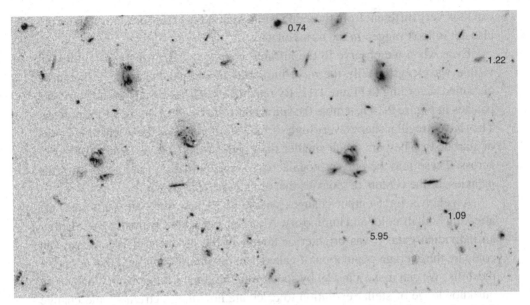

Fig. 9.14. A section of the Hubble Ultra Deep Field in blue (B, left) and red (i, right) light. Redshifts of selected galaxies are marked. Note the normal-looking spiral at $z \lesssim 1$ to the right of the center, and the merging group to the left – Space Telescope Science Institute.

thousandth of the solar level, although about 30% of the clouds were even more metal-poor with $Z \sim Z_\odot/3000$. If many stars were formed early in small galaxies, the products of their nucleosynthesis would be spread throughout the Lyα forest. How much of the heavy elements should they make? Adding up the ultraviolet light from the galaxies of Figure 9.17, we find that by $z \sim 2.5$ they alone would produce enough metals to give an average abundance $\sim Z_\odot/30$ if mixed evenly through the baryons. So why do we see only $0.001 Z_\odot$ in the Lyα forest, which contains most of the baryons? This is known as the *missing metals* problem. For lack of definite information, we conclude that the missing metals are hidden in highly ionized states in hot gas that is too diffuse to cool.

9.4 The first galaxies

Figure 9.14 shows part of the *Hubble Ultra Deep Field*, a region of the sky where deep images have been taken with the Hubble Space Telescope in four colors, roughly the U, B, i, and z passbands of Tables 1.2 and 1.3. Some of the galaxies in this field are small nearby objects, but others are known to have redshifts up to $z \sim 5$. These look more irregular and asymmetric than present-day systems. But we see them by light that was emitted in the ultraviolet, by their young massive stars, and Figure 5.10 shows that the ultraviolet images of nearby normal galaxies can look very different from those in visible light. Local star-forming galaxies are also apt to look ragged in the ultraviolet.

Even when we observe in the infrared to map out light that was emitted at visible wavelengths, only the most luminous of the galaxies at $z \sim 1$ can be classified according to Figure 1.11. By redshift $z \gtrsim 2$, hardly any of the luminous patches in Figure 9.14 resemble the spirals and ellipticals of our nearby Universe. They have irregular shapes, very high surface brightness, and the bright blue colors of starbursts. They are much smaller than present-day galaxies, only 0.1″–0.2″ across. These may be only protogalactic fragments that will merge to form the galaxies, or the centers of galaxies that are not yet full-grown.

A galaxy's light output changes with time as new stars are born and age, affecting both its color and luminosity. Although we cannot normally test whether an individual galaxy was brighter or fainter in the past than it is now, we can compare the average population of galaxies at the present day with that at higher redshifts. We can make a model by specifying the star-forming history, and using our knowledge of stellar evolution to calculate how its spectrum should change with time. This amounts to finding the evolutionary term $e(z)$ in Equation 8.46. We then test the model by calculating how many galaxies of a given type should be found. For example, if galaxy disks have built up their stars at a steady rate, then they should always have roughly the same number of young blue stars, while the red stars build up over time. If elliptical galaxies made all their stars at redshifts $z \gtrsim 3$, we can see from the starburst model of Figure 6.18 that their stars should have been both bluer and brighter in the past.

Figure 9.15 represents a look backward over at least half of the Universe's history, showing the numbers of galaxies at each luminosity per *comoving* volume. The number of red galaxies, with spectra showing only old stars, is about the same as today. We know that the population of stars that makes up each galaxy at $z \sim 1$ will have faded by the present, as in Figure 6.18. So these galaxies must make new stars, or grow by eating their companions as we discussed in Section 7.1; or else new bright red galaxies have been produced. For example, a blue galaxy that ceases forming stars will become red.

At $z \sim 1$ the number of blue galaxies that are actively forming new stars was larger at each luminosity than it is today. Each one was about three times brighter, or else there were more blue galaxies in the past. We cannot blame stellar fading: today's blue galaxies have formed new stars in the past gigayear, or they would now be red, not blue. The extra starbirth is not caused by starbursts in galaxy mergers; fewer than 10% of the galaxies that contribute to Figure 9.15 are merging. Instead, normal galaxies like our own must have made stars more rapidly than they do now. These systems should redden over time, as old stars make up more of their mass; and indeed the average color of a star-forming galaxy is now redder than it was at $z \sim 1$.

We can compare the observed rate of starbirth in a galaxy with what is needed to build up the stars that we see over the lifetime of the Universe; a starburst galaxy is one that makes stars much faster than this average rate. We saw in Section 2.1

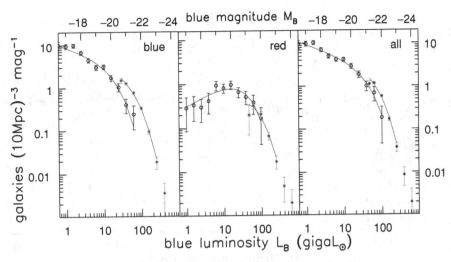

Fig. 9.15. Numbers of galaxies per comoving 10 Mpc cube between absolute magnitude $M(B)$ and $M(B) + 1$: open circles are for nearby galaxies with $0.2 < z < 0.4$, and filled dots for those at $1 < z < 1.2$. The number of red galaxies at each luminosity has changed little, but luminous blue galaxies are far more common at $z \approx 1$. As in Figure 1.16, very luminous galaxies are mainly red, while most of the the dimmer ones are blue – DEEP2: S. Faber *et al.*, astroph/0506044.

that our Milky Way's disk has formed stars at a steady pace over the past few gigayears, while small galaxies more often make their stars in spurts (Section 4.4). Figure 6.20 shows that the biggest galaxies, with $L_r > 8 \times 10^{10} L_\odot$ or twice the Milky Way's luminosity, are now forming stars at only 1%–10% of their average rates over cosmic history. Today, <1% of galaxies with more than $2 \times 10^{10} \mathcal{M}_\odot$ of stars (about a third as much as the Milky Way) have a global starburst, although bursts confined to the center are more common. But at $z \sim 0.7$, fully 40% of such galaxies are making stars significantly faster than their average rate.

The spectrum of a galaxy that forms its stars in bursts will resemble those in Figure 6.18, with the addition of some red light from the older stars. At $z \sim 0.7$ almost half of large galaxies show 'post-starburst' spectra, with deep Balmer absorption lines; they must have stopped making stars quite abruptly about 0.5–2 Gyr earlier. This is as we expect for galaxies where starbirth took place in spurts of $\sim$100 Myr, separated by quieter periods of a gigayear or more. Today, only a few percent of large galaxies have post-starburst spectra.

9.4.1 Lyman break galaxies

To study galaxies at high redshift, we must first find a good way to pick them out from our images. All gas-rich galaxies show a break in the spectrum at 912 Å: neutral gas around the star and within the galaxy absorbs much of the light at shorter wavelengths. In objects at $z \gtrsim 4$, intergalactic gas of the Lyα forest removes light below 1216 Å. The *Lyman break galaxies* are distant star-forming

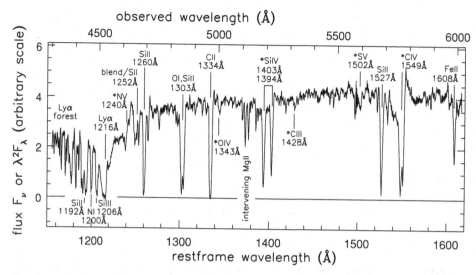

Fig. 9.16. The spectrum of the Lyman break galaxy cB58 at $z = 2.723$. Lines character-istic of hot stellar photospheres are marked *; near 5100 Å, we see MgII absorption from gas at lower redshift. The spectral lines are similar to the starburst in Figure 5.24, but this galaxy is bluer with F_ν approximately constant; that starburst has roughly $F_\lambda \propto \lambda^{-1}$ or $F_\nu \propto \nu^{-1}$ – C. Steidel.

systems found by searching for these spectral signatures. At $z \sim 3$, the break at 912 Å shifts to wavelengths $\lambda \approx 3650$ Å, in the near-ultraviolet U band. So galaxies at this redshift will almost disappear from U-band images, while the color $B - V$ in visible light is still blue or neutral. At redshifts $z \sim 4 - 5$ a star-forming galaxy is not seen in the B band but is still bright in V. At $z \sim 6$ it becomes dark in the I band at 8000 Å as the break at 1216 Å moves into that region.

Figure 9.16 shows cB58, a Lyman break galaxy at redshift $z = 2.7$, corre-sponding to 2.4 Gyr after the Big Bang. The ultraviolet spectral lines are similar to those of the nearby starburst galaxy of Figure 5.24, but this galaxy is bluer. There is no strong Lyα emission line. Although hot stars radiate profusely here, the Lyα photons scatter repeatedly from hydrogen atoms before they leave the galaxy; during their lengthy travels they are easily absorbed by even small amounts of dust. The CIV absorption line has an abrupt edge on the red side, but blends smoothly away to the blue. This *P Cygni* profile is characteristic of the expanding atmospheres of massive hot stars.

The metal lines tell us that the young stars have metal abundance $0.4 Z_\odot \lesssim Z \lesssim Z_\odot$; those heavy elements were made by previous generations of stars. In the gas we can measure the abundances of several elements; oxygen is at 40% of the solar level while iron reaches only $\sim 10\%$, although more may be hidden in dust grains. This is the same pattern as we see in old stars in the Milky Way in Figure 4.17. Oxygen is produced in Type II supernovae, the explosions of short-lived massive stars, while iron is released from Type Ia supernovae which involve low-mass long-lived stars. So oxygen is added to the gas within $\sim$10 Myr after the first significant star formation, whereas iron will be released only $\sim$1 Gyr later. Together with a low level of nitrogen, also made in low-mass stars (see Section 4.2), this ratio suggests that cB58 made most of its stars in the past 250 Myr.

We see blueshifted absorption lines showing that gas is streaming outward; this is typical for Lyman break galaxies. In cB58 the wind moves outward at 250 km s^{-1}, carrying off about as much mass as the galaxy turns into new stars. At least some of this metal-bearing material will escape to intergalactic space. Stars are made from dense gas, so it is not surprising to detect radio-frequency lines from molecules such as CO. In cB58 we find $\sim$10$^9 \mathcal{M}_\odot$ of molecular gas, enough to form stars at its current rate for only about 50 Myr. This gas is dense enough to absorb 80% of the ultraviolet light of the young stars, re-radiating energy $L_{FIR} \sim 10^{11} L_\odot$ in the far-infrared.

Observing these distant galaxies in visible or near-infrared light, we receive energy emitted in the ultraviolet, which comes only from their most massive and recently born stars. To infer the pace of starbirth, we must guess at the *initial mass function*, giving the proportions of massive and less-massive stars. In cB58, the P Cygni profile of the CIV line and the shapes of other lines tell us that, at least among its massive stars, this galaxy has roughly the same initial mass function as the Milky Way. We can measure the ultraviolet luminosity L_λ(1500 Å) in the hybrid units of 3.86×10^{26} W Å^{-1}; recall that $L_\odot = 3.86 \times 10^{26}$ W is the Sun's bolometric luminosity. Then, if the initial mass function has the Salpeter form of Equation 2.5 for $\mathcal{M} \gtrsim 0.1 \mathcal{M}_\odot$, stars are produced at the rate

$$\dot{\mathcal{M}}_\star \sim (3\text{--}5) \times 10^{-7} L_\lambda(1500 \text{ Å}) \mathcal{M}_\odot \text{ yr}^{-1}. \tag{9.24}$$

Making $\mathcal{M}_\odot$ yr^{-1} in stars corresponds to L_λ(1500 Å) $\sim 3 \times 10^6 L_\odot$Å^{-1}, or a luminosity $\lambda L_\lambda \sim 4 \times 10^9 L_\odot$. For a more realistic mass function that includes fewer low-mass stars, $\dot{\mathcal{M}}_\star$ is only half as large for a given ultraviolet luminosity.

The best estimate of starbirth would be to add the rate given by Equation 9.24 to that given by Equation 7.11, which measures the stellar light intercepted by dust and re-radiated in the far-infrared. Typically, the more vigorous the starbirth, the larger the fraction of the young stars' light intercepted by dust. Most of the Lyman break galaxies studied at $z \sim 3$ make (10–50)$\mathcal{M}_\odot$ yr^{-1} of new stars; cB58 produces $\sim$40$\mathcal{M}_\odot$ yr^{-1}. This is modest compared with 200$\mathcal{M}_\odot$ yr^{-1} in the nearby ultraluminous merging system Arp 220 or the submillimeter galaxies (see below). Today $\sim$5$\mathcal{M}_\odot$ yr^{-1} of stars are born in a large spiral galaxy like the Milky Way, while rates for typical local starbursts range up to 30$\mathcal{M}_\odot$ yr^{-1}.

Lyman break galaxies are very bright because of their short-lived massive stars: they are typically several times more luminous in visible light than an $L_\star$ galaxy defined by Figure 1.16. Figure 6.18 shows that, between roughly 50 Myr and 1 Gyr after a starburst, the Balmer jump near 4000 Å increases in strength. Using it to date the current episode of starbirth, we find that the fastest star-formers have typically been making stars for only $30-100\,\mathrm{Myr}$, while the more sedate galaxies have continued for $\gtrsim 1\,\mathrm{Gyr}$. The color of light emitted at visible wavelengths gives information on older stars, which represent most of the mass. We find that Lyman break galaxies are not massive: a few have assembled $5 \times 10^{10}\mathcal{M}_\odot$ of stars, roughly the same as the Milky Way today, but most are small systems like cB58 with $\mathcal{M}_\star \sim 10^{10}\mathcal{M}_\odot$.

Because the light must be spread out in wavelength, spectroscopy of these faint galaxies is very demanding. Some of the best-observed high-redshift galaxies appear many times brighter because of *gravitational lensing*: their light is bent as it passes through a galaxy cluster on its way to us (see Section 7.4). Light from cB58 is amplified about 30-fold in this way. Otherwise, spectroscopy of the most distant galaxies is a task for the large 8–10 m telescopes.

Very bright Lyman break galaxies are seen to $z \gtrsim 6$; BD38 at $z = 5.515$ is an example. The spectrum looks much like that of cB58 in Figure 9.16, although we see it barely 1 Gyr after the Big Bang. It was then making $140\mathcal{M}_\odot\,\mathrm{yr}^{-1}$ of new stars, in an episode that had already lasted $\sim$200 Myr. These are not the galaxy's first stars; the strength of their spectral lines shows that they already are half as metal-rich as the Sun. Observations with the Spitzer Space Telescope at 4.5 μm show the red light of $(1-6) \times 10^{10}\mathcal{M}_\odot$ of stars with ages of $600-700\,\mathrm{Myr}$. These systems are quite small: BD38 is one of the largest, but is only 1.6 kpc in radius. In contrast to the very gas-rich objects that we discuss below, an active nucleus of any kind is rare; we see them in only a few percent of Lyman break galaxies.

Problem 9.11 At optical and near-infrared wavelengths, the flux F_λ from each square arcsecond of the night sky increases roughly as $F_\lambda \propto \lambda^{2.5}$. To measure the energy that a galaxy emits near the Lyα line at 1216 Å, we must observe near $1216(1+z)$ Å. Explain why the measured flux F_λ from each square arcsecond of the galaxy's image decreases as $(1+z)^5$, so that $F_\lambda(\mathrm{galaxy})/F_\lambda(\mathrm{sky}) \propto (1+z)^{-7.5}$. Ly$\alpha$ emission from protogalaxies at $z \sim 5$ is hard to see without an 8–10 m telescope.

9.4.2 Hidden stars: submillimeter galaxies and molecular gas

Even locally, galaxies with intense star formation hide most of the blue and ultraviolet light of their young stars behind dusty gas. As we saw in Section 7.1, only a few percent of this light escapes from a starburst galaxy like M82 or Arp 220. On average, about two-thirds of the ultraviolet light escapes from local galaxies; the rest is absorbed to warm the grains of dust, and re-radiated in the infrared. At $z \sim 1$, *most* stars were made in dusty places: the energy of starlight emerges mainly as the infrared light of luminous infrared galaxies (LIRGs) with $L_{\mathrm{FIR}} > 10^{11}L_\odot$, making $\gtrsim 50\mathcal{M}_\odot\,\mathrm{yr}^{-1}$ of new stars (see Equation 7.11). At higher redshifts, dusty starbursts become even more common. Just as some quasars have large quantities of molecular gas around their centers, so many of the most luminous star-forming galaxies have nuclear activity. The active nuclei are conspicuous in X-rays, which can shine out through the dusty gas, but star formation still contributes most of their energy output.

We saw in Section 7.1 that galaxy mergers can trigger a starburst. These are rare today, and contribute little to the total of star formation. Locally, fewer than 1% of Milky-Way-sized galaxies are locked in close interaction and likely to merge soon with a comparably large system. Even at $z \sim 1$, roughly a third of LIRGs are normal-looking spiral galaxies; a quarter are irregulars; in another quarter, the light is concentrated into a compact center. Only 20% are clearly undergoing major mergers, though these are the most powerful sources. But, at redshifts $z = 2$–3, the far-infrared light comes predominantly from *ultraluminous* infrared galaxies (ULIRGs), with $L_{FIR} > 10^{12} L_\odot$, forming more than $200 \mathcal{M}_\odot \, \text{yr}^{-1}$ of new stars. Many of the ULIRGs show signs of recent or ongoing merger. It is not simply that our telescopes are not powerful enough to see dimmer objects: at these wavelengths, adding up the light of the ULIRGs accounts for almost all of the background radiation shown in Figure 1.19.

In observing dusty galaxies, Equation 8.46 comes to our aid: the term $k(z)$ is so large and negative that the measured flux of a galaxy at $\lambda \sim 1$ mm is nearly constant over the range $1 \lesssim z \lesssim 10$. Some of the most vigorously star-forming galaxies were found first by searching at 850 µm for the redshifted emission of warm dust. In an optical image, these *submillimeter galaxies* are faint with irregular and complex shapes, and often seem to be merging with a neighbor. At these redshifts, lines from the $3 \rightarrow 2$ and $4 \rightarrow 3$ transitions of the CO molecule (see Table 1.8) are shifted to around 3 mm, where they can be observed from the ground.

One object studied in this way, J02399 at $z = 2.8$, lies behind a galaxy cluster where gravitational lensing makes it appear 2.5 times brighter. With $L_{FIR} \sim 10^{13} L_\odot$, Equation 7.11 indicates that it is making $600 \mathcal{M}_\odot \, \text{yr}^{-1}$ of new stars. At this rate, its supply of $6 \times 10^{10} \mathcal{M}_\odot$ of molecular gas will last only 100 Myr. The gas lies in a ring at radius 8 kpc, rotating at $420 \, \text{km s}^{-1}$, so the mass inside is $3 \times 10^{11} \mathcal{M}_\odot$. Recall from Section 2.3 and Problem 6.6 that most of the mass in the central few kiloparsecs of a galaxy is stars and gas, not dark matter. If J02399 is similar, the galaxy has built $\sim 2 \times 10^{11} \mathcal{M}_\odot$ of stars – too many for the current burst, unless by coincidence we see it on the point of exhausting its gas.

The most luminous galaxies discovered at 850 µm have redshifts $1.5 < z < 3.5$. Like the very powerful quasars of Figure 8.13, but unlike bright star-forming galaxies (look ahead to Figure 9.17), their numbers peak at $z \sim 2$. Most contain $\sim 2 \times 10^{10} \mathcal{M}_\odot$ of molecular gas, far more than the Lyman break galaxies, and ten times as much as the Milky Way. But this is only enough to fuel their starbirth for 20–40 Myr. The gas usually lies in a ring or disk within 2 kpc of the center rotating at 400–$500 \, \text{km s}^{-1}$, enclosing a mass $\gtrsim 10^{11} \mathcal{M}_\odot$. They contain $\sim 3 \times 10^{10} \mathcal{M}_\odot$ of young stars, and possibly many more old stars. The gas and young stars alone amount to $5 \times 10^{10} \mathcal{M}_\odot$, which is roughly the mass of normal (baryonic) matter in the Milky Way today. But less-luminous objects also contribute much of the total submillimeter emission, and those lie mostly at $z \lesssim 1.5$, in a pattern more like that of Figure 9.17. The ALMA (Atacama Large Millimeter Array) radio telescope observing at 0.4–4 mm will make sensitive and detailed maps of dusty starbursts at $3 \lesssim z \lesssim 12$.

Problem 9.12 Dust grains in a starburst galaxy are heated to a temperature $T \sim 50\,\mathrm{K}$. For a blackbody at temperature T, the luminosity $L_\nu \propto \nu n(\nu)$, where $n(\nu)$ is given by Equation 1.35. But because the grains are small, with sizes $\lesssim 1\,\mu\mathrm{m}$, they radiate inefficiently at longer wavelengths: the dust emission follows $L_\nu \propto \nu^3 n(\nu)$. Show that, for $\lambda \gg 300\,\mu\mathrm{m}$, we have $L_\nu \propto \nu^4$.

If we observe at 100 GHz or 3 mm, the power $F_\nu(\nu)\Delta\nu$ that we receive between frequencies ν and $\nu + \Delta\nu$ was emitted between $\nu(1+z)$ and $(\nu + \Delta\nu)(1+z)$. Equation 8.37 tells us that the redshift decreases the energy received from each square arcsecond by a factor of $(1+z)^4$. Show that, near this frequency, the flux F_ν from each square arcsecond of a starburst galaxy is almost *constant* for redshifts $5 \lesssim z \lesssim 20$. Once we have large telescopes with sensitive detectors in this spectral region, we should be able to see extremely distant star-forming galaxies.

9.4.3   Old, red, and dead?

Figure 6.20 shows that the most optically-luminous galaxies today are making hardly any new stars. When did the earliest galaxies finish building their stellar bodies? Locally, almost all red galaxies are red because they lack young blue stars; only a few, like the starburst M82, are red because dust hides those stars. Recent observations from space, where the infrared sky is dark, show that this is reversed at $z = 2$–3. In one recent sample, only 3 of 13 red galaxies at $z \sim 2.5$ really seem to be 'dead': observed at 8μm in the infrared, they are dim while the dusty galaxies blaze brightly with re-radiated starlight. Comparing their spectra with models like Figure 6.18 shows that these are massive galaxies, with $\mathcal{M}_\star > 3 \times 10^{10}\,\mathcal{M}_\odot$, where hardly any stars have been born for the past $\sim 2\,\mathrm{Gyr}$. They are making new stars at less than 0.1% of the average rate needed to build the galaxy in the 2.6 Gyr since the Big Bang.

The earliest 'red and dead' galaxies yet observed are at $z \sim 6$. They are too faint for us to take their spectra, but we can look at their light in broad bands. Some of them show increased light at wavelengths longer than 3 μm. The 4000 Å break of Figure 6.17 is so strong that most of the stars must be 200–600 Myr old – they were born already at $z \sim 7$–13! These galaxies have $\mathcal{M}_\star \sim (1$–$4) \times 10^{10}\,\mathcal{M}_\odot$ – they have made stars equivalent to 20%–50% of the Milky Way.

By $z \sim 1.5$ or 4.2 Gyr after the Big Bang, red galaxies with roughly the same stellar mass as the Milky Way, $\mathcal{M}_\star > 5 \times 10^{10}\,\mathcal{M}_\odot$, probably contain more than half the stars in the galaxies. A few real monsters even have $\mathcal{M}_\star > 10^{11}\,\mathcal{M}_\odot$ at

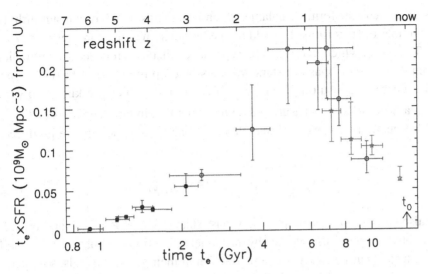

Fig. 9.17. Star formation per *comoving* Mpc^3, derived from light emitted at ultraviolet wavelengths, calculated for the benchmark cosmology. Stars show points measured in the ultraviolet by Galex, and filled circles are from Lyman break galaxies. The mass of stars formed in any time interval is proportional to the area under the points – E. Stanway and A. Barger.

$z \approx 1.5$. Since there is roughly six times more dark matter in the Universe than baryonic material, these objects must have dark halos of $\sim 10^{12} \mathcal{M}_\odot$. According to the cold dark matter models of Section 8.5, such massive halos should be very rare at early times. Their stars are about as metal-rich as the Sun, and have ages around 2 Gyr, so the last significant starbirth was at $z \sim 3$. Forming $(300\text{--}500)\mathcal{M}_\odot\,\text{yr}^{-1}$ of new stars, it would take only 300 Myr (a typical timescale for starbursts) for the 'hidden' systems of the last subsection to build up these masses. The most luminous of the dusty starbursts discovered at 850 μm might have developed first into these 'red and dead' galaxies, and then into today's luminous ellipticals.

9.4.4 The star-forming history of the Universe

Using variants of Equation 9.24, we can estimate how fast the ultraviolet-luminous galaxies formed their stars. According to Figure 9.17 most stars in the Universe were born at $z \sim 1$, between 5 and 8 Gyr after the Big Bang. Star-forming galaxies seem to have flourished later than the luminous active nuclei of Figure 8.13, which reach their peak at $z \sim 2$. It also shows less extreme variation: starbirth today is only about five times less vigorous than it was at its peak, whereas at $z \sim 4$ it is at roughly half the peak level.

At $z \gtrsim 0.7$ much of the ultraviolet radiation of young stars is absorbed by dust, and Figure 9.17 includes a correction based on the color of the ultraviolet light. But virtually no ultraviolet light escapes from the dustiest systems such as the submillimeter galaxies, so they will be missed from the plot. We have also missed fainter star-forming galaxies, even if they host much of the starbirth. But the number of stars formed should not be less than that given in Figure 9.17.

The rate at which those stars have produced metals is related to the ultraviolet luminosity of their massive stars. We must add up the yield of heavy elements from supernovae resulting from stars of each mass, which are known to within about a factor of two. Measuring the ultraviolet flux in the same units of $L_\odot \, \text{Å}^{-1}$ as for Equation 9.24, we find that elements heavier than helium are produced at the rate

$$\dot{\mathcal{M}}_Z \sim 8 \times 10^{-9} L_\lambda (1500 \, \text{Å}) \mathcal{M}_\odot \, \text{yr}^{-1}. \tag{9.25}$$

We also had to assume here that high-mass stars, with $\mathcal{M} \gtrsim 6\mathcal{M}_\odot$, are formed in the relative proportions specified by the Salpeter function of Equation 2.5. Luckily, all such stars burn roughly the same fraction of their gas into metals, so Equation 9.25 is not very sensitive to the initial mass function adopted.

According to Figure 9.17, roughly a quarter of all the stars had formed by $z = 2$, and they should have released enough metals to give an average abundance $\sim Z_\odot/30$ if mixed evenly throughout the baryons. At the end of the previous section we saw that this level is already higher than what we observe. If many small star-forming galaxies contributed to reionization, then even more 'missing metals' must be hidden in diffuse gas.

Is our picture of the life of distant galaxies consistent with what we see in the Local Group, where we can take spectra of individual bright stars to reconstruct the star-forming history? About half of its stellar mass is in the bulges of M31 and the Milky Way, and in Chapter 4 we saw that most of these stars are probably old enough to predate $z \sim 1.5$. The Milky Way's disk has formed stars steadily over the past few gigayears. The disk of M31 is more luminous but has fewer young stars, suggesting that the pace of starbirth has slackened over that period. Averaging over the Local Group, starbirth appears to have been at its most vigorous 8–13 Gyr in the past, perhaps somewhat earlier than the peak in Figure 9.17. However, the atmospheres of the stars that we see locally all contain elements heavier than helium; very few have less than 10^{-3} of the Sun's metal abundance. We have found no 'fossils' from the earliest stellar generation, formed out of hydrogen and helium alone. Thus further studies both of nearby and of distant galaxies are needed in order to tell us whether Figure 9.17 fairly represents cosmic starbirth. New instruments that can observe at longer wavelengths, such as ALMA and the planned James Webb Space Telescope, will help us to pierce through the dust to explore the birth of the galaxies.

Appendix A

Units and conversions

Table A.1 Units and prefixes

magnitudes	see Section 1.1
arcsecond	$1'' = (1/60) \times 1'$; $1/206\,265$ radians
arcminute	$1' = (1/60) \times 1°$
angstrom	$\text{Å} = 10^{-10}\,\text{m} = 0.1\,\text{nm}$
nanometer	$\text{nm} = 10^{-9}\,\text{m}$
micron	$\mu\text{m} = 10^{-6}\,\text{m} = 10^{-4}\,\text{cm}$
centimeter	$\text{cm} = 10^{-2}\,\text{m}$
jansky	$\text{Jy} = 10^{-26}\,\text{W}\,\text{m}^{-2}\,\text{Hz}^{-1}$
joule	$\text{J} = 10^7\,\text{erg or } 10^7\,\text{g}\,\text{cm}^2\,\text{s}^{-2}$
watt	$\text{W} = 10^7\,\text{erg}\,\text{s}^{-1}$
micro	μ $1\,\mu\text{s} = 10^{-6}\,\text{s}$: microsecond
milli	m $1\,\text{mJy} = 10^{-3}\,\text{Jy}$: millijansky
kilo	k $1\,\text{km} = 10^3\,\text{m}$: kilometer
mega	M $1\,\text{Mpc} = 10^6\,\text{pc}$: megaparsec
giga	G $1\,\text{Gyr} = 10^9\,\text{yr}$: gigayear

Table A.2 Conversion factors

Sound speed in atomic hydrogen	$c_s = \sqrt{k_B T/m_p} = 9\,\text{km}\,\text{s}^{-1} \times \sqrt{T/10^4\,\text{K}}$
Surface density	$\mathcal{M}_\odot\,\text{pc}^{-2} = 1.25 \times 10^{20}\,\text{H atoms cm}^{-2}$
Volume density	$\mathcal{M}_\odot\,\text{pc}^{-3} = 6.7 \times 10^{-23}\,\text{g}\,\text{cm}^{-3}$
	or $44\,\text{H atoms cm}^{-3}$
Surface brightness	$L_\odot\,\text{pc}^{-2} = 27\,\text{mag arcsec}^{-2}$ in B
Luminosity	$M_B = -20,\quad L_B = 1.6 \times 10^{10}\,L_\odot$
	$M_B = -18,\quad L_B = 2.5 \times 10^9\,L_\odot$
	$M_B = -16,\quad L_B = 3.9 \times 10^8\,L_\odot$
Speed	$1\,\text{km}\,\text{s}^{-1} = 1.023\,\text{pc Myr}^{-1}$
Gravitational constant	$G = 4.5 \times 10^{-3}$ if mass is in $\mathcal{M}_\odot$,
	distance in pc, time in Myr
Vector products	$\mathbf{A} \times (\mathbf{B} \times \mathbf{C}) = (\mathbf{A} \cdot \mathbf{C})\mathbf{B} - (\mathbf{A} \cdot \mathbf{B})\mathbf{C}$
	$(\mathbf{A} \times \mathbf{B}) \cdot (\mathbf{C} \times \mathbf{D}) = (\mathbf{A} \cdot \mathbf{C})(\mathbf{B} \cdot \mathbf{D}) - (\mathbf{A} \cdot \mathbf{D})(\mathbf{B} \cdot \mathbf{C})$

Table A.3 Physical constants

Gravitational constant	$G = 6.67 \times 10^{-8}\,\mathrm{cm^3\,s^{-2}\,g^{-1}}$ or $6.67 \times 10^{-11}\,\mathrm{N\,m^2\,kg^{-2}}$
Speed of light	$c = 2.997\,924\,58 \times 10^{10}\,\mathrm{cm\,s^{-1}}$ or $2.997\,924\,58 \times 10^{8}\,\mathrm{m\,s^{-1}}$
Planck's constant	$h_P = 6.626 \times 10^{-27}\,\mathrm{erg\,s}$ or $6.626 \times 10^{-34}\,\mathrm{J\,s}$
Photon energy	$\nu h_P = 4.136 \times (\nu/10^{15}\,\mathrm{Hz})\,\mathrm{eV}$ or $1.240 \times (1\,\mu\mathrm{m}/\lambda)\,\mathrm{eV}$
Boltzmann constant	$k_B = 1.381 \times 10^{-16}\,\mathrm{erg\,K^{-1}}$ or $1.381 \times 10^{-23}\,\mathrm{J\,K^{-1}}$ or $0.862\,\mathrm{MeV}/10^{10}\,\mathrm{K}$
Blackbody constant	$a_B = 8\pi^5 k_B^4/(15c^2 h_P^3) = 7.566 \times 10^{-15}\,\mathrm{erg\,cm^{-3}\,K^{-4}}$ or $7.566 \times 10^{-16}\,\mathrm{J\,m^{-3}\,K^{-4}}$
Stefan–Boltzmann constant	$\sigma_{SB} = c a_B/4 = 2\pi^5 k_B^4/(15c^2 h_P^3) = 5.671 \times 10^{-8}\,\mathrm{W\,m^{-2}\,K^{-4}}$ or $5.671 \times 10^{-5}\,\mathrm{erg\,s^{-1}\,cm^{-2}\,K^{-4}}$
Charge on electron	$e = 1.602 \times 10^{-19}\,\mathrm{coulomb}$ or $4.803 \times 10^{-10}\,\mathrm{esu}$
Electron-volt	$\mathrm{eV} = 1.602 \times 10^{-12}\,\mathrm{erg}$ or $1.602 \times 10^{-19}\,\mathrm{J}$
Electron mass	$m_e = 9.11 \times 10^{-28}\,\mathrm{g}$ or $9.11 \times 10^{-31}\,\mathrm{kg}$ $m_e c^2 = 0.511\,\mathrm{MeV}$
Proton mass	$m_p = 1.673 \times 10^{-24}\,\mathrm{g}$ or $1.673 \times 10^{-27}\,\mathrm{kg}$ $m_p c^2 = 938.3\,\mathrm{MeV}$
Neutron mass	$(m_n - m_p)c^2 = 1.293\,\mathrm{MeV}$
Thomson cross-section	$\sigma_T = (8\pi/3)[e^2/m_e c^2]^2$ (cgs) or $(8\pi/3)[e^2/(4\pi\epsilon_0 m_e c^2)]^2$ (SI) $6.652 \times 10^{-25}\,\mathrm{cm^2}$ or $6.652 \times 10^{-29}\,\mathrm{m^2}$
Fine structure constant	$\alpha = 2\pi e^2/(ch_P)$ (cgs) or $e^2/(2\epsilon_0 ch_P)$ (SI) 7.297×10^{-3} or $1/137.04$
SI electromagnetic constants	$\mu_0 = 4\pi \times 10^{-7}\,\mathrm{H\,m^{-1}}$ $\epsilon_0 = 1/(\mu_0 c^2) = 8.854 \times 10^{-12}\,\mathrm{C^2\,m^{-2}\,N^{-1}}$

Table A.4 Astronomical constants

Tropical year (1900)	$\mathrm{yr} = 3.155\,693 \times 10^{7}\,\mathrm{s}$
Astronomical unit	$\mathrm{AU} = 1.496 \times 10^{13}\,\mathrm{cm}$ or $1.496 \times 10^{8}\,\mathrm{km}$
Light-year	$\mathrm{ly} = 9.46 \times 10^{17}\,\mathrm{cm}$ or $9.46 \times 10^{12}\,\mathrm{km}$
Parsec	$\mathrm{pc} = (648\,000/\pi)\,\mathrm{AU}$ or $206\,265\,\mathrm{AU}$, $3.09 \times 10^{18}\,\mathrm{cm}$ or 3.26 light-years
Solar radius	$R_\odot = 6.96 \times 10^{10}\,\mathrm{cm}$ or $6.96 \times 10^{5}\,\mathrm{km}$
Solar mass	$\mathcal{M}_\odot = 1.99 \times 10^{33}\,\mathrm{g}$ or $1.99 \times 10^{30}\,\mathrm{kg}$
Solar luminosity (bolometric)	$L_\odot = 3.86 \times 10^{33}\,\mathrm{erg\,s^{-1}}$ or $3.86 \times 10^{26}\,\mathrm{W}$
Sun's effective temperature	$T_{\mathrm{eff}} = 5780\,\mathrm{K}$
Sun's surface gravity	$g_\odot = 2.74 \times 10^{4}\,\mathrm{cm\,s^{-2}}$ or $274\,\mathrm{m\,s^{-2}}$
Solar absolute magnitude	$M_{B,\odot} = +5.48$ $M_{V,\odot} = +4.83$ $M_{K,\odot} = +3.31$ $M_{\mathrm{bol},\odot} = +4.75$
Earth's mass	$M_E = 5.98 \times 10^{27}\,\mathrm{g}$ or $5.98 \times 10^{24}\,\mathrm{kg}$
Earth's radius	$R_E = 6.38 \times 10^{8}\,\mathrm{cm}$ or $6.38 \times 10^{3}\,\mathrm{km}$
Earth's surface gravity	$g_E = 980.7\,\mathrm{cm\,s^{-2}}$ or $9.807\,\mathrm{m\,s^{-2}}$
Earth's orbit (sidereal year)	$3.155\,815 \times 10^{7}\,\mathrm{s}$
Average Earth–Moon distance	$3.84 \times 10^{5}\,\mathrm{km}$
Hubble 'constant'	$H_0 = 100h\,\mathrm{km\,s^{-1}\,Mpc^{-1}}$; $0.4 \lesssim h \lesssim 0.8$
Hubble time	$t_H = 1/H_0 = 9.78 h^{-1}$ gigayears $c/H_0 = 2.99 h^{-1}$ gigaparsecs
Critical density	$\rho_{\mathrm{crit}} = 1.9 \times 10^{-26} h^2\,\mathrm{kg\,m^{-3}}$ or $2.8 \times 10^{11} h^2\,\mathcal{M}_\odot\,\mathrm{Mpc^{-3}}$

Table A.5 Frequently used symbols

γ	Lorentz factor $1/\sqrt{1 - V^2/c^2}$
λ	wavelength
ν	frequency
ρ	volume density of mass
σ	velocity dispersion, standard deviation, or
	comoving radius coordinate (cosmology)
Σ	surface density of mass
$\Phi(\mathbf{x}, t)$	gravitational potential energy per unit mass
$\Phi(L)$	luminosity function
$\Omega(R)$	angular speed in circular orbit at radius R
$\Omega(t)$	ratio of cosmic density to critical value ρ_{crit}: present value Ω_0
Ω_m, Ω_B	present-day ratio of density in matter or in baryons to critical value
Ω_Λ	present-day ratio of 'dark energy' density to critical value
$a(t)$	$\mathcal{R}(t)/\mathcal{R}(t_0)$: dimensionless scale factor for cosmic expansion
$\mathcal{E}$	energy
E	energy per unit mass
F_λ, F_ν	flux of energy per unit wavelength or frequency
$f(\mathbf{x}, \mathbf{v}, t)$	distribution function: density of particles at $\mathbf{x}, \mathbf{v}$ in phase space
H_0	Hubble 'constant': present value of parameter $H(t) = \dot{\mathcal{R}}(t)/\mathcal{R}(t)$
h	H_0 in units of 100 km s^{-1} Mpc^{-1}
HI	atomic hydrogen
HII	ionized hydrogen
H$_2$	molecular hydrogen
$I(\mathbf{x})$	surface brightness (units of mag arcsec^{-2} or $L_\odot$ pc^{-2})
L	luminosity: $L_\odot$ is the Sun's luminosity
L_*	$2 \times 10^{10} L_\odot$, typical luminosity of bright galaxy: see Equation 1.24
$\mathcal{L}$	angular momentum
$\mathbf{L}$	angular momentum per unit mass (vector)
M	absolute magnitude
m	apparent magnitude, or mass
$\mathcal{M}$	mass: $\mathcal{M}_\odot$ is the solar mass
$\mathcal{M}/L$	mass-to-light ratio: units $\mathcal{M}_\odot/L_\odot$
$\mathcal{N}$	surface density: number of stars or atoms
n	volume density: number of stars or atoms
r	radius (in three-dimensional space)
R	radius (two-dimensional) or distance from point in disk to Galactic center
R_0	distance from Sun to Galactic center
$\mathcal{R}(t)$	scale length for Universe at time t after the Big Bang
t	time
T	temperature
V_r	radial velocity: motion away from or toward the observer
$V(R)$	linear speed in circular orbit at radius R
V_{max}	peak rotation speed
Z	mass fraction of $metals$, elements heavier than H and He
z	redshift or distance above Galactic midplane

Table A.6 Astronomical buzzwords

Early-type star	hot: early in spectral sequence OBAFGKM
Late-type star	cool: late in spectral sequence OBAFGKM
Dwarf star	main-sequence star (except for 'white dwarf')
Early-type galaxy	E or S0: 'early' in Hubble sequence
Late-type galaxy	spiral or irregular: 'late' in Hubble sequence
Dwarf galaxy	luminosity $L \lesssim 10^9 L_\odot$
Metals	elements heavier than helium
Redshift z	Doppler shift $(\lambda_{obs} - \lambda_e)/\lambda_e$
Radial velocity V_r	motion away from or toward the observer
Tangential velocity V_t	motion perpendicular to the observer
Scale length or height	distance over which density falls by factor of e

Appendix B

Bibliography

We have drawn for our presentation on the following graduate texts:

J. Binney & S. Tremaine, 1987, *Galactic Dynamics* (Princeton University Press, Princeton, New Jersey), on the dynamics of galaxies and star clusters; F. Combes, P. Boissé, A. Mazure, & A. Blanchard, *Galaxies and Cosmology*, 2nd edition (English translation, 2002; Springer, Heidelberg, Germany), covering similar ground to our text; and J. Binney & M. Merrifield, 1998, *Galactic Astronomy*, 3rd edition (Princeton University Press, Princeton, New Jersey), which gives a comprehensive review of observations of our Galaxy and others.

On a similar level to this book, see S. Phillipps, 2005, *The Structure and Evolution of Galaxies* (Wiley, Chichester, UK) and P. Schneider, *Extragalactic Astronomy and Cosmology: An Introduction* (English translation, 2006; Springer, Heidelberg/Berlin, Germany).

At a more elementary level, see M. H. Jones & R. J. Lambourne (eds.), 2003, *An Introduction to Galaxies and Cosmology* (Cambridge University Press and the Open University, Cambridge, UK).

More specialized references are given under chapter headings below.

Chapter 1

Introductions to stellar structure on an undergraduate level include D. A. Ostlie & B. W. Carroll, 1996, *An Introduction to Modern Stellar Astrophysics* (Addison-Wesley, Reading, Massachusetts); A. C. Phillips, 1994, *The Physics of Stars* (Wiley, Chichester, UK); and D. Prialnik, 2000, *An Introduction to the Theory of Stellar Structure and Evolution* (Cambridge University Press, Cambridge, UK).

Graduate texts include C. J. Hansen & S. D. Kawaler, 1994, *Stellar Interiors: Physical Principles, Structure, and Evolution* (Springer, New York); D. Arnett, 1996, *Supernovae and Nucleosynthesis* (Princeton University Press, Princeton, New Jersey) for stellar evolution beyond the main sequence; and M. Salaris and S. Cassisi, 2005, *Evolution of Stars and Stellar Populations* (Wiley, Chichester, UK).

Hubble's original account of galaxy classification is given in E. Hubble, 1936, *The Realm of the Nebulae* (Yale University Press; reprinted by Dover, New York); it is illustrated in A. Sandage, 1961, *The Hubble Atlas of Galaxies* (Carnegie Institute of Washington, Washington, DC). Modern treatments include S. van den Bergh, 1998, *Galaxy Morphology and Classification* (Cambridge University Press, Cambridge, UK).

Chapter 2

Two undergraduate texts on interstellar gas and dust are J. E. Dyson & D. A. Williams, 1997, *The Physics of the Interstellar Medium*, 2nd edition, and D. C. B. Whittet, 1992, *Dust in the Galactic Environment* (both from Institute of Physics Publishing, London and Bristol, UK). On the graduate level, see J. Lequeux, 2004, *The Interstellar Medium* (English translation, 2004; Springer, Berlin and Heidelberg, Germany).

Chapter 3

The standard graduate text is J. Binney & S. Tremaine, 1987, *Galactic Dynamics* (Princeton University Press, Princeton, New Jersey).

Chapter 4

S. van den Bergh, 2000, *The Galaxies of the Local Group* (Cambridge University Press, Cambridge, UK); on element production in the Big Bang and afterwards, see B. E. Pagel, 1997, *Nucleosynthesis and Chemical Evolution of Galaxies* (Cambridge University Press, Cambridge, UK).

Chapter 5

Texts on array detectors in astronomy include G. H. Rieke, 1994, *Detection of Light: from the Ultraviolet to the Submillimeter* (Cambridge University Press, Cambridge, UK); for a wider wavelength range, see P. Léna, F. Lebrun, & F. Mignard, *Observational Astrophysics*, 2nd edition (English translation, 1998; Springer, Berlin, Germany).

For spectroscopy, see C. R. Kitchin, 1995, *Optical Astronomical Spectroscopy* (Institute of Physics Publishing, Bristol, UK); D. F. Gray, 2005, *The Observation and Analysis of Stellar Photospheres*, 3rd edition (Cambridge University Press, Cambridge, UK) is a graduate-level text.

On statistics and observational uncertainties, see P. R. Bevington & D. K. Robinson, 1992, *Data Reduction and Error Analysis for the Physical Sciences*, 2nd edition (McGraw-Hill, New York); and R. Lupton, 1993, *Statistics in Theory and Practice* (Princeton University Press, Princeton, New Jersey).

On radio astronomy, see B. Burke & F. Graham-Smith, 2002, *An Introduction to Radio Astronomy*, 2nd edition (Cambridge University Press, Cambridge, UK); and G. L. Verschuur & K. I. Kellermann, eds., 1988, *Galactic and Extragalactic Radio Astronomy*, 2nd edition (Springer, New York).

Chapter 6

I. Stewart, 1990, *Does God Play Dice? The Mathematics of Chaos* (Blackwell, Cambridge, Massachusetts) gives a clear discussion of mathematical chaos, written for the general reader.

Chapter 7

On gravitational lensing, at the graduate level, see P. Schneider, J. Ehlers, & E. E. Falco, 1992, *Gravitational Lenses* (Springer, New York).

Chapter 8

For a descriptive introduction to cosmology, see T. Padmanabhan, 1998, *After the First Three Minutes* (Cambridge University Press, Cambridge, UK).

B. Ryden, 2003, *Introduction to Cosmology* (Addison-Wesley, San Francisco) is a very clear undergraduate text. See also A. Liddle, 2003, *An Introduction to Modern Cosmology*, 2nd edition (John Wiley & Sons, Chichester, UK); and M. Lachièze-Rey, *Cosmology: A First Course* (English translation, 1995; Cambridge University Press, Cambridge, UK).

Recent and comprehensive graduate texts are J. A. Peacock, 1999, *Cosmological Physics* (Cambridge University Press, Cambridge, UK); and M. S. Longair, 1998, *Galaxy Formation* (Springer, Berlin, Germany).

Chapter 9

For comprehensive reviews of active nuclei, see B. M. Peterson, 1997, *Active Galactic Nuclei* (Cambridge University Press, Cambridge, UK); and A. J. Kembhavi & J. V. Narlikar, 1999, *Quasars and Active Galactic Nuclei* (Cambridge University Press, Cambridge, UK).

For relevant physics, an undergraduate text using SI units is M. S. Longair, *High Energy Astrophysics*, 2nd edition (Cambridge University Press, Cambridge, UK): 1992, Volume 1, *Particles, Photons and their Detection*; 1994, Volume 2, *Stars, the Galaxy and the Interstellar Medium*.

On the graduate level, and in cgs units, are F. H. Shu, 1991, *The Physics of Astrophysics*, Volume 1, *Radiation* (University Science Books, Mill Valley, California); and J. H. Krolik, 1999, *Active Galactic Nuclei* (Princeton University Press, Princeton, New Jersey).

Appendix C

Hints for problems

Problem 1.14 See Table 1.4 for the Sun's absolute magnitude M_V, and remember that $M_B = M_V + (B - V)$, while $M_I = M_V - (V - I)$.

Problem 1.20 The energy in the background radiation is 10^{54} J Mpc^{-3}. The Sun radiates 4×10^{26} W or 10^{43} J Gyr^{-1}; so, from Equation 1.25, galaxies emit about 2×10^{51} J Mpc^{-3} Gyr^{-1}. Even though the galaxies were brighter in the past, all the starlight radiated since the Big Bang falls far short of the energy in the cosmic background radiation.

Problem 2.4 If the rate of starbirth at time t is $\mathcal{B}(t)$, then, for stars that spend time $\tau_{MS} < \tau_{gal}$ on the main sequence, the initial luminosity function $\Psi(M_V)$ is related to the present-day luminosity function Φ_{MS} by

$$\Psi(M_V) = \Phi_{MS}(M_V) \times \int_0^{\tau_{gal}} \mathcal{B}(t)\, dt \left/ \int_{\tau_{gal}-\tau_{MS}}^{\tau_{gal}} \mathcal{B}(t) dt \right. .$$

The faster $\mathcal{B}(t)$ declines with time, the more of these short-lived stars must be born to yield the numbers that we see today.

Problem 2.5 Taking $L \propto \mathcal{M}^{3.5}$ for each star, the number N of stars and their total luminosity L are

$$N = \xi_0 \int_{\mathcal{M}_l}^{\mathcal{M}_u} \left(\frac{\mathcal{M}}{\mathcal{M}_\odot}\right)^{-2.35} \frac{d\mathcal{M}}{\mathcal{M}_\odot}, \qquad L = L_\odot \xi_0 \int_{\mathcal{M}_l}^{\mathcal{M}_u} \left(\frac{\mathcal{M}}{\mathcal{M}_\odot}\right)^{-2.35+3.5} \frac{d\mathcal{M}}{\mathcal{M}_\odot}.$$

The integral for N and that for the total mass both diverge as the lower limit $\mathcal{M}_l \rightarrow 0$, while that for L becomes large as $\mathcal{M}_u$ increases. Almost all the light of a young cluster comes from the few most massive stars. At ages beyond 2–3 Gyr light comes mainly from red giants, which reach roughly the same luminosity independently of the star's initial mass.

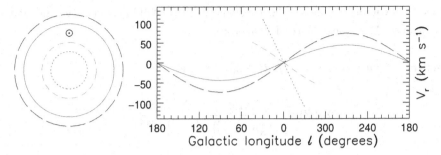

Fig. C.1. Radial velocity V_r of gas on four rings, at radii $R = 4, 6, 10$, and 12 kpc, with circular speed $V(R) = 220$ km s^{-1}. The Sun $\odot$ is at $R_0 = 8$ kpc.

Problem 2.7 See astro-ph/0309416.

Problem 2.11 Because of the Malmquist bias, the stars in your sample are brighter than the average for the whole sky. If you do not correct for the bias, and simply assume that the stars of your sample are an average selection of those in the sky, you will overestimate their distances. The average values that you infer for any other properties that are linked to the luminosity will also be wrong.

Problem 2.12 Blue stars with $m_V = 20$ must be far from the midplane, where disk stars are rare. The red stars are dim nearby dwarfs; at this apparent magnitude, a red giant would be halfway to M31!

Problem 2.13 $L_{eye} = 0.08L_\odot$, corresponding via Equation 1.6 to $\mathcal{M}_{eye} \approx 0.6\mathcal{M}_\odot$. After 3 Gyr, only stars below $\mathcal{M}_u \approx 1.5\mathcal{M}_\odot$ are still on the main sequence. Between $\mathcal{M}_l = 0.2\mathcal{M}_\odot$ and $\mathcal{M}_u$, $6.08\xi_0$ stars were made; their total mass is $2.54\xi_0\mathcal{M}_\odot$, so $\xi_0 = 3.95 \times 10^6$. There are $N_{eye} = 1.05\xi_0 \approx 4 \times 10^6$ main-sequence stars with $\mathcal{M}_{eye} \lesssim \mathcal{M} \lesssim \mathcal{M}_u$.

Table 1.1 shows that, for low-mass stars, the red giant phase lasts about a third as long as the main-sequence life. There are $0.1\xi_0$ stars that live between 2.25 and 3 Gyr on the main sequence ($1.5\mathcal{M}_\odot \lesssim \mathcal{M} \lesssim 1.8\mathcal{M}_\odot$); adding them makes little difference to N_{eye}. A star with $L < L_{eye}$ can be seen only to $r_{max} = 3$ pc$(L/L_{eye})^{0.5} \approx 3$ pc$(\mathcal{M}/\mathcal{M}_{eye})^{2.5}$. The number of stars between $\mathcal{M}$ and $\mathcal{M} + \Delta\mathcal{M}$ within the sphere $r_{max}(\mathcal{M})$ is proportional to $(\mathcal{M}/\mathcal{M}_{eye})^{7.5-2.35}$, decreasing rapidly as $\mathcal{M} < \mathcal{M}_{eye}$.

Problem 2.15 See Figure C.1. At $(l = 120°, V > 0)$ we see local motions in gas near the Sun, not Galactic rotation.

Problem 2.20 Near the center where the density is close to $\rho_H(0)$, Equation 2.19 gives $V(r) \to rV_H/(\sqrt{3}a_H)$. At large radius $V(R) \to V_H$: see Figure 5.19. Far beyond a_H, the mass $\mathcal{M}(r)$ rises linearly with radius. In a real galaxy, the dark halo does not extend forever; at some radius its density must start to fall below that of Equation 2.19. But Equation 2.18 tells us that, in a spherical halo, the orbital speed at radius r depends only on the mass within r. As long as the density is close to that of Equation 2.19 within radius r, we can use Equation 2.20 to calculate $V(r)$.

Problem 3.2 On setting $s^2 = z^2/(R^2 + a_P^2)$ we have

$$\Sigma_P(R) = \frac{3\mathcal{M}a_P^2}{2\pi\left(R^2 + a_P^2\right)^{2}} \int_0^\infty \frac{ds}{(1 + s^2)^{5/2}}.$$

Write $s = \tan\phi$ to show that the integral is 2/3.

Problem 3.10 $\mathcal{E} = m\Phi(1 - \alpha/2)$, $\mathcal{L} = m\sqrt{\alpha K}\,r^{1-\alpha/2}$.
$\Delta\mathcal{L}_{\text{tot}} = 0$ when $\Delta r_2 = -\Delta r_1 (m_1/m_2)(r_1/r_2)^{-\alpha/2}$; the energy change is then

$$\Delta\mathcal{E}_{\text{tot}} = \alpha K(1 - \alpha/2)m_1\,\Delta r_1\,r_1^{-\alpha-1}[1 - (r_1/r_2)^{1+\alpha/2}].$$

Problem 3.12 Substituting $s = \tan\phi$ shows that the integral

$$\int_0^\infty \frac{s^2\,ds}{(1 + s^2)^3} = \frac{\pi}{16}.$$

Problem 3.20 Differentiating the effective potential gives

$$r^4\frac{d\Phi_{\text{eff}}}{dr} = L^2\left(\frac{3G\mathcal{M}_{\text{BH}}}{c^2} - r\right) + G\mathcal{M}_{\text{BH}}r^2,$$

which is zero in circular orbit: $L^2 > 0$ so $r > 3G\mathcal{M}/c^2$. For stability,

$$r^5\frac{d^2\Phi_{\text{eff}}}{dr^2} = L^2\left(r - \frac{6G\mathcal{M}_{\text{BH}}}{c^2}\right) > 0.$$

Problem 3.22 See Figure C.2. From Equation 3.71,

$$\kappa^2(R) = \Omega^2(R) + \frac{V_H^2}{a_H^2}\frac{1}{1 + R^2/a_H^2}.$$

Problem 2.20 gives $\Omega^2(R) \to V_H^2/(3a_H^2)$ at the center, so $\kappa \to 2\Omega$, as expected when the density is constant. At large radius $\Omega \to V_H/R$ and $\kappa^2 \to 2\Omega^2$.

Problem 3.25 Note that $2\,d^2\phi/dy^2 \cdot d\phi/dy = d(d\phi/dy)^2/dy$, so multiply the equation by $d\phi/dy$ and integrate to find $(d\phi/dy)^2$. Then, recall that we set $\phi = 0$ and $d\phi/dy = 0$ at $y = 0$, so that

$$y(\phi) = \int_0^\phi \frac{d\psi}{\sqrt{1 - e^{-\psi}}}.$$

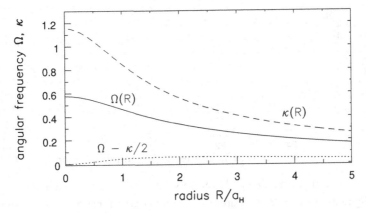

Fig. C.2. For a circular orbit at radius R in the 'dark-halo' potential, the angular velocity Ω (solid curve), the epicyclic frequency κ (dashed curve), and $\Omega - \kappa/2$ (dotted curve). The units are V_H/a_H.

Setting $u = e^{-\psi/2}$, and then $t = \text{sech}\, u$, and integrating yields $e^{-\phi/2} = \text{sech}(y/2)$. At large z, $n(z) = n_0 e^{-\phi} \to 4n_0 e^{-|z|/z_0}$; the midplane density n_0 is four times lower than the inward extrapolation of the exponential.

Problem 4.1 The brightest blue stars in this region have $V - I \approx 0 \approx M_V$; they are main-sequence stars, of late B or early A types. The brightest red stars are K giants, not supergiants. Just as we found few supergiants among the solar-neighborhood stars of Figure 2.2, such rare very luminous stars are missing from this small patch of the LMC's disk.

Problem 4.4 The Jacobi radius $r_J = 0.01$ AU, while the average distance between Earth and Moon $r_{EM} = 0.0026$ AU. The ratio of the gravitational forces from Earth and Sun is $(m/\mathcal{M})(1\,\text{AU}/r_{EM})^2 \approx 0.5$.

Problem 4.5 Since $m \ll \mathcal{M}$, the mass center C is at the halo center. What rotation rate Ω must you choose to follow m in its circular orbit? In Equation 4.7, you can find $\partial \Phi_H/\partial x$ from $\mathcal{M}(<r)$.

Problem 4.10 We have yet another reason to think that gas is entering the Galaxy. The Milky Way's disk produces $(3–5)\mathcal{M}_\odot$ of new stars each year (Section 2.1) which will use up our $(5–10) \times 10^9 \mathcal{M}_\odot$ of disk gas within 1–3 Gyr. Dying stars return ($\mathcal{M}_\odot$ to $2\mathcal{M}_\odot$) of gas per year, which is not enough to avoid exhausting the supply. If gas flows into the disk from outside, we avoid the uncomfortable conclusion that we see the Milky Way and other spiral galaxies at the very end of their star-forming lives.

Problem 5.1 At the 0.6 m telescope: $1.1''$ pixel; $37'$ field. At 4 m prime focus: $0.54''$ pixel; $18'$ field. For 4 m $f/7.5$: $0.167''$ pixel; $5.7'$ field. If seeing is $0.8''$, the image at $f/7.5$ will be no sharper than at prime focus; the entire galaxy fits into the field of view only at the 0.6 m telescope.

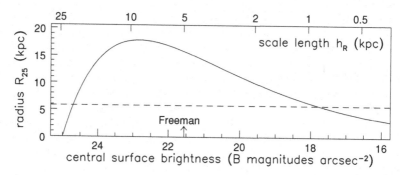

Fig. C.3. For an exponential disk of luminosity $2.5 \times 10^{10} L_{B,\odot}$, the radius R_{25} where the surface brightness $I(B) = 25$ mag arcsec^{-2} is plotted as a function of scale length h_R or central surface brightness $I_B(0)$.

Problem 5.2 Look back at Problem 1.14.

Problem 5.3 NGC 7331 is a larger and brighter galaxy than our Milky Way. For $d = 13.7$ Mpc, $R_{25} \approx 22$ kpc. At 8 kpc radius the surface brightness is $40 L_{I,\odot}$ pc^{-2}, about twice that near the Sun: Problem 2.8. Why did we choose $H_0 = 60$ km s^{-1} Mpc^{-1} for this problem, rather than our usual value of 75? For nearby galaxies, those with $V_r \lesssim 2000$ km s^{-1}, peculiar motions (see Section 1.4) can be large compared with the cosmic expansion. Looking at these galaxies, we would measure a higher or lower value for the Hubble constant. Using Cepheid variable stars (see Section 4.1; Hughes *et al.* 1998 *ApJ* **501**, 32) places NGC 7331 at 15.5 Mpc; the Tully–Fisher relation (see Problem 5.11) gives a similar distance, which corresponds to an 'effective H_0' near 55 km s^{-1} Mpc^{-1}. Section 8.4 discusses how the uneven gravitational pull of clustered galaxies induces peculiar motions.

Problem 5.4 See Figure C.3. The central brightness is $140 L_{B,\odot}$ pc^{-2}, and $M_B = -20.5$ is equivalent to $2.5 \times 10^{10} L_{B,\odot}$.

Problem 5.5 By Equation 1.35, the energy received between wavelength λ and $\lambda + \Delta\lambda$ from each steradian of a blackbody at temperature T is

$$F_\lambda \, \Delta\lambda = \frac{2h_p^2 c^2}{\lambda^5} \frac{\Delta\lambda}{e^{h_P c/(\lambda k_B T)} - 1} \, \mathrm{W\,m^{-2}}.$$

F_λ peaks at 10 μm when $T = 300$ K; see Equation 1.5. When $\lambda = 10$ μm, at $T = 100$ K the factor $e^{h_P c/\lambda k_B T} - 1$ is $\sim 1.5 \times 10^4$ times larger than it is at 300 K, so F_λ is reduced by the same multiple.

Problem 5.6 $\lambda/D = 21$ cm/73 m corresponds to $\sim 10'$, but structures larger than about half this size are already significantly 'resolved out' of interferometric maps.

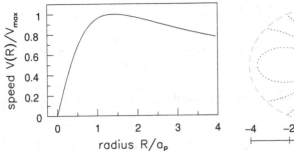
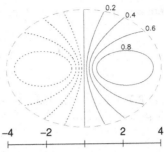

Fig. C.4. Left, the rotation curve in the Plummer potential; $V(R)$ is in units of V_{max}, radius R is in units of a_P. Right, a 'spider diagram' of contours of radial velocity $V_r - V_{sys}$ in units of $\sin i \, V_{max}$, for gas in a disk tilted at $i = 30°$ from face-on; negative contours are shown dotted.

Problem 5.8 $V^2(r) = GMr^2/(r^2 + a_P^2)^{3/2}$. Where $V(r)$ starts to fall with radius, we find closed loops in the spider diagram of Figure C.4.

Problem 5.11 For NGC 7331, Figure 5.20 shows that $V_{max} = 250 \, km \, s^{-1}$; by Equation 5.6, this corresponds to $L_I \approx 10^{11} L_\odot$. That would imply a distance $d \approx 16 \, Mpc$ for the galaxy. But the Tully–Fisher relation is not exact; we see in Figure 5.23 that some galaxies are more luminous than average for their rotation speed, and others are dimmer.

Problem 5.13 All the frequencies fall as $1/R$. A two-armed spiral with pattern speed Ω_p can live between radii R_+ and R_-, where $\Omega_p = \Omega \pm \kappa/2$, respectively; $\kappa = \sqrt{2}\Omega$, so $R_+/R_- = (\sqrt{2} + 1)/(\sqrt{2} - 1) \approx 5.8$. The radii where $\Omega_p = \Omega \pm \kappa/4$ are in the ratio $(2\sqrt{2} + 1)/(2\sqrt{2} - 1) \approx 1.6$.

Problem 5.14 From Problem 3.2, the Plummer radius for M33's nucleus is $a_P \lesssim 0.6 \, pc$, about $1/17$ of that for the globular cluster in Problem 3.13. The velocity dispersion is twice as large, so the nuclear cluster is no more than $2.4^2/16.6 \approx 1/4$ times as massive, or about $5 \times 10^5 \mathcal{M}_\odot$.

Problem 7.19 Follow the method of Problem 3.2, setting $s = \tan\phi$ to show that $\int_0^\infty (1 + s^2)^{-1} \, ds = \pi/2$.

Problem 7.21 For the Plummer and dark-halo models we have

$$\theta_E(P) = \frac{a_P}{d_{Lens}} \sqrt{\frac{\Sigma(0)}{\Sigma_{crit}} - 1}; \quad \theta_E(DH) = \frac{2a_H}{d_{Lens}} \sqrt{\frac{\Sigma(0)}{\Sigma_{crit}}} \sqrt{\frac{\Sigma(0)}{\Sigma_{crit}} - 1}.$$

Problem 8.6 If $\xi(r) = \xi_0 r^{-(3+n)}$ then

$$P(k) = 4\pi \xi_0 \int_0^\infty \frac{\sin(kr)}{kr} r^{-(1+n)} \, dr = 4\pi \xi_0 k^n \int_0^\infty t^{-(2+n)} \sin t \, dt.$$

Problem 8.7 We can write the potential $\Phi(\mathbf{x})$ and the fluctuation in density $\bar{\rho}\delta(\mathbf{x})$ in terms of their Fourier transforms as

$$\Phi(\mathbf{x}) = \frac{1}{(2\pi)^3} \int \Phi_{\mathbf{k}} e^{-i\mathbf{k}\cdot\mathbf{x}} \, d^3\mathbf{k} \ \text{and} \ \delta(\mathbf{x}) = \frac{1}{(2\pi)^3} \int \delta_{\mathbf{k}} e^{-i\mathbf{k}\cdot\mathbf{x}} \, d^3\mathbf{k}.$$

In the *random-phase* approximation, the $\delta_{\mathbf{k}}$ are independent random variables. Since δ is real, $\delta_{-\mathbf{k}} = \delta_{\mathbf{k}}^*$, the complex conjugate. Applying Poisson's equation for each $\mathbf{k}$-component in turn gives $-k^2\Phi_{\mathbf{k}} = 4\pi G\bar{\rho}\delta_{\mathbf{k}}$.

Now we must link $\delta_{\mathbf{k}}$ to $P(k)$. The correlation function $\xi(r)$ is the average at all possible points $\mathbf{x}$ of $\delta(\mathbf{x})\delta(\mathbf{x}+\mathbf{r})$; when counting galaxies, think of $\delta(\mathbf{x})$ as being nonzero only very near the position where one is present. If the cosmos is isotropic, it will not matter in which direction we take our step $\mathbf{r}$ away from the initial point $\mathbf{x}$. So we can write

$$\xi(r) = \frac{1}{(2\pi)^3} \int |\delta_{\mathbf{k}}^2| e^{-i\mathbf{k}\cdot\mathbf{r}} \, d^3\mathbf{k};$$

On comparing this with Equation 8.3, we see that $P(k) = |\delta_{\mathbf{k}}^2|$.

Using Parseval's theorem, and remembering that $\delta_{\mathbf{k}}$ depends only on the magnitude of k, the fluctuation in density $\bar{\rho}\delta(\mathbf{x})$ follows

$$\delta^2 = \frac{1}{(2\pi)^3} \int \delta_{\mathbf{k}}\delta_{\mathbf{k}}^* \, d^3\mathbf{k} = \frac{1}{(2\pi)^3} \int |\delta_{\mathbf{k}}^2| 4\pi k^2 \, dk = \int \frac{k^3 P(k)}{2\pi^2} \frac{dk}{k}.$$

So wavenumbers near k contribute an amount $\Delta_k^2 = k^3 P(k)/(2\pi^2)$ to the fluctuations $\langle\delta^2\rangle$ in the density, as stated in Equation 8.4. By similar reasoning, the same range of wavenumbers gives rise to fluctuations in the potential Φ amounting to $\Delta_\Phi^2 = k^3|\Phi_{\mathbf{k}}^2|/(2\pi^2)$. Poisson's equation tells us that $k^4|\Delta_\Phi^2| \propto |\Delta_k^2| \propto k^3 P(k)$. If $P(k) \propto k$ then $|\Delta_\Phi^2|$ does not vary with k, so we have equal fluctuations in Φ on all spatial scales.

Index

Printed in the United States
by Baker & Taylor Publisher Services